MEDICAL IMAGING

PRINCIPLES AND PRACTICES

MEDICAL IMAGING

PRINCIPLES AND PRACTICES

Edited by
Mostafa Analoui
Joseph D. Bronzino
Donald R. Peterson

CRC Press
Taylor & Francis Group
Boca Raton London New York

CRC Press is an imprint of the
Taylor & Francis Group, an **informa** business

CRC Press
Taylor & Francis Group
6000 Broken Sound Parkway NW, Suite 300
Boca Raton, FL 33487-2742

Printed in the United States of America on acid-free paper
Version Date: 20120619

International Standard Book Number: 978-1-4398-7102-7 (Hardback)

Library of Congress Cataloging-in-Publication Data

Medical imaging : principles and practices / edited by Mostafa Analoui, Joseph D. Bronzino, and Donald R. Peterson.
p. ; cm.
Includes bibliographical references and index.
ISBN 978-1-4398-7102-7 (hardcover : alk. paper)
I. Analoui, Mostafa. II. Bronzino, Joseph D., 1937- III. Peterson, Donald R.
[DNLM: 1. Diagnostic Imaging. WN 180]

616.07'543--dc23

2012024329

Visit the Taylor & Francis Web site at
http://www.taylorandfrancis.com

and the CRC Press Web site at
http://www.crcpress.com

Contents

Introduction

While the discovery of x-ray is commonly considered as a pivotal event, ushering in modern medical imaging, there have been various optical-based *in vitro* and *in vivo* modalities routinely used for centuries. Owing to the exponential technological innovations of the late nineteenth and twentieth centuries, imaging had an explosive growth, primarily in the emergence of new modalities and associated tools. During this period, a number of emerging technologies played key roles in advances in biomedical engineering and clinical practice. The contribution of these technologies can be summarized into (1) enabling objective and quantitative measurement of *in vivo* phenomena, (2) enhancing our understanding of disease process and interaction with therapeutic agents, and (3) new discoveries in the pathophysiology of disease and new diagnosis. Medical imaging is among the technologies that have strongly influenced all these three aspects and beyond. The discovery of x-ray, as a landmark event, enabled us to see the "invisible," opening a new era in medical diagnostics. More important, it opened up the thinking around the interaction of electromagnetic signal with human tissue and the utility of its selective absorption, scattering, diffusion, and reflection as a tool for understanding the physiology, evolution of disease, and therapy.

It is based on such a fundamental concept that all imaging modalities developed and utilized, that is, interaction of internal/external energy and its differential behavior in healthy and diseased tissues. X-ray initiated this journey, which has led to a broad range of solutions for 2-D and 3-D visualization of anatomic structures and functions. *In vitro* utilization of such tools in biomedical research has been even broader compared to clinical applications.

This book offers (1) a selective review of key imaging modalities focusing on those with established clinical utilization and (2) examples of quantitative tools for image analysis, modeling, and interpretation. The first two chapters provide a detailed overview of x-ray imaging and computed tomography. Projectional radiography and computed tomography cover a large portion of clinical imaging in practice. Although the fundamentals are quite similar to its early discovery, significant and evolutionary changes have been made in the past 115 years, which are reflected in these two chapters.

Magnetic resonance imaging (MRI) covers a broad range of tools and instrumentations based on differential magnetic properties of tissue and its relaxivity. MRI and its associated techniques have become one of the fastest-growing modalities since the last quarter of the twentieth century. The chapter on MRI covers fundamental concepts in signal acquisition and processing, followed by an overview of functional MRI (fMRI) and chemical shift imaging. Considering the depth of technical materials, an additional chapter covers topics in magnetic resonance microscopy.

While the concept of making the patient radioactive for diagnostic purposes seemed counterintuitive for a number of years, this is the basic concept behind nuclear imaging. The physics of instrumentation and signal collection have been covered in the chapter on nuclear medicine, along with their application in clinical practice.

Chapter 8 deals with electrical impedance tomography (EIT). EIT is among a number of promising emerging technologies that are identifying their unique application for *in vivo* imaging, along with the challenges for gaining clinical and regulatory acceptance.

The second half of the book covers selective topics on quantitative tools and approaches for analysis and interpretation of images acquired primarily by MRI and PET. In Chapter 9, the authors offer an overview of recent developments in the use of MRI for vascular imaging and its key advantages over ultrasound-based approaches. Chapter 10 covers dynamic contrast-enhanced MRI (DCE-MRI), a critical method for the analysis of oncologic tumor and its response to antiangiogenic class of therapeutics. The focus of Chapter 11 is modeling of myocardial deformation toward better understanding of cardiac function and associated anomalies. Perhaps one of the most promising utilities of MR in joint disease is in osteoarthritis (OA), which is covered in depth in Chapter 12. This chapter offers a comprehensive overview of OA, various MR approaches for imaging knee, and methods for quantitative image analysis. Chapter 13 offers a unique perspective in the utilization of quantitative PET for its applications in drug discovery and development, which is rapidly becoming an indispensable tool for clinical and research applications. Chapter 14 addresses the key issues around organizing and searching multimodality imaging data set, which is becoming increasingly important yet a challenging issue in clinical imaging. This chapter is intended to highlight complexities associated with high-volume data management, access, and sharing.

It must be noted that this book offers a selective overview of medical imaging and a comprehensive coverage perhaps requires multiple volumes. The editor hopes that these selected topics will provide readers the depth and breadth of this field and the challenges lying ahead of the technical and clinical community of researchers and practitioners.

Mostafa Analoui
New York, New York

Editors

Mostafa Analoui is the head of Healthcare and Life Sciences at The Livingston Group, New York. Previously he was the senior director at Pfizer Global Research and Development. Dr. Analoui is actively involved in investment, management, and scientific/business development of nanotechnology, drug discovery/development, diagnostic imaging, and global strategies.

While at Pfizer, he was the site head for Global Clinical Technology in Groton and New London, a division focusing on emerging technologies for development and validation of biomarkers and diagnostics for drug development. Prior to joining Pfizer, Dr. Analoui was the director of Oral and Maxillofacial Imaging Research, associate professor of radiology at Indiana University, and associate professor of biomedical engineering and electrical and computer engineering at Purdue University. He was also president and CEO of Therametric Technology Inc. He received his PhD from Purdue University, followed by a postdoctoral fellowship at IBM TJ Watson Research Center in New York.

In addition to industry leadership in biomedical and technology fields, he consults and lectures in the United States, Europe, and Asia. He has also served on various scientific, regulatory, and business advisory committees and boards, including NIH, NSF, PhRMA, NASA, and OECD. Dr. Analoui has authored over 130 publications, including journal articles, book chapters, and technical reports. He is a senior member of IEEE, SPIE, and RSNA.

He currently serves as a board member of VirtualScopics (Nasdaq: VSCP), Calando Pharmaceutical (Nasdaq: ARWR), BEACON (Biomedical Engineering Alliance and Consortium), and NanoBusiness Commercialization Association.

Joseph D. Bronzino earned a BSEE from Worcester Polytechnic Institute, Worcester, Massachusetts, in 1959, an MSEE from the Naval Postgraduate School, Monterey, California, in 1961, and a PhD in electrical engineering from Worcester Polytechnic Institute in 1968. He is presently the Vernon Roosa Professor of Applied Science, an endowed chair at Trinity College, Hartford, Connecticut, and president of the Biomedical Engineering Alliance and Consortium (BEACON), which is a nonprofit organization consisting of academic and medical institutions as well as corporations dedicated to the development of new medical technology. To accomplish this goal, BEACON facilitates collaborative research, industrial partnering, and the development of emerging companies.

Dr. Bronzino is the author of over 200 journal articles and 15 books, including *Technology for Patient Care* (C.V. Mosby, 1977), *Computer Applications for Patient Care* (Addison-Wesley, 1982), *Biomedical Engineering: Basic Concepts and Instrumentation* (PWS Publishing Co., 1986), *Expert Systems: Basic Concepts* (Research Foundation of State University of New York, 1989), *Medical Technology and Society: An Interdisciplinary Perspective* (MIT Press and McGraw-Hill, 1990), *Management of Medical Technology* (Butterworth/Heinemann, 1992), *The Biomedical Engineering Handbook* (CRC Press, 1st Edition, 1995; 2nd Edition, 2000; 3rd Edition, 2006), *Introduction to Biomedical Engineering* (Academic Press, 1st Edition, 1999; 2nd Edition, 2006), *Biomechanics: Principles and Applications* (CRC Press,

2002), *Biomaterials: Principles and Applications* (CRC Press, 2002), *Tissue Engineering* (CRC Press, 2002), and *Biomedical Imaging* (CRC Press, 2002).

Dr. Bronzino is a fellow of IEEE and the American Institute of Medical and Biological Engineering (AIMBE), an honorary member of the Italian Society of Experimental Biology, past chairman of the Biomedical Engineering Division of the American Society for Engineering Education (ASEE), a charter member of the Connecticut Academy of Science and Engineering (CASE), a charter member of the American College of Clinical Engineering (ACCE), a member of the Association for the Advancement of Medical Instrumentation (AAMI), past president of the IEEE-Engineering in Medicine and Biology Society (EMBS), past chairman of the IEEE Health Care Engineering Policy Committee (HCEPC), and past chairman of the IEEE Technical Policy Council in Washington, DC. He is a member of Eta Kappa Nu, Sigma Xi, and Tau Beta Pi. He is also a recipient of the IEEE Millennium Medal for "his contributions to biomedical engineering research and education" and the Goddard Award from WPI for Outstanding Professional Achievement in 2005. He is presently editor-in-chief of the Academic Press/Elsevier BME Book Series.

Donald R. Peterson earned his PhD in biomedical engineering and his MS in mechanical engineering at the University of Connecticut. He has been an active member of the BME faculty since 1999, offering courses in biomechanics, biodynamics, biofluid mechanics, and ergonomics, and during the past year he has served as BME Graduate Program Committee chair. He is an assistant professor in the School of Medicine, where he is the director of the Biodynamics Laboratory and the Bioengineering Center at the University of Connecticut Health Center. His research involves the modeling of human interactions with existing and developmental devices such as powered and nonpowered tools, dental instruments, computer input devices, musical instruments, sports equipment, and spacesuit and space tool development for NASA. Dr. Peterson has written more than 45 scholarly publications appearing in journals and textbooks.

Contributors

Amir A. Amini
Department of Electrical and Computer Engineering
University of Louisville
Louisville, Kentucky

Edward Ashton
VirtualScopics, Inc.
Rochester, New York

Niranjan Balu
Department of Radiology
University of Washington
Seattle, Washington

D.C. Barber
Department of Cardiovascular Science
Sheffield University
Yorkshire, United Kingdom

R. Boellaard
Department of Nuclear Medicine and PET Research
VU University Medical Center
Amsterdam, the Netherlands

Thomas F. Budinger
Department of Electrical Engineering and Computer Sciences
University of California, Berkeley
Berkeley, California

Gador Canton
Department of Radiology
University of Washington
Seattle, Washington

Huijun Chen
Department of Radiology
University of Washington
Seattle, Washington

Wei Chen
Center for Magnetic Resonance Research
University of Minnesota Medical School
Minneapolis, Minnesota

David A. Chesler
Department of Radiology
Massachusetts General Hospital
Boston, Massachusetts

Steven Conolly
Department of Biomedical Engineering
Stanford University
Stanford, California

Barbara Y. Croft
National Institutes of Health
Bethesda, Maryland

Ian A. Cunningham
Victoria Hospital
The John P. Robarts Research Institute
University of Western Ontario
London, Ontario, Canada

Thomas Deserno
Department of Medical Informatics
Aachen University
Aachen, Germany

K. Whittaker Ferrara
Riverside Research Institute
Riverside, California

Richard L. Goldberg
Department of Biomedical Engineering
University of North Carolina
Chapel Hill, North Carolina

Xiaoping Hu
Center for Magnetic Resonance Research
University of Minnesota Medical School
Minneapolis, Minnesota

N. Huber
Biomedical Engineering Group
University of Sussex
Brighton, United Kingdom

G. Allan Johnson
Departments of Biomedical Engineering and Physics
Duke University Medical Center
Durham, North Carolina

Philip F. Judy
Department of Radiology
Brigham and Women's Hospital
Harvard Medical School
Boston, Massachusetts

William S. Kerwin
Department of Radiology
University of Washington
Seattle, Washington

Kenneth K. Kwong
Department of Radiology
Massachusetts General Hospital
Boston, Massachusetts

M. Lubberink
Department of Biomedical Engineering and PET Center
Uppsala University
Uppsala, Sweden

Albert Macovski
Departments of Electrical Engineering and Radiology
Stanford University
Stanford, California

R.P. Maguire
Novartis Institutes for Biomedical Research
Basel, Switzerland

Jack G. Mottley
Department of Electrical and Computer Engineering
University of Rochester
Rochester, New York

Maqbool Patel
Center for Magnetic Resonance Research
University of Minnesota Medical School
Minneapolis, Minnesota

John Pauly
Department of Electrical Engineering
Stanford University
Stanford, California

John Schenck
General Electric Corporate Research and Development Center
Schenectady, New York

Vijay Shah
VirtualScopics, Inc.
Rochester, New York

Stephen W. Smith
Department of Biomedical Engineering
Duke University
Durham, North Carolina

José G. Tamez-Peña
Tec de Monterrey
Nuevo León, México

Saara Totterman
Qmetrics Technologies, LLC
Rochester, New York

Benjamin M.W. Tsui
Department of Radiology and Radiological Sciences
Johns Hopkins University
Baltimore, Maryland

Kamil Ugurbil
Center for Magnetic Resonance Research
University of Minnesota Medical School
Minneapolis, Minnesota

Henry F. VanBrocklin
Department of Radiology
University of California, Berkeley
Berkeley, California

Hui Wang
Department of Electrical Engineering
University of Louisville
Louisville, Kentucky

Jinnan Wang
Philips Research North America
Briarcliff Manor, New York

W. Wang
Biomedical Engineering Group
University of Sussex
Brighton, United Kingdom

Martin J. Yaffe
Imaging/Bioengineering Research
Sunnybrook Health Sciences Centre
and
Departments of Medical Biophysics and Radiology
University of Toronto
Toronto, Ontario, Canada

Chun Yuan
Department of Radiology
University of Washington
Seattle, Washington

Xiaohong Zhou
Duke University Medical Center
Durham, North Carolina

1

Mammography

Martin J. Yaffe
Sunnybrook Health Sciences Centre
University of Toronto

1.1 Introduction

Mammography is an x-ray imaging procedure for examination of the breast. It is used primarily not only for the detection and diagnosis of breast cancer, but also for presurgical localization of suspicious areas and in the guidance of needle biopsies.

Breast cancer is a major killer of women. Approximately 254,650 women were diagnosed with breast cancer in the United States in 2009 and 40,170 women died of this disease (ACS, 2009). Its cause is not currently known; however, it has been demonstrated that *survival* is greatly improved if disease is detected at an *early stage* (Smart et al., 1993; Tabar et al., 1993). Mammography is at present the most effective means of detecting early-stage breast cancer. It is used both for investigating symptomatic patients (diagnostic mammography) and for *screening* of asymptomatic women in selected age groups.

Breast cancer is detected on the basis of four types of signs on the mammogram:

1. The characteristic morphology of a tumor mass
2. Certain presentations of mineral deposits as specks called microcalcifications
3. Architectural distortion of normal tissue patterns caused by the disease
4. Asymmetry between corresponding regions of images of the left and right breast

1.2 Principles of Mammography

The mammogram is an x-ray shadowgram formed when x-rays from a quasi-point source irradiate the breast and the transmitted x-rays are recorded by an *image receptor*. Because of the spreading of the x-rays from the source, structures are magnified as they are projected onto the image receptor. The signal is a result of differential attenuation of x-rays along paths passing through the structures of the breast.

The essential features of image quality are summarized in Figure 1.1. This is a one-dimensional profile of x-ray transmission through a simplified computer model of the breast (Fahrig et al., 1992), illustrated in Figure 1.2. A region of reduced transmission corresponding to a structure of interest such as a tumor, a calcification or normal *fibroglandular* tissue is shown. The imaging system must have sufficient *spatial resolution* to delineate the edges of fine structures in the breast. Structural detail as small as 50 μm must be adequately resolved. Variation in x-ray attenuation among tissue structures in the breast gives rise to *contrast*. The detectability of structures providing subtle contrast is impaired, however, by an overall random fluctuation in the profile, referred to as mottle or *noise*. Because the breast is sensitive to ionizing radiation, which at least for high doses is known to cause breast cancer, it is desirable to use the lowest radiation dose compatible with excellent image quality. The components of the imaging system will be described and their design will be related to the imaging performance factors discussed in this section.

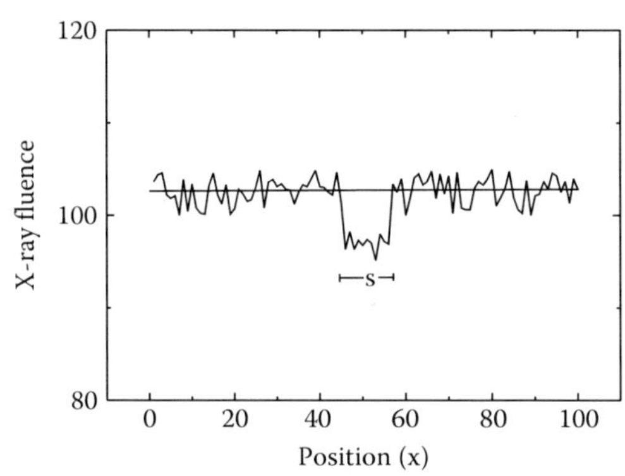

FIGURE 1.1 Profile of a simple x-ray projection image, illustrating the role of contrast, spatial resolution, and noise in mammographic image quality.

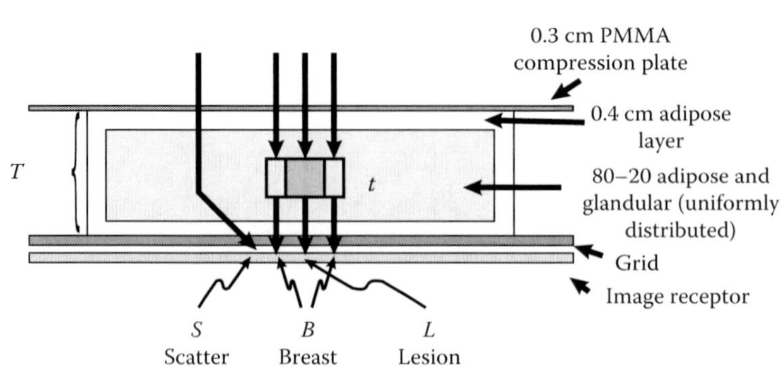

FIGURE 1.2 Simplified computer model of the mammographic image acquisition process.

1.3 Physics of Image Formation

In the model of Figure 1.2, an "average" breast composed of 80% adipose tissue and 20% fibroglandular tissue is considered (Yaffe et al., 2009). For the simplified case of monoenergetic x-rays of energy, E, the number of x-rays recorded in a fixed area of the image is proportional to

$$N_B = N_0(E)e^{-\mu T} \tag{1.1}$$

in the "background" and

$$N_L = N_0(E)e^{-[\mu(T - t) + \mu' t]} \tag{1.2}$$

in the shadow of the lesion or other structure of interest. In Equations 1.1 and 1.2, $N_0(E)$ is the number of x-rays that would be recorded in the absence of tissue in the beam, μ and μ' are the attenuation coefficients of the breast tissue and the lesion, respectively, T is the thickness of the breast, and t is the thickness of the lesion.

The difference in x-ray transmission gives rise to *subject* contrast which can be defined as

$$C_0 = \frac{N_B - N_L}{N_B + N_L} \tag{1.3}$$

$$\frac{1 - e^{-[\mu' - \mu t]}}{1 + e^{-[\mu' - \mu t]}} \tag{1.4}$$

For the case of monoenergetic x-rays and temporarily ignoring scattered radiation, contrast would depend only on the thickness of the lesion and the difference between its attenuation coefficient and that of the background material. These are not valid assumptions; x-ray beams are polyenergetic and some scattered radiation is always recorded by the image receptor, so in actuality, contrast also depends to some extent on μ and T.

Shown in Figure 1.3 are x-ray attenuation coefficients measured versus energy on samples of three types of materials found in the breast: adipose tissue, normal fibroglandular breast tissue, and infiltrating ductal carcinoma (one type of breast tumor) (Johns and Yaffe, 1987). Both the attenuation coefficients themselves and their difference ($\mu' - \mu$) decrease with increasing E. As shown in Figure 1.4, which is based on Equation 1.4, this causes C_s to fall as x-ray energy increases. Note that the subject contrast of even small calcifications in the breast is greater than that for a tumor because of the greater difference in attenuation coefficient between calcium and breast tissue.

For a given image recording system (also referred to as the detector), a proper exposure requires a specific value of x-ray energy transmitted by the breast and incident on the receptor, that is, a specific value of N_B. The breast entrance air kerma required to produce an image is, therefore, proportional to

$$N_0 = N_B(E)e^{+\mu T} \tag{1.5}$$

Because μ decreases with energy, the required entrance air kerma for constant signal at the image receptor, N_B, will increase if E is reduced to improve image contrast. A better measure of the risk of radiation-induced breast cancer than air kerma is the mean glandular dose (MGD) (BEIR, 1990). MGD is calculated as the product of the entrance air kerma and a factor, obtained experimentally or by Monte Carlo radiation transport calculations, which converts from incident air kerma to dose (Wu et al., 1991, 1994; Boone, 1999; Dance et al., 2000). The conversion factor increases with E so that MGD does not fall

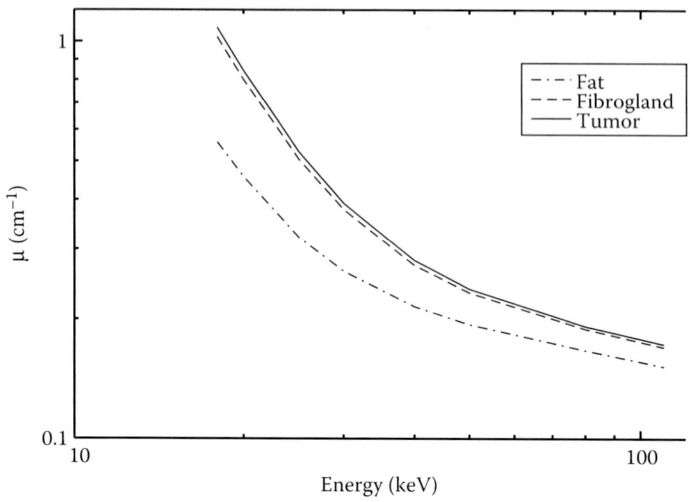

FIGURE 1.3 Measured x-ray linear attenuation coefficients of breast fibroglandular tissue, breast fat, and infiltrating ductal carcinoma plotted versus x-ray energy.

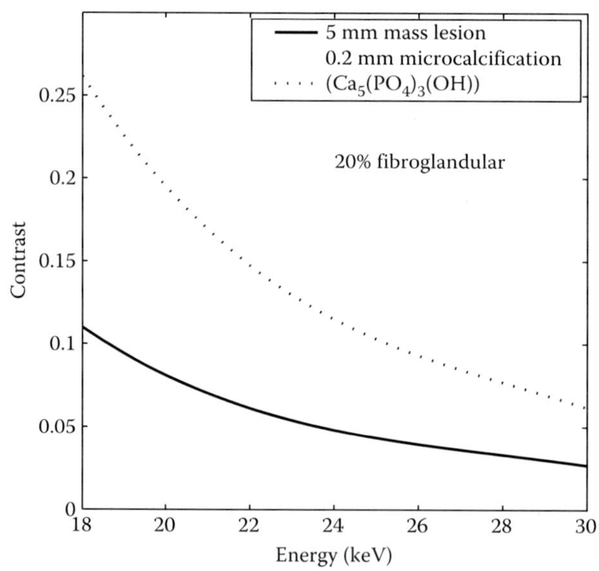

FIGURE 1.4 Dependence of mammographic subject contrast on x-ray energy.

as quickly with energy as does entrance air kerma. The trade-off between image contrast and radiation dose necessitates important compromises in establishing mammographic operating conditions.

1.4 Equipment

In a mammography unit, an x-ray tube and an image receptor are mounted on opposite sides of a mechanical assembly or gantry. Because the breast must be imaged from different aspects and to accommodate patients of different height, the assembly can be adjusted in a vertical axis and rotated about a horizontal axis as shown in Figure 1.5.

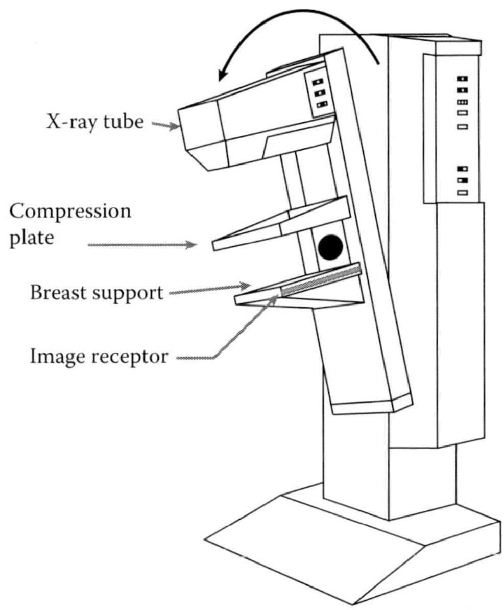

FIGURE 1.5 Schematic diagram of a dedicated mammography machine.

Most general radiography equipment is designed such that the image field is centered below the x-ray source. In mammography, the system's geometry is arranged as in Figure 1.6a where a vertical line from the x-ray source grazes the chest wall of the patient and intersects orthogonally with the edge of the image receptor closest to the patient. If the x-ray beam were centered over the breast as in Figure 1.6b, some of the tissue near the chest wall would be projected inside of the patient where it could not be recorded.

Radiation leaving the x-ray tube passes through a metallic spectral-shaping filter, a beam-defining aperture, and a plate which compresses the breast. Those rays transmitted through the breast are incident on an antiscatter "grid." The x-rays that successfully pass through the grid then strike the image receptor, where they interact, and deposit most of their energy locally. A fraction of the x-rays pass through the receptor without interaction and impinge upon a sensor which is used to activate the automatic exposure control mechanism of the unit.

1.4.1 X-Ray Source

Practical monoenergetic x-ray sources are not available and the x-rays used in mammography arise from bombardment of a metal target by electrons in a hot-cathode vacuum tube. The x-rays are emitted from the target over a spectrum of energies, ranging up to the peak kilovoltage applied to the x-ray tube. Typically, the x-ray tube employs a rotating anode design in which electrons from the cathode strike the anode *target* material at a small angle (0–16°) from normal incidence (Figure 1.7). Over 99% of the energy from the electrons is dissipated as heat in the anode. The angled surface and the distribution of the electron bombardment along the circumference of the rotating anode disk allows the energy to be spread over a larger area of target material while presenting a much smaller effective *focal spot* as viewed from the imaging plane. On modern equipment, the typical "nominal" focal spot size for normal contact mammography is 0.3 mm while the smaller spot used primarily for magnification is 0.1 mm. The specifications for x-ray focal spot size tolerance, established by NEMA (National Electrical Manufacturers Association) in their standard, NEMA XR 5-1992 (R1999), or the IEC (International Electrotechnical Commission) allow the *effective focal spot size* to be considerably larger than these nominal sizes. For example, the NEMA specification allows the effective focal spot size to be 0.45 mm in width and 0.65 mm in length for a nominal 0.3 mm spot and 0.15 mm in each dimension for a nominal 0.1 mm spot.

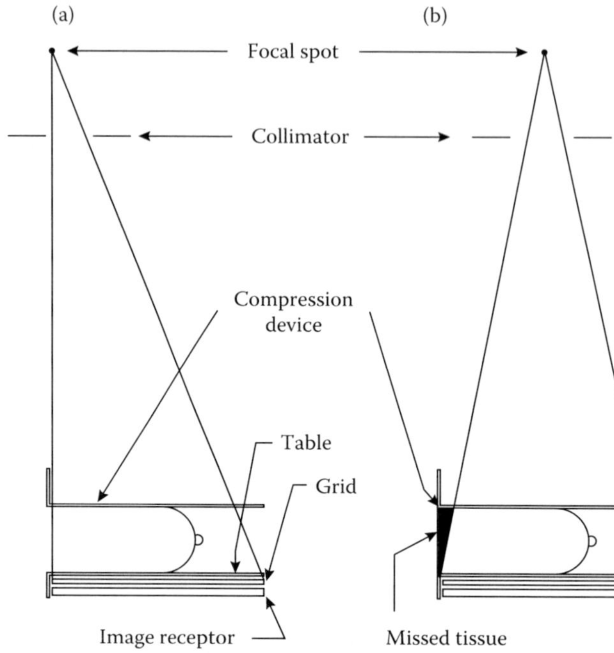

FIGURE 1.6 Geometric arrangement of system components in mammography. (a) Correct alignment provides good tissue coverage; (b) incorrect alignment causes tissue near the chest wall not to be imaged.

The nominal focal spot size is defined relative to the effective spot size at a "reference axis." As shown in Figure 1.7, this reference axis, which may vary from manufacturer to manufacturer, is normally specified at some mid-point in the image. The effective size of the focal spot will monotonically increase from the anode side to the cathode side of the imaging field. Normally, x-ray tubes are arranged such that the cathode side of the tube is adjacent to the patient's chest wall, since the highest intensity of x-rays is available at the cathode side, and the attenuation of x-rays by the patient is generally greater near the chest wall of the image.

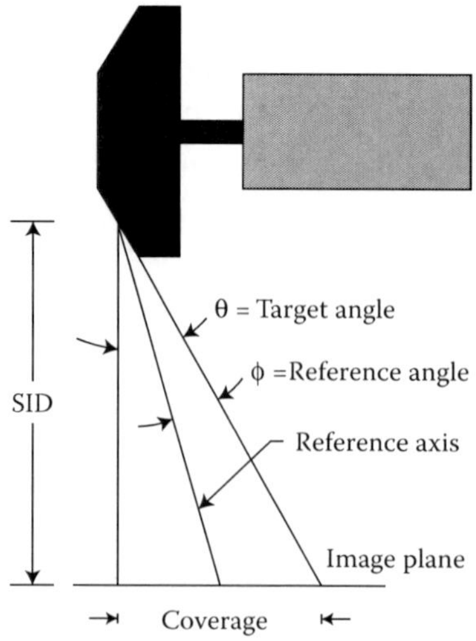

FIGURE 1.7 Angled-target x-ray source provides improved heat loading but causes effective focal spot size to vary across the image.

The spatial resolution capability of the imaging system is partially determined by the effective size of the focal spot and by the degree of magnification of the anatomy at any plane in the breast. This is illustrated in Figure 1.8 where, by similar triangles, the unsharpness region due to the finite size of the focal spot is linearly related to the effective size of the spot and to the ratio of OID to SOD, where SOD is the source–object distance and OID is the object–image receptor distance. Because the breast is a three-dimensional structure, this ratio and, therefore, the unsharpness will vary for different planes within the breast.

The size of the focal spot determines the heat loading capability of the x-ray tube target. For smaller focal spots, the current through the x-ray tube must be reduced, necessitating increased exposure times and the possibility of loss of resolution due to motion of anatomical structures. Loss of geometric resolution can be controlled in part by minimizing OID/SOD, that is, by designing the equipment with greater source–breast distances, by minimizing space between the breast and the image receptor and by compressing the breast to reduce its overall thickness.

Magnification is often used intentionally to improve the signal-to-noise ratio (SNR) or, in the case of digital mammography, the resolution of the image. This is accomplished by elevating the breast above the image receptor, in effect reducing SOD and increasing OID. Under these conditions, resolution is invariably limited by focal spot size and use of a small spot for magnification imaging (typically a nominal size of 0.1 mm) is critical.

Because monoenergetic x-rays are not available, one attempts to define a spectrum providing energies which give a reasonable compromise between radiation dose and image contrast. The spectral shape can be controlled by adjustment of the kilovoltage, choice of the target material, and the type and thickness of metallic filter placed between the x-ray tube and the breast.

Based on models of the imaging problem in mammography, it has been suggested that the optimum energy for imaging lies between 18 and 23 keV, depending on the thickness and composition of the breast (Beaman and Lillicrap, 1982). It has been found that for the breast of typical thickness and composition, the characteristic x-rays from molybdenum at 17.4 and 19.6 keV provide good

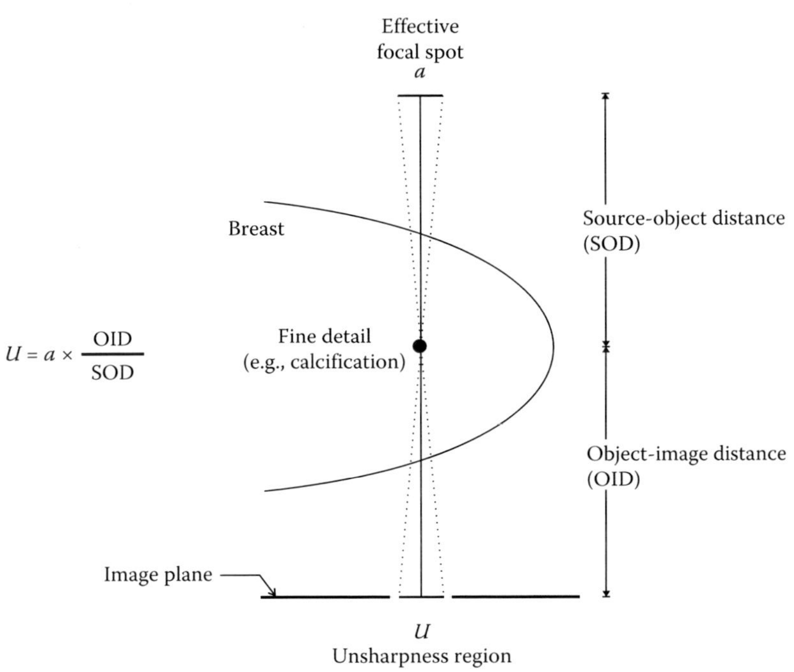

FIGURE 1.8 Dependence of focal spot unsharpness on focal spot size and magnification factor.

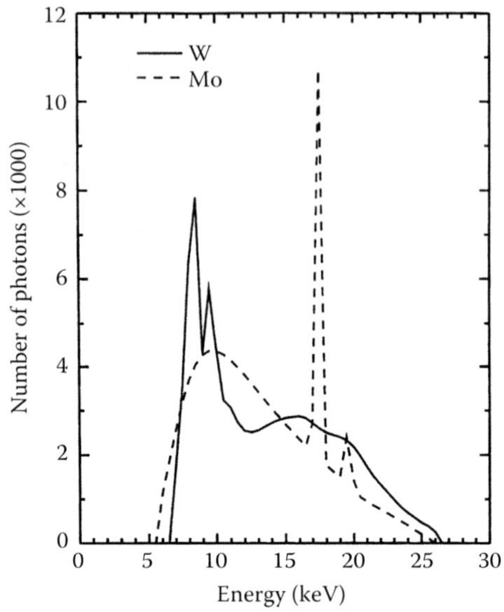

FIGURE 1.9 Comparison of tungsten and molybdenum target x-ray spectra.

imaging performance. For this reason, molybdenum target x-ray tubes are used on the vast majority of mammography machines.

Most mammography tubes use beryllium exit windows between the evacuated tube and the outside world as glass or other metals used in general-purpose tubes would provide excessive attenuation of the useful energies for mammography. Figure 1.9 compares tungsten target and molybdenum target spectra for beryllium window x-ray tubes. Under some conditions, tungsten may provide appropriate image quality for mammography; however, it is essential that the intense emission of L radiation from tungsten be filtered from the beam before it is incident upon the breast, because extremely high doses to the skin would result from this radiation without useful contribution to the mammogram.

1.4.2 Filtration of X-Ray Beam

In conventional radiology, filters made of aluminum or copper are used to provide selective removal of low x-ray energies from the beam before it is incident upon the patient. In mammography, particularly when a molybdenum anode x-ray tube is employed, a molybdenum filter 20–35 μm thick is generally used. This filter attenuates x-rays both at low energies and those above its own K-absorption edge allowing the molybdenum characteristic x-rays from the target to pass through the filter with relatively high efficiency. As illustrated in Figure 1.10, this K-edge filtration results in a spectrum enriched with x-ray energies in the range of 17–20 keV.

Although this spectrum is relatively well suited for imaging the breast of average attenuation, slightly higher energies are desirable for imaging dense thicker breasts. Because the molybdenum target spectrum is so heavily influenced by the characteristic x-rays, an increase in the kilovoltage alone does not substantially change the shape of the spectrum. The beam can be "hardened," however, by employing filters of higher atomic number than molybdenum. For example, rhodium (atomic no. 45) has a K absorption edge at 23 keV, providing strong attenuation both for x-rays above this energy and for those at substantially lower energies. Used with a molybdenum target x-ray tube and slightly increased kV, it provides a spectrum with increased penetration (reduced dose) compared to the Mo/Mo combination.

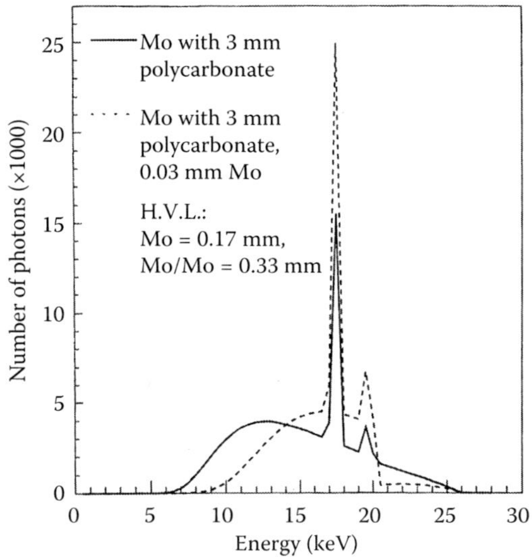

FIGURE 1.10 Molybdenum target spectrum unfiltered (solid line) and filtered by 0.03 mm Mo foil (dashed line).

It is possible to go further in optimizing imaging performance, by "tuning" the effective spectral energy by using other target materials in combination with appropriate K-edge filters (Jennings et al., 1993). One manufacturer employs an x-ray tube incorporating both molybdenum and rhodium targets, where the electron beam can be directed toward one or the other of these materials (Heidsieck et al., 1991). With this system, the filter material (rhodium, molybdenum, etc.) can be varied to suit the target which has been selected. Similarly, work has been reported on K-edge filtration of tungsten spectra (Desponds et al., 1991), where the lack of pronounced K characteristic peaks provides more flexibility in spectral shaping with filters. With the increasing use of digital mammography and tomosynthesis (described below), the optimization shifts toward the use of more penetrating beams, for example, rhodium target with a rhodium filter or tungsten target with an aluminum or silver filter.

1.4.3 Compression Device

There are several reasons for applying firm (but not necessarily painful) compression to the breast during the examination. Compression causes the different tissues to be spread out, minimizing superposition from different planes, and thereby improving a conspicuity of structures. As will be discussed later, scattered radiation can degrade contrast in the mammogram. The use of compression decreases the ratio of scattered to directly transmitted radiation reaching the image receptor. Compression also decreases the distance from any plane within the breast to the image receptor (i.e., OID) and in this way reduces geometric unsharpness. The compressed breast provides lower overall attenuation to the incident x-ray beam, allowing radiation dose to be reduced. The compressed breast also provides more uniform attenuation over the image. This reduces the exposure range that must be recorded by the imaging system, allowing more flexibility in choice of films to be used. Finally, compression provides a clamping action which reduces anatomical motion during the exposure reducing this source of image unsharpness.

It is important that the compression plate allows the breast to be compressed parallel to the image receptor, and that the edge of the plate at the chest wall be straight and aligned with both the focal spot and image receptor to maximize the amount of breast tissue which is included in the image (see Figure 1.6).

1.4.4 Antiscatter Grid

Lower x-ray energies are used for mammography than for other radiological examinations. At these energies, the probability of photoelectric interactions within the breast is significant. Nevertheless, the probability of Compton scattering of x-rays within the breast is still quite high. Scattered radiation recorded by the image receptor has the effect of creating a quasi-uniform haze on the image and causes the subject contrast to be reduced to

$$C_S = \frac{C_0}{1 + \text{SPR}} \tag{1.6}$$

where C_0 is the contrast in the absence of scattered radiation, given by Equation 1.4 and SPR is the scatter-to-primary (directly transmitted) x-ray ratio at the location of interest in the image.

In the absence of an antiscatter device, 37–50% of the total radiation incident on the image receptor would have experienced a scattering interaction within the breast, that is, the scatter to primary ratio would be 0.6–1.0. In addition to contrast reduction, the recording of scattered radiation uses up part of the dynamic range of the image receptor (a limiting factor primarily in film mammography) and adds statistical noise to the image.

Antiscatter *grids* have been designed for mammography. These are composed of linear lead (Pb) septa separated by a rigid interspace material. Generally, the grid septa are not strictly parallel but focused (toward the x-ray source). Because the primary x-rays all travel along direct lines from the x-ray source to the image receptor while the scatter diverges from points within the breast, the grid presents a smaller acceptance aperture to scattered radiation than to primary and thereby discriminates against scattered radiation. Grids are characterized by their *grid ratio* (ratio of the path length through the interspace material to the interseptal width) which typically ranges from 3.5:1 to 5:1. When a grid is used, the SPR typically is reduced by a factor of about 5, leading in most cases to a substantial improvement in image contrast (Wagner, 1991).

On modern mammography equipment, the grid is an integral part of the system, and during x-ray exposure, is moved to blur the image of the grid septa to avoid a distracting pattern in the mammogram. It is important that this motion be uniform and of sufficient amplitude to avoid nonuniformities in the image, particularly for short exposures that occur when the breast is relatively lucent.

Because of absorption of primary radiation by the septa and by the interspace material, part of the primary radiation transmitted by the patient does not arrive at the image receptor. In addition, by removing some of the scattered radiation, the grid causes the overall radiation fluence to be reduced from that which would be obtained in its absence. In film mammography, to obtain a radiograph of proper optical density (blackness), the entrance air kerma to the patient must be increased by a factor known as the *Bucky factor* to compensate for these losses. Typical Bucky factors are in the range of 2–3.

A linear grid does not provide scatter rejection for those quanta traveling in planes parallel to the septa. One manufacturer employs a crossed grid that consists of septa that run in orthogonal directions to address this limitation. The improved scatter rejection is accomplished at doses comparable to those required with a linear grid, because the interspace material of the crossed grid is air rather than a solid. To avoid artifacts, the grid is moved in a very precise way during the exposure to ensure a uniform blurring of the image of the grid itself.

1.4.5 Screen-Film Mammography

1.4.5.1 Fluorescent Screens

When first introduced, mammography was carried out using direct exposure radiographic film in order to obtain the high spatial resolution required. Since the mid-1970s, high-resolution fluorescent screens have been used to convert the x-ray pattern from the breast into an optical image. These screens are used

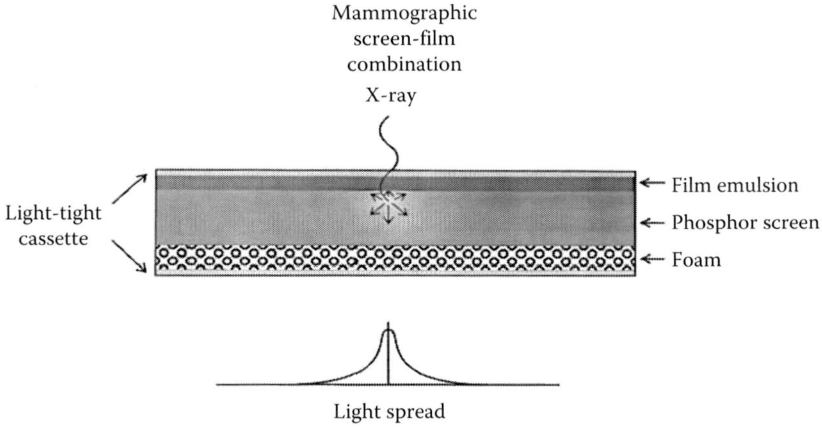

FIGURE 1.11 Design of a screen-film image receptor for mammography.

in conjunction with single-coated radiographic film, and the configuration is shown in Figure 1.11. With this arrangement, the x-rays pass through the cover of a light-tight cassette and the film to impinge upon the screen. Absorption is exponential, so that a large fraction of the x-rays are absorbed near the entrance surface of the screen. The phosphor crystals, which absorb the energy, produce light in an isotropic distribution. Because the film emulsion is pressed tightly against the entrance surface of the screen, the majority of the light quanta have only a short distance to travel to reach the film. Light quanta traveling longer distances have an opportunity to spread laterally (see Figure 1.11), and in this way degrade the spatial resolution. To discriminate against light quanta traveling along these longer oblique paths, the phosphor material of the screen is generally treated with a dye which absorbs much of this light, giving rise to a sharper image. A typical phosphor used for mammography is gadolinium oxysulfide (Gd_2O_2S). Although the K-absorption edge of gadolinium occurs at too high an energy to be useful in mammography, the phosphor material is dense (7.44 g/cm^3) so that the *quantum efficiency* (the fraction of incident x-rays which interact with the screen) is good (about 60%). The signal produced by the detector (in a phosphor, the amount of light produced) is related to the number of x-rays that interact with it (the product of the number of incident x-rays and the quantum efficiency) and the gain provided by the detector. In a phosphor, this gain results from the conversion of the kinetic energy of relatively high energy interacting x-ray quanta into much lower energy optical quanta. For example, one x-ray quantum at 24 keV carries enough kinetic energy to liberate about 8000 light quanta. The process is <100% efficient. At 15% *conversion efficiency*, about 1200 light quanta are emitted per interacting x-ray.

1.4.5.2 Film

The photographic film emulsion for mammography is designed with a characteristic curve such as that shown in Figure 1.12 which is a plot of the optical density (blackness) provided by the processed film versus the logarithm of the x-ray exposure to the screen. Film provides nonlinear input–output transfer characteristics. The local gradient of this curve controls the display contrast presented to the radiologist. Where the curve has a shallow gradient, a given increment of kerma provides little change in optical density, rendering structures imaged in this part of the curve difficult to visualize. Where the curve is steep, the film provides excellent image contrast. The range of kerma over which contrast is appreciable is referred to as the *latitude* of the film. Because the film is constrained between two optical density values, the *base* ± *fog* density of the film, where no intentional x-ray exposure has resulted, and the maximum density provided by the emulsion, there is a compromise between maximum gradient of the film and the latitude which it provides. For this reason, some regions of the mammogram will generally be underexposed or overexposed, that is, rendered with suboptimal contrast.

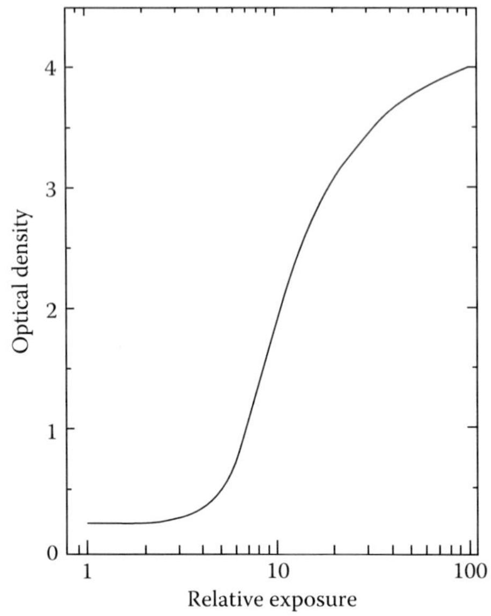

FIGURE 1.12 Characteristic curve of a mammographic screen-film image receptor.

1.4.5.3 Film Processing

Mammography film is processed in an automatic processor similar to that used for general radiographic films. It is important that the development temperature, time, and rate of replenishment of the developer chemistry be compatible with the type of film emulsion used and be designed to maintain good contrast of the film.

1.4.6 Digital Mammography

There are several technical factors associated with screen-film mammography which limit the ability to display the finest or most subtle details, and produce images with the most efficient use of radiation dose to the patient. In screen-film mammography, the film must act as an image acquisition detector as well as a storage and display device. Because of its sigmoidal shape, the range of x-ray exposures over which the film display gradient is significant, that is, the image latitude, is limited. If a tumor is located in either a relatively lucent or more opaque region of the breast, then the contrast displayed to the radiologist may be inadequate because of the limited gradient of the film. This is particularly a concern in patients whose breasts contain large amounts of fibroglandular tissue, the so-called dense breast.

Another limitation of film mammography is the effect of structural noise due to the granularity of the film emulsion used to record the image. This impairs the detectibility of microcalcifications and other fine structures within the breast. While Poisson's quantum noise is unavoidable, it should be possible to virtually eliminate structural noise by technical improvements. Existing screen-film mammography also suffers because of the inefficiency of grids in removing the effects of scattered radiation and with compromises in spatial resolution versus quantum efficiency inherent in the screen-film image receptor.

Many of the limitations of conventional mammography can be effectively overcome with a *digital mammography* imaging system, in which image acquisition, display, and storage are performed independently, allowing optimization of each. For example, acquisition can be performed with low noise, highly linear x-ray detectors, while since the image is stored digitally, it can be displayed with contrast independent of the detector properties and defined by the needs of the radiologist. Whatever image processing techniques are found useful, ranging from simple contrast enhancement to histogram modification and spatial frequency filtering, they could conveniently be applied.

The first commercial digital mammography systems were introduced in 2000 and at the time of writing over 80% of mammography systems in the United States have been converted to digital. The key technical elements that differ from screen-film mammography are the x-ray detector and the display device. The detector should have the following characteristics:

1. Efficient absorption of the incident radiation beam
2. Linear response over a wide range of incident radiation intensity
3. Low intrinsic noise
4. Spatial resolution on the order of 10 cycles/mm (50 μm sampling)
5. Can accommodate at least an 18 × 24 cm and preferably a 24 × 30 cm field size
6. Acceptable imaging time and heat loading of the x-ray tube
7. Robust and stable performance over environmental conditions and time

Two main approaches have been taken in detector development—area detectors and slot detectors. In the former, the entire image is acquired simultaneously, while in the latter only a portion of the image is acquired at one time and the full image is obtained by scanning the x-ray beam and detector across the breast. Area detectors offer convenient fast image acquisition and could be used with conventional x-ray machines, but may still require a grid, while slot systems are slower, require a scanning x-ray beam, but use relatively simple detectors and have excellent intrinsic efficiency at scatter rejection.

In addition, detectors can be distinguished by the x-ray absorber material, by the process with which the x-ray signal is converted to electronic charge and by the readout mechanism. Four designs now in clinical use will be described.

1.4.6.1 Phosphor Flat Panel Detector

In this type of detector system, x-rays are absorbed by a cesium iodide (CsI) phosphor layer and produce light. The phosphor crystal is evaporated to form needle-like structures and these tend to channel the light through the phosphor while reducing lateral spreading. The phosphor is deposited directly on a matrix of about 2000^2 photodiodes with thin film transistor (TFT) switches fabricated on a large area amorphous silicon plate (Figure 1.13) (Shaw et al., 2004). Each photodiode/TFT combination forms a detector element (del) generally corresponding to one pixel of the digital image. The electronic signal is read out on a series of data lines as the switches in each column of the array are activated. The signals from those lines are then multiplexed and digitized to form the image.

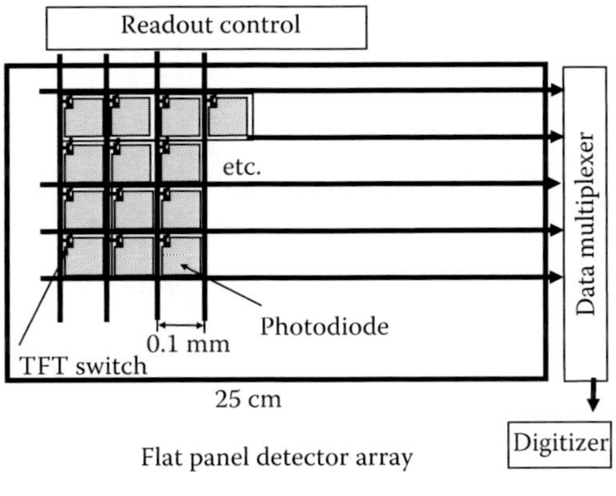

FIGURE 1.13 Schematic representation of a flat-panel phosphor-based digital mammography system. The readout assembly is shown here. A thallium-activated CsI phosphor is evaporated on this surface to serve as the x-ray absorber.

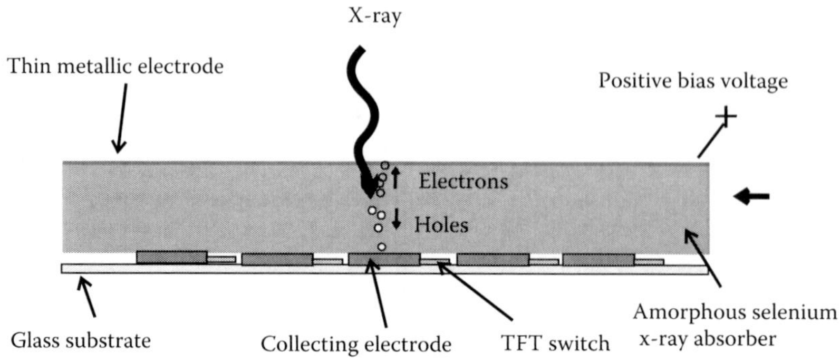

FIGURE 1.14 Cross-sectional view of an amorphous selenium flat panel detector system. The readout array is similar to that illustrated in Figure 1.13.

1.4.6.2 Amorphous Selenium Flat Panel Detector

Here, as in the previous system, the detector is based on large-area flat plate containing a matrix of TFT readout switches. In this case, however, a layer of amorphous selenium is evaporated onto the plate and a thin metal electrode is evaporated over the selenium (Figure 1.14). A bias field of a few volts per μm of detector thickness is applied across the detector. When x-rays are absorbed in the selenium, the energetic photoelectrons and Compton recoil electrons liberated in the material lose their energy by interacting with neighboring atoms and creating about 20 electron-hole pairs for each keV of energy deposited. These drift across the electric field and charges are deposited on landing electrodes that form the dels and replace the photodiodes that were employed in the flat-panel phosphor detector (Yorker et al., 2002). The charge signal from each del is read out in the same manner as described for that system. One potential advantage of this detector technology is that unlike phosphors where the light spreads between the point of creation and where it is detected, here the electrical charge is guided by the applied external field so that it the blurring due to spreading can be minimized. This allows the construction of a detector that is thick enough to have high x-ray absorption efficiency without the loss of spatial resolution that would occur as the thickness of the absorber in phosphor detectors is increased.

1.4.6.3 Photostimulable Phosphor Detector

In these systems, the x-ray absorber is a flat plate formed of a photostimulable phosphor material. When exposed to x-rays, electrons that are liberated during deceleration of the photoelectrons and Compton recoil electrons can become trapped in the phosphor. The number of electron traps filled is directly related to the total x-ray energy absorbed in the phosphor. The plate is placed in a reader device and scanned with a red HeNe laser beam which stimulates the traps to release the electrons (Fetterly and Schueler, 2003). The transition of these electrons through energy levels in the phosphor crystal results in the emission of blue light, which is measured as a function of the laser position on the plate to form the image signal as illustrated schematically in Figure 1.15.

Photostimulable phosphor detectors often have limited spatial resolution because the light from the readout laser scatters and discharges signal from traps that are laterally displaced from the point where the laser beam enters the phosphor. In addition, there is a fundamental compromise required in setting the electron trapping efficiency of the phosphor and this leads to a reduction in signal formation sensitivity. If the trapping efficiency is low, the electrons freed by the x-ray interaction will not be trapped and there will be loss of signal. On the other hand, if it is increased, then electrons that are detrapped during readout will be retrapped rather than creating the light.

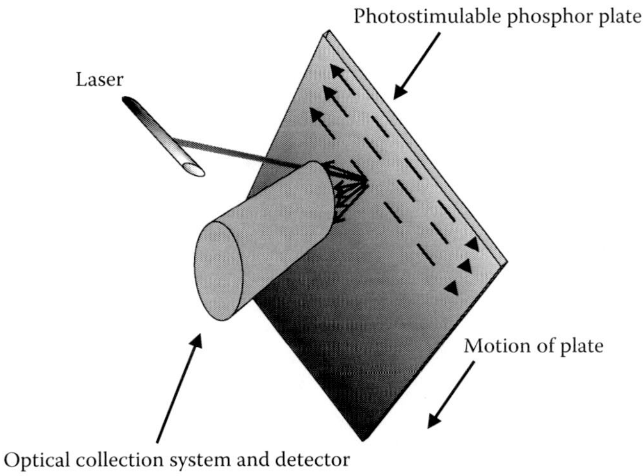

FIGURE 1.15 Readout system for photostimulable phosphor plates. The laser beam scans across the plate in one dimension as the plate is gradually moved through the reader in the orthogonal dimension. Typically red light is used to detrap electrons and blue light is emitted and collected to form the image.

1.4.6.4 X-Ray Quantum Counting Detectors

The detectors described previously all produce their signal by integrating the charge produced by many x-rays. Even if the x-rays were monoenergetic, there would be a statistical variation in the amount of charge deposited per quantum. Furthermore, these x-rays span an energy distribution, again causing fluctuation in the charge produced per quantum (Swank, 1973). These two sources of fluctuation form a type of image noise over and above the fundamental Poisson noise associated with the random interaction of x-rays in the detector. (Noise is discussed in the next section.) Additionally, when the detector responds to a spectrum of x-rays, the lower energy x-rays produce more contrast, but those at higher energy produce more charge. So the detector implicitly weights the signal with quanta that contain less diagnostic information. Quantum counting detectors are designed to produce one pulse of charge per x-ray quantum absorbed and that pulse is simply counted by the detector circuitry (Åslund et al., 2007). In this way, an equal signal (i.e., one count) is produced, regardless of the actual energy deposited by the quantum. A potential advantage of this design is an improvement in SNR per radiation dose or a dose reduction without loss of SNR.

A more thorough review of detectors for digital mammography is given in Yaffe (2010) and of detectors for digital x-ray imaging in Yaffe and Rowlands (1997).

1.4.7 Noise and Dose

Noise in mammography results primarily from two sources—the random absorption of x-rays in the detector and structural or electronic noise associated with the detector. The first, commonly known as quantum noise, is governed by Poisson's statistics so that for a given area of image the standard deviation in the number of x-rays recorded, σ_N is equal to the square root of the mean number, N, recorded. In other words, the noise in the image is dependent on both the amount of radiation striking the imaging system per unit area and the quantum efficiency of the imaging system. The SNR, $N/\sigma_N = N^{1/2}$, increases as the number of quanta used to form the image increases.

Structural noise is associated with the inherent granularity of the film emulsion in film imaging or the spatial nonuniformity of the detector sensitivity in digital systems. In digital systems other than CR systems, much of the structural noise, also referred to as fixed-pattern noise, can be removed from the image by correction. This is accomplished by producing a low-noise image (large exposure) of a uniform x-ray attenuator to record the pattern of detector sensitivity and any other nonuniformities in the x-ray field and then using this as a "flat-field" mask to correct subsequent images of the breast. Electronic noise arises from the circuitry employed to read out digital images.

For phosphor-based systems as used in screen-film mammography and in some digital mammography systems, in order to maintain high spatial resolution, the screen must be made relatively thin to avoid lateral diffusion of light and this limits the quantum efficiency that can be achieved. The loss of gain could be offset by increasing the conversion efficiency or, in film imaging, by employing a more sensitive film. In film mammography, both the brightness and contrast of the image are closely linked with the light exposure to the film. Therefore, there is a desired degree of exposure to the film. Increasing the gain by increasing the conversion efficiency could result in a film of the required optical density, but obtained using an insufficient number of x-rays to ensure adequate SNR. Similarly, the structural noise due to film granularity increases as more sensitive films are used, so that again film speed must be limited to maintain high image quality. For current high-quality mammographic imaging employing an antiscatter grid, with films exposed to a mean optical density of at least 1.6, the mean glandular dose to a 5.3 cm thick compressed breast is approximately 2.4 milligray. A full examination (two views of each breast) then yields a dose of approximately 4.8 mGy to each breast. For CR systems, it has been found in many cases that doses should be comparable to those used in screen-film mammography. For other types of digital mammography systems, the doses can typically be reduced by about 25–30% (Hendrick et al., 2010). For quantum counting detector systems, a greater dose decrease is often possible.

1.4.8 Automatic Control of Exposure Parameters

It is difficult for the technologist to estimate the attenuation of the breast by inspection, and, therefore, modern mammography units are equipped with automatic exposure control (AEC). For screen-film mammography, the AEC radiation sensors are located behind the image receptor so that they do not cast a shadow on the image. The sensors measure the x-ray fluence transmitted through both the breast and the receptor and provide a signal which can be used to discontinue the exposure when a certain preset amount of radiation has been received by the image receptor. The location of the sensor must be adjustable so that it can be placed behind the appropriate region of the breast in order to obtain proper image density. For film systems, the AEC devices must be calibrated so that approximately constant image optical density results independent of variations in breast attenuation, kilovoltage setting, or field size. With modern equipment, automatic exposure control is generally microprocessor-based so that relatively sophisticated corrections can be made during the exposure for the above effects and for *reciprocity law* failure of the film.

On digital mammography systems other than those using photostimulable phosphor detectors, the detector itself is generally employed as the sensor for automatic control. This provides much more flexibility as algorithms can be used to apply different weightings to signals from different areas of the image field and to ignore regions that are not relevant to the diagnosis. In digital mammography, brightness and contrast can be adjusted while viewing the image at a workstation. Therefore, the automatic control is often configured to terminate the exposure when a given SNR is obtained at a particular location in the image field (e.g., the most attenuating part of the breast).

1.4.8.1 Automatic Control of Kilovoltage and Target/Filter Combination

Many modern mammography units also incorporate automatic control of the filter/kilovoltage combination or, on systems with multitarget x-ray tubes, the target/filter/kilovoltage combination. Penetration through the breast depends on both breast thickness and composition. For a breast that is dense, it is possible that a very long exposure time would be required to achieve adequate film blackening or detector SNR. This results in high dose to the breast and possibly blur due to anatomical motion. It is possible to sense the compressed breast thickness and the transmitted exposure rate and to employ an algorithm to automatically choose the x-ray target and/or beam filter as well as the kilovoltage.

1.5 Quality Control

Mammography is one of the most technically demanding radiographic procedures, and in order to obtain optimal results, all components of the system must be operating properly. Recognizing this, many jurisdictions have implemented and administered mammography accreditation programs. These should evaluate not only the technical performance, but also the qualifications and continuing education of personnel in facilities performing mammography.

In order to verify proper operation, a rigorous quality control program should be in effect. As an example, a program of tests (summarized in Table 1.1) and methods for performing them are contained in the quality control manuals for screen-film mammography published by the American College of

TABLE 1.1 Mammographic Quality Control Minimum Test Frequencies

Test	Performed by	Minimum Frequency
Darkroom cleanliness	Radiologic technologist	Daily
Processor quality control		Daily
Screen cleanliness		Weekly
Viewboxes and viewing conditions		Weekly
Phantom images		Weekly
Visual check list		Monthly
Repeat analysis		Quarterly
Analysis of fixer retention in film		Quarterly
Darkroom fog		Semiannually
Screen-film contact		Semiannually
Compression		Semiannually
Mammographic unit assembly evaluation	Medical physicist	Annually
Collimation assessment		Annually
Focal spot size performance		Annually
kVp accuracy/reproducibility		Annually
Beam quality assessment (half-value layer)		Annually
Automatic exposure control (AEC) system performance assessment		Annually
Uniformity of screen speed		Annually
Breast entrance exposure and mean glandular dose		Annually
Image quality—phantom evaluation		Annually
Artifact assessment		Annually
Radiation output rate		Annually
Viewbox luminance and room illuminance		Annually
Compression release mechanism		Annually

Source: Hendrick RE et al. 1999. *Mammography Quality Control Manuals (Radiologist, Radiologic Technologist, Medical Physicist)*. American College of Radiology, Reston, VA. With permission.

Radiology (Hendrick et al., 1999). Programs for quality assurance in digital mammography have been or are being developed by The National Health Service in the United Kingdom, the European Community, the Flemish Screening Program in Belgium, the Norwegian government, The International Atomic Energy Agency (IAEA, 2011), and The American College of Radiology.

1.6 Breast CT, Digital Breast Tomosynthesis, and Contrast Imaging

All mammography, film or digital, is based on projection of x-rays through the three-dimensional breast onto a two-dimensional detector. Superposition of structures in some cases creates complex images in which cancers can be missed; conversely, the superposition can create the appearance of cancers in the projection image where no cancer actually exists. This is a fundamental limitation in all projection radiography and has been overcome in other parts of the body by creating three-dimensional images (sets of tomographic slices) by reconstruction from projections. Systems for dedicated breast computed tomography (CT) are beginning to appear (Boone et al., 2006; Ning et al., 2006). These are based on a table in which the woman lies prone with the breast pendant through an aperture. An x-ray tube and detector array rotate about the breast below the table to acquire image projections.

Another approach, called digital breast tomosynthesis (DBT), illustrated in Figure 1.16, employs the platform of the digital mammography system with one major modification; the x-ray source can rotate over a limited angular range about the stationary compressed breast and detector. By acquiring a set of from 9 to 30 low-dose projection images over a limited angular range of between ±10° and ±20° about the normal to the breast, either iterative techniques or filtered back projection algorithms can be used to reconstruct a set of tomographic slice images (Wu et al., 2003). These slices through the breast can be viewed as a movie loop. While tomosynthesis images do contain artifacts associated with the limited-angle data set, the images provide high spatial resolution in the x-ray plane and moderate (~1 mm) resolution in the *z* (depth in the breast) plane at doses only slightly higher than those employed for two-dimensional mammography.

Some breast cancers will not be seen on mammography, CT, or DBT because they lack the physical differences in attenuation properties that give rise to x-ray contrast. In such cases, it is possible that an

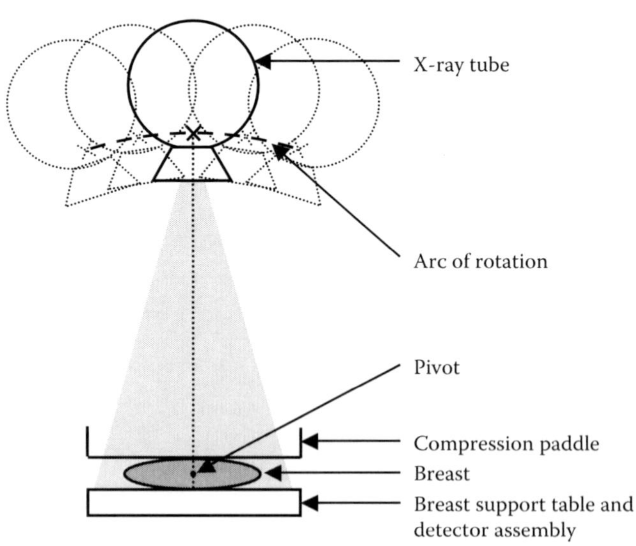

FIGURE 1.16 Concept of tomosynthesis. Tomographic images through the breast are reconstructed from a set of two-dimensional angulated x-ray projection images.

x-ray imaging technique that exploits the phenomenon of tumor angiogenesis, the development of new blood vessels to supply a tumor, may provide a signal that can reveal the presence and extent of the cancer. Angiogenic vessels tend to be poorly formed and leaky. If an iodinated contrast agent is injected into a vein in the arm, the contrast agent will leak and pool near the tumor and then will be washed out. The iodine signals are easily picked up in the mammographic image to reveal the abnormality. Because of the high atomic number of iodine, this type of imaging provides good contrast even at relatively high energies, where the x-rays easily penetrate the noniodinated areas, allowing doses to be kept low. Conspicuity of the lesion can be further enhanced by acquiring images before and after injection of the agent and displaying the result subtraction of the logarithms of the two images (Skarpathiotakis et al., 2002; Jong et al., 2003) or by acquiring images at energies below and above the K edge of iodine, and performing the appropriate weighted subtraction (Lewin et al., 2003; Dromain et al., 2006, 2009). In both cases, the iodine signal is largely isolated from that due to soft tissues in the breast, facilitating visualization of the tumor.

1.7 Stereotactic Biopsy Devices

Stereoscopic x-ray imaging techniques are currently used for the guidance of needle "core" biopsies. These procedures can be used to investigate suspicious mammographic or clinical findings without the need for surgical excisional biopsies, resulting in reduced patient risk, discomfort, and cost. In these stereotactic biopsies, the gantry of a mammography machine is modified to allow angulated views of the breast (typically ±15° from normal incidence) to be achieved. From measurements obtained from these images, the three-dimensional location of a suspicious lesion is determined and a needle equipped with a spring-loaded cutting device can be accurately placed in the breast to obtain the tissue samples. While this procedure can be performed on an upright mammography unit, special dedicated systems are available to allow its performance with the patient lying prone on a table. The accuracy of sampling the appropriate tissue depends critically on the alignment of the system components and the quality of the images produced. A thorough review of stereotactic imaging, including recommended quality control procedures, is given by Hendrick and Parker (1994).

1.8 Summary

Mammography is a technically demanding imaging procedure which can help reduce mortality from breast cancer. To be successful at this purpose, both the technology and technique used for imaging must be optimized. This requires careful system design and attention to quality control procedures. Digital mammography is gradually replacing screen-film mammography and three-dimensional techniques for mammography may provide improved accuracy for breast cancer detection and diagnosis.

Defining Terms

Conversion efficiency: The efficiency of converting the energy from x-rays absorbed in a phosphor material into that of emitted light quanta.

Fibroglandular tissue: A mixture of tissues within the breast composed of the functional glandular tissue and the fibrous supporting structures.

Focal spot: The area of the anode of an x-ray tube from which the useful beam of x-rays is emitted. Also known as the target.

Grid: A device consisting of evenly spaced lead strips which functions like a venetian blind in preferentially allowing x-rays traveling directly from the focal spot without interaction in the patient to pass through, while those whose direction has been diverted by scattering in the patient strike the slats of the grid and are rejected. Grids improve the contrast of radiographic images at the price of increased dose to the patient.

Image receptor: A device which records the distribution of x-rays to form an image. In mammography, the image receptor is generally composed of a light-tight cassette containing a fluorescent screen, which absorbs x-rays and produces light, coupled to a sheet of photographic film.

Quantum efficiency: The fraction of incident x-rays which interact with a detector or image receptor.

Screening: Examination of asymptomatic individuals to detect disease.

Survival: An epidemiological term giving the fraction of individuals diagnosed with a given disease alive at a specified time after diagnosis, for example, "10-year survival."

References

ACS American Cancer Society. Breast Cancer Facts and Figures 2009/2010. http://www.acsevents.org/downloads/STT/F861009_final%209-08-09.pdf accessed May 25, 2010.

Åslund M, Cederström B, Lundqvist M, and Danielsson M. 2007. Physical characterization of a scanning photon counting digital mammography system based on Si-strip detectors. *Med. Phys.* 34(6):1918–1925.

Beaman SA and Lillicrap SC. 1982. Optimum x-ray spectra for mammography. *Phys. Med. Biol.* 27:1209–1220.

BEIR. 1990. *Health Effects of Exposure to Low Levels of Ionizing Radiation (BEIR V)*. National Academy Press, Washington DC, pp. 163–170.

Boone JM. 1999. Glandular breast dose for monoenergetic and high-energy x-ray beams: Monte Carlo assessment. *Radiology* 213:23–37.

Boone JM, Nelson T, Kwan A et al. 2006. Computed tomography of the breast: Design, fabrication, characterization, and initial clinical testing. *Med. Phys.* 33:2185.

Dance DR, Skinner CL, Young KC, Beckett JR, and Kotre CJ. 2000. Additional factors for the estimation of mean glandular breast dose using the UK mammography dosimetry protocol. *Phys. Med. Biol.* 45:3225–3240.

Desponds L, Depeursinge C, Grecescu M et al. 1991. Image of anode and filter material on image quality and glandular dose for screen-film mammography. *Phys. Med. Biol.* 36:1165–1182.

Dromain C, Balleyguier C, Adlera G et al. 2009. Contrast-enhanced digital mammography. *Eur. J. Radiol.* 69:34–42.

Dromain C, Balleyguier C, Muller S et al. 2006. Evaluation of tumor angiogenesis of breast carcinoma using contrast-enhanced digital mammography. *AJR Am. J. Roentgenol.* 187(5):W528–37.

Fahrig R, Maidment ADA, and Yaffe MJ. 1992. Optimization of peak kilovoltage and spectral shape for digital mammography. *Proc. SPIE* 1651:74–83.

Fetterly KA and Schueler BA. 2003. Performance evaluation of a dual-side read dedicated mammography computed radiography system. *Med. Phys.* 30:1843–1853.

Heidsieck R, Laurencin G, Ponchin A et al. 1991. Dual target x-ray tubes for mammographic examinations: Dose reduction with image quality equivalent to that with standard mammographic tubes. *Radiology* 181(P):311.

Hendrick RE et al. 1994. *Mammography Quality Control Manuals (Radiologist, Radiologic Technologist, Medical Physicist)*. American College of Radiology, Reston, VA.

Hendrick RE et al. 1999. *Mammography Quality Control Manuals (Radiologist, Radiologic Technologist, Medical Physicist)*. American College of Radiology, Reston, VA.

Hendrick RE and Parker SH. 1994. Stereotaxic imaging. In *A Categorical Course in Physics: Technical Aspects of Breast Imaging*. 3rd Edition, eds. A.G. Haus and M.J. Yaffe, pp. 263–274. RSNA Publications, Oak Brook, IL.

Hendrick RE, Pisano ED, Averbuch A et al. 2010. Technical comparison of digital mammography to screen-film mammography in the American College of Radiology Imaging Network (ACRIN) Digital Mammographic Imaging Screening Trial (DMIST). *AJR Am. J. Roentgenol.* 194(2):362–369.

IAEA Quality Assurance Programme for Digital Mammography. 2011. IAEA Human Health Series No.17 International Atomic Energy Agency, Vienna.

Jennings RJ, Quinn PW, Gagne RM, and Fewell TR. 1993. Evaluation of x-ray sources for mammography. *Proc. SPIE* 1896:259–268.

Johns PC and Yaffe MJ. 1987. X-ray characterization of normal and neoplastic breast tissues. *Phys. Med. Biol.* 32:675–695.

Jong RA, Yaffe MJ, Skarpathiotakis M et al. 2003. Contrast-enhanced digital mammography: Initial clinical experience. *Radio. Sep.* 228(3):842–850.

Lewin JM, Isaacs PK, Vance V, and Larke FJ. 2003. Dual-energy contrast-enhanced digital subtraction mammography: Feasibility. *Radiology* 229(1):261–268.

Ning R, Conover D, Yu Y et al. 2006. A novel cone beam breast CT scanner: Preliminary system evaluation. *Proc. SPIE* 6142:614211.

Shaw J, Albagli D, Wei C-Y, and Granfors P. 2004. Enhanced a-Si/CsI–based flat panel x-ray detector for mammography. In *Medical Imaging: Physics of Medical Imaging*, eds. M.J. Yaffe and M.J. Flynn, pp. 370–378. *Proceedings of SPIE* Vol. 5368 SPIE, Bellingham, WA, 2004.

Skarpathiotakis M, Yaffe MJ, Bloomquist AK et al. 2002. Development of contrast digital mammography. *Med. Phys.* 29(10):2419–26.

Smart CR, Hartmann WH, Beahrs OH et al. 1993. Insights into breast cancer screening of younger women: Evidence from the 14-year follow-up of the Breast Cancer Detection Demonstration Project. *Cancer* 72:1449–1456.

Swank RK. 1973. Absorption and noise in x-ray phosphors. *J. Appl. Phys.* 44:4199–4203.

Tabar L, Duffy SW, and Burhenne LW. 1993. New Swedish breast cancer detection results for women aged 40–49. *Cancer* 72(Suppl):1437–1448.

Wagner AJ. 1991. Contrast and grid performance in mammography. In *Screen Film Mammography: Imaging Considerations and Medical Physics Responsibilities,* eds. G.T. Barnes and G.D. Frey, pp. 115–134. Medical Physics Publishing, Madison, WI.

Wu T, Stewart A, Stanton M et al. 2003. Tomographic mammography using a limited number of low-dose cone-beam projection images. *Med. Phys.* 30:365.

Wu X, Barnes GT, and Tucker DM. 1991. Spectral dependence of glandular tissue dose in screen-film mammography. *Radiology* 179:143–148.

Wu X, Gingold EL, Barnes GT, and Tucker DM. 1994. Normalized average glandular dose in molybdenum target-rhodium filter and rhodium target-rhodium filter mammography. *Radiology* 193:83–89.

Yaffe MJ. Detectors for digital mammography. 2010. In: *Digital Mammography Series: Medical Radiology Subseries: Diagnostic Imaging Bick*, Ulrich; Diekmann, Felix (Eds), Springer, XVI, 220 p.

Yaffe MJ and Rowlands JA. 1997. X-ray detectors for digital radiography. *Phys Med Biol.* 42:1–39.

Yaffe MJ, Boone JM, Packard N et al. 2009. The myth of the 50–50 breast. *Med. Phys.* 36:5437–5443.

Yorker JG, Jeromin LS, Lee DL et al. 2002. Characterization of a full-field digital mammography detector based on direct x-ray conversion in selenium. *Proc. SPIE* 4682:21.

Further Information

Barnes GT and Frey GD (eds). 1991. *Screen Film Mammography: Imaging Considerations and Medical Physics Responsibilities.* Medical Physics Publishing, Madison, WI. Considerable practical information related to obtaining and maintaining high quality mammography is provided here.

Haus AG (ed). 1993. *Film Processing in Medical Imaging.* Medical Physics Publishing. Madison, WI. This book deals with all aspects of medical film processing with particular emphasis on mammography.

Haus AG and Yaffe MJ. 1994. *A Categorical Course in Physics: Technical Aspects of Breast Imaging.* RSNA Publications, Oak Brook, IL. In this syllabus to a course presented at the Radiological Society of North America, all technical aspects of mammography are addressed by experts and a clinical overview is presented in language understandable by the physicist or biomedical engineer.

Yaffe MJ et al. 1993. *Recommended Specifications for New Mammography Equipment: ACR-CDC Cooperative Agreement for Quality Assurance Activities in Mammography.* ACR Publications, Reston VA.

2

Computed Tomography

Ian A. Cunningham
University of Western Ontario

Philip F. Judy
Harvard Medical School

2.1 Instrumentation

Ian A. Cunningham

The development of **computed tomography (CT)** in the early 1970s revolutionized medical radiology. For the first time, physicians were able to obtain high-quality tomographic (cross-sectional) images of internal structures of the body. Over the next ten years, 18 manufacturers competed for the exploding world CT market. Technical sophistication increased dramatically, and even today, CT continues to mature, with new capabilities being researched and developed.

Computed tomographic images are reconstructed from a large number of measurements of **x-ray transmission** through the patient (called **projection data**). The resulting images are tomographic "maps" of the x-ray linear **attenuation** coefficient. The mathematical methods used to **reconstruct** CT images from projection data are discussed in the next section. In this section, the hardware and instrumentation in a modern scanner are described.

The first practical CT instrument was developed in 1971 by **Dr. G.N. Hounsfield** in England and was used to image the brain (Hounsfield, 1980). The projection data were acquired in approximately 5 min, and the tomographic image was reconstructed in approximately 20 min. Since then, CT technology has developed dramatically, and CT has become a standard imaging procedure for virtually all parts of the body in thousands of facilities throughout the world. Projection data are typically acquired in approximately 1 sec, and the image is reconstructed in 3 to 5 sec. One special-purpose scanner described below acquires the projection data for one tomographic image in 50 msec. A typical modern CT scanner is shown in Figure 2.1, and typical CT images are shown in Figure 2.2.

The fundamental task of CT systems is to make an extremely large number (approximately 500,000) of highly accurate measurements of x-ray transmission through the patient in a precisely controlled geometry. A basic system generally consists of a **gantry**, a patient table, a **control console**, and a

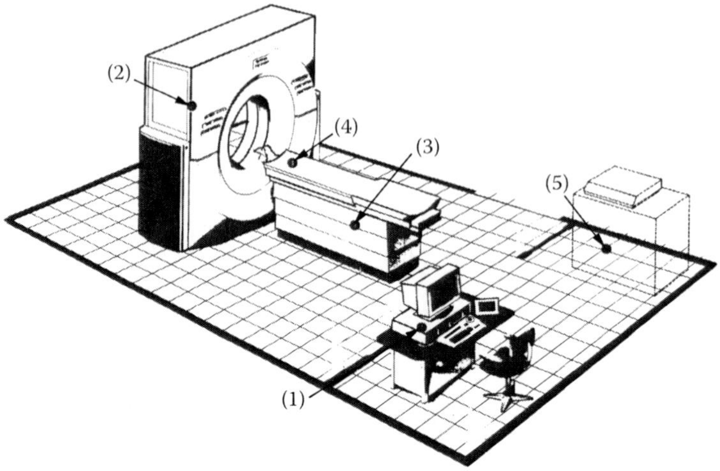

FIGURE 2.1 Schematic drawing of a typical CT scanner installation, consisting of (1) control console, (2) gantry stand, (3) patient table, (4) head holder, and (5) laser imager. (Courtesy of Picker International, Inc.)

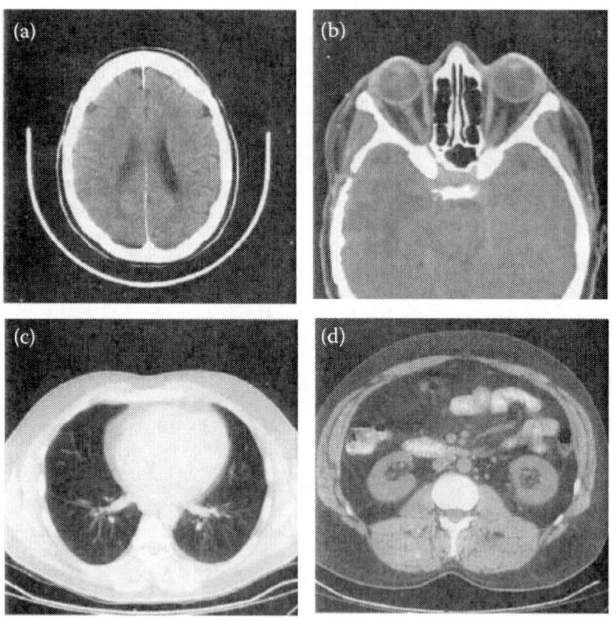

FIGURE 2.2 Typical CT images of (a) brain, (b) head showing orbits, (c) chest showing lungs, and (d) abdomen.

computer. The gantry contains the **x-ray source**, **x-ray detectors**, and the **data-acquisition system (DAS)**.

2.1.1 Data-Acquisition Geometries

Projection data may be acquired in one of several possible geometries described below, based on the scanning configuration, scanning motions, and detector arrangement. The evolution of these geometries is descried in terms of "generations," as illustrated in Figure 2.3, and reflects the historical

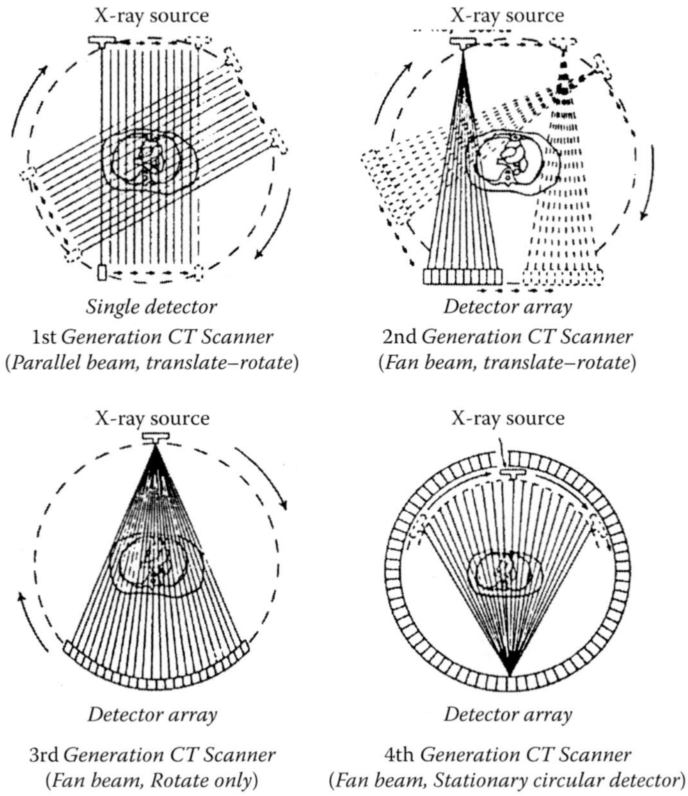

FIGURE 2.3 Four generations of CT scanners illustrating the parallel- and fan-beam geometries. (Taken from Robb R.A. 1982. CRC *Crit. Rev. Biomed. Eng.* 7: 265.)

development (Newton and Potts, 1981; Seeram, 1994). Current CT scanners use either third-, fourth-, or fifth-generation geometries, each having their own pros and cons.

2.1.1.1 First Generation: Parallel-Beam Geometry

Parallel-beam geometry is the simplest technically and the easiest with which to understand the important CT principles. Multiple measurements of x-ray transmission are obtained using a single highly collimated x-ray **pencil beam** and detector. The beam is translated in a linear motion across the patient to obtain a projection profile. The source and detector are then rotated about the patient isocenter by approximately 1 degree, and another projection profile is obtained. This translate-rotate scanning motion is repeated until the source and detector have been rotated by 180 degrees. The highly collimated beam provides excellent rejection of radiation scattered in the patient; however, the complex scanning motion results in long (approximately 5-min) **scan times.** This geometry was used by Hounsfield in his original experiments (Hounsfield, 1980) but is not used in modern scanners.

2.1.1.2 Second Generation: Fan Beam, Multiple Detectors

Scan times were reduced to approximately 30 sec with the use of a **fan beam** of x-rays and a linear **detector array**. A translate-rotate scanning motion was still employed; however, a larger rotate increment could be used, which resulted in shorter scan times. The reconstruction algorithms are slightly more complicated than those for first-generation algorithms because they must handle fan-beam projection data.

2.1.1.3 Third Generation: Fan Beam, Rotating Detectors

Third-generation scanners were introduced in 1976. A fan beam of x-rays is rotated 360 degrees around the isocenter. No translation motion is used; however, the fan beam must be wide enough to completely contain the patient. A curved detector array consisting of several hundred independent detectors is mechanically coupled to the x-ray source, and both rotate together. As a result, these rotate-only motions acquire projection data for a single image in as little as 1 sec. Third-generation designs have the advantage that thin tungsten septa can be placed between each detector in the array and focused on the x-ray source to reject **scattered radiation**.

2.1.1.4 Fourth Generation: Fan Beam, Fixed Detectors

In a fourth-generation scanner, the x-ray source and fan beam rotate about the isocenter, while the detector array remains stationary. The detector array consists of 600 to 4800 (depending on the manufacturer) independent detectors in a circle that completely surrounds the patient. Scan times are similar to those of third-generation scanners. The detectors are no longer coupled to the x-ray source and hence cannot make use of **focused septa** to reject scattered radiation. However, detectors are calibrated twice during each rotation of the x-ray source, providing a self-calibrating system. Third-generation systems are calibrated only once every few hours.

Two detector geometries are currently used for fourth-generation systems (1) a rotating x-ray source inside a fixed detector array and (2) a rotating x-ray source outside a nutating detector array. Figure 2.4 shows the major components in the gantry of a typical fourth-generation system using a fixed-detector array. Both third- and fourth-generation systems are commercially available, and both have been highly successful clinically. Neither can be considered an overall superior design.

2.1.1.5 Fifth Generation: Scanning Electron Beam

Fifth-generation scanners are unique in that the x-ray source becomes an integral part of the system design. The detector array remains stationary, while a high-energy electron beams is electronically swept along a semicircular tungsten strip **anode**, as illustrated in Figure 2.5. X-rays are produced at the point where the electron beam hits the anode, resulting in a source of x-rays that rotates about the patient with no moving parts (Boyd et al., 1979). Projection data can be acquired in approximately 50 msec, which is fast enough to image the beating heart without significant motion artifacts (Boyd and Lipton, 1983).

An alternative fifth-generation design, called the dynamic spatial reconstructor (DSR) scanner, is in use at the Mayo Clinic (Ritman, 1980, 1990). This machine is a research prototype and is not available commercially. It consists of 14 x-ray tubes, scintillation screens, and video cameras. **Volume CT** images can be produced in as little as 10 msec.

2.1.1.6 Spiral/Helical Scanning

The requirement for faster scan times, and in particular for fast multiple scans for **three-dimensional imaging**, has resulted in the development of spiral (**helical**) scanning systems (Kalendar et al., 1990). Both third- and fourth-generation systems achieve this using self-lubricating slip-ring technology (Figure 2.6) to make the electrical connections with rotating components. This removes the need for power and signal cables which would otherwise have to be rewound between scans and allows for a continuous rotating motion of the x-ray fan beam. Multiple images are acquired while the patient is translated through the gantry in a smooth continuous motion rather than stopping for each image. Projection data for multiple images covering a volume of the patient can be acquired in a single breath hold at rates of approximately one **slice** per second. The reconstruction algorithms are more sophisticated because they must accommodate the spiral or helical path traced by the x-ray source around the patient, as illustrated in Figure 2.7.

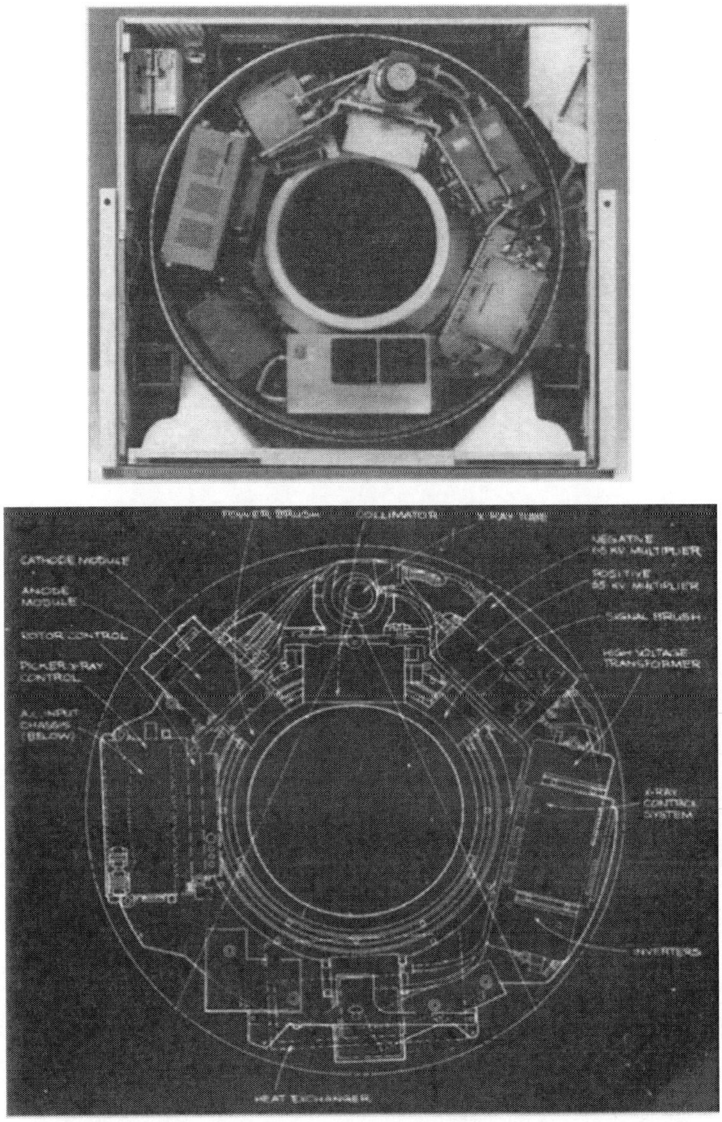

FIGURE 2.4 The major internal components of a fourth-generation CT gantry are shown in a photograph with the gantry cover removed (upper) and identified in the line drawing (lower). (Courtesy of Picker International, Inc.)

2.1.2 X-Ray System

The x-ray system consists of the x-ray source, detectors, and a data-acquisition system.

2.1.2.1 X-Ray Source

With the exception of one fifth-generation system described above, all CT scanners use bremsstrahlung x-ray tubes as the source of radiation. These tubes are typical of those used in diagnostic imaging and produce x-rays by accelerating a beam of electrons onto a target anode. The anode area from which x-rays are emitted, projected along the direction of the beam, is called the **focal spot**. Most systems have two possible focal spot sizes, approximately 0.5×1.5 mm and 1.0×2.5 mm. A collimator assembly is

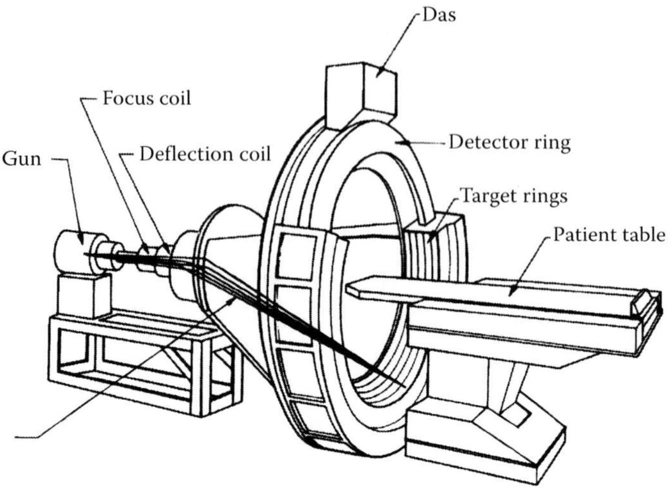

FIGURE 2.5 Schematic illustration of a fifth-generation ultrafast CT system. Image data are acquired in as little as 50 msec, as an electron beam is swept over the strip anode electronically. (Courtesy of Imatron, Inc.)

FIGURE 2.6 Photograph of the slip rings used to pass power and control signals to the rotating gantry. (Courtesy of Picker International, Inc.)

used to control the width of the fan beam between 1.0 and 10 mm, which in turn controls the width of the imaged slice.

The power requirements of these tubes are typically 120 kV at 200 to 500 mA, producing x-rays with an energy spectrum ranging between approximately 30 and 120 keV. All modern systems use high-frequency generators, typically operating between 5 and 50 kHz (Brunnett et al., 1990). Some spiral systems use a stationary generator in the gantry, requiring high-voltage (120-kV) slip rings, while others use a rotating generator with lower-voltage (480-V) slip rings. Production of x-rays in bremsstrahlung tubes is an inefficient process, and hence most of the power delivered to the tubes results in heating

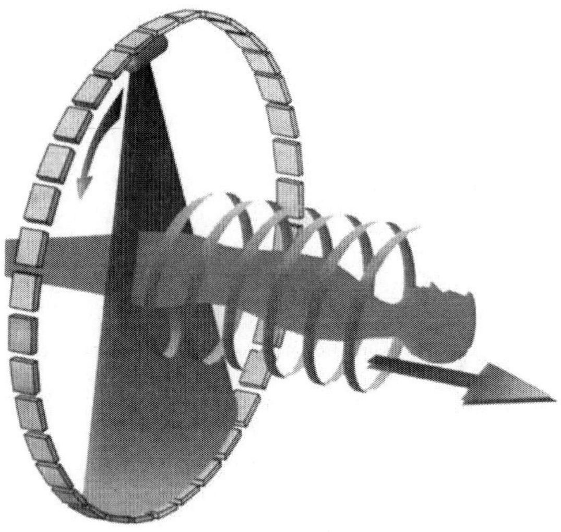

FIGURE 2.7 Spiral scanning causes the focal spot to follow a spiral path around the patient as indicated. (Courtesy of Picker International, Inc.)

of the anode. A heat exchanger on the rotating gantry is used to cool the tube. **Spiral scanning**, in particular, places heavy demands on the heat-storage capacity and cooling rate of the x-ray tube.

The intensity of the x-ray beam is attenuated by **absorption** and scattering processes as it passes through the patient. The degree of attenuation depends on the energy spectrum of the x-rays as well as on the average atomic number and mass density of the patient tissues. The transmitted intensity is given by

$$I_t = I_o e^{-\int_0^L \mu(x)\mathrm{d}x} \tag{2.1}$$

where I_o and I_t are the incident and transmitted beam intensities, respectively; L is the length of the x-ray path; and $m(x)$ is the **x-ray linear attenuation coefficient**, which varies with tissue type and hence is a function of the distance x through the patient. The integral of the attenuation coefficient is therefore given by

$$\int_0^L \mu(x)\mathrm{d}x = -\frac{1}{L}\ln(I_t/I_0) \tag{2.2}$$

The reconstruction algorithm requires measurements of this integral along many paths in the fan beam at each of many angles about the isocenter. The value of L is known, and I_o is determined by a system calibration. Hence values of the integral along each path can be determined from measurements of I_t.

2.1.2.2 X-Ray Detectors

X-ray detectors used in CT systems must (a) have a high overall efficiency to minimize the patient radiation dose, have a large dynamic range, (b) be very stable with time, and (c) be insensitive to temperature variations within the gantry. Three important factors contributing to the detector efficiency are geometric efficiency, quantum (also called capture) efficiency, and conversion efficiency (Villafanaet et al., 1987). Geometric efficiency refers to the area of the detectors sensitive to radiation as a fraction of the total exposed area. Thin septa between detector elements to remove scattered radiation, or other insensitive regions, will degrade this value. Quantum efficiency refers to the fraction of incident x-rays on the detector that are absorbed and contribute to the measured signal. Conversion efficiency refers to the ability to accurately convert the absorbed x-ray signal into an electrical signal (but is not the same as the energy conversion efficiency). Overall efficiency is the product of the three, and it generally lies between 0.45 and 0.85. A value of less than 1 indicates a nonideal detector system and results in a required increase in patient radiation dose if image quality is to be maintained. The term dose efficiency sometimes has been used to indicate overall efficiency.

Modern commercial systems use one of two detector types: solid-state or gas ionization detectors.

Solid-State Detectors. Solid-state detectors consist of an array of scintillating crystals and photodiodes, as illustrated in Figure 2.8. The scintillators generally are either cadmium tungstate (CdWO4) or a ceramic material made of rare earth oxides, although previous scanners have used bismuth germanate crystals with photomultiplier tubes. Solid-state detectors generally have very high quantum and conversion efficiencies and a large dynamic range.

Gas Ionization Detectors. Gas ionization detectors, as illustrated in Figure 2.9, consist of an array of chambers containing compressed gas (usually xenon at up to 30 atm pressure). A high voltage is applied to tungsten septa between chambers to collect ions produced by the radiation. These detectors have excellent stability and a large dynamic range; however, they generally have a lower quantum efficiency than solid-state detectors.

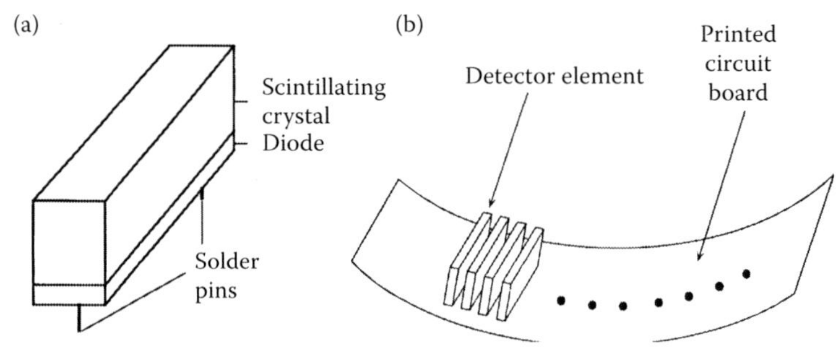

FIGURE 2.8 (a) A solid-state detector consists of a scintillating crystal and photodiode combination. (b) Many such detectors are placed side by side to form a detector array that may contain up to 4800 detectors.

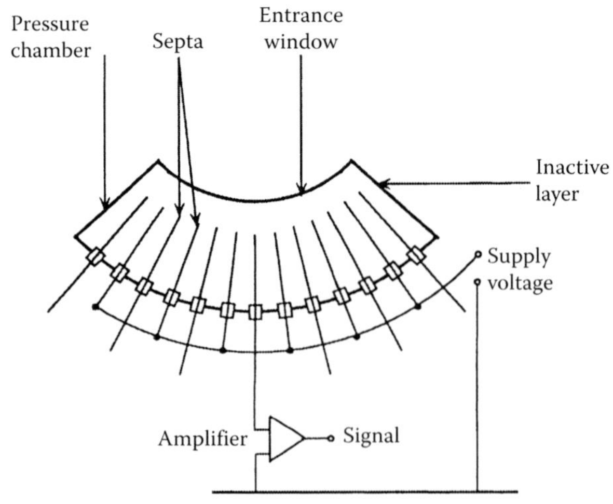

FIGURE 2.9 Gas ionization detector arrays consist of high-pressure gas in multiple chambers separated by thin septa. A voltage is applied between alternating septa. The septa also act as electrodes and collect the ions created by the radiation, converting them into an electrical signal.

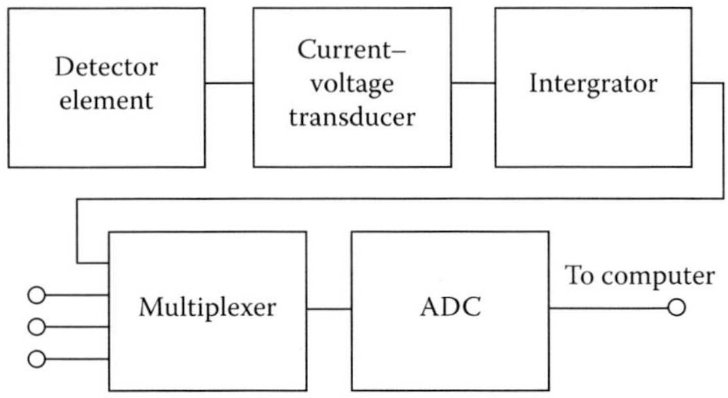

FIGURE 2.10 The data-acquisition system converts the electrical signal produced by each detector to a digital value for the computer.

2.1.2.3 Data-Acquisition System

The transmitted fraction I_t/I_0 in Equation 2.2 through an obese patient can be less than 10 to 4. Thus it is the task of the data-acquisition system (DAS) to accurately measure I_t over a dynamic range of more than 104, encode the results into digital values, and transmit the values to the system computer for reconstruction. Some manufacturers use the approach illustrated in Figure 2.10, consisting of precision preamplifiers, current-to-voltage converters, analog integrators, multiplexers, and analog-to-digital converters. Alternatively, some manufacturers use the preamplifier to control a synchronous voltage-to-frequency converter (SVFC), replacing the need for the integrators, multiplexers, and analog-to-digital converters (Brunnett et al., 1990). The logarithmic conversion required in Equation 2.2 is performed with either an analog logarithmic amplifier or a digital lookup table, depending on the manufacturer.

Sustained data transfer rates to the computer are as high as 10 Mbytes/sec for some scanners. This can be accomplished with a direct connection for systems having a fixed detector array. However, third-generation slip-ring systems must use more sophisticated techniques. At least one manufacturer uses optical transmitters on the rotating gantry to send data to fixed optical receivers (Siemens, 1989).

2.1.2.4 Computer System

Various computer systems are used by manufacturers to control system hardware, acquire the projection data, and reconstruct, display, and manipulate the tomographic images. A typical system is illustrated in Figure 2.11, which uses 12 independent processors connected by a 40-Mbyte/sec multibus. Multiple custom array processors are used to achieve a combined computational speed of 200 MFLOPS (million floating-point operations per second) and a reconstruction time of approximately 5 sec to produce an image on a 1024 × 1024 pixel display. A simplified UNIX operating system is used to provide a multitasking, multiuser environment to coordinate tasks.

2.1.3 Patient Dose Considerations

The patient dose resulting from CT examinations is generally specified in terms of the CT dose index (CTDI) (Felmlee et al., 1989; Rothenberg and Pentlow, 1992), which includes the dose contribution from radiation scattered from nearby slices. A summary of CTDI values, as specified by four manufacturers, is given in Table 2.1.

2.1.4 Summary

Computed tomography revolutionized medical radiology in the early 1970s. Since that time, CT technology has developed dramatically, taking advantage of developments in computer hardware and detector technology. Modern systems acquire the projection data required for one tomographic image in approximately 1 sec and present the reconstructed image on a 1024 × 1024 matrix display within a few seconds.

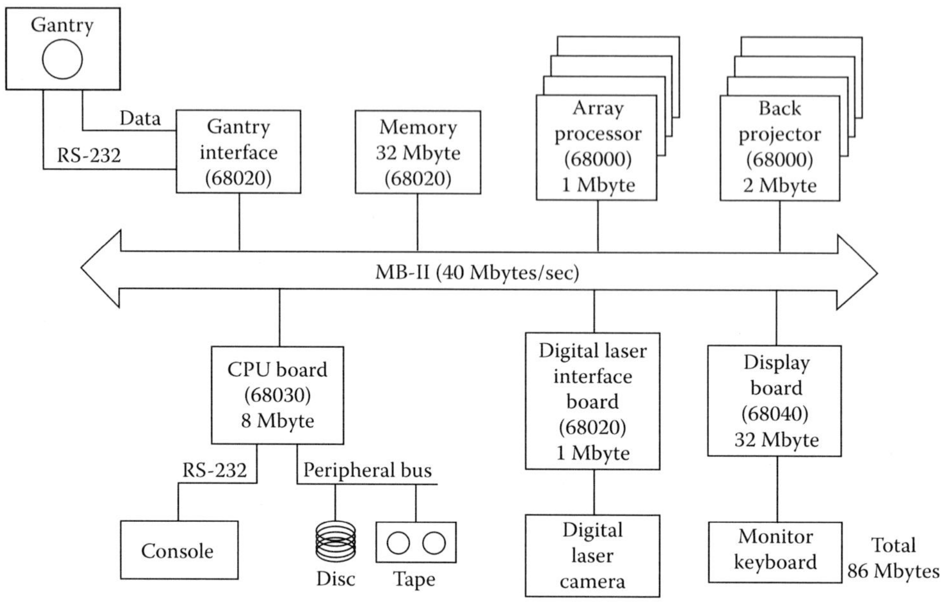

FIGURE 2.11 The computer system controls the gantry motions, acquires the x-ray transmission measurements, and reconstructs the final image. The system shown here uses 1,268,000-family CPUs. (Courtesy of Picker International, Inc.)

TABLE 2.1 Summary of the CT Dose Index (CTDI) Values at Two Positions (Center of the Patient and Near the Skin) as Specified by Four CT Manufacturers for Standard Head and Body Scans

Manufacturer	Detector	kVp	mA	Scan Time (sec)	CTDI, Center (mGy)	CTDI, Skin (mGy)
A, head	Xenon	120	170	2	50	48
A, body	Xenon	120	170	2	14	25
A, head	Solid state	120	170	2	40	40
A, body	Solid state	120	170	2	11	20
B, head	Solid state	130	80	2	37	41
B, body	Solid state	130	80	2	15	34
C, head	Solid state	120	500	2	39	50
C, body	Solid state	120	290	1	12	28
D, head	Solid state	120	200	2	78	78
D, body	Solid state	120	200	2	9	16

The images are high-quality tomographic "maps" of the x-ray linear attenuation coefficient of the patient tissues.

Defining Terms

Absorption: Some of the incident x-ray energy is absorbed in patient tissues and hence does not contribute to the transmitted beam.

Anode: A tungsten bombarded by a beam of electrons to produce x-rays. In all but one fifth-generation system, the anode rotates to distribute the resulting heat around the perimeter. The anode heat-storage capacity and maximum cooling rate often limit the maximum scanning rates of CT systems.

Attenuation: The total decrease in the intensity of the primary x-ray beam as it passes through the patient, resulting from both scatter and absorption processes. It is characterized by the linear attenuation coefficient.

Computed tomography (CT): A computerized method of producing x-ray tomographic images. Previous names for the same thing include computerized tomographic imaging, computerized axial tomography (CAT), computer-assisted tomography (CAT), and reconstructive tomography (RT).

Control console: The control console is used by the CT operator to control the scanning operations, image reconstruction, and image display.

Cormack, Dr. Allan MacLeod: A physicist who developed mathematical techniques required in the reconstruction of tomographic images. Dr. Cormack shared the Nobel Prize in Medicine and Physiology with Dr. G.N. Hounsfield in 1979 (Cormack, 1980).

Data-acquisition system (DAS): Interfaces the x-ray detectors to the system computer and may consist of a preamplifier, integrator, multiplexer, logarithmic amplifier, and analog-to-digital converter.

Detector array: An array of individual detector elements. The number of detector elements varies between a few hundred and 4800, depending on the acquisition geometry and manufacturer. Each detector element functions independently of the others.

Fan beam: The x-ray beam is generated at the focal spot and so diverges as it passes through the patient to the detector array. The thickness of the beam is generally selectable between 1.0 and 10 mm and defines the slice thickness.

Focal spot: The region of the anode where x-rays are generated.

Focused septa: Thin metal plates between detector elements which are aligned with the focal spot so that the primary beam passes unattenuated to the detector elements, while scattered x-rays which normally travel in an altered direction are blocked.

Gantry: The largest component of the CT installation, containing the x-ray tube, collimators, detector array, DAS, other control electronics, and the mechanical components required for the scanning motions.

Helical scanning: The scanning motions in which the x-ray tube rotates continuously around the patient while the patient is continuously translated through the fan beam. The focal spot therefore traces a helix around the patient. Projection data are obtained which allow the reconstruction of multiple contiguous images. This operation is sometimes called spiral, volume, or three-dimensional CT scanning.

Hounsfield, Dr. Godfrey Newbold: An engineer who developed the first practical CT instrument in 1971. Dr. Hounsfield received the McRobert Award in 1972 and shared the Nobel Prize in Medicine and Physiology with Dr. A.M. Cormack in 1979 for this invention (Hounsfield, 1980).

Image plane: The plane through the patient that is imaged. In practice, this plane (also called a slice) has a selectable thickness between 1.0 and 10 mm centered on the image plane.

Pencil beam: A narrow, well-collimated beam of x-rays.

Projection data: The set of transmission measurements used to reconstruct the image.

Reconstruct: The mathematical operation of generating the tomographic image from the projection data.

Scan time: The time required to acquire the projection data for one image, typically 1.0 sec.

Scattered radiation: Radiation that is removed from the primary beam by a scattering process. This radiation is not absorbed but continues along a path in an altered direction.

Slice: See Image plane.

Spiral scanning: See Helical scanning.

Three-dimensional imaging: See Helical scanning.

Tomography: A technique of imaging a cross-sectional slice.

Volume CT: See Helical scanning.

X-ray detector: A device that absorbs radiation and converts some or all of the absorbed energy into a small electrical signal.

X-ray linear attenuation coefficient *m*: Expresses the relative rate of attenuation of a radiation beam as it passes through a material. The value of *m* depends on the density and atomic number of the material and on the x-ray energy. The units of *m* are cm^{-1}.

X-ray source: The device that generates the x-ray beam. All CT scanners are rotating-anode bremsstrahlung x-ray tubes except one-fifth generation system, which uses a unique scanned electron beam and a strip anode.

X-ray transmission: The fraction of the x-ray beam intensity that is transmitted through the patient without being scattered or absorbed. It is equal to I_t/I_o in Equation 2.2, can be determined by measuring the beam intensity both with (I_t) and without (I_o) the patient present, and is expressed as a fraction. As a rule of thumb, n^2 independent transmission measurements are required to reconstruct an image with an $n \times n$ sized pixel matrix.

References

Body D.P. et al. 1979. A proposed dynamic cardiac 3D densitometer for early detection and evaluation of heart disease. *IEEE Trans. Nucl. Sci.* 26: 2724–2727.

Boyd D.P. and Lipton M.J. 1983. Cardiac computed tomography. *Proc. IEEE* 71: 198–307.

Brunnett C.J., Heuscher D.J., Mattson R.A., and Vrettos C.J. 1990. CT Design Considerations and Specifications. Picker International, CT Engineering Department, Ohio.

Cormack A.M. 1980. Nobel award address: early two-dimensional reconstruction and recent topics stemming from it. *Med. Phys.* 7(4): 277.

Felmlee J.P., Gray J.E., Leetzow M.L., and Price J.C. 1989. Estimated fetal radiation dose from multislice CT studies. *Am. Roent. Ray. Soc.* 154: 185.

Hounsfield G.N. 1980. Nobel award address: computed medical imaging. *Med. Phys.* 7(4): 283.

Kalendar W.A., Seissler W., Klotz E. et al. 1990. Spiral volumetric CT with single-breath-hold technique, continuous transport, and continuous scanner rotation. *Radiology* 176: 181.

Newton T.H. and Potts D.G. (eds). 1981. *Radiology of the Skull and Brain: Technical Aspects of Computed Tomography.* St. Louis, Mosby.

Picker. 1990. *Computed Dose Index PQ2000 CT Scanner.* Ohio, Picker International.

Ritman E.L. 1980. Physical and technical considerations in the design of the DSR, and high temporal resolution volume scanner. *AJR* 134: 369.

Ritman E.L. 1990. Fast computed tomography for quantitative cardiac analysis—state of the art and future perspectives. *Mayo Clin. Proc.* 65: 1336.

Robb R.A. 1982. X-ray computed tomography: an engineering synthesis of multiscientific principles. *CRC Crit. Rev. Biomed. Eng.* 7: 265.

Rothenberg L.N. and Pentlow K.S. 1992. Radiation dose in CT. *RadioGraphics* 12: 1225.

Seeram E. 1994. *Computed Tomography: Physical Principles, Clinical Applications and Quality Control.* Philadelphia, Saunders.

Siemens. 1989. *The Technology and Performance of the Somatom Plus.* Erlangen, Germany, Siemens Aktiengesellschaft, Medical Engineering Group.

Villafana T., Lee S.H., and Rao K.C.V.G. (eds). 1987. *Cranial Computed Tomography.* New York, McGraw-Hill.

Further Information

A recent summary of CT instrumentation and concepts is given by E. Seeram in *Computed Tomography: Physical Principles, Clinical Applications and Quality Control.* The author summarizes CT from the perspective of the nonmedical, nonspecialist user. A summary of average CT patient doses is described by

Rothenberg and Pentlow (1992) in *Radiation Dose in CT*. Research papers on both fundamental and practical aspects of CT physics and instrumentation are published in numerous journals, including *Medical Physics*, *Physics in Medicine and Biology*, *Journal of Computer Assisted Tomography*, *Radiology*, *British Journal of Radiology*, and the IEEE Press. A comparison of technical specifications of CT systems provided by the manufacturers is available from ECRI to help orient the new purchaser in a selection process. Their Product Comparison System includes a table of basic specifications for all the major international manufactures.

2.2 Reconstruction Principles

Philip F. Judy

Computed tomography (CT) is a two-step process: (1) the transmission of an x-ray beam is measured through all possible straight-line paths as in a plane of an object, and (2) the attenuation of an x-ray beam is estimated at points in the object. Initially, the transmission measurements will be assumed to be the results of an experiment performed with a narrow monoenergetic beam of x-rays that are confined to a plane. The designs of devices that attempt to realize these measurements are described in the preceding section. One formal consequence of these assumptions is that the logarithmic transformation of the measured x-ray intensity is proportional to the line integral of attenuation coefficients. In order to satisfy this assumption, computer processing procedures on the measurements of x-ray intensity are necessary even before image reconstruction is performed. These linearization procedures will reviewed after background.

Both analytical and iterative estimations of linear x-ray attenuation have been used for transmission CT reconstruction. Iterative procedures are of historic interest because an iterative reconstruction procedure was used in the first commercially successful CT scanner (EMI, Mark I, Hounsfield, 1973). They also permit easy incorporation of physical processes that cause deviations from the linearity. Their practical usefulness is limited. The first EMI scanner required 20 min to finish its reconstruction. Using the identical hardware and employing an analytical calculation, the estimation of attenuation values was performed during the 4.5-min data acquisition and was made on a 160×160 matrix. The original iterative procedure reconstructed the attenuation values on an 80×80 matrix and consequently failed to exploit all the spatial information inherent in transmission data.

Analytical estimation, or direct reconstruction, uses a numerical approximation of the inverse Radon transform (Radon, 1917). The direct reconstruction technique (convolution-backprojection) presently used in x-ray CT was initially applied in other areas such as radio astronomy (Bracewell and Riddle, 1967) and electron microscopy (Crowther et al., 1970; Ramachandran and Lakshminarayana, 1971). These investigations demonstrated that the reconstructions from the discrete spatial sampling of bandlimited data led to full recovery of the cross-sectional attenuation. The random variation (noise) in x-ray transmission measurements may not be bandlimited. Subsequent investigators (Herman and Rowland, 1973; Shepp and Logan, 1974; Chesler and Riederer, 1975) have suggested various bandlimiting windows that reduce the propagation and amplification of noise by the reconstruction. These issues have been investigated by simulation, and investigators continue to pursue these issues using a computer phantom (e.g., Guedon and Bizais, 1994, and references therein) described by Shepp and Logan. The subsequent investigations of the details of choice of reconstruction parameters has had limited practical impact because real variation of transmission data is bandlimited by the finite size of the focal spot and radiation detector, a straightforward design question, and because random variation of the transmission tends to be uncorrelated. Consequently, the classic precedures suffice.

2.2.1 Image Processing: Artifact and Reconstruction Error

An artifact is a reconstruction defect that is obviously visible in the image. The classification of an image feature as an artifact involves some visual criterion. The effect must produce an image feature that is greater than the random variation in image caused by the intrinsic variation in transmission measurements. An artifact not recognized by the physician observer as an artifact may be reported as a lesion. Such false-positive reports could lead to an unnecessary medical procedure, for example, surgery to remove an imaginary tumor. A reconstruction error is a deviation of the reconstruction value from its expected value. Reconstruction errors are significant if the application involves a quantitative measurement, not a common medical application. The reconstruction errors are characterized by identical material at different points in the object leading to different reconstructed attenuation values in the image which are not visible in the medical image.

Investigators have used computer simulation to investigate artifact (Herman, 1980) because image noise limits the visibility of their visibility. One important issue investigated was required spatial sampling of transmission slice plane (Crawford and Kak, 1979; Parker et al., 1982). These simulations provided a useful guideline in design. In practice, these aliasing artifacts are overwhelmed by random noise, and designers tend to oversample in the slice plane. A second issue that was understood by computer simulation was the partial volume artifact (Glover and Pelc, 1980). This artifact would occur even for mononergetic beams and finite beam size, particularly in the axial dimension. The axial dimension of the beams tend to be greater (about 10 mm) than their dimensions in the slice plane (about 1 mm). The artifact is created when the variation of transmission within the beam varies considerably, and the exponential variation within the beam is summed by the radiation detector. The logarithm transformation of the detected signal produces a nonlinear effect that is propagated throughout the image by the reconstruction process. Simulation was useful in demonstrating that isolated features in the same cross-section act together to produce streak artifacts. Simulations have been useful to illustrate the effects of patient motion during the data-acquisition streaks off high-contrast objects.

2.2.2 Projection Data to Image: Calibrations

Processing of transmission data is necessary to obtain high-quality images. In general, optimization of the projection data will optimize the reconstructed image. Reconstruction is a process that removes the spatial correlation of attenuation effects in the transmitted image by taking advantage of completely sampling the possible transmissions. Two distinct calibrations are required: registration of beams with the reconstruction matrix and linearization of the measured signal.

Without loss of generalization, a projection will be considered a set of transmissions made along parallel lines in the slice plane of the CT scanner. Without loss of generalization means that essential aspects of all calibration and reconstruction procedures required for fan-beam geometries are captured by the calibration and reconstruction procedures described for parallel projections. One line of each projection is assumed to pass through the center of rotation of data collection. Shepp et al. (1979) showed that errors in the assignment of that center-of-rotation point in the projections could lead to considerable distinctive artifacts and that small errors (0.05 mm) would produce these effects. The consequences of these errors have been generalized to fan-beam collection schemes, and images reconstructed from 180-degree projection sets were compared with images reconstructed from 360-degree data sets (Kijewski and Judy, 1983). A simple misregistration of the center of rotation was found to produce blurring of image without the artifact. These differences may explain the empirical observation that most commercial CT scanners collect a full 360-degree data set even though 180 degree of data will suffice.

The data-acquisition scheme that was designed to overcome the limited sampling inherent in third-generation fan-beam systems by shifting detectors a quarter sampling distance while opposite 180-degree projection is measured, has particularly stringent registration requirements. Also, the fourth-generation scanner does not link the motion of the x-ray tube and the detector; consequently, the center of rotation is determined as part of a calibration procedure, and unsystematic effects lead to artifacts that mimic noise besides blurring the image.

Misregistration artifacts also can be mitigated by feathering. This procedure requires collection of redundant projection data at the end of the scan. A single data set is produced by linearly weighting the redundant data at the beginning and end of the data collection (Parker et al., 1982). These procedures have be useful in reducing artifacts from gated data collections (Moore et al., 1987).

The other processing necessary before reconstruction of project data is linearization. The formal requirement for reconstruction is that the line integrals of some variable be available; this is the variable that ultimately is reconstructed. The logarithm of x-ray transmission approximates this requirement. There are physical effects in real x-ray transmissions that cause deviations from this assumption. X-ray beams of sufficient intensity are composed of photons of different energies. Some photons in the beam interact with objects and are scattered rather than absorbed. The spectrum of x-ray photons of different attenuation coefficients means the logarithm of the transmission measurement will not be proportional to the line integral of the attenuation coefficient along that path, because an attenuation coefficient cannot even be defined. An effective attenuation coefficient can only be defined uniquely for a spectrum for a small mass of material that alters that intensity. It has to be small enough not to alter the spectrum (McCullough, 1979).

A straightforward approach to this nonunique attenuation coefficient error, called hardening, is to assume that the energy dependence of the attenuation coefficient is constant and that differences in attenuation are related to a generalized density factor that multiplies the spectral dependence of attenuation. The transmission of an x-ray beam then can be estimated for a standard material, typically water, as a function of thickness. This assumption is that attenuations of materials in the object, the human body, differ because specific gravities of the materials differ. Direct measurements of the transmission of an actual x-ray beam may provide initial estimates that can be parameterized. The inverse of this function provides the projection variable that is reconstructed. The parameters of the function are usually modified as part of a calibration to make the CT image of a uniform water phantom flat.

Such a calibration procedure does not deal completely with the hardening effects. The spectral dependence of bone differs considerably from that of water. This is particularly critical in imaging of the brain, which is contained within the skull. Without additional correction, the attenuation values of brain are lower in the center than near the skull.

The detection of scattered energy means that the reconstructed attenuation coefficient will differ from the attenuation coefficient estimated with careful narrow-beam measurements. The x-rays appear more penetrating because scattered x-rays are detected. The zero-ordered scatter, a decrease in the attenuation coefficient by some constant amount, is dealt with automatically by the calibration that treats hardening. First-order scattering leads to a widening of the x-ray beam and can be dealt with by a modification of the reconstruction kernel.

2.2.3 Projection Data to Image: Reconstruction

The impact of CT created considerable interest in the formal aspects of reconstruction. There are many detailed descriptions of direct reconstruction procedures. Some are presented in textbooks used in graduate courses for medical imaging (Barrett and Swindell, 1981; Cho et al., 1993). Herman (1980) published a textbook that was based on a two-semester course that dealt exclusively with reconstruction principles, demonstrating the reconstruction principles with simulation.

The standard reconstruction method is called convolution-backprojection. The first step in the procedure is to convolve the projection, a set of transmissions made along parallel lines in the slice plane, with a reconstruction kernel derived from the inverse Radon transform. The choice of kernel is dictated by bandlimiting issues (Herman and Rowland, 1973; Shepp and Logan, 1974; Chesler and Riederer, 1975). It can be modified to deal with the physical aperture of the CT system (Bracewell, 1977), which might include the effects of scatter. The convolved projection is then backprojected onto a two-dimensional image matrix. Backprojection is the opposite of projection; the value of the projection is added to each point along the line of the projection. This procedure makes sense in the continuous description, but in the discrete world of the computer, the summation is done over the image matrix.

Consider a point of the image matrix; very few, possibly no lines of the discrete projection data intersect the point. Consequently, to estimate the projection value to be added to that point, the procedure must interpolate between two values of sampled convolve projection. The linear interpolation scheme is a significant improvement over nearest project nearest to the point. More complex schemes get confounded with choices of reconstruction kernel, which are designed to accomplish standard image processing in the image, for example, edge enhancement.

Scanners have been developed to acquire a three-dimensional set of projection data (Kalender et al., 1990). The motion of the source defines a spiral motion relative to the patient. The spiral motion defines an axis. Consequently, only one projection is available for reconstruction of the attenuation values in the plane. This is the back-projection problem just discussed; no correct projection value is available from the discrete projection data set. The solution is identical: a projection value is interpolated from the existing projection values to estimate the necessary projections for each plane to be reconstructed. This procedure has the advantage that overlapping slices can be reconstructed without additional exposure, and this eliminates the risk that a small lesion will be missed because it straddles adjacent slices. This data-collection scheme is possible because systems that continuously rotate have been developed. The spiral scan motion is realized by moving the patient through the gantry. Spiral CT scanners have made possible the acquisition of an entire data set in a single breath hold.

References

Barrett H.H. and Swindell W. 1981. *Radiological Imaging: The Theory and Image Formation, Detection, and Processing*, Vol. 2. New York, Academic Press.

Bracewell R.N. and Riddle A.C. 1967. Inversion of fan-beam scans in radio astronomy. *The Astrophys. J.* 150: 427–434.

Chesler D.A. and Riederer S.J. 1975. Ripple suppression during reconstruction in transverse tomography. *Phys. Med. Biol.* 20(4): 632–636.

Cho Z., Jones J.P., and Singh M. 1993. *Foundations of Medical Imaging*. New York, Wiley & Sons, Inc.

Crawford C.R. and Kak A.C. 1979. Aliasing artifacts in computerized tomography. *Appl. Opt.* 18: 3704–3711.

Glover G.H. and Pelc N.J. 1980. Nonlinear partial volume artifacts in x-ray computed tomography. *Med. Phys.* 7: 238–248.

Guedon J.-P. and Bizais. 1994. Bandlimited and harr filtered back-projection reconstruction. *IEEE Trans. Med. Imag.* 13(3): 430–440.

Herman G.T. and Rowland S.W. 1973. Three methods for reconstruction objects for x-rays—a comparative study. *Comp. Graph. Imag. Process.* 2: 151–178.

Herman G.T. 1980. *Image Reconstruction from Projection: The Fundamentals of Computerized Tomography*. New York, Academic Press.

Hounsfield G.N. 1973. Computerized transverse axial scanning (tomography): Part I. *Brit. J. Radiol.* 46: 1016–1022.

Kalender W.A., Seissler W., Klotz E. et al. 1990. Spiral volumetric CT with single-breath-hold technique, continuous transport, and continuous scanner rotation. *Radiology* 176: 181–183.

Kijewski M.F. and Judy P.F. 1983. The effect of misregistration of the projections on spatial resolution of CT scanners. *Med. Phys.* 10: 169–175.

McCullough E.C. 1979. Specifying and evaluating the performance of computed tomographic (CT) scanners. *Med. Phys.* 7: 291–296.

Moore S.C., Judy P.F., Garnic J.D. et al. 1983. The effect of misregistration of the projections on spatial resolution of CT scanners. *Med. Phys.* 10: 169–175.

3

Magnetic Resonance Imaging

Steven Conolly
Stanford University

Albert Macovski
Stanford University

John Pauly
Stanford University

John Schenck
General Electric Corporate Research and Development Center

Kenneth K. Kwong
Massachusetts General Hospital

David A. Chesler
Massachusetts General Hospital

Xiaoping Hu
University of Minnesota Medical School

Wei Chen
University of Washington

Maqbool Patel
University of Minnesota Medical School

Kamil Ugurbil
University of Minnesota Medical School

3.1 Acquisition and Processing

Steven Conolly, Albert Macovski, and John Pauly

Magnetic resonance imaging (MRI) is a clinically important medical imaging modality due to its exceptional soft-tissue contrast. MRI was invented in the early 1970s [1]. The first commercial scanners appeared about 10 years later. Noninvasive MRI studies are now supplanting many conventional invasive procedures. A 1990 study [2] found that the principal applications for MRI are examinations of the head (40%), spine (33%), bone and joints (17%), and the body (10%). The percentage of bone and joint studies was growing in 1990.

Although typical imaging studies range from 1 to 10 min, new fast imaging techniques acquire images in less than 50 msec. MRI research involves fundamental tradeoffs between resolution, imaging time, and signal-to-noise ratio (SNR). It also depends heavily on both gradient and receiver coil hardware innovations.

In this section we provide a brief synopsis of basic nuclear magnetic resonance (NMR) physics. We then derive the *k*-**space** analysis of MRI, which interprets the received signal as a scan of the Fourier transform of the image. This powerful formalism is used to analyze the most important imaging sequences. Finally, we discuss the fundamental contrast mechanisms for MRI.

3.1.1 Fundamentals of MRI

Magnetic resonance imaging exploits the existence of induced nuclear magnetism in the patient. Materials with an odd number of protons or neutrons possess a weak but observable nuclear magnetic moment. Most commonly protons (^1H) are imaged, although carbon (^{13}C), phosphorous (^{31}P), sodium (^{23}Na), and fluorine (^{19}F) are also of significant interest. The nuclear moments are normally randomly oriented, but they align when placed in a strong magnetic field. Typical field strengths for imaging range between 0.2 and 1.5 T, although spectroscopic and functional imaging work is often performed with higher field strengths. The nuclear **magnetization** is very weak; the ratio of the induced magnetization to the applied fields is only 4×10^{-9}. The collection of nuclear moments is often referred to as magnetization or **spins**.

The static nuclear moment is far too weak to be measured when it is aligned with the strong static magnetic field. Physicists in the 1940s developed resonance techniques that permit this weak moment to be measured. The key idea is to measure the moment while it oscillates in a plane perpendicular to the static field [3,4]. First one must tip the moment away from the static field. When perpendicular to the static field, the moment feels a torque proportional to the strength of the static magnetic field. The torque always points perpendicular to the magnetization and causes the spins to oscillate or precess in a plane perpendicular to the static field. The frequency of the rotation ω_0 is proportional to the field:

$$\omega_0 = -\gamma B_0$$

where γ, the **gyromagnetic ratio**, is a constant specific to the nucleus, and B_0 is the magnetic field strength. The direction of B_0 defines the z-axis. The **precession** frequency is called the **Larmor frequency**. The negative sign indicates the direction of the precession.

Since the precessing moments constitute a time-varying flax, they produce a measurable voltage in a loop antenna arranged to receive the x and y components of induction. It is remarkable that in MRI we are able to directly measure induction from the precessing nuclear moments of water protons.

Recall that to observe this precession, we first need to tip the magnetization away from the static field. This is accomplished with a weak rotating radiofrequency (RF) field. It can be shown that a rotating RF field introduces a fictitious field in the z direction of strength ω/γ. By tuning the frequency of the RF field to ω_0, we effectively delete the B_0 field. The RF slowly nutates the magnetization away from the z-axis. The Larmor relation still holds in this "rotating frame," so the frequency of the nutation is γB_1, where B_1 is the amplitude of the RF field. Since the coils receive x and y (transverse) components of induction, the signal is maximized by tipping the spins completely into the transverse plane. This is accomplished by a $\pi/2$ RF pulse, which requires $\gamma B_1 \tau = \pi/2$, where τ is the duration of the RF pulse. Another useful RF pulse rotates spins by π radians. This can be used to invert spins. It also can be used to refocus transverse spins that have dephased due to B_0 field inhomogeneity. This is called a **spin echo** and is widely used in imaging.

NMR has been used for decades in chemistry. A complex molecule is placed in a strong, highly uniform magnetic field. Electronic shielding produces microscopic field variations within the molecule so that geometrically isolated nuclei rotate about distinct fields. Each distinct magnetic environment produces a peak in the spectra of the received signal. The relative size of the spectral peaks gives the ratio of nuclei in each magnetic environment. Hence the NMR spectrum is extremely useful for elucidating molecular structure.

The NMR signal from a human is due predominantly to water protons. Since these protons exist in identical magnetic environments, they all resonate at the same frequency. Hence the NMR signal is simply proportional to the volume of the water. They key innovation for MRI is to impose spatial variations on the magnetic field to distinguish spins by their location. Applying a magnetic field gradient causes each region of the volume to oscillate at a distinct frequency. The most effective nonuniform field is a linear gradient where the field and the resulting frequencies vary linearly with distance along the object being studied. Fourier analysis of the signal obtains a map of the spatial distribution of spins. This argument is formalized below, where we derive the powerful *k*-space analysis of MRI [5,6].

3.1.1.1 *k*-Space Analysis of Data Acquisition

In MRI, we receive a volume integral from an array of oscillators. By ensuring that the phase, "signature" of each oscillator is unique, one can assign a unique location to each spin and thereby reconstruct an image. During signal reception, the applied magnetic field points in the z direction. Spins precess in the xy plane at the Larmor frequency. Hence a spin at position $\mathbf{r} = (x,y,z)$ has a unique phase θ that describes its angle relative to the y axis in the xy plane:

$$s(t) \propto \frac{\mathrm{d}}{\mathrm{d}t}\int_{V} M(\mathbf{r})\mathrm{e}^{-\mathrm{i}\theta(\mathbf{r},t)/\mathrm{d}r} \tag{3.1}$$

where $B_z(\mathbf{r},t)$ is the z component of the instantaneous, local magnetic flux density. This formula assumes there are no x and y field components.

A coil large enough to receive a time-varying flux uniformly from the entire volume produces an EMF proportional to

$$\theta(\mathbf{r},t) = -\gamma\int_{0}^{t} B_z(\mathbf{r},\tau)d\tau \tag{3.2}$$

where $M(\mathbf{r})$ represents the equilibrium moment density at each point $\mathbf{r}$.

The key idea for imaging is to superimpose a linear field gradient on the static field B_0. This field points in the direction z, and its magnitude varies linearly with a coordinate direction. For example, an x gradient points in the z direction and varies along the coordinate x. This is described by the vector field $xG_x\hat{\mathbf{z}}$, where $\hat{\mathbf{z}}$ is the unit vector in the z direction. In general, the gradient is $(xG_x + yG_y + zG_z)\hat{\mathbf{z}}$, which can be written compactly as the dot product $\mathbf{G} \cdot \mathbf{r}\hat{\mathbf{z}}$. These gradient field components can vary with time, so the total z field is

$$B_z(\mathbf{r},t) = B_0 + \mathbf{G}(t) \cdot \mathbf{r} \tag{3.3}$$

In the presence of this general time-varying gradient, the received signal is

$$s(t) \propto \frac{\mathrm{d}}{\mathrm{d}t}\int_{V} \mathrm{e}^{\mathrm{i}\gamma B_0 t} M(\mathbf{r})\mathrm{e}^{-\mathrm{i}\gamma\int_0^t \mathbf{G}(\tau)\cdot\mathbf{r}\,\mathrm{d}\tau}\,\mathrm{d}r \tag{3.4}$$

The center frequency γB_0 is always much larger than the bandwidth of the signal. Hence the derivative operation is approximately equivalent to multiplication by $-\mathrm{i}\omega_0$. The signal is demodulated by the waveform $\mathrm{e}^{\mathrm{i}\gamma B_0 t}$ to obtain the "baseband" signal:

$$s(t) \propto -\mathrm{i}\omega_0\int_{V} M(\mathbf{r})\mathrm{e}^{-\mathrm{i}\gamma\int_0^t \mathbf{G}(\tau)\cdot\mathbf{r}\,\mathrm{d}\tau}\,\mathrm{d}r \tag{3.5}$$

It will be helpful to define the term $\mathbf{k}(t)$:

$$\mathbf{k}(t) = \gamma \int_0^t \mathbf{G}(\tau)\,d\tau \tag{3.6}$$

Then we can rewrite the received baseband signal as

$$S(t) \propto \int_V M(\mathbf{r}) e^{-i\mathbf{k}(t)\cdot\mathbf{r}}\, dr \tag{3.7}$$

which we can now identify as the spatial Fourier transform of $M(\mathbf{r})$ evaluated at $\mathbf{k}(t)$. That is, $S(t)$ scans the spatial frequencies of the function $M(\mathbf{r})$. This can be written explicitly as

$$S(t) \propto M(\mathbf{k}(t)) \tag{3.8}$$

where $M(\mathbf{k})$ is the three-dimensional Fourier transform of the object distribution $M(\mathbf{r})$. Thus we can view MRI with linear gradients as a "scan" of k-space or the spatial Fourier transform of the image. After the desired portion of k-space is scanned, the image $M(\mathbf{r})$ is reconstructed using an inverse Fourier transform.

3.1.1.1.1 2D Imaging

Many different gradient waveforms can be used to scan k-space and to obtain a desired image. The most common approach, called *two-dimensional Fourier transform imaging* (**2D FT**), is to scan through k-space along several horizontal lines covering a rectilinear grid in 2D k-space. See Figure 3.1 for a schematic of the k-space traversal. The horizontal grid lines are acquired using 128 to 256 excitations separated by a time TR, which is determined by the desired contrast, RF flip angle, and the T_1 of the desired components of the image. The horizontal-line scans through k-space are offset in k_y by a variable area y-gradient pulse, which happens before data acquisition starts. These variable offsets in k_y are

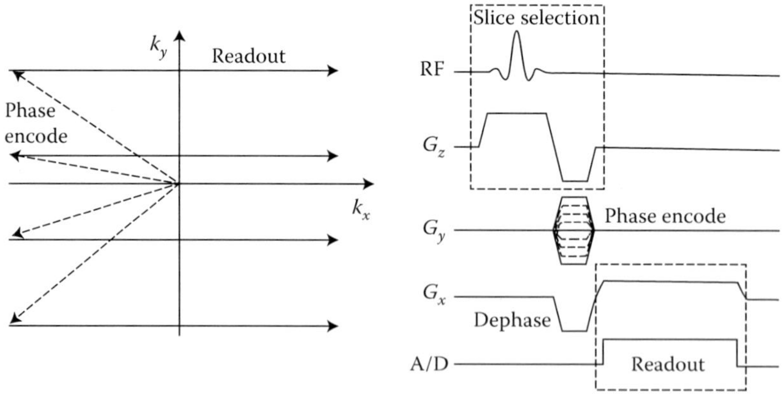

FIGURE 3.1 The drawing on the left illustrates the scanning pattern of the 2D Fourier transform imaging sequence. On the right is a plot of the gradient and RF waveforms that produce this pattern. Only four of the N_y horizontal k-space lines are shown. The phase-encode period initiates each acquisition at a different k_y and at—k_x(max). Data are collected during the horizontal traversals. After all N_y k-space lines have been acquired, a 2D FFT reconstructs the image. Usually 128 or 256 k_y lines are collected, each with 256 samples along k_x. The RF pulse and the z gradient waveform together restrict the image to a particular slice through the subject.

called *phase encodes* because they affect the phase of the signal rather than the frequency. Then for each k_y phase encode, signal is acquired while scanning horizontally with a constant x gradient.

3.1.1.1.2 Resolution and Field of View

The fundamental image characteristics of resolution and field of view (FOV) are completely determined by the characteristics of the k-space scan. The extent of the coverage of k-space determines the resolution of the reconstructed image. The resolution is inversely proportional to the highest spatial frequency acquired:

$$\frac{1}{\Delta x} = \frac{k_x(\max)}{\pi} = \frac{\gamma G_x T}{2\pi} \tag{3.9}$$

$$\frac{1}{\Delta y} = \frac{k_y(\max)}{\pi} = \frac{\gamma G_y T_{\text{phase}}}{\pi} \tag{3.10}$$

where G_x is the readout gradient amplitude and T is the readout duration. The time T_{phase} is the duration of the phase-encode gradient G_y. For proton imaging on a 1.5-T imaging system, a typical gradient strength is $G_x = 1$ G/cm. The signal is usually read for about 8 msec. For water protons, $\gamma = 26{,}751$ rad/sec/G, so the maximum excursion in k_x is about 21 rad/mm. Hence we cannot resolve an object smaller than 0.3 mm in width. From this one might be tempted to improve the resolution dramatically using very strong gradients or very long readouts. But there are severe practical obstacles, since higher resolution increases the scan time and also degrades the image SNR.

In the phase-encode direction, the k-space data are sampled discretely. This discrete sampling in k-space introduces replication in the imaqge domain [7]. If the sampling in k-space is finer than 1/FOV, then the image of the object will not fold back on itself. When the k-space samling is coarser than 1/FOV, the image of the object does fold back over itself. This is termed *aliasing*. Aliasing is prevented in the readout dirction by the sampling filter.

3.1.1.1.3 Perspective

For most imaging systems, diffraction limits the resolution. That is, the resolution is limited to the wavelength divided by the angle subtended by the receiver aperture, which means that the ultimate resolution is approximately the wavelength itself. This is true for imaging systems based on optics, ultrasound, and x-rays (although there are other important factors, such as quantum noise, in x-ray).

MRI is the only imaging system for which the resolution is independent of the wavelength. In MRI, the wavelength is often many meters, yet submillimeter resolution is routinely achieved. The basic reason is that no attempt is made to focus the radiation pattern to the individual pixel or voxel (volume element), as is done in all other imaging modalities. Instead, the gradients create spatially varying magnetic fields so that individual pixels emit unique waveform signatures. These signals are decoded and assigned to unique positions. An analogous problem is isolating the signals from two transmitting antenna towers separated by much less than a wavelength. Directive antenna arrays would fail because of diffraction spreading. However, we can distinguish the two signals if we use the a priori knowledge that the two antennas transmit at different frequencies. We can receive both signals with a wide-angle antenna and then distinguish the signals through frequency-selective filtering.

3.1.1.1.4 SNR Considerations

The signal strength is determined by the EMF induced from each voxel due to the processing moments. The magnetic moment density is proportional to the polarizing field B_0. Recall that the EMF is proportional to the rate of change of the coil flux. The derivative operation multiples the signal by the Larmor frequency, which is proportional to B_0, so the received signal is proportional to B_0^2 times the volume of the voxel V_v.

In a well-designed MRI system, the dominant noise source is due to thermally generated currents within the conductive tissues of the body. These currents create a time-varying flux which induces noise voltages in the receiver coil. Other noise sources include the thermal noise from the antenna and from the first amplifier. These subsystems are designed so that the noise is negligible compared with the noise from the patient. The noise received is determined by the total volume seen by the antenna pattern V_n and the effective resistivity and temperature of the conductive tissue. One can show [8] that the standard deviation of the noise from conductive tissue varies linearly with B_0. The noise is filtered by an integration over the total acquisition time T_{acq}, which effectively attenuates the noise standard deviation by $\sqrt{T_{acq}}$. Therefore, the SNR varies as

$$\mathrm{SNR} \propto \frac{B_0^2 V_v}{B_0 V_n / \sqrt{T_{acq}}} = B_0 \sqrt{T_{acq}} (V_v / V_n) \tag{3.11}$$

The noise volume V_n is the effective volume based on the distribution of thermally generated currents. For example, when imaging a spherical object of radius r, the noise standard deviation varies as $r^{5/2}$ [9]. The effective resistance depends strongly on the radius because currents near the outer radius contribute more to the noise flux seen by the receiver coil.

To significantly improve the SNR, most systems use *surface coils*, which are simply small coils that are just big enough to see the desired region of the body. Such a coil effectively maximizes the voxel-volume to noise-volume ratio. The noise is significantly reduced because these coils are sensitive to currents from a smaller part of the body. However, the field of view is somewhat limited, so "phased arrays" of small coils are now being offered by the major manufacturers [10]. In the phased array, each coil sees a small noise volume, while the combined responses provide the wide coverage at a greatly improved SNR.

3.1.1.1.5 *Fast Imaging*

The 2D FT scan of k-space has the disadvantage that the scan time is proportional to the number of phase encodes. It is often advantageous to trade off SNR for a shorter scan time. This is especially true when motion artifacts dominate thermal noise. To allow for a flexible tradeoff of SNR for imaging time, more than a single line in k-space must be covered in a single excitation. The most popular approach, called echo-planar imaging (EPI), traverses k-space back and forth on a single excitation pulse. The k-space trajectory is drawn in Figure 3.2.

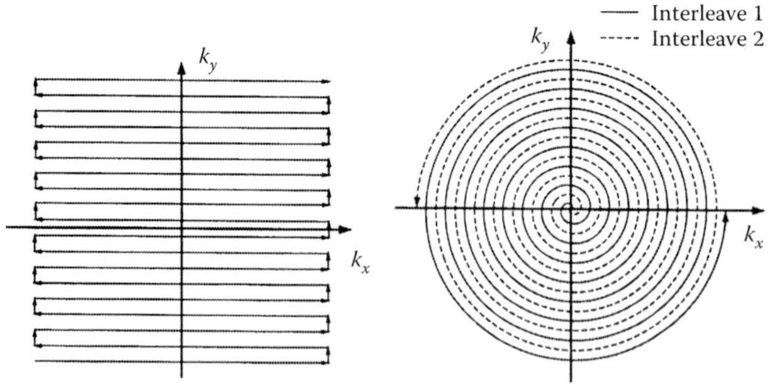

FIGURE 3.2 Alternative methods for the rapid traversal of k-space. On the left is the echo planar trajectory. Data are collected during the horizontal traversals. When all N_y horizontal lines in k-space have been acquired, the data are sent to a 2D FFT to reconstruct the image. On the right is an interleaved spiral trajectory. The data are interpolated to a 2D rectilinear grid and then Fourier transformed to reconstruct the image. These scanning techniques allow for imaging within a breathhold.

It is important that the tradeoff be flexible so that you can maximize the imaging time given the motion constraints. For example, patients can hold their breath for about 12 sec. So a scan of 12 sec duration gives the best SNR given the breath-hold constraint. The EPI trajectory can be interleaved to take full advantage of the breath-hold interval. If each acquisition takes about a second, 12 interleaves can be collected. Each interleaf acquires every twelfth line in k-space.

Another trajectory that allows for a flexible tradeoff between scan time and SNR is the spiral trajectory. Here the trajectory starts at the origin in k-space and spirals outward. Interleaving is accomplished by rotating the spirals. Figure 3.2 shows two interleaves in a spiral format. Interleaving is very helpful for reducing the hardware requirements (peak amplitude, peak slew rate, average dissipation, etc.) for the gradients amplifiers. For reconstruction, the data are interpolated to a 2D rectilinear grid and then Fourier-transformed. Our group has found spiral imaging to be very useful for imaging coronary arteries within a breath-hold scan [11]. The spiral trajectory is relatively immune to artifacts due to the motion of blood.

3.1.2 Contrast Mechanisms

The tremendous clinical utility of MRI is due to the great variety of mechanisms that can be exploited to create image contrast. If magnetic resonance images were restricted to water density, MRI would be considerably less useful, since most tissues would appear identical. Fortunately, many different MRI contrast mechanisms can be employed to distinguish different tissues and disease processes.

The primary contrast mechanisms exploit *relaxation* of the magnetization. The two types of relaxations are termed **spin–lattice relaxation**, characterized by a relaxation time T_1, and **spin–spin relaxation**, characterized by a relaxation time T_2.

Spin–lattice relaxation describes the rate of recovery of the z component of magnetization toward equilibrium after it has been disturbed by RF pulses. The recovery is given by

$$M_z(t) = M_0(1 - e^{-t/T_1}) + M_z(0)e^{-t/T_1} \tag{3.12}$$

where M_0 is the equilibrium magnetization. Differences in the T_1 time constant can be used to produce image contrast by exciting all magnetization and then imaging before full recovery has been achieved. This is illustrated on the left in Figure 3.3. An initial $\pi/2$ RF pulse destroys all the longitudinal

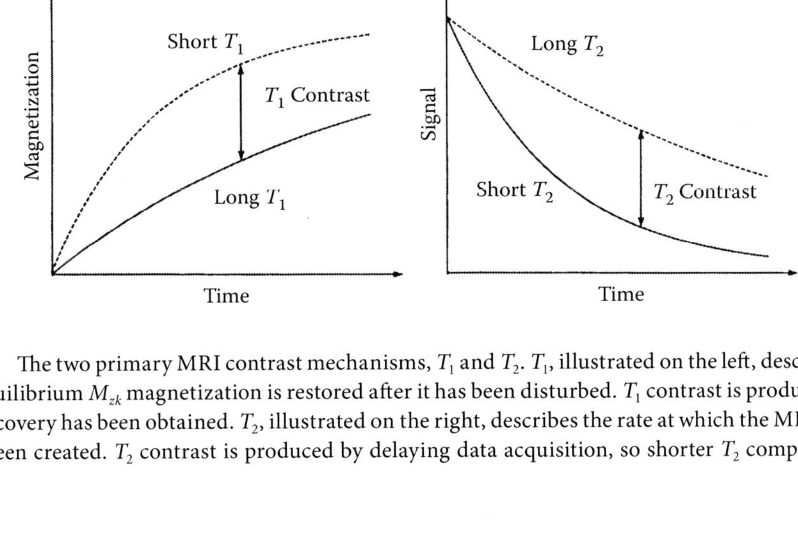

FIGURE 3.3 The two primary MRI contrast mechanisms, T_1 and T_2. T_1, illustrated on the left, describes the rate at which the equilibrium M_{zk} magnetization is restored after it has been disturbed. T_1 contrast is produced by imaging before full recovery has been obtained. T_2, illustrated on the right, describes the rate at which the MRI signal decays after it has been created. T_2 contrast is produced by delaying data acquisition, so shorter T_2 components produce less signal.

magnetization. The plots show the recovery of two different T_1 components. The short T_1 component recovers faster and produces more signal. This gives a T_1-weighted image.

Spin–spin relaxation describes the rate at which the NMR signal decays after it has been created. The signal is proportional to the transverse magnetization and is given by

$$M_{xy}(t) = M_{xy}(0)e^{-t/T_2} \tag{3.13}$$

Image contrast is produced by delaying the data acquisition. The decay of two different T_2 species is plotted on the right in Figure 3.3. The signal from the shorter T_2 component decays more rapidly, while that of the longer T_2 component persists. At the time of data collection, the longer T_2 component produces more signal. This produces a T_2-weighted image.

Figure 3.4 shows examples of these two basic types of contrast. These images are of identical axial sections through the brain of a normal volunteer. The image of the left was acquired with an imaging method that produces T_1 contrast. The very bright ring of subcutaneous fat is due to its relatively short T_1. White matter has a shorter T_1 than gray matter, so it shows up brighter in this image. The image on the right was acquired with an imaging method that produces T_2 contrast. Here the cerebrospinal fluid in the ventricles is bright due to its long T_2. White matter has a shorter T_2 than gray matter, so it is darker in this image.

There are many other contrast mechanisms that are important in MRI. Different chemical species have slightly different resonant frequencies, and this can be used to image one particular component. It is possible to design RF and gradient waveforms so that the image shows only moving spins. This is of great utility in MR angiography, allowing the noninvasive depiction of the vascular system. Another contrast mechanism is called T_2. This relaxation parameter is useful for functional imaging. It occurs when there is a significant spread of Larmor frequencies within a voxel. The superposition signal is attenuated faster than T_2 due to destructive interference between the different frequencies.

In addition to the intrinsic tissue contrast, artificial MRI contrast agents also can be introduced. These are usually administered intravenously or orally. Many different contrast mechanisms can be exploited, but the most popular agents decrease both T_1 and T_2. One agent approved for clinical use is gadolinium DPTA. Decreasing T_1 causes faster signal recovery and a higher signal on a T_1-weighted image. The contrast-enhanced regions then show up bright relative to the rest of the image.

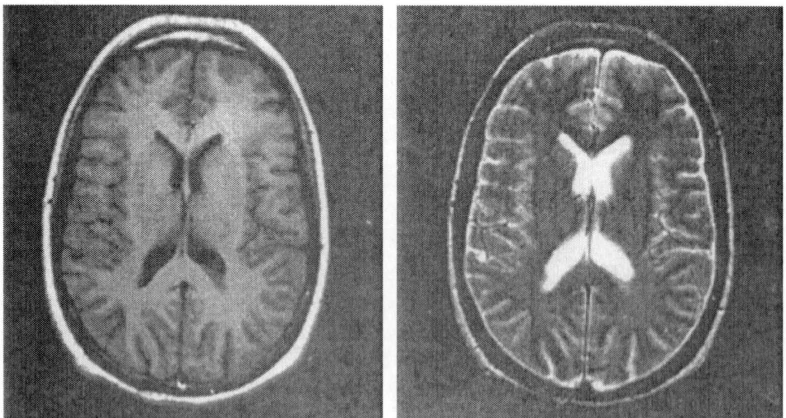

FIGURE 3.4 Example images of a normal volunteer demonstrating T_1 contrast on the left and T_2 contrast on the right.

Defining Terms

D FT: A rectilinear trajectory through k-space. This popular acquisition scheme requires several (usually 128 to 256) excitations separated by a time TR, which is determined by the desired contrast, RF flip angle, and the T_1 of the desired components of the image.

Gyromagnetic ratio γ: An intrinsic property of a nucleus. It determines the Larmor frequency through the relation $\omega_0 = -\gamma B_0$.

k**-space**: The reciprocal of object space, k-space describes MRI data acquisition in the spatial Fourier transform coordinate system.

Larmor frequency ω_0: The frequency of precession of the spins. It depends on the product of the applied flux density B_0 and on the gyromagnetic ratio γ. The Larmor frequency is $\omega_0 = -\gamma B_0$.

Magnetization M: The macroscopic ensemble of nuclear moments. The moments are induced by an applied magnetic field. At body temperatures, the amount of magnetization is linearly proportional ($M_0 = 4 \times 10^{-9} H_0$) to the applied magnetic field.

Precession: The term used to describe the motion of the magnetization about an applied magnetic field. The vector locus traverses a cone. The precession frequency is the frequency of the magnetization components perpendicular to the applied field. The precession frequency is also called the *Larmor frequency* ω_0.

Spin echo: The transverse magnetization response to a π RF pulse. The effects of field inhomogeneity are refocused at the middle of the spin echo.

Spin–lattice relaxation T_1: The exponential rate constant describing the decay of the z component of magnetization toward the equilibrium magnetization. Typical values in the body are between 300 and 3000 msec.

Spin–spin relaxation T_2: The exponential rate constant describing the decay of the transverse components of magnetization (M_x and M_y).

Spins M: Another name for magnetization.

References

1. Lauterbur P.C. 1973. *Nature* 242: 190.
2. Evens R.G. and Evens J.R.G. 1991. *AJR* 157: 603.
3. Bloch F., Hansen W.W., and Packard M.E. 1946. *Phys. Rev.* 70: 474.
4. Bloch F. 1946. *Phys. Rev.* 70: 460.
5. Twieg D.B. 1983. *Med. Phys.* 10: 610.
6. Ljunggren S. 1983. *J. Magn. Reson.* 54: 338.
7. Bracewell R.N. 1978. *The Fourier Transform and Its Applications*. New York, McGraw-Hill.
8. Hoult D.I. and Lauterbur P.C. 1979. *J. Magn. Reson.* 34: 425.
9. Chen C.N. and Hoult D. 1989. *Biomedical Magnetic Resonance Technology*. New York, Adam Hilger.
10. Roemer P.B., Edelstein W.A., Hayes C.E. et al. 1990. *Magn. Reson. Med.* 16: 192.
11. Meyer C.H., Hu B.S., Nishimura D.G., and Macovski A. 1992. *Magn. Reson. Med.* 28: 202.

3.2 Hardware/Instrumentation

John Schenck

This section describes the basic components and the operating principles of MRI scanners. Although scanners capable of making diagnostic images of the human internal anatomy through the use of **magnetic resonance imaging** (MRI) are now ubiquitous devices in the radiology departments of hospitals in the United States and around the world, as recently as 1980 such scanners were available only in a handful of research institutions. Whole-body superconducting magnets became available in the early

1980s and greatly increased the clinical acceptance of this new imaging modality. Market research data indicate that between 1980 and 1996 more than 100 million clinical MRI scans were performed worldwide. By 1996 more than 20 million MRI scans were being performed each year.

MRI scanners use the technique of **nuclear magnetic resonance (NMR)** to induce and detect a very weak radio frequency signal that is a manifestation of **nuclear magnetism**. The term *nuclear magnetism* refers to weak magnetic properties that are exhibited by some materials as a consequence of the nuclear **spin** that is associated with their atomic nuclei. In particular, the proton, which is the nucleus of the hydrogen atom, possesses a nonzero nuclear spin and is an excellent source of NMR signals. The human body contains enormous numbers of hydrogen atoms—especially in water (H_2O) and lipid molecules. Although biologically significant NMR signals can be obtained from other chemical elements in the body, such as phosphorous and sodium, the great majority of clinical MRI studies utilize signals originating from protons that are present in the lipid and water molecules within the patient's body.

The patient to be imaged must be placed in an environment in which several different magnetic fields can be simultaneously or sequentially applied to elicit the desired NMR signal. Every MRI scanner utilizes a strong static field magnet in conjunction with a sophisticated set of **gradient coils** and radiofrequency coils. The gradients and the radiofrequency components are switched on and off in a precisely timed pattern, or **pulse sequence**. Different pulse sequences are used to extract different types of data from the patient. MR images are characterized by excellent contrast between the various forms of soft tissues within the body. For patients who have no ferromagnetic foreign bodies within them, MRI scanning appears to be perfectly safe and can be repeated as often as necessary without danger (Shellock and Kanal, 1998). This provides one of the major advantages of MRI over conventional x-ray and computed tomographic (CT) scanners. The NMR signal is not blocked at all by regions of air or bone within the body, which provides a significant advantage over ultrasound imaging. Also, unlike the case of nuclear medicine scanning, it is not necessary to add radioactive tracer materials to the patient.

3.2.1 Fundamentals of MRI Instrumentation

Three types of magnetic fields—main fields or static fields (B_2), gradient fields, an radiofrequency (RF) fields (B_1)—are required in MRI scanners. In practice, it is also usually necessary to use coils or magnets that produce shimming fields to enhance the spatial uniformity of the static field B_0. Most MRI hardware engineering is concerned with producing and controlling these various forms of magnetic fields. The ability to construct NMR instruments capable of examining test tube-sized samples has been available since shortly after World War II. The special challenge associated with the design and construction of medical scanners was to develop a practical means of scaling these devices up to sizes capable of safely and comfortably accommodating an entire human patient. Instruments capable of human scanning first became available in the late 1970s. The successful implementation of MRI requires a two-way flow of information between analog and digital formats (Figure 3.5). The main magnet, the gradient and RF coils, and the gradient and RF power supplies operate in the analog domain. The digital domain is centered on a general-purpose computer (Figure 3.6) that is used to provide control information (signal timing and amplitude) to the gradient and RF amplifiers, to process time-domain MRI signal data returning from the receiver, and to drive image display and storage systems. The computer also provides miscellaneous control functions, such as permitting the operator to control the position of the patient table.

3.2.2 Static Field Magnets

The main field magnet (Thomas, 1993) is required to produce an intense and highly uniform, **static magnetic field** over the entire region to be imaged. To be useful for imaging purposes, this field must be extremely uniform in space and constant in time. In practice, the spatial variation of the main field of a whole-body scanner must be less than about 1 to 10 parts per million (ppm) over a region approximately 40 cm in diameter. To achieve these high levels of homogeneity requires careful attention to magnet

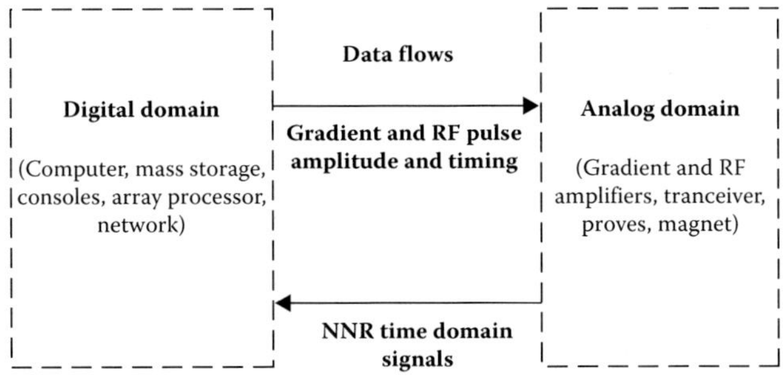

FIGURE 3.5 Digital and analog domains for MRI imaging. MRI involves the flow of data and system commands between these two domains. (Courtesy of W.M. Leue. Reprinted with permission from Schenck, J.F., and W.M. Leue, Instrumentation: Magnets, coils, and hardware, in S.W. Atlas (Ed.), *Magnetic Resonance Imaging of the Brain and Spine*, 2nd ed., pp. 1–27, Lippincott-Raven, Philadelphia, 1996.)

design and to manufacturing tolerances. The temporal drift of the field strength is normally required to be less than 0.1 ppm/h.

Two units of magnetic field strength are now in common use. The gauss (G) has a long historical usage and is firmly embedded in the older scientific literature. The tesla (T) is a more recently adopted unit, but is a part of the SI system of units and, for this reason, is generally preferred. The tesla is a much larger unit than the gauss—1 T corresponds to 10,000 G. The magnitude of the earth's magnetic field is about

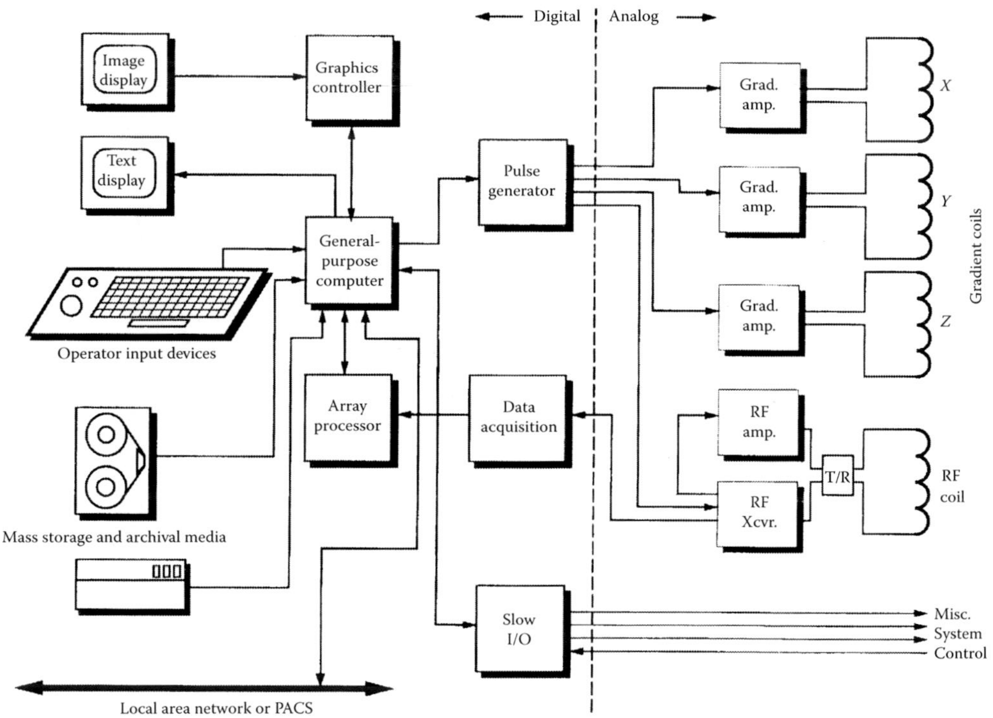

FIGURE 3.6 Block diagram for an MRI scanner. A general-purpose computer is used to generate the commands that control the pulse sequence and to process data during MR scanning. (Courtesy of W.M. Leue. Reprinted with permission from Schenck, J.F., and W.M. Leue, Instrumentation: Magnets, coils, and hardware, in S.W. Atlas (Ed.), *Magnetic Resonance Imaging of the Brain and Spine*, 2nd ed., pp. 1–27, Lippincott-Raven, Philadelphia, 1996.)

0.05 mT (5000 G). The static magnetic fields of modern MRI scanners arc most commonly in the range of 0.5 to 1.5 T; useful scanners, however, have been built using the entire range from 0.02 to 8 T. The signal-to-noise ratio (SNR) is the ratio of the NMR signal voltage to the ever-present noise voltages that arise within the patient and within the electronic components of the receiving system. The SNR is one of the key parameters that determine the performance capabilities of a scanner. The maximum available SNR increases linearly with field strength. The improvement in SNR as the field strength is increased is the major reason that so much effort has gone into producing high-field magnets for MRI systems.

Magnetic fields can be produced by using either electric currents or permanently magnetized materials as sources. In either case, the field strength falls off rapidly away from the source, and it is not possible to create a highly uniform magnetic field on the outside of a set of sources. Consequently, to produce the highly uniform field required for MRI, it is necessary to more or less surround the patient with a magnet. The main field magnet must be large enough, therefore, to effectively surround the patient; in addition, it must meet other stringent performance requirements. For these reasons, the main field magnet is the most important determinant of the cost, performance, and appearance of an MRI scanner. Four different classes of main magnets—(1) permanent magnets, (2) electromagnets, (3) resistive magnets, and (4) superconducting magnets—have been used in MRI scanners.

3.2.2.1 Permanent Magnets and Electromagnets

Both these magnet types use magnetized materials to produce the field that is applied to the patient. In a permanent magnet, the patient is placed in the gap between a pair of permanently magnetized pole faces. Electromagnets use a similar configuration, but the pole faces are made of soft magnetic materials, which become magnetized only when subjected to the influence of electric current coils that are wound around them. Electromagnets, but not permanent magnets, require the use of an external power supply. For both types of magnets, the magnetic circuit is completed by use of a soft iron yoke connecting the pole faces to one another (Figure 3.7). The gap between the pole faces must be large enough to contain the patient as well as the gradient and RF coils. The permanent magnet materials available for use in MRI scanners include high-carbon iron, alloys such as Alnico, ceramics such as barium ferrite, and rare earth alloys such as samarium cobalt.

Permanent magnet scanners have some advantages: they produce a relatively small fringing field and do not require power supplies. However, they tend to be very heavy (up to 100 T) can produce only

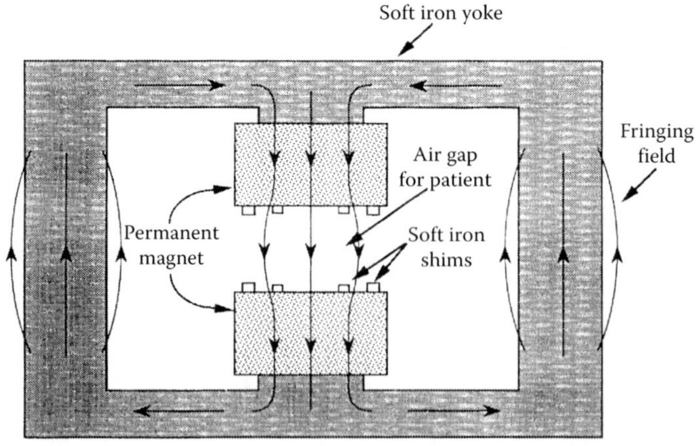

FIGURE 3.7 Permanent magnet. The figure shows a schematic cross-section of a typical permanent magnet configuration. Electromagnets have a similar construction but are energized by current-carrying coils wound around the iron yoke. Soft magnetic shims are used to enhance the homogeneity of the field. (Reprinted with permission from Schenck, J.F., and W.M. Leue, Instrumentation: Magnets, coils, and hardware, in S.W. Atlas (Ed.), *Magnetic Resonance Imaging of the Brain and Spine*, 2nd ed., pp. 1–27, Lippincott-Raven, Philadelphia, 1996.)

relatively low fields—on the order of 0.3 T or less. They are also subject to temporal field drift caused by temperature changes. If the pole faces are made from an electrically conducting material, eddy currents induced in the pole faces by the pulsed gradient fields can limit performance as well. A recently introduced alloy of neodymium, boron, and iron (usually referred to as *neodymium iron*) has been used to make lighter-weight permanent magnet scanners.

3.2.2.2 Resistive Magnets

The first whole-body scanners, manufactured in the late 1970s and early 1980s, used four to six large coils of copper or aluminum wire surrounding the patient. These coils are energized by powerful (40–100 kW) direct-current (dc) power supplies. The electrical resistance of the coils leads to substantial joule heating, and the use of cooling water flowing through the coils is necessary to prevent overheating. The heat dissipation increases rapidly with field strength, and it is not feasible to build resistive magnets operating at fields much higher than 0.15–0.3 T. At present, resistive magnets are seldom used except for very low field strength (0.02–0.06 T) applications.

3.2.2.3 Superconducting Magnets

Since the early 1980s, the use of cryogenically cooled superconducting magnets (Wilson, 1983) has been the most satisfactory solution to the problem of producing the static magnet field for MRI scanners. The property of exhibiting absolutely no electrical resistance near absolute zero has been known as an exotic property of some materials since 1911. Unfortunately, the most common of these materials, such as lead, tin, and mercury, exhibit a phase change back to the normal state at relatively low magnetic field strengths and cannot be used to produce powerful magnetic fields. In the 1950s, a new class of materials (type II superconductors) was discovered. These materials retain the ability to carry loss-free electric currents in very high fields. One such material, an alloy of niobium and titanium, has been used in most of the thousands of superconducting whole-body magnets that have been constructed for use in MRI scanners (Figure 3.8). The widely publicized discovery in 1986 of another class of materials which remain superconducting at much higher temperatures than any previously known material has not yet lead to any material capable of carrying sufficient current to be useful in MRI scanners.

Figure 3.9 illustrates the construction of a typical superconducting whole-body magnet. In this case, six coils of superconducting wire are connected in a series and carry an intense current—on the order of 200 A—to produce the 1.5-T magnetic field at the magnet's center. The diameter of the coils is about

FIGURE 3.8 Superconducting magnet. This figure shows a 1.5-T whole-body superconducting magnet. The nominal warm bore diameter is 1 m. The patient to be imaged, as well as the RF and gradient coils, are located within this bore. (Courtesy of General Electric Medical Systems. Reprinted with permission from Schenck, J.F., and W.M. Leue, Instrumentation: Magnets, coils, and hardware, in S.W. Atlas (Ed.), *Magnetic Resonance Imaging of the Brain and Spine*, 2nd ed., pp. 1–27, Lippincott-Raven, Philadelphia, 1996.)

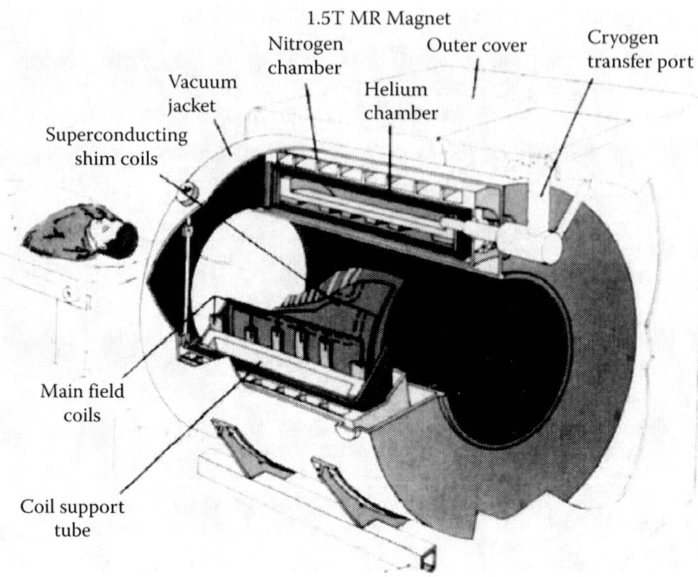

FIGURE 3.9 Schematic drawing of a superconducting magnet. The main magnet coils and the superconducting shim coils are maintained at liquid helium temperature. A computer-controlled table is used to advance the patient into the region of imaging. (Reprinted with permission from Schenck, J.F., and W.M. Leue, Instrumentation: Magnets, coils, and hardware, in S.W. Atlas (Ed.), *Magnetic Resonance Imaging of the Brain and Spine*, 2nd ed., pp. 1–27, Lippincott-Raven, Philadelphia, 1996.)

1.3 m, and the total length of wire is about 65 km (40 miles). The entire length of this wire must be without any flaws—such as imperfect welds—that would interrupt the superconducting properties. If the magnet wire has no such flaws, the magnet can be operated in the persistent mode—that is, once the current is established, the terminals may be connected together, and a constant persistent current flow indefinitely so long as the temperature of the coils is maintained below the superconducting transition temperature. This temperature is about 10 K for niobium–titanium wire. The coils are kept at this low temperature by encasing them in a double-walled cryostat (analogous to a Thermos bottle) that permits them to be immersed in liquid helium at a temperature of 4.2 K. The gradual boiling of liquid helium caused by inevitable heat leaks into the cryostat requires that the helium be replaced on a regular schedule. Many magnets now make use of cryogenic refrigerators that reduce or eliminate the need for refilling the liquid helium reservoir. The temporal stability of superconducting magnets operating in the persistent mode is truly remarkable—magnets have operated for years completely disconnected from power supplies and maintained their magnetic field constant to within a few parts per million. Because of their ability to achieve very strong and stable magnetic field strengths without undue power consumption, superconducting magnets have become the most widely used source of the main magnetic fields for MRI scanners.

3.2.2.4 Magnetic Field Homogeneity

The necessary degree of spatial uniformity of the field can be achieved only by carefully placing the coils at specific spatial locations. It is well known that a single loop of wire will produce, on its axis, a field that is directed along the coil axis and that can be expressed as a sum of spherical harmonic fields. The first term in this sum is constant in space and represents the desired field that is completely independent of position. The higher-order terms represent contaminating field inhomogeneities that spoil the field uniformity. More than a century ago, a two-coil magnet system—known as the *Helmholtz pair*—was developed which produced a much more homogeneous field at its center than is produced by a single current loop. This design is based on the mathematical finding that when two coaxial coils of the same radius are separated by a distance equal to their radius, the first nonzero contaminating term in the harmonic expansion is of the fourth order. This results in an increased region of the field homogeneity, which, although it is useful in many applications, is far too small to be useful in MRI scanners. However, the principle of eliminating low-order harmonic fields can be extended by using additional coils. This is the method now used to increase the volume of field homogeneity to values that are useful for MRI. For example, in the commonly used six-coil system, it is possible to eliminate all the error fields through the twelfth order.

In practice, manufacturing tolerances and field perturbations caused b extraneous magnetic field sources—such as steel girders in the building surrounding the magnet—produce additional inhomogeneity in the imaging region. These field imperfections are reduced by the use of shimming fields. One approach—*active shimming*—uses additional coils (either resistive coils, superconducting coils, or some of each) which are designed to produce a magnetic field corresponding to a particular term in the spherical harmonic expansion. When the magnet is installed, the magnetic field is carefully mapped, and the currents in the shim coils are adjusted to cancel out the terms in the harmonic expansion to some prescribed high order. The alternative approach—*passive shimming*—utilizes small permanent magnets that are placed at the proper locations along the inner walls of the magnet bore to cancel out contaminating fields. If a large object containing magnetic materials—such as a power supply—is moved in the vicinity of superconducting magnets, it may be necessary to reset the shimming currents or magnet locations to account for the changed pattern of field inhomogeneity.

3.2.2.5 Fringing Fields

A large, powerful magnet produces a strong magnetic field in the region surrounding it as well as in its interior. This fringing field can produce undesirable effects such as erasing magnetic tapes (and credit cards). It is also a potential hazard to people with implanted medical devices such as cardiac pacemakers. For safety purposes, it is general practice to limit access to the region where the fringing field becomes intense. A conventional boundary for this region is the "5-gaussline," which is about 10 to 12 m from the center of an unshielded 1.5-T magnet. Magnetic shielding—in the form of iron plates (passive shielding) or external coils carrying current in the direction opposite to the main coil current (active shielding)—is frequently used to restrict the region in which the fringing field is significant.

3.2.3 Gradient Coils

Three gradient fields, one each for the *x*, *y*, and *z* directions of a Cartesian coordinate system, are used to code position information into the MRI signal and to permit the imaging of thin anatomic slices (Thomas, 1993). Along with their larger size, it is the use of these gradient coils that distinguishes MRI scanners from the conventional NMR systems such as those used in analytical chemistry. The direction of the static field, along the axis of the scanner, is conventionally taken as the *z* direction, and it is only the Cartesian component of the gradient field in this direction that produces a significant contribution to the resonant behavior of the nuclei. Thus, the three relevant gradient fields are $B_z = G_xX$, $B_z = G_yX$, and $B_z = G_zX$. MRI scans are carried out by subjecting the spin system to a sequence of pulsed gradient and RF fields. Therefore, it is necessary to have three separate coils—one for each of the relevant gradient fields—each with its own power supply and under independent computer control. Ordinarily, the most practical method for constructing the gradient coils is to wind them on a cylindrical coil form that surrounds the patient and is located inside the warm bore of the magnet. The *z* gradient field can be produced by sets of circular coils wound around the cylinder with the current direction reversed for coils on the opposite sides of the magnet center ($z = 0$). To reduce deviations from a perfectly linear B_z gradient field, a spiral winding can be used with the direction of the turns reversed at $z = 0$ and the spacing between windings decreasing away from the coil center (Figure 3.10). A more complex current pattern is required to produce the transverse (*x* and *y*) gradients. As indicated in Figure 3.11, transverse gradient fields are produced by windings which utilize a four-quadrant current pattern.

FIGURE 3.10 *Z*-gradient coil. The photograph shows a spiral coil wound on a cylindrical surface with an overwiding near the end of the coil. (Courtesy of R.J. Dobberstein, General Electric Medical Systems. Reprinted with permission from Schenck, J.F., and W.M. Leue, Instrumentation: Magnets, coils, and hardware, in S.W. Atlas (Ed.), *Magnetic Resonance Imaging of the Brain and Spine*, 2nd ed., pp. 1–27, Lippincott-Raven, Philadelphia, 1996.)

The generation of MR images requires that a rapid sequence of time-dependent gradient fields (on all three axes) be applied to the patient. For example, the commonly used technique of **spin-warp imaging** (Edelstein et al., 1980) utilizes a slice-selection gradient pulse to select the spins in a thin (3–10 mm) slice of the patient and then applies readout and phase-encoding gradients in the two orthogonal directions to encode two-dimensional spatial information into the NMR signal. This, in turn, requires that the currents in the three gradient coils be rapidly switched by computer-controlled power supplies. The rate at which gradient currents can be switched is an important determinant of the imaging capabilities of a scanner. In typical scanners, the gradient coils have an electrical resistance of about 1 Ω and an inductance of about 1 mH, and the gradient field can be switched from 0 to 10 mT/m (1 G/cm) in about

FIGURE 3.11 Transverse gradient coil. The photograph shows the outer coil pattern of an actively shielded transverse gradient coil. (Courtesy of R.J. Dobberstien, General Electric Medical Systems. Reprinted with permission from Schenck, J.F., and W.M. Leue, Instrumentation: Magnets, coils, and hardware, in S.W. Atlas (Ed.), *Magnetic Resonance Imaging of the Brain and Spine*, 2nd ed., pp. 1–27, Lippincott-Raven, Philadelphia, 1996.)

0.5 msec. The current must be switched from 0 to about 100 A in this interval, and the instantaneous voltage on the coils, *Ldi*/*dt*, is on the order of 200 V. The power dissipation during the switching interval is about 20 kW. In more demanding applications, such as are met in cardiac MRI, the gradient field may be as high as 4–5 mT/m and switched in 0.2 msec or less. In this case, the voltage required during gradient switching is more than 1 kV. In many pulse sequences, the switching duty cycle is relatively low, and coil heating is not significant. However, fast-scanning protocols use very rapidly switched gradients at a high duty cycle. This places very strong demands on the power supplies, and it is often necessary to use water cooling to prevent overheating the gradient coils.

3.2.4 Radiofrequency Coils

Radiofrequency (RF) coils are components of every scanner and are used for two essential purposes—transmitting and receiving signals at the resonant frequency of the protons within the patient (Schenck, 1993). The precession occurs at the **Larmor frequency** of the protons, which is proportional to the static magnetic field. At 1T this frequency is 42.58 MHz. Thus in the range of field strengths currently used in whole-body scanners, 0.02 to 4 T, the operating frequency ranges from 0.85 to 170.3 MHz. For the commonly used 1.5-T scanners, the operating frequency is 63.86 MHz. The frequency of MRI scanners overlaps the spectral region used for radio and television broadcasting. As an example, the frequency of a 1.5-T scanner is within the frequency hand 60 to 66 MHz, which is allocated to television channel 3. Therefore, it is not surprising that the electronic components in MRI transmitter and receiver chains closely resemble corresponding components in radio and television circuitry. An important difference between MRI scanners and broadcasting systems is that the transmitting and receiving antennas of broadcast systems operate in the far field of the electromagnetic wave. These antennas are separated by many wavelengths. On the other hand, MRI systems operate in the near field, and the spatial separation of the sources and receivers is much less than a wavelength. In far-field systems, the electromagnetic energy is shared equally between the electric and magnetic components of the wave. However, in the near field of magnetic dipole sources, the field energy is almost entirely in the magnetic component of the electromagnetic wave. This difference accounts for the differing geometries that are most cost effective for broadcast and MRI antenna structures.

Ideally, the RF field is perpendicular to the static field, which is in the z direction. Therefore, the RF field can be linearly polarized in either the x or y direction. However, the most efficient RF field results from quadrature excitation, which requires a coil that is capable of producing simultaneous x and y fields with a 90-degree phase shift between them. Three classes of RF coils—body coils, head coils, and surface coils—are commonly used in MRI scanners. These coils are located in the space between the patient and the gradient coils. Conducting shields just inside the gradient coils are used to prevent electromagnetic coupling between the RF coils and the rest of the scanner. Head and body coils are large enough to surround the legion being imaged and are designed to produce an RF magnetic field that is uniform across the region to be imaged. Body coils are usually constructed on cylindrical coil forms and have a large enough diameter (50 to 60 cm) to entirely surround the patient's body. Coils are designed only for head imaging (Figure 3.12) have a smaller diameter (typically 28 cm). Surface coils are smaller coils designed to image a restricted region of the patient's anatomy. They come in a wide variety of shapes and sizes. Because they can be closely applied to the region of interest, surface coils can provide SNR advantages over head and body coils for localized regions, but because of their asymmetric design, they do not have uniform sensitivity.

A common practice is to use separate coils for the transmitter and receiver functions. This permits the use of a large coil—such as the body coil—with a uniform excitation pattern as the transmitter and a small surface coil optimized to the anatomic region—such as the spine—being imaged. When this two-coil approach is used, it is important to provide for electronically decoupling of the two coils because they are tuned at the same frequency and will tend to have harmful mutual interactions.

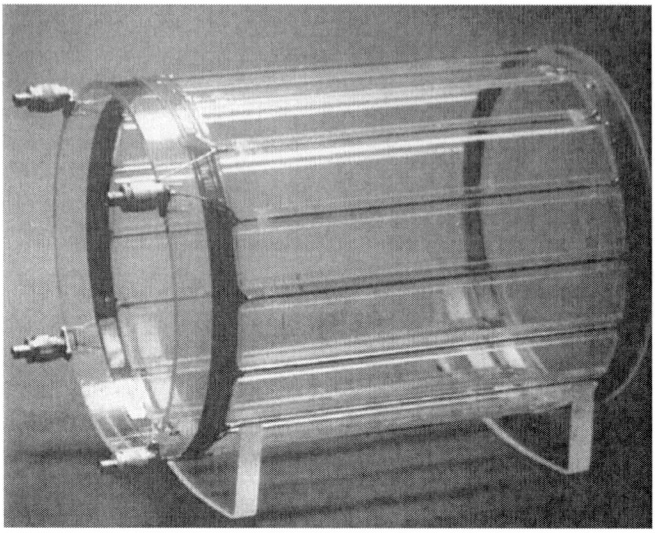

FIGURE 3.12 Birdcage resonator. This is a head coil designed to operate in a 4-T scanner at 170 MHz. Quadrature excitation and receiver performance are achieved by using two adjacent ports with a 90-degree phase shift between them. (Reprinted with permission from Schenck, J.F., and W.M. Leue, Instrumentation: Magnets, coils, and hardware, in S.W. Atlas (Ed.), *Magnetic Resonance Imaging of the Brain and Spine*, 2nd ed., pp. 1–27, Lippincott-Raven, Philadelphia, 1996.)

3.2.5 Digital Data Processing

A typical scan protocol calls for a sequence of tailored RF and gradient pulses with duration controlled in steps of 0.1 msec. To achieve sufficient dynamic range in control of pulse amplitudes, 12- to 16-bit digital-to-analog converters are used. The RF signal at the Larmor frequency (usually in the range from 1 to 200 MHz) is mixed with a local oscillator to produce a baseband signal which typically has a **bandwidth** of 16–32 kHz. The data-acquisition system must digitize the baseband signal at the Nyquist rate, which requires sampling the detected RF signal at a rate one digital data point every 5–20 msec. Again, it is necessary to provide sufficient dynamic range. Analog-to-digital converters with 16–18 bits are used to produce the desired digitized signal data. During the data acquisition, information is acquired at a rate on the order of 800 kilobytes per second, and each image can contain up to a megabyte of digital data. The array processor (AP) is a specialized computer that is designed for the rapid performance of specific algorithms, such as the **fast Fourier transform (FFT)**, which are used to convert the digitized time-domain data to image data. Two-dimensional images are typically displayed as 256 × 128, 256 × 256, or 512 × 512 **pixel** arrays. The images can be made available for viewing within about 1 sec after data acquisition. Three-dimensional imagining data, however, require more computer processing, and this results in longer delays between acquisition and display.

A brightness number, typically containing 16 bits of gray-scale information, is calculated for each pixel element of the image, and this corresponds to the signal intensity originating in each **voxel** of the object. To make the most effective use of the imaging information, sophisticated display techniques, such as multi-image displays, rapid sequential displays (cine loop), and three-dimensional renderings of anatomic surfaces, are frequently used. These techniques are often computationally intensive and require the use of specialized computer hardware. Interfaces to microprocessor-based workstations are frequently used to provide such additional display and analysis capabilities. MRI images are available as digital data; therefore, there is considerable utilization of local area networks (LANs) to distribute information throughout the hospital, and long-distance digital transmission of the images can be used for purposes of teleradiology.

3.2.6 Current Trends in MRI

At present, there is a substantial effort directed at enhancing the capabilities and cost-effectiveness of MR imagers. The activities include efforts to reduce the cost of these scanners, improve image quality, reduce scan times, and increase the number of useful clinical applications. Examples of these efforts include the development of high-field scanners, the advent of MRI-guided therapy, and the development of niche scanners that are designed for specific anatomical and clinical applications. Scanners have been developed that are dedicated to low-field imaging of the breast and other designs are dedicated to orthopedic applications such as the knees, wrists, and elbows. Perhaps the most promising incipient application of MRI is to cardiology. Scanners are now being developed to permit studies of cardiac wall motion, cardiac perfusion, and the coronary arteries in conjunction with cardiac stress testing. These scanners emphasize short magnet configurations to permit close monitoring and interaction with the patient, and high strength rapidly switched gradient fields.

Conventional spin-warp images typically require several minutes to acquire. The fast spin echo (FSE) technique can reduce this to the order of 20 sec, and gradient-echo techniques can reduce this time to a few seconds. The echo-planar technique (EPI) (Wehrli, 1990; Cohen and Weisskoff, 1991) requires substantially increased gradient power and receiver bandwidth but can produce images in 40 to 60 msec. Scanners with improved gradient hardware that are capable of handling higher data-acquisition rates are now available.

For most of the 1980s and 1990s, the highest field strength commonly used in MRI scanners was 1.5 T. To achieve better SNRs, higher-field scanners, operating at fields up to 4 T, were studied experimentally. The need for very high-field scanners has been enhanced by the development of functional brain MRI. This technique utilizes magnetization differences between oxygenated and deoxygenated hemoglobin, and this difference is enhanced at higher field strengths. It has now become possible to construct 3- and 4-T and even 8-T (Robitaille et al., 1998), whole-body scanners of essentially the same physical size (or footprint) as conventional 1.5-T systems. Along with the rapidly increasing clinical interest in functional MRI, this is resulting in a considerable increase in the use of high-field systems.

For the first decade or so after their introduction, MRI scanners were used almost entirely to provide diagnostic information. However, there is now considerable interest in systems capable of performing image-guided, invasive surgical procedures. Because MRI is capable of providing

FIGURE 3.13 Open magnet for MEI-guided therapy. This open-geometry superconducting magnet provides a surgeon with direct patient access and the ability to interactively control the MRI scanner. This permits imaging to be performed simultaneously with surgical interventions.

excellent soft-tissue contrast and has the potential for providing excellent positional information with submillimeter accuracy, it can be used for guiding biopsies and stereotactic surgery. The full capabilities of MRI-guided procedures can only be achieved if it is possible to provide surgical access to the patient simultaneously with the MRI scanning. This has lead to the development of new system designs, including the introduction of a scanner with a radically modified superconducting magnet system that permits the surgeon to operate at the patient's side within the scanner (Figure 3.13) (Schenck et al., 1995; Black et al., 1997). These systems have lead to the introduction of magnetic field-compatible surgical instruments, anesthesia stations, and patient monitoring equipment (Schenck, 1996).

Defining Terms

Bandwidth: The narrow frequency range, approximately 32 kHz, over which the MRI signal is transmitted. The bandwidth is proportional to the strength of the readout gradient field.

Echo-planar imaging (EPI): A pulse sequence used to produce very fast MRI scans. EPI times can be as short as 50 msec.

Fast Fourier transform (FFT): A mathematical technique used to convert data sampled from the MRI signal into image data. This version of the Fourier transform can be performed with particular efficiency on modern array processors.

Gradient coil: A coil designed to produce a magnetic field for which the field component B: varies linearly with position. Three gradient coils, one each for the x, y, and z directions, are required MRI. These coils are used to permit slice selection and to encode position information into the MRI signal.

Larmor frequency: The rate at which the magnetic dipole moment of a particle precesses in an applied magnetic field. It is proportional to the field strength and is 42.58 MHz for protons in a 1-T magnetic field.

Magnetic resonance imaging (MRI): A technique for obtaining images of the internal anatomy based on the use of nuclear magnetic resonance signals. During the 1980s, it became a major modality for medical diagnostic imaging.

Nuclear magnetic resonance (NMR): A technique for observing and studying nuclear magnetism. It is based on partially aligning the nuclear spins by use of a strong, static magnetic field, stimulating these spins with a radiofrequency field oscillating at the Larmor frequency, and detecting the signal that is induced at this frequency.

Nuclear magnetism: The magnetic properties arising from the small magnetic dipole moments possessed by the atomic nuclei of some materials. This form of magnetism is much weaker than the more familiar form that originates from the magnetic dipole moments of the atomic electrons.

Pixel: A single element or a two-dimensional array of image data.

Pulse sequence: A series of gradient and radiofrequency pulses used to organize the nuclear spins into a pattern that encodes desired imaging information into the NMR signal.

Quadrature excitation and detection: The use of circularly polarized, rather than linearly polarized, radio frequency fields to excite an detect the NMR signal. It provides a means of reducing the required excitation power by 1/2 and increasing the signal-to-noise ratio by 2.

Radiofrequency (RF) coil: A coil designed to excite and/or detect NMR signals. These coils can usually be tuned to resonate at the Larmor frequency of the nucleus being studied.

Spin: The property of a particle, such as an electron or nucleus, that leads to the presence of an intrinsic angular momentum and magnetic moment.

Spin-warp imagining: The pulse sequence used in the most common method of MRI imaging. It uses a sequence of gradient field pulses to encode position information into the NMR signal and applies Fourier transform mathematics to this signal to calculate the image intensity value for each pixel.

Static magnetic field: The field of the main magnet that is used to magnetize the spins and to drive their Larmor precession.

Voxel: The volume element associated with a pixel. The voxel volume is equal to the pixel area multiplied by the slice thickness.

References

Black P.Mc.L., Moriarty T., Alexander E. III. et al. 1997. Development and implementation of intraoperative magnetic resonance imaging and its neurosurgical applications. *Neurosurgery* 41: 831.

Cohen M.S. and Weisskoff R.M. 1991. Ultra-fast imaging. *Magn. Reson. Imag.* 9: 1.

Edelstein W.A., Hutchinson J.M.S., Johnson G., and Redpath T.W. 1980. Spin-warp NMR imaging and applications to human whole-body imaging. *Phys. Med. Biol.* 25: 751.

Robitaille P.-M.L., Abdujalil A.M., Kangarlu A. et al. 1998. Human magnetic resonance imaging at 8 T. *NMR Biomed.* 11: 263.

Schenck J.F. 1993. Radiofrequency coils: types and characteristics. In M.I. Bronskill and P. Sprawls (Eds.), *The Physics of MRI, Medical Physics Monograph No. 21*, pp. 98–134. Woodbury, NY, American Institute of Physics.

Schenck J.F. 1996. The role of magnetic susceptibility in magnetic resonance imaging: magnetic field compatibility of the first and second kinds. *Med. Phys.* 23: 815.

Schenck J.F. and Leue W.M. 1996. Instrumentation: magnets coils and hardware. In S.W. Atlas (Ed.), *Magnetic Resonance Imaging of the Brain and Spine*, 2nd ed., pp. 1–27. Philadelphia, Lippincott-Raven.

Schenck J.F., Jolesz A., Roemer P.B. et al. 1995. Superconducting open-configuration MR imaging system for image-guided therapy. *Radiology* 195: 805.

Shellock F.G. and Kanal E. 1998. *Magnetic Resonance: Bioeffects, Safety and Patient Management*, 2nd ed. Philadelphia, Saunders.

Thomas S.R. 1993. Magnet and gradient coils: types and characteristics. In M.J. Bronskill and P. Sprawls (Eds.), *The Physics of MRI, Medical Physics Monograph No. 21*, pp. 56–97. Woodbury, NY, American Institute of Physics.

Wehrli F.W. 1990. Fast scan magnetic resonance: principles and applications. *Magn. Reson. Q.* 6: 165.

Wilson M.N. 1983. *Superconducting Magnets*. Oxford, Clarendon Press.

Further Information

There are several journals devoted entirely to MR imaging. These include *Magnetic Resonance in Medicine, JMRI—Journal of Magnetic Resonance Imaging*, and *NMR in Biomedicine*, all three of which are published by Wiley-Liss, 605 Third Avenue, New York, NY 10158. Another journal dedicated to this field is *Magnetic Resonance Imaging* (Elsevier Publishing, 655 Avenue of the Americas, New York, NY 10010). The clinical aspects of MRI are covered extensively in *Radiology* (Radiological Society of North America, 2021 Spring Road, Suite 600, Oak Brook, IL 60521), *The American Journal of Radiology* (American Roentgen Ray Society, 1891 Preston White Drive, Reston, VA 20191), as well as in several other journals devoted to the practice of radiology. There is a professional society, now known as the International Society for Magnetic Resonance in Medicine (ISMRM), devoted to the medical aspects of magnetic resonance. The main offices of this society are at 2118 Milvia, *Suite* 201, Berkeley, CA 94704. This society holds an annual meeting that includes displays of equipment and the presentation of approximately 2800 technical papers on new developments in the field. The annual *Book of Abstracts* of this meeting provides an excellent summary of current activities in the field. Similarly, the annual meeting of the Radiological Society of North America (RSNA) provides extensive coverage of MRI that is particularly strong on the clinical applications. The RSNA is located at 2021 Spring Road, Suite 600, Oak Brook, IL 60521.

Several book-length accounts of MRI instrumentation and techniques are available. *Biomedical Magnetic Resonance Technology* (Adam Higler, Bristol, 1989) by Chen and D.I. Hoult, *The Physics of MRI* (Medical Physics Monograph 21, American Institute of Physics, Woodbury, NY, 1993), edited by M.J. Bronskill and P. Sprawls, and *Electromagnetic Analysis and Design in Magnetic Resonance Imaging* (CRC Press, Boca Raton, FL, 1998) by J.M. Jin each contain thorough accounts of instrumentation and the physical aspects of MRI. There are many books that cover the clinical aspects of MRI. Of particular interest are *Magnetic Resonance Imaging*, 3rd edition (Mosby, St. Louis, 1999), edited by D.D. Stark and W.G. Bradley, Jr., and *Magnetic Resonance Imaging of the Brain and Spine*, 2nd ed. (Lipincott-Raven, Philadelphia, 1996), edited by S.W. Atlas.

3.3 Functional MRI

Kenneth K. Kwong and David A. Chesler

Functional magnetic resonance imaging (fMRI), a technique that images intrinsic blood signal change with magnetic resonance (MR) imagers, has in the last 3 years become one of the most successful tools used to study blood flow and perfusion in the brain. Since changes in neuronal activity are accompanied by focal changes in cerebral blood flow (CBF), blood volume (CBV), blood oxygenation, and metabolism, these physiologic changes can be used to produce functional maps of mental operations.

There are two basic but completely different techniques used in fMRI to measure CBF. The first one is a classic steady-state perfusion technique first proposed by Detre et al. [1], who suggested the use of saturation or inversion of incoming blood signal to quantify absolute blood flow [1–5]. By focusing on blood flow *change* and not just steady-state blood flow, Kwong et al. [6] were successful in imaging brain visual functions associated with quantitative perfusion change. There are many advantages in studying blood flow change because many common baseline artifacts associated with MRI absolute flow techniques can be subtracted out when we are interested only in changes. And one obtains adequate information in most functional neuroimaging studies with information of flow change alone.

The second technique also looks at change of a blood parameter—blood oxygenation *change* during neuronal activity. The utility of the change of blood oxygenation characteristics was strongly evident in Turner's work [7] with cats with induced hypoxia. Turner et al. found that with hypoxia, the MRI signal from the cats' brains went down as the level of deoxyhemoglobin rose, a result that was an extension of an earlier study by Ogawa et al. [8,9] of the effect of deoxyhemoglobin on MRI signals in animals' veins. Turner's new observation was that when oxygen was restored, the cats' brain signals climbed up and went *above* their baseline levels. This was the suggestion that the vascular system overcompensated by bringing more oxygen, and with more oxygen in the blood, the MRI signal would rise beyond the baseline.

Based on Turner's observation and the perfusion method suggested by Detre et al., movies of human visual cortex activation utilizing both the perfusion and blood oxygenation techniques were successfully acquired in May of 1991 (Figure 3.14) at the Massachusetts General Hospital with a specially equipped superfast 1.5-T system known as an *echo-planar imaging* (EPI) MRI system [10]. fMRI results using intrinsic blood contrast were first presented in public at the Tenth Annual Meeting of the Society of Magnetic Resonance in Medicine in August of 1991 [6,11]. The visual cortex activation work was carried out with flickering goggles, a photic stimulation protocol employed by Belliveau et al. [12] earlier to acquire the MRI functional imaging of the visual cortex with the injection of the contrast agent gadolinium-DTPA. The use of an external contrast agent allows the study of change in blood volume. The intrinsic blood contrast technique, sensitive to blood flow and blood oxygenation, uses no external contrast material. Early model calculation showed that signal due to blood perfusion change would only be around 1% above baseline, and the signal due to blood oxygenation change also was quite small. It was quite a pleasant surprise that fMRI results turned out to be so robust and easily detectable.

The blood oxygenation–sensitive MRI signal change, coined *blood oxygenation level dependent* (BOLD) by Ogawa et al. [8,9,13], is in general much larger than the MRI perfusion signal change during

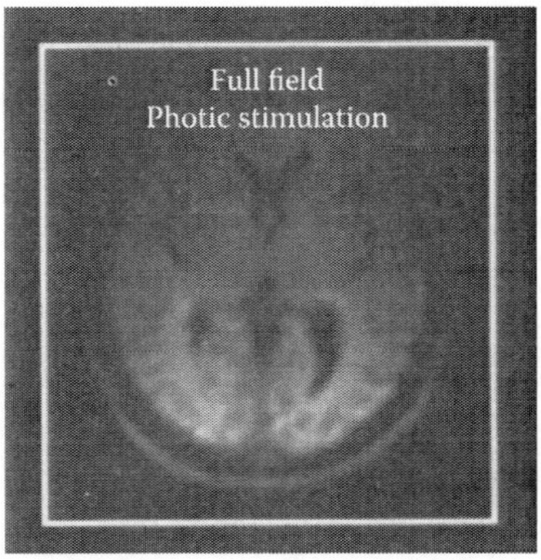

FIGURE 3.14 Functional MR image demonstrating activation of the primary visual cortex (V1). Image acquired on May 9, 1991 with a blood oxygenation–sensitive MRI gradient-echo (GE) technique.

brain activation. Also, while the first intrinsic blood contrast fMRI technique was demonstrated with a superfast EPI MRI system, most centers doing fMRI today are only equipped with conventional MRI systems, which are really not capable of applying Detre's perfusion method. Instead, the explosive growth of MR functional neuroimaging [14–33] in the last three years relies mainly on the measurement of blood oxygenation change, utilizing a MR parameter called T_2. Both high speed echo planar (EPI) and conventional MR have now been successfully employed for functional imaging in MRI systems with magnet field strength ranging from 1.5 to 4.0 T.

3.3.1 Advances in Functional Brain Mapping

The popularity of fMRI is based on many factors. It is safe and totally noninvasive. It can be acquired in single subjects for a scanning duration of several minutes, and it can be repeated on the same subjects as many times as necessary. The implementation of the blood oxygenation sensitive MR technique is universally available. Early neuroimaging, work focused on time-resolved MR topographic mapping of human primary visual (VI) (Figures 3.15 and 3.16), motor (MI), somatosensory (S1), and auditory (A1) cortices during task activation. Today, with BOLD technique combined with EPI, one can acquire 20 or more contiguous brain slices covering the whole head (3 × 3 mm in plane and 5 mm slice thickness) every 3 sec for a total duration of several minutes. Conventional scanners can only acquire a couple of slices at a time. The benefits of whole-head imaging are many. Not only can researchers identify and test their hypotheses on known brain activation centers, they can also search for previous unknown or unsuspected sites. High resolution work done with EPI has a resolution of 1.5 × 1.5 mm in plane and a slice thickness of 3 mm. Higher spatial resolution has been reported in conventional 1.5-T MR systems [34].

Of note with Figure 3.16 is that with blood oxygenation–sensitive MR technique, one observers an undershoot [6,15,35] in signal in V1 when the light stimulus is turned off. The physiologic mechanism underlying the undershoot is still not well understood.

The data collected in the last 3 years have demonstrated that fMRI maps of the visual cortex correlate well with known retinotopic organization [24,36]. Higher visual regions such as V5/MT [37] and motor-cortex organization [6,14,27,38] have been explored successfully. Preoperative planning work (Figure 3.17) using motor stimulation [21,39,40] has helped neurosurgeons who attempt to preserve primary areas from tumors to be resected. For higher cognitive functions, several fMRI language studies have already demonstrated known language-associated regions [25,26,41,42] (Figure 3.18). There is more detailed modeling work on the mechanism of functional brain mapping by blood-oxygenation change [43–46]. Postprocessing techniques that would help to alleviate the serious problem of motion/displacement artifacts are available [47].

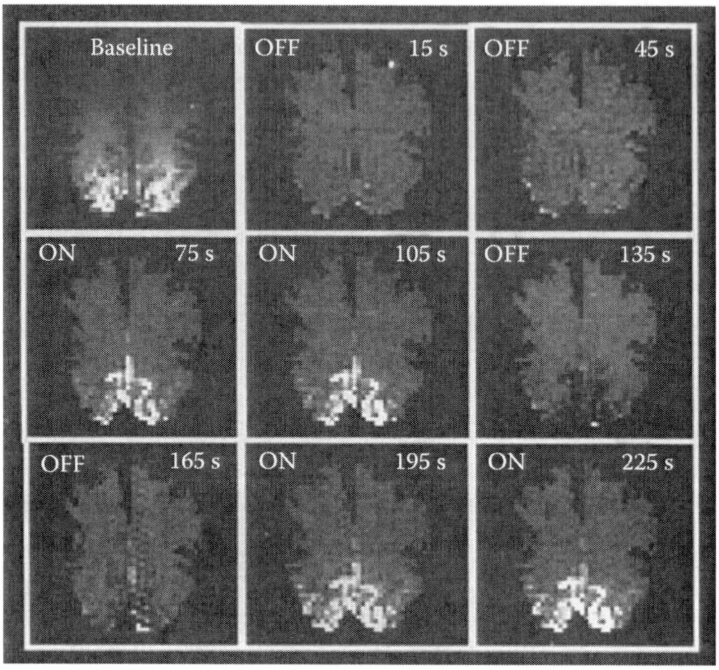

FIGURE 3.15 Movie of fMRI mapping of primary visual cortex (V1) activation during visual stimulation. Images are obliquely aligned along the calcarie fissures with the occipital pole at the bottom. Images were acquired at 3-sec intervals using a blood oxygenation–sensitive MRI sequence (80 images total). A baseline image acquired during darkness (*upper left*) was subtracted from subsequent images. Eight of these subtraction images are displayed, chosen when the image intensities reached a steady-state signal level during darkness (OFF) and during 8-Hz photic stimulation (ON). During stimulation, local increases in signal intensity are detected in the posteromedial regions of the occipital lobes along the calcarine fissures.

3.3.2 Mechanism

Flow-sensitive images show increased perfusion with stimulation, while blood oxygenation–sensitive images show changes consistent with an increase in venous blood oxygenation. Although the precise biophysical mechanisms responsible for the signal changes have yet to be determined, good hypotheses exist to account for our observations.

Two fundamental MRI relaxation rates, T_1 and T_2, are used to describe the fMRI signal. T_1 is the rate at which the nuclei approach thermal equilibrium, and perfusion change can be considered as an additional T_1 change. T_2 represents the rate of the decay of MRI signal due to magnetic field inhomogeneities, and the change of T_2 is used to measure blood-oxygenation change.

T_2 changes reflect the interplay between changes in cerebral blood flow, volume, and oxygenation. As hemoglobin becomes deoxygenated, it becomes more paramagnetic than the surrounding tissue [48] and thus creates a magnetically inhomogeneous environment. The observed *increased* signal on T_2-weighted images during activation reflects a decrease in deoxyhemoglobin content, that is, an increase in venous blood oxygenation. Oxygen delivery, cerebral blood flow, and cerebral blood volume all increase with neuronal activation. Because CBF (and hence oxygen-delivery) changes exceed CBV changes by 2 to 4 times [49], while blood–oxygen extraction increases only slightly [50,51], the total paramagnetic blood deoxyhemoglobin content within brain tissue voxels will decrease with brain activation. The resulting decrease in the tissue-blood magnetic susceptibility difference leads to less intravoxel dephasing

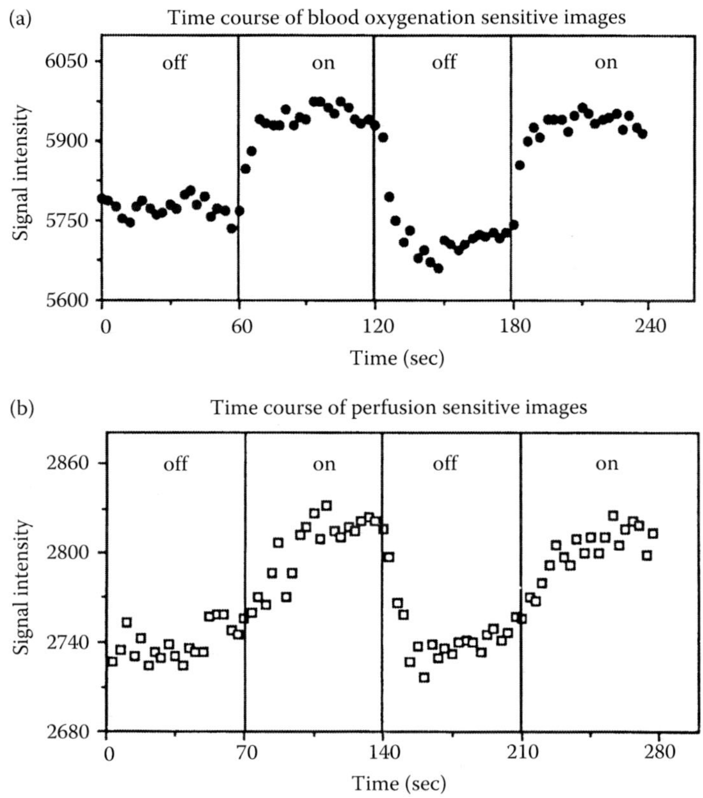

FIGURE 3.16 Signal intensity changes for a region of interest (~60 mm^2) within the visual cortex during darkness and during 8-Hz photic stimulation. Results using oxygenation–sensitive (a) and flow-sensitive (b) techniques are shown. The flow-sensitive data were collected once every 3.5 sec, and the oxygenation-sensitive data were collected once every 3 sec. Upon termination of photic stimulation, an undershoot in the oxygenation-sensitive signal intensity is observed.

within brain tissue voxels and hence *increased* signal on T_2-weighted images [6,14,15,17]. These results independently confirm PET observations that activation-induced changes in blood flow and volume are accompanied by little or no increases in tissue oxygen consumption [50–52].

Since the effect of volume susceptibility difference $\Delta\chi$ is more pronounced at high field strength [53], higher-field imaging magnets [17] will increase the observed T_2 changes.

Signal changes can also be observed on T_1-weighted MR images. The relationship between T_1 and regional blood flow was characterized by Detre et al. [1]:

$$\frac{\mathrm{d}M}{\mathrm{d}t} = \frac{M_0 - M}{T_1} + fM_\mathrm{b} - \frac{f}{\lambda}M \tag{3.14}$$

where M is tissue magnetization and M_b is incoming blood signal. M_0 is proton density, f is the flow in mL/gm/unit time, and λ is the brain–blood partition coefficient of water (~0.95 mL/g). From this equation, the brain tissue magnetization M relaxes with an apparent T_1 time constant $T_{1\mathrm{app}}$ given by

$$\frac{f}{\lambda} = \frac{1}{T_{1\mathrm{app}}} - \frac{1}{T_1} \tag{3.15}$$

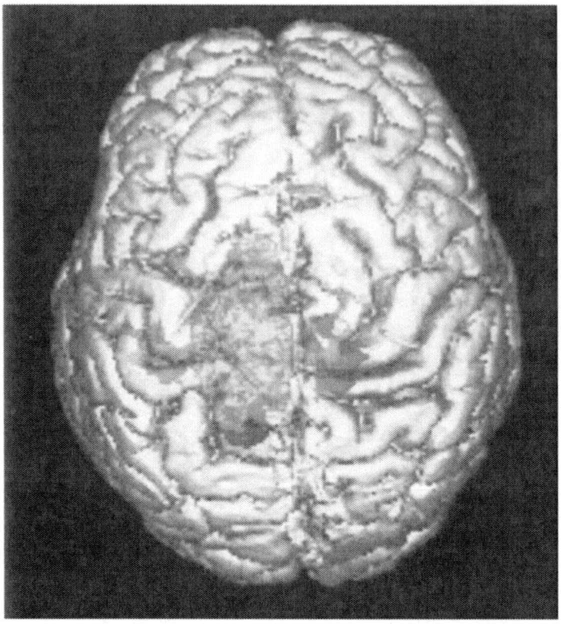

FIGURE 3.17 Functional MRI mapping of motor cortex for preoperative planning. This three-dimensional rendering of the brain represents fusion of functional and structural anatomy. Brain is viewed from the top. A tumor is shown in the left hemisphere, near the midline. The other areas depict sites of functional activation during movement of the right hand, right foot, and left foot. The right foot cortical representation is displaced by tumor mass effect from its usual location. (Courtesy of Dr. Brad Buchbinder.)

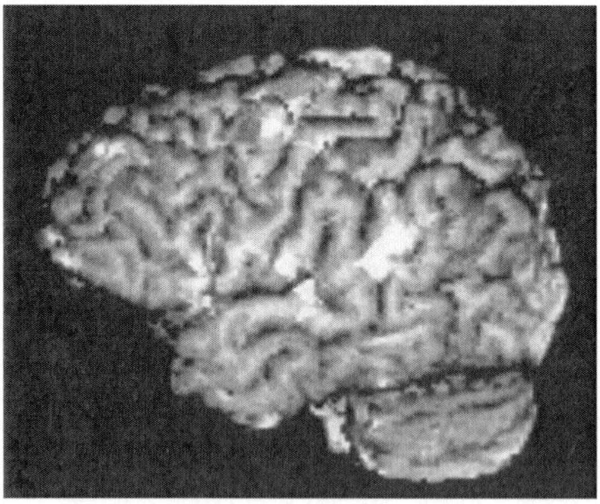

FIGURE 3.18 Left hemisphere surface rendering of functional data (EPI, gradient-echo, ten oblique coronal slices extending to posterior sylvian fissure) and high-resolution anatomic image obtained on a subject (age 33 years) during performance of a same-different (visual matching) task of pairs of words or nonwords (false font strings). Foci of greatest activation for this study are located in dominant perisylvian cortex, that is, inferior frontal gyrus (Broca's area), superior temporal gyrus (Wernicke's area), and inferior parietal lobule (angular gyrus). Also active in this task are sensorimotor cortex and prefrontal cortex. The perisylvian sites of activation are known to be key nodes in a left hemisphere language network. Prefrontal cortex probably plays a more general, modulatory role in attentional aspects of the task. Sensorimotor activation is observed in most language studies despite the absence of overt vocalization. (Courtesy of Dr. Randall Benson.)

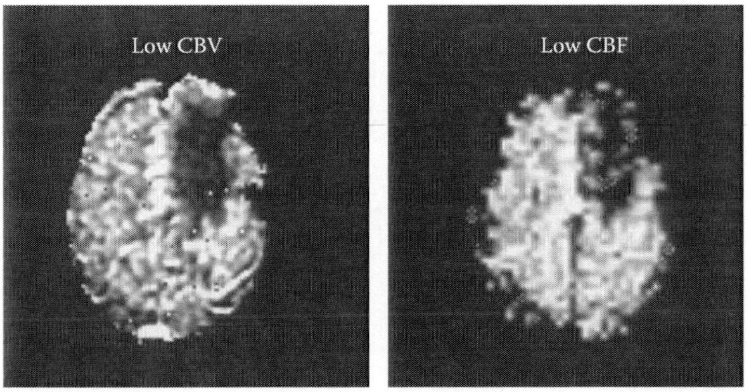

FIGURE 3.19 Functional MRI cerebral blood flow (CBF) index (*right*) of a low-flow brain tumor (dark region right of the midline) generated by the subtraction of a flow-nonsensitive image from a flow-sensitive image. This low-flow region matches well with a cerebral blood volume (CBV) map (*left*) of the tumor region generated by the injection of a bolus of MRI contrast agent Gd-DTPA, a completely different and established method to measure hemodynamics with MRI.

where the $T_{1\text{app}}$ is the observed (apparent) longitudinal relaxation time with flow effects included. T_1 is the true tissue longitudinal relaxation time in the absence of flow. If we assume that the true tissue T_1 remains constant with stimulation, a change in blood flow Δf will lead to a change in the observed $T_{1\text{app}}$:

$$\Delta \frac{1}{T_{1\text{app}}} = \Delta \frac{f}{\lambda} \tag{3.16}$$

Thus the MRI signal change can be used to estimate the change in blood flow.

From Equation 3.14, if the magnetization of blood and tissue always undergoes a similar T_1 relaxation, the flow effect would be minimized. This is a condition that can be approximated by using a flow-nonsensitive T_1 technique inverting *all* the blood coming into the imaged slice of interest. This flow-nonsensitive sequence can be subtracted from a flow-sensitive T_1 technique to provide an index of CBF without the need of external stimulation [54,55] (Figure 3.19). Initial results with tumor patients show that such flow-mapping techniques are useful for mapping out blood flow of tumor regions [55].

Other flow techniques under investigation include the continuous inversion of incoming blood at the carotid level [1] or the use of a single inversion pulse at the carotid level (EPIstar) inverting the incoming blood [56,57]. Compared with the flow-nonsensitive and flow-sensitive methods, the blood-tagging techniques at the carotid level are basically similar concepts except that the MRI signal of tagged blood is expected to be smaller by a factor that depends on the time it takes blood to travel from the tagged site to the imaged slice of interest [55]. The continuous-inversion technique also has a significant problem of magnetization transfer [1] that contaminates the flow signal with a magnetization transfer signal that is several times larger. On the other hand, the advantage of the continuous inversion is that it can under optimal conditions provide a flow contrast larger than all the other methods by a factor of *e* [55].

3.3.3 Problem and Artifacts in fMRI: The Brain–Vein Problem? The Brain-Inflow Problem?

The artifacts arising from large vessels pose serious problems to the interpretation of oxygenation sensitive fMRI data. It is generally believed that microvascular changes are specific to the underlying region of neuronal activation. However, MRI gradient echo (GE) is sensitive to vessels of all dimensions [46,58], and there is concern that macrovascular changes distal to the site of neuronal activity can be induced [20]. This has been known as the *brain–vein problem*. For laboratories not equipped with EPI, gradient echo (GE) sensitive to variations in T_2 and magnetic susceptibility are the only realistic sequences available for fMRI acquisition, so the problem is particularly acute.

In addition, there is a non-deoxyhemoglobin-related problem, especially acute in conventional MRI. This is the inflow problem of fresh blood that can be time-locked to stimulation [28,29,59]. Such non-parenchymal and macrovascular responses can introduce error in the estimate of activated volumes.

3.3.4 Techniques to Reduce the Large Vessel Problems

In dealing with the inflow problems, EPI has special advantages over conventional scanners. The use of long repetition times (2 to 3 sec) in EPI significantly reduces the brain-inflow problem. Small-flip-angle methods in conventional MRI scanners can be used to reduce inflow effect [59]. Based on inflow modeling, one observes that at an angle smaller than the Ernst angle [60], the inflow effect drops much faster than the tissue signal response to activation. Thus one can effectively remove the inflow artifacts with small-flip-angle techniques.

A new exciting possibility is to add small additional velocity-dephasing gradients to suppress slow in-plane vessel flow [60,61]. Basically, moving spins lose signals, while stationary spins are unaffected. The addition of these velocity-dephasing gradients drops the overall MRI signal (Figure 3.20). The hypothesis that large vessel signals are suppressed while tissue signals remain intact is a subject of ongoing research.

Another advantage with EPI is that another oxygenation-sensitive method such as the EPI T_2-weighted spin-echo (T2SE) is also available. T2SE methods are sensitive to the MRI parameter T_2, which is affected by microscopic susceptibility and hence blood oxygenation. Theoretically, T2SE methods are far less sensitive to large vessel signals [1,6,46,58]. For conventional scanners, T2SE methods take too long to perform and therefore are not practical options.

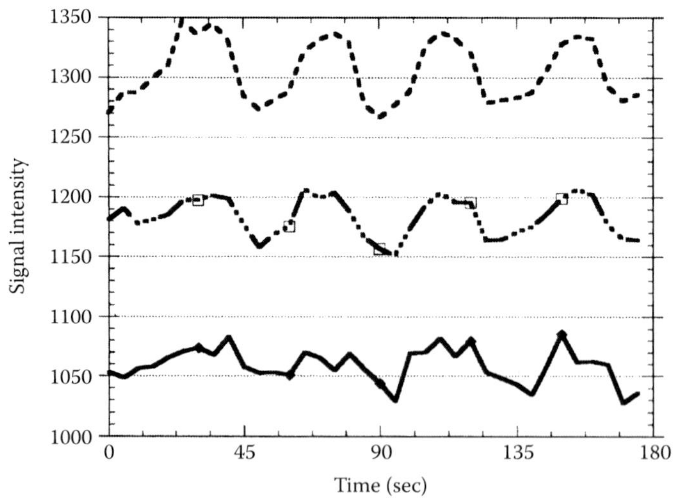

FIGURE 3.20 The curves represent time courses of MRI response to photic stimulation (off-on-off-on …) with different levels of velocity-dephasing gradients turned on to remove MRI signals coming from the flowing blood of large vessels. The top curve had no velocity-dephasing gradients turned on. The bottom curve was obtained with such strong velocity-dephasing gradients turned on that all large vessel signals were supposed to have been eliminated. The middle curve represents a moderate amount of velocity-dephasing gradients, a tradeoff between removing large vessel signals and retaining a reasonable amount of MRI signal to noise.

The flow model [1] based on T_1-weighted sequences and independent of deoxyhemoglobin is also not so prone to large vessel artifacts, since the T_1 model is a model of perfusion at the tissue level.

Based on the study of volunteers, the average T_2-weighted GE signal percentage change at V1 was $2.5 \pm 0.8\%$. The average oxygenation-weighted T2SE signal percentage change was $0.7 \pm 0.3\%$. The average perfusion-weighted and T_1-weighted MRI signal percentage change was $1.5 \pm 0.5\%$. These results demonstrate that T2SE and T_1 methods, despite their ability to suppress large vessels, are not competitive with T_2 effect at 1.5 T. However, since the microscopic effect detected by T2SE scales up with field strength [62], we expect the T2SE to be a useful sequence at high field strength such as 3 or 4 T. Advancing field strength also should benefit T_1 studies due to better signal-to-noise and to the fact that T_1 gets longer at higher field strength.

While gradient-echo sequence has a certain ambiguity when it comes to tissue versus vessels, its sensitivity at current clinical field strength makes it an extremely attractive technique to identify activation sites. By using careful paradigms that rule out possible links between the primary activation site and secondary sites, one can circumvent many of the worries of "signal from the primary site draining down to secondary sites." A good example is as follows: photic stimulation activates both the primary visual cortex and the extrastriates. To show that the extrastriates are not just a drainage from the primary cortex, one can utilize paradigms that activate the primary visual cortex but not the extrastriate, and vice versa. There are many permutations of this [37]. This allows us to study the higher-order functions umambiguously even if we are using gradient-echo sequences.

The continuous advance of MRI mapping techniques utilizing intrinsic blood-tissue contrast promises the development of a functional human neuroanatomy of unprecedented spatial and temporal resolution.

References

1. Detre J., Leigh J., Williams D., and Koretsky A. 1992. *Magn. Reson. Med.* 23: 37.
2. Williams D.S., Detre J.A., Leigh J.S., and Koretsky A.P. 1992. *Proc. Natl Acad. Sci. USA* 89: 212.
3. Zhang W., Williams D.S., and Detre J.A. 1992. *Magn. Reson. Med.* 25: 362.
4. Zhang W., Williams D.S., and Koretsky A.P. 1993. *Magn. Reson. Med.* 29: 416.
5. Dixon W.T., Du L.N., Faul D. et al. 1986. *Magn. Reson. Med.* 3: 454.
6. Kwong K.K., Belliveau J.W., Chesler D.A. et al. 1992. *Proc. Natl Acad. Sci. USA* 89: 5675.
7. Turner R., Le Bihan D., Moonen C.T. et al. 1991. *Magn. Reson. Med.* 22: 159.
8. Ogawa S., Lee T.M., Kay A.R., and Tank D.W. 1990. *Proc. Natl Acad. Sci. USA* 87: 9868.
9. Ogawa S. and Lee T.M. 1990. *Magn. Reson. Med.* 16: 9.
10. Cohen M.S. and Weisskoff R.M. 1991. *Magn. Reson. Imag.* 9: 1.
11. Brady T.J. 1991. *Society of Magnetic Resonance in Medicine*, San Francisco, CA Vol. 2.
12. Belliveau J.W., Kennedy D.N. Jr, McKinstry R.C. et al. 1991. *Science* 254: 716.
13. Ogawa S., Lee T.M., Nayak A.S., and Glynn P. 1990. *Magn. Reson. Med.* 14: 68.
14. Bandettini P.A., Wong E.C., Hinks R.S. et al. 1992. *Magn. Reson. Med.* 25: 390.
15. Ogawa S., Tank D.W., Menon R. et al. 1992. *Proc. Natl Acad. Sci. USA* 89: 5951.
16. Frahm J., Bruhn H., Merboldt K., and Hanicke W. 1992. *J. Magn. Reson. Imag.* 2: 501.
17. Turner R., Jezzard P., Wen H. et al. 1992. *Society of Magnetic Resonance in Medicine Eleventh Annual Meeting*, Berlin.
18. Blamire A., Ogawa S., Ugurbil K. et al. 1992. *Proc. Natl Acad. Sci. USA* 89: 11069.
19. Menon R., Ogawa S., Tank D., and Ugurbil K. 1993. *Magn. Reson. Med.* 30: 380.
20. Lai S., Hopkins A., Haacke E. et al. 1993. *Magn. Reson. Med.* 30: 387.
21. Cao Y., Towle V.L., Levin D.N. et al. 1993. *Society of Magnetic Resonance in Medicine Meeting.*
22. Connelly A., Jackson G.D., Frackowiak R.S.J. et al. 1993. *Radiology* 188: 125.
23. Kim S.G., Ashe J., Georgopouplos A.P. et al. 1993. *J. Neurophys.* 69: 297.

24. Schneider W., Noll D.C., and Cohen J.D. 1993. *Nature* 365: 150.
25. Hinke R.M., Hu X., Stillman A.E. et al. 1993. *Neurol. Rep.* 4: 675.
26. Binder J.R., Rao S.M., Hammeke T.A. et al. 1993. *Neurology* (suppl. 2): 189.
27. Rao S.M., Binder J.R., Bandettini P.A. et al. 1993. *Neurology* 43: 2311.
28. Gomiscek G., Beisteiner R., Hittmair K. et al. 1993. *MAGMA* 1: 109.
29. Duyn J., Moonen C., de Boer R. et al. 1993. *Society of Magnetic Resonance in Medicine, th Annual Meeting*, New York.
30. Hajnal J.V., Collins A.G., White S.J. et al.. 1993. *Magn. Reson. Med.* 30: 650.
31. Hennig J., Ernst T., Speck O. et al. 1994. *Magn. Reson. Med.* 31: 85.
32. Constable R.T., Kennan R.P., Puce A. et al. 1994. *Magn. Reson. Med.* 31: 686.
33. Binder J.R., Rao S.M., Hammeke T.A. et al. 1994. *Ann. Neurol.* 35: 662.
34. Frahm J., Merboldt K., and Hänicke W. 1993. *Magn. Reson. Med.* 29: 139.
35. Stern C.E., Kwong K.K., Belliveau J.W. et al. 1992. *Society of Magnetic Resonance in Medicine Annual Meeting*, Berlin, Germany.
36. Belliveau J.W., Kwong K.K., Baker J.R. et al. 1992. *Society of Magnetic Resonance in Medicine Annual Meeting*, Berlin, Germany.
37. Tootell R.B.H., Kwong K.K., Belliveau J.W. et al. 1993. *Investigative Ophthalmology and Visual Science*, p. 813.
38. Kim S.-G., Ashe J., Hendrich K. et al. 1993. *Science* 261: 615.
39. Buchbinder B.R., Jiang H.J., Cosgrove G.R. et al. 1994. *ASNR* 162.
40. Jack C.R., Thompson R.M., Butts R.K. et al. 1994. *Radiology* 190: 85.
41. Benson R.R., Kwong K.K., Belliveau J.W. et al. 1993. *Soc. Neurosci.*
42. Benson R.R., Kwong K.K., Buchbinder B.R. et al. 1994. *Society of Magnetic Resonance*, San Francisco.
43. Ogawa S., Menon R., Tank D. et al. 1993. *Biophys. J.* 64: 803.
44. Ogawa S., Lee T.M., and Barrere B. 1993. *Magn. Reson. Med.* 29: 205.
45. Kennan R.P., Zhong J., and Gore J.C. 1994. *Magn. Reson. Med.* 31: 9.
46. Weisskoff R.M., Zuo C.S., Boxerman J.L., and Rosen B.R. 1994. *Magn. Reson. Med.* 31: 601.
47. Bandettini P.A., Jesmanowicz A., Wong E.C., and Hyde J.S. 1993. *Magn. Reson. Med.* 30: 161.
48. Thulborn K.R., Waterton J.C., Matthews P.M., and Radda G.K. 1982. *Biochim. Biophys. Acta* 714: 265.
49. Grubb R.L., Raichle M.E., Eichling J.O., and Ter-Pogossian M.M. 1974. *Stroke* 5: 630.
50. Fox P.T. and Raichle M.E. 1986. *Proc. Natl Acad. Sci. USA* 83: 1140.
51. Fox P.T., Raichle M.E., Mintun M.A., and Dence C. 1988. *Science* 241: 462.
52. Prichard J., Rothman D., Novotny E. et al. 1991. *Proc. Natl Acad. Sci. USA* 88: 5829.
53. Brooks R.A. and Di Chiro G. 1987. *Med. Phys.* 14: 903.
54. Kwong K., Chesler D., Zuo C. et al. 1993. *Society of Magnetic Resonance in Medicine, th Annual Meeting*, New York, p. 172.
55. Kwong K.K., Chesler D.A., Weisskoff R.M., and Rosen B.R. 1994. *Society of Magnetic Resonance*, San Francisco.
56. Edelman R., Sievert B., Wielopolski P. et al. 1994. *JMRI* 4 (P): 68.
57. Warach S., Sievert B., Darby D. et al. 1994. *JMRI* 4: S8.
58. Fisel C.R., Ackerman J.L., Buxton R.B. et al. 1991. *Magn. Reson. Med.* 17: 336.
59. Frahm J., Merboldt K., and Hanicke W. 1993. *Society of Magnetic Resonance in Medicine, th Annual Meeting*, New York, p. 1427.
60. Kwong K.K., Chesler D.A., Boxerman J.L. et al. 1994. *Society of Magnetic Resonance*, San Francisco.
61. Song W., Bandettini P., Wong E., and Hyde J. 1994. Personal communication.
62. Zuo C., Boxerman J., and Weisskoff R. 1992. *Society of Magnetic Resonance in Medicine, th Annual Meeting*, Berlin, p. 866.

3.4 Chemical-Shift Imaging: An Introduction to Its Theory and Practice

Xiaoping Hu, Wei Chen, Maqbool Patel, and Kamil Ugurbil

Over the past two decades, there has been a great deal of development in the application of nuclear magnetic resonance (NMR) to biomedical research and clinical medicine. Along with the development of magnetic resonance imaging [1], *in vivo* magnetic resonance spectroscopy (MRS) is becoming a research tool for biochemical studies of humans as well as a potentially more specific diagnostic tool, since it provides specific information on individual chemical species in living systems. Experimental studies in animals and humans have demonstrated that MRS can be used to study the biochemical basis of disease and to follow the treatment of disease.

Since biologic subjects (e.g., humans) are heterogeneous, it is necessary to spatially localize the spectroscopic signals to a well-defined volume or region of interest (VOI or ROI, respectively) in the intact body. Toward this goal, various localization techniques have been developed (see Reference 2 for a recent review). Among these techniques, chemical-shift imaging (CSI) or spectroscopic imaging [3–6] is an attractive technique, since it is capable of producing images reflecting the spatial distribution of various chemical species of interest. Since the initial development of CSI in 1982 [3], further developments have been made to provide better spatial localization and sensitivity, and the technique has been applied to numerous biomedical problems.

In this section we will first present a qualitative description of the basic principles of chemical-shift imaging and subsequently present some practical examples to illustrate the technique. Finally, a summary is provided in the last subsection.

3.4.1 General Methodology

In an NMR experiment, the subject is placed in a static magnetic field B_0. Under the equilibrium condition, nuclear spins with nonzero magnetic moment are aligned along B_0, giving rise to an induced bulk magnetization. To observe the bulk magnetization, it is tipped to a direction perpendicular to B_0 (transverse plane) with a radiofrequency (RF) pulse that has a frequency corresponding to the resonance frequency of the nuclei. The resonance frequency is determined by the product of the gyromagnetic ratio of the nucleus γ and the strength of the static field, that is, γB_0, and is called the *Larmor frequency*. The Larmor frequency also depends on the chemical environment of the nuclei, and this dependency gives rise to chemical shifts that allow one to identify different chemical species in an NMR spectrum. Upon excitation, the magnetization in the transverse plane (perpendicular to the main B_0 field direction) oscillates with the Larmor frequencies of all the different chemical species and induces a signal in a receiving RF coil; the signal is also termed the *free induction decay (FID)*. The FID can be Fourier transformed with respect to time to produce a spectrum in frequency domain.

In order to localize an NMR signal from an intact subject, spatially selective excitation and/or spatial encoding are usually utilized. Selective excitation is achieved as follows: in the excitation, an RF pulse with a finite bandwidth is applied in the presence of a linear static magnetic field gradient. With the application of the gradient, the Larmor frequency of spins depends linearly on the spatial location along the direction of the gradient. Consequently, only the spins in a slice whose resonance frequency falls into the bandwidth of the RF pulse are excited.

The RF excitation rotates all or a portion of the magnetization to the transverse plane, which can be detected by a receiving RF coil. Without spatial encoding, the signal detected is the integral of the signals over the entire excited volume. In CSI based on Fourier imaging, spatial discrimination is achieved by phase encoding. Phase encoding is accomplished by applying a gradient pulse after the excitation and before the data acquisition. During the gradient pulse, spins precess at Larmor frequencies that vary linearly along the direction of the gradient and accrue a phase proportional to the position along the phase-encoding gradient as well as the strength and the duration of the gradient pulse. This acquired spatially encoded phase is typically expressed as $\vec{k} \cdot \vec{r} = \int \gamma \vec{g}(t) \cdot \vec{r} dt$, where γ is the gyromagnetic ratio; $\vec{r}$ is the vector designating spatial location; $\vec{g}(t)$ defines the magnitude, the direction, and the time dependence of the magnetic field gradient applied during the phase-encoding; and the integration is performed over time when the phase-encoding gradient is on. Thus, in one-dimensional phase encoding, if the phase encoding is along, for example, the y axis, the phase acquired becomes $k \times y = \int \gamma g_y(t) \times y \, dt$. The acquired signal $S(t)$ is the integral of the spatially distributed signals modulated by a spatially dependent phase, given by the equation

$$S(t) = \int \rho(\vec{r},t) e^{(i\vec{k}\cdot\vec{r})} d^3r \tag{3.17}$$

where ρ is a function that describes the spatial density and the time evolution of the transverse magnetization of all the chemical species in the sample. This signal mathematically corresponds to a sample of the Fourier transform along the direction of the gradient. The excitation and detection process is repeated with various phase-encoding gradients to obtain many phase-encoded signals that can be inversely Fourier-transformed to resolve an equal number of pixels along this direction. Taking the example of one-dimensional phase-encoding along the y axis to obtain a one-dimensional image along this direction of n pixels, the phase encoding gradient is incremented n times so that n FIDs are acquired, each of which is described as

$$S(t,n) = \int \rho^*(y,t) e^{(1nk_0y)} dy \tag{3.18}$$

where ρ^* is already integrated over the x and z directions, and k_0 is the phase-encoding increment; the latter is decided on using the criteria that the full field of view undergo a 360-degree phase difference when $n = 1$, as dictated by the sampling theorem. The time required for each repetition (TR), which is dictated by the longitudinal relaxation time, is usually on the order of seconds.

In CSI, phase encoding is applied in one, two, or three dimensions to provide spatial localization. Meanwhile, selective excitation also can be utilized in one or more dimensions to restrict the volume to be resolved with the phase encodings. For example, with selective excitation in two dimensions, CSI in one spatial dimension can resolve voxels within the selected column. In multidimensional CSI, all the phase-encoding steps along one dimension need to be repeated for all the steps along the others. Thus, for three dimensions with M, N, and L number of phase encoding steps, one must acquire $M \times N \times L$ number of FIDS:

$$S(t,m,n,l) = \int \rho(\vec{r},t) e^{i(mkx_0x + nky_0y + lkz_0z)} d^3\vec{r} \tag{3.19}$$

where m, n, and l must step through M, N, and L in integer steps, respectively. As a result, the time needed for acquiring a chemical-shift image is proportional to the number of pixels desired and may be very long. In practice, due to the time limitation as well as the signal-to-noise ratio (SNR) limitation, chemical-shift imaging is usually performed with relatively few spatial encoding steps, such as 16 × 16 or 32 × 32 in a two-dimensional experiment.

The data acquired with the CSI sequence need to be properly processed before the metabolite information can be visualized and quantitated. The processing consists of spatial reconstruction and spectral processing. Spatial reconstruction is achieved by performing discrete inverse Fourier transformation, for each of the spatial dimensions, with respect to the phase-encoding steps. The spatial Fourier transform is applied for all the points of the acquired FID. For example, for a data set from a CSI in two spatial dimensions with 32×32 phase-encoding steps and 1024 sampled data points for each FID, a 32×32 two-dimensional inverse Fourier transform is applied to each of the 1024 data points. Although the nominal spatial resolution achieved by the spatial reconstruction is determined by the number of phase-encoding steps and the field of view (FOV), it is important to note that due to the limited number of phase-encoding steps used in most CSI experiments, the spatial resolution is severely degraded by the truncation artifacts, which results in signal "bleeding" between pixels. Various methods have been developed to reduce this problem [7–14].

The localized FIDs derived from the spatial reconstruction are to be further processed by spectral analysis. Standard procedures include Fourier transformation, filtering, zero-filling, and phasing. The localized spectra can be subsequently presented for visualization or further processed to produce quantitative metabolite information. The presentation of the localized spectra in CSI is not a straightforward task because there can be thousands of spectra. In one-dimensional experiments, localized spectra are usually presented in a stack plot. In two-dimensional experiments, localized spectra are plotted in small boxes representing the extent of the pixels, and the plots can be overlaid on corresponding anatomic image for reference. Spectra from three-dimensional CSI experiments are usually presented slice by slice, each displaying the spectra as in the two-dimensional case.

To derive metabolite maps, peaks corresponding to the metabolites of interest need to be quantified. In principle, the peaks can be quantified using the standard methods developed for spectral quantification [15–17]. The most straightforward technique is to calculate the peak areas by integrating the spectra over the peak of interest if it does not overlap with other peaks significantly. In integrating all the localized spectra, spectral shift due to B_0 inhomogeneity should be taken into account. A more robust approach is to apply spectral fitting programs to each spectrum to obtain various parameters of each peak. The fitted area for the peak of interest can then be used to represent the metabolite signal. The peak areas are then used to generate metabolite maps, which are images with intensities proportional to the localized peak area. The metabolite map can be displayed by itself as a gray-scale image or color-coded image or overlaid on a reference anatomic image.

3.4.2 Practical Examples

To illustrate the practical utility of CSI, we present two representative CSI studies in this section. The sequence for the first study is shown in Figure 3.21. This is a three-dimensional sequence in which phase encoding is applied in all three directions and no slice selection is used. Such a sequence is usually used with a surface RF coil whose spatial extent of sensitivity defines the field of view. In this sequence, the FID is acquired immediately after the application of the phase-encoding gradient to minimize the decay of the transverse magnetization, and the sequence is suitable for imaging metabolites with short transverse relaxation time (e.g., ATP).

With the sequence shown in Figure 3.21, a phosphorus-31 CSI study of the human brain was conducted using a quadrature surface coil. A nonselective RF pulse with an Ernest angle (40 degrees) optimized for the repetition time was used for the excitation. Phase-encoding gradients were applied for a duration of 500 μ sec; the phase-encoding gradients were incremented according to a FOV of $25 \times 25 \times 20$ cm^3. Phase-encoded FIDs were acquired with 1024 complex data points over a sampling window of 204.8 msec; the corresponding spectral width was 5000 Hz. To reduce intervoxel signal contamination, a technique that utilizes variable data averaging to introduce spatial filtering during the data acquisition for optimal signal-to-noise ratio is employed [7–10], resulting in spherical voxels with diameter of 3 cm (15 cc volume). The data were acquired with a *TR* of 1 sec, and the total acquisition time was approximately 28 min.

The acquired data were processed to generate three-dimensional voxels, each containing a localized phosphorus spectrum, in a $17 \times 13 \times 17$ matrix. In Figure 3.22a–c, spectra in three slices of the

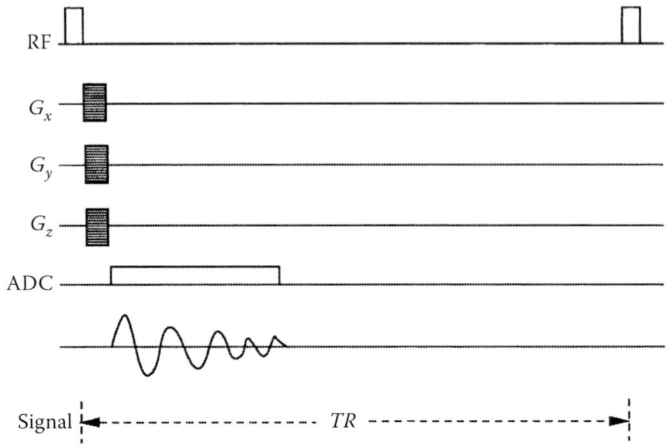

FIGURE 3.21 Sequence diagram for a three-dimensional chemical shift imaging sequence using a nonselective RF pulse.

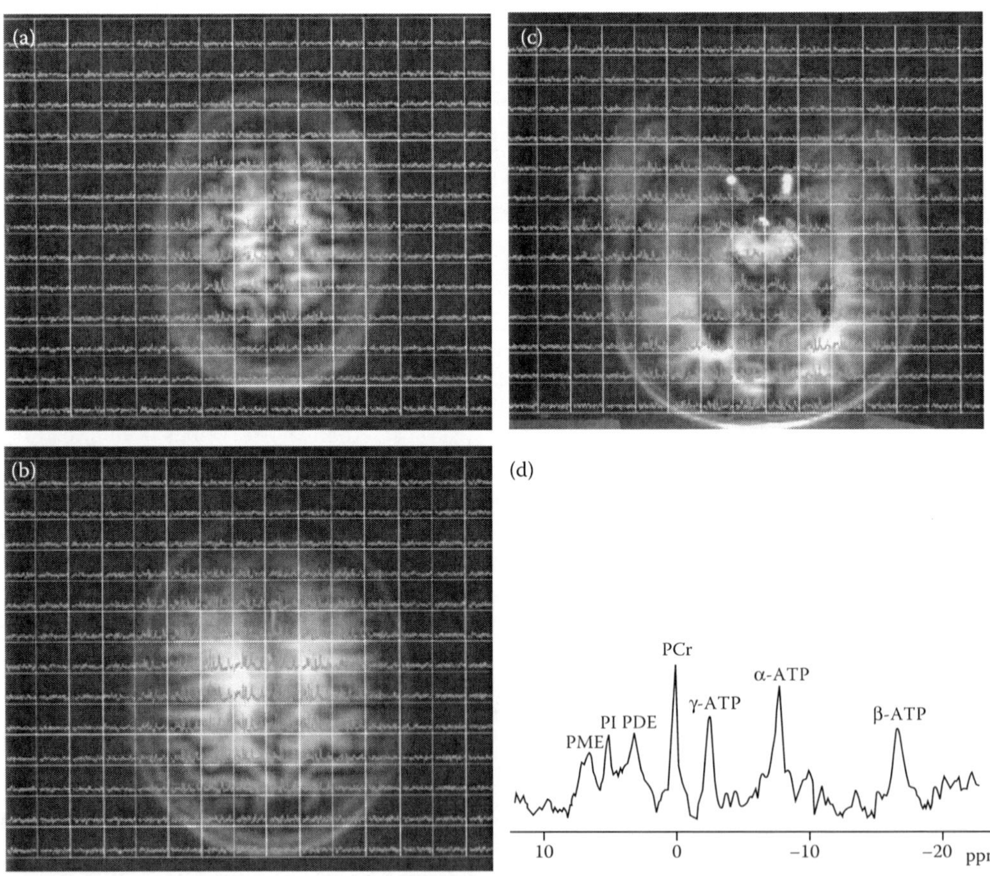

FIGURE 3.22 (a–c) Boxed plot of spectra in three slices from the three-dimensional ^{31}P CSI experiment overload on corresponding anatomic images. The spectral extent displayed is from 10 to 20 ppm. A 20-Hz line broadening is applied to all the spectra. (d) Representative spectrum from the three-dimensional ^{31}P CSI shown in (b). Metabolite peaks are labeled.

three-dimensional CSI are presented; these spectra are overlaid on the corresponding anatomic images obtained with a T_1-weighted imaging sequence. One representative spectrum of the brain is illustrated in Figure 3.22d, where the peaks corresponding to various metabolites are labeled. It is evident that the localized phosphorus spectra contain a wealth of information about several metabolites of interest, including adenosine triphosphate (ATP), phosphocreatine (PCr), phosphomonoester (PME), inorganic phosphate (P_i), and phosphodiester (PDE). In pathologic cases, focal abnormalities in phosphorus metabolites have been detected in patients with tumor, epilepsy, and other diseases [18–25].

The second study described below is performed with the sequence depicted in Figure 3.23. This is a two-dimensional spin-echo sequence in which a slice is selectively excited by a 90-degree excitation pulse. The 180-degree refocusing pulse is selective with a slightly broader slice profile. Here the phase-encoding gradients are applied before the refocusing pulse; they also can be placed after the 180-degree pulse or split to both sides of the 180-degree pulse. This sequence was used for a proton CSI experiment. In proton CSI, a major problem arises from the strong water signal that overwhelms that of the metabolites. In order to suppress the water signal, many techniques have been devised [26–29]. In this study, a three-pulse CHESS [26] technique was applied before the application of the excitation pulse as shown in Figure 3.23. The CSI experiment was performed on a 1.4-cm slice with 32 × 32 phase encodings over a 22 × 22 cm^2 FOV. The second half of the spin-echo was acquired with 512 complex data points over a sampling window of 256 msec, corresponding to a spectral width of 2000 Hz. Each phase-encoding FID was acquired twice for data averaging. The repetition time (*TR*) and the echo time (*TE*) used were 1.2 sec and 136 msec, respectively. The total acquisition time was approximately 40 min.

Another major problem in proton CSI study of the brain is that the signal from the subcutaneous lipid usually is much stronger than those of the metabolites, and this strong signal leaks into pixels within the brain due to truncation artifacts. To avoid lipid signal contamination, many proton CSI studies of the brain are performed within a selected region of interest excluding the subcutaneous fat [30–34]. Recently, several techniques have been proposed to suppress the lipid signal and consequently suppress the lipid signal contamination. These include the use of WEFT [27] and the use of outer-volume signal suppression [34]. In the example described below, we used a technique that utilizes the spatial location of the lipid to extrapolate data in the *k*-space to reduce the signal contamination due to truncation [35].

In Figure 3.24, the results from the proton CSI study are presented. In panel (a), the localized spectra are displayed. Note that the spectra in the subcutaneous lipid are ignored because they are all off the scale. The nominal spatial resolution is approximately 0.66 cc. A spectrum from an individual pixel in this study is presented in Figure 3.24b with metabolite peaks indicated. Several metabolite peaks, such as those corresponding to the *N*-acetyl aspartate (NAA), creatine/phosphocreatine (Cr/PCr), and choline (Cho), are readily identified. In addition, there is still a noticeable amount of residual lipid signal

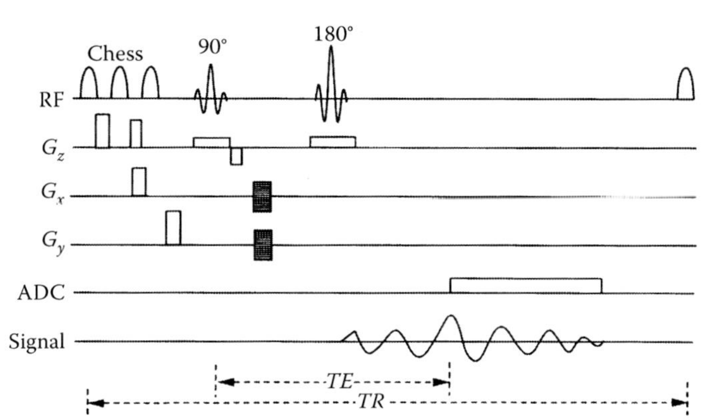

FIGURE 3.23 A two-dimensional spin-echo CSI sequence with chemical selective water suppression (CHESS) for proton study.

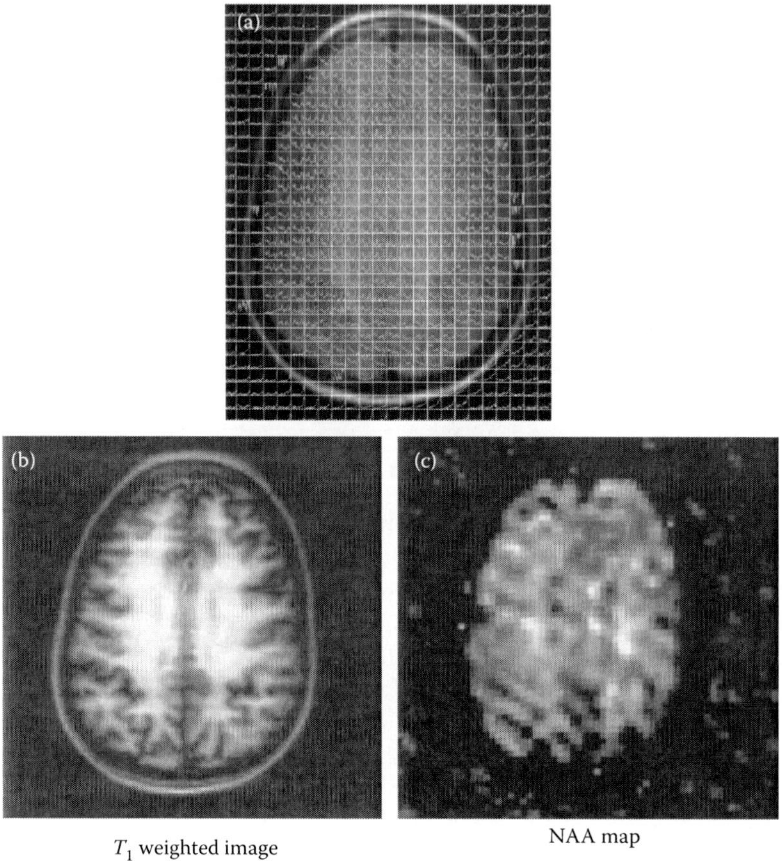

FIGURE 3.24 (a) Boxed plot of spectra for the two-dimensional proton study overlaid on the anatomic image. A spectral range of 1.7 to 3.5 ppm is used in the plot to show Cho, PCr/Cr, and NAA. A 5-Hz line broadening is applied in the spectral processing. (b) A representative spectrum from the two-dimensional CSI in panel (a). Peaks corresponding to Cho, PCr/Cr, and NAA are indicated. (c) A map of the area under the NAA peak obtained by spectral fitting. The anatomic image is presented along with the metabolite map for reference. The spatial resolution of the metabolite image can be appreciated from the similarities between the two images. The lipid suppression technique has successfully eliminated the signal contamination from the lipid in the skull.

contamination despite the use of the data extrapolation technique. Without the lipid suppression technique, the brain spectra would be severely contaminated by the lipid signal, making the detection of the metabolite peaks formidable. The peak of NAA in these spectra is fitted to generate the metabolite map shown in panel (c). Although the metabolite map is not corrected for coil sensitivity and other factors and only provides a relative measure of the metabolite concentration in the brain, it is a reasonable measure of the NAA distribution in the brain slice. The spatial resolution of the CSI study can be appreciated from the brain structure present in the map. In biomedical research, proton CSI is potentially the most promising technique, since it provides best sensitivity and spatial resolution. Various *in vivo* applications of proton spectroscopy can be found in the literature [36].

3.4.3 Summary

CSI is a technique for generating localized spectra that provide a wealth of biochemical information that can be used to study the metabolic activity of living system and to detect disease associated biochemical changes. This section provides an introduction to the technique and illustrates it by two representative examples. More specific topics concerning various aspects of CSI can be found in the literature.

Acknowledgments

The authors would like to thank Dr. Xiao-Hong Zhu for assisting data acquisition and Mr. Gregory Adriany for hardware support. The studies presented here are supported by the National Institute of Health (RR08079).

References

1. Lauterbur P.C. 1973. Image formation by induced local interactions: examples employing nuclear magnetic resonance. *Nature* 242: 190.
2. Alger J.R. 1994. Spatial localization for in vivo magnetic resonance spectroscopy: concepts and commentary. In R.J. Gillies (ed.), *NMR in Physiology and Biomedicine*, pp. 151–168. San Diego, CA, Academic Press.
3. Brown T.R., Kincaid M.B., and Ugurbil K. 1982. NMR chemical shift imaging in three dimensions. *Proc. Natl Acad. Sci. USA* 79: 3523.
4. Maudsley A.A., Hilal S.K., Simon H.E., and Perman W.H. 1983. Spatially resolved high resolution spectroscopy by "four dimensional" NMR. *J. Magn. Reson.* 51: 147.
5. Haselgrove J.C., Subramanian V.H., Leigh J.S. Jr. et al. 1983. in vivo one-dimensional imaging of phosphorous metabolites by phosphorus-31 nuclear magnetic resonance. *Science* 220: 1170.
6. Maudsley A.A., Hilal S.K., Simon H.E., and Wittekoek S. 1984. in vivo MR spectroscopic imaging with P-31. *Radiology* 153: 745.
7. Garwood M., Schleich T., Ross B.D. et al. 1985. A modified rotating frame experiment based on a Fourier window function: application to in vivo spatially localized NMR spectroscopy. *J. Magn. Reson.* 65: 239.
8. Garwood M., Robitalle P.M., and Ugurbil K. 1987. Fourier series windows on and off resonance using multiple coils and longitudinal modulation. *J. Magn. Reson.* 75: 244.
9. Mareci T.H. and Brooker H.R. 1984. High-resolution magnetic resonance spectra from a sensitive region defined with pulsed gradients. *J. Magn. Reson.* 57: 157.
10. Brooker H.R., Mareci T.H., and Mao J.T. 1987. Selective Fourier transform localization. *Magn. Reson. Med.* 5: 417.
11. Hu X., Levin D.N., Lauterbur P.C., and Spraggins T.A. 1988. SLIM: spectral localization by imaging. *Magn. Reson. Med.* 8: 314.
12. Liang Z.P. and Lauterbur P.C. 1991. A generalized series approach to MR spectroscopic imaging. *IEEE Trans. Med. Imag.* MI-10: 132.
13. Hu X. and Stillman A.E. 1991. Technique for reduction of truncation artifact in chemical shift images. *IEEE Trans. Med. Imag.* MI-10 3: 290.
14. Hu X., Patel M.S., and Ugurbil K. 1993. A new strategy for chemical shift imaging. *J. Magn. Reson.* B103: 30.
15. van den Boogaart A., Ala-Korpela M., Jokisaari J., and Griffiths J.R. 1994. Time and frequency domain analysis of NMR data compared: an application to 1D 1H spectra of lipoproteins. *Magn. Reson. Med.* 31: 347.
16. Ernst T., Kreis R., and Ross B. 1993. Absolute quantification of water and metabolites in human brain: I. Compartments and water. *J. Magn. Reson.* 102: 1.
17. Kreis R., Ernst T., and Ross B. 1993. Absolute quantification of water and metabolites in human brain. II. Metabolite concentration. *J. Magn. Reson.* 102: 9.
18. Lenkinski R.E., Holland G.A., Allman T. et al. 1988. Integrated MR imaging and spectroscopy with chemical shift imaging of P-31 at 1.5 T: Initial clinical experience. *Radiology* 169: 201.

19. Hugg J.W., Matson G.B., Twieg D.B. et al. ^{31}P MR spectroscopic imaging of normal and pathological human brains. *Magn. Reson. Imag.* 10: 227.
20. Vigneron D.B., Nelson S.J., Murphy-Boesch J. et al. 1990. Chemical shift imaging of human brain: axial, sagittal, and coronal ^{31}P metabolite images. *Radiology* 177: 643.
21. Hugg J.W., Laxer K.D., Matson G.B. et al. 1992. Lateralization of human focal epilepsy by ^{31}P magnetic resonance spectroscopic imaging. *Neurology* 42: 2011.
22. Meyerhoff D.J., Maudsley A.A., Schafer S., and Weiner M.W. 1992. Phosphorous-31 magnetic resonance metabolite imaging in the human body. *Magn. Reson. Imag.* 10: 245.
23. Bottomley P.A., Hardy C., and Boemer P. 1990. Phosphate metabolite imaging and concentration measurements in human heart by nuclear magnetic resonance. *J. Magn. Reson. Med.* 14: 425.
24. Robitaille P.M., Lew B., Merkle H. et al. 1990. Transmural high energy phosphate distribution and response to alterations in workload in the normal canine myocardium as studied with spatially localized ^{31}P NMR spectroscopy. *Magn. Reson. Med.* 16: 91.
25. Ugurbil K., Garwood M., Merkle H. et al. 1989. Metabolic consequences of coronary stenosis: transmurally heterogeneous myocardial ischemia studied by spatially localized ^{31}P NMR spectroscopy. *NMR Biomed.* 2: 317.
26. Hasse A., Frahm J., Hanicker H., and Mataei D. 1985. ^{1}H NMR chemical shift selective (CHESS) imaging. *Phys. Med. Biol.* 30: 341.
27. Patt S.L. and Sykes B.D. 1972. T_1 water eliminated Fourier transform NMR spectroscopy. *Chem. Phys.* 56: 3182.
28. Moonen C.T.W. and van Zijl P.C.M. 1990. Highly effective water suppression for in vivo proton NMR spectroscopy (DRYSTEAM). *J. Magn. Reson.* 88: 28.
29. Ogg R., Kingsley P., and Taylor J.S. 1994. WET: A T_1 and B_1 insensitive water suppression method for in vivo localized ^{1}H. *NMR Spectroscopy* B104: 1.
30 Lampman D.A., Murdoch J.B., and Paley M. 1991. In vivo proton metabolite maps using MESA 3D technique. *Magn. Reson. Med.* 18: 169.
31. Luyten P.R., Marien A.J.H., Heindel W. et al. 1990. Metabolic imaging of patients with intracranial tumors: ^{1}H MR spectroscopic imaging and PET. *Radiology* 176: 791.
32. Arnold D.L., Matthews P.M., Francis G.F. et al. 1992. Proton magnetic resonance spectroscopic imaging for metabolite characterization of demyelinating plaque. *Ann. Neurol.* 31: 319.
33. Duijin J.H., Matson G.B., Maudsley A.A. et al. 1992. Human brain infarction: proton MR spectroscopy. *Radiology* 183: 711.
34. Duyn J.H., Gillen J., Sobering G. et al. 1993. Multisection proton MR spectroscopic imaging of the brain. *Radiology* 188: 277.
35. Patel M.S. and Hu X. 1994. Selective data extrapolation for chemical shift imaging. *Soc. Magn. Reson. Abstr.* 3: 1168.
36. Rothman D.L. 1994. ^{1}H NMR studies of human brain metabolism and physiology. In R.J. Gillies (ed.), *NMR in Physiology and Biomedicine*, pp. 353–372. San Diego, CA, Academic Press.

4

Nuclear Medicine

Barbara Y. Croft
National Institutes of Health

Benjamin M.W. Tsui
Johns Hopkins University

4.1 Instrumentation

Barbara Y. Croft

Nuclear medicine can be defined as the practice of making patients radioactive for diagnostic and therapeutic purposes. The radioactivity is injected intravenously, rebreathed, or ingested. It is the internal circulation of radioactive material that distinguishes nuclear medicine from diagnostic radiology and radiation oncology in most of its forms. This section will examine only the diagnostic use and will concentrate on methods for detecting the radioactivity from outside the body without trauma to the patient. Diagnostic nuclear medicine is successful for two main reasons: (1) it can rely on the use of very small amounts of materials (picomolar concentrations in chemical terms) thus usually not having any effect on the processes being studied, and (2) the radionuclides being used can penetrate tissue and be detected outside the patient. Thus the materials can trace processes or "opacify" organs without affecting their function.

4.1.1 Parameters for Choices in Nuclear Medicine

Of the various kinds of emanations from radioactive materials, photons alone have a range in tissue great enough to escape so that they can be detected externally. Electrons or beta-minus particles of high energy can create bremsstrahlung in interactions with tissue, but the radiation emanates from the site of the interaction, not the site of the beta ray's production. Positrons or beta-plus particles annihilate with electrons to create gamma rays so that they can be detected. For certain radionuclides, the emanation being detected is x-rays, in the 50- to 100-keV energy range.

The half-lives of materials in use in nuclear medicine range from a few minutes to weeks. The half-life must be chosen with two major points in mind: the time course of the process being studied and the radiation dose to the target organ, that is, that organ with the highest concentration over the longest

TABLE 4.1 Gamma Ray Detection

Type of Sample	Activity (μCi)	Energy (keV)	Type of Instrument
Patient samples, for example, blood, urine	0.001	0–5000	Gamma counter with annular NaI(TI) detector, 1 or 2 PMTs, external Pb shielding
Small organ function <30 cm field of view at 60 cm distance	5–200	20–1500	2–4-in. NaI(TI) detector with flared Pb collimator
Static image of body part, for example, liver, lung	0.2–30	50–650	Rectilinear scanner with focused Pb collimator
Dynamic image of body part, for example, xenon in airways	2–30	80–300	Anger camera and parallel-hole Pb collimator
Static tomographic image of body part	See Section 4.1		

time (the cumulated activity or area underneath the activity versus time curve). In general, it is desired to stay under 5 **rad** to the target organ.

The choice of the best energy range to use is also based on two major criteria: the energy that will penetrate tissue but can be channeled by heavy metal shielding and collimation and that which will interact in the detector to produce a pulse. Thus the ideal energy is dependent on the detector being used and the kind of examination being performed. Table 4.1 describes the kinds of gamma-ray detection, the activity and energy ranges, and an example of the kind of information to be gained. The lesser amounts of activity are used in situations of lesser spatial resolution and of greater sensitivity. Positron imaging is omitted because it is treated elsewhere.

Radiation dose is affected by all the emanations of a radionuclide, not just the desirable ones, thus constricting the choice of nuclide further. There can be no alpha radiation used in diagnosis; the use of materials with primary beta radiation should be avoided because the beta radiation confers a radiation dose without adding to the information being gained. For imaging, in addition, even if there is a primary gamma ray in the correct energy window for the detector, there should be no large amount of radiation, either of primary radiation of higher energy, because it interferes with the image collimation, or of secondary radiation of a very similar energy, because it interferes with the perception of the primary radiation emanating from the site of interest.

For imaging using heavy-metal collimation, the energy range is constrained to be that which will emanate from the human body and which the collimation can contain, or about 50 to 500 keV.

Detectors must be made from materials that exhibit some detectable change when ionizing radiation is absorbed and that are of a high enough atomic number and density to make possible stopping large percentages of those gamma rays emanating (high sensitivity). In addition, because the primary gamma rays are not the only rays emanating from the source—a human body and therefore a distributed source accompanied by an absorber—there must be energy discrimination in the instrument to prevent the formation of an image of the scattered radiation. To achieve pulse size proportional to energy, and therefore to achieve identification of the energy and source of the energy, the detector must be a proportional detector. This means that Geiger-Muller detection, operating in an all-or-none fashion, is not acceptable.

Gaseous detectors are not practical because their density is not great enough. Liquid detectors (in which any component is liquid) are not practical because the liquid can spill when the detector is positioned; this problem can be compensated for if absolutely necessary, but it is better to consider it from the outset. Another property of a good detector is its ability to detect large numbers of gamma rays per time unit. With detection capabilities to separate 100,000 counts per second or a dead time of 2 msec, the system is still only detecting perhaps 1,000 counts per square centimeter per second over a 10 × 10 cm area. The precision of the information is governed by **Poisson statistics**, so the imprecision in information collected for 1 sec in a square centimeter is ±3% at the 1 standard deviation level. Since we would hope for better spatial resolution than 1 cm^2, the precision is obviously worse than this. This points to the need for fast detectors, in addition to the aforementioned sensitivity. The more detector that

TABLE 4.2 Ways of Imaging Using Lead Collimation

Moving probe; rectilinear scanner
Array of multiple crystals; autofluoroscope, "fly-eye" camera
Two moving probes: dual-head rectilinear scanner
Large single-crystal system: Anger camera
Two crystals on opposite sides of patient for two views using Anger logic
Large multiple-crystal systems using Anger logic SPECT
Other possibilities

surrounds the patient, the more sensitive the system will be. Table 4.2 lists in order from least sensitive to most sensitive some of the geometries used for imaging in nuclear medicine. This generally is also a listing from the older methods to the more recent.

For the purposes of this section, we shall consider that the problems of counting patient and other samples and of detecting the time course of activity changes in extended areas with probes are not our topic and confine ourselves to the attempts made to image distributions of gamma-emitting radionuclides in patients and research subjects. The previous section treats the three-dimensional imaging of these distributions; this section will treat detection of the distribution in a planar fashion or the image of the projection of the distribution onto a planar detector.

4.1.2 Detection of Photon Radiation

Gamma rays are detected when atoms in a detector are ionized and the ions are collected either directly as in gaseous or semiconductor systems or by first conversion of the ionized electrons to light and subsequent conversion of the light to electrons in a photomultiplier tube (P-M tube or PMT). In all cases there is a voltage applied across some distance that causes a pulse to be created when a photon is absorbed.

The gamma rays are emitted according to Poisson statistics because each decaying nucleus is independent of the others and has an equal probability of decaying per unit time. Because the uncertainty in the production of gamma rays is therefore on the order of magnitude of the square root of the number of gamma rays, the more gamma rays that are detected, the less the proportional uncertainty will be. Thus sensitivity is a very important issue for the creation of images, since the rays will be detected by area. To get better resolution, one must have the numbers of counts and the apparatus to resolve them spatially. Having the large numbers of counts also means the apparatus must resolve them temporally.

The need for **energy resolution** carries its own burden. Depending on the detector, the energy resolution may be easily achieved or not (Table 4.3). In any case, the attenuation and scattering inside the body means that there will be a range of gamma rays emitted, and it will be difficult to tell those scattered through very small angles from those not scattered at all. This affects the spatial resolution of the instrument.

TABLE 4.3 Detector Substances and Size Considerations, Atomic Number of the Attenuator, Energy Resolution Capability

PMT connected
NaI(TI): up to 50 cm across; 63; 5–10%
Plastic scintillators: unlimited; 6; only Compton absorption for gamma rays used in imaging
CsI(TI): < 3 × 3 cm; 53, 55; poorer than NaI(TI)
BiGermanate: < 3 × 3 cm; 83; poorer than NaI(TI)
Semiconductors: Liquid nitrogen operation and liquid nitrogen storage
GeLi: < 3 × 3 cm; 32; < 1%
SiLi: < 3 × 3 cm; 14; < 1%

TABLE 4.4 Calculation of Number of Counts Achieved with Anger Camera

	Cpm	Cps
Activity	0.001	
mCi/cm^3		
counts/sec		3.7×10^7
counts/min	2.22×10^9	
2π geometry	1.11×10^9	1.85×10^7
Attenuated by tissue of 0.12/cm attenuation and 3 cm thick	7.44×10^8	1.29×10^7
X Camera efficiency of 0.0006	4.64×10^5	7744
Good uptake in liver = 5 mCi/1000 g = 0.005 mCi/g	2.32×10^6	3.8×10^4
Thyroid uptake of Tc-99m = (2 mCi/37 g)	4.6×10^5	7.7×10^3
2% = 0.001 mCi/G		

The current practice of nuclear medicine has defined the limits of the amount of activity that can be administered to a patient by the amount of radiation dose. Since planar imaging with one detector allows only 2 pi detection at best and generally a view of somewhat less because the source is in the patient and the lead collimation means that only rays that are directed from the decay toward the crystal will be detected, it is of utmost importance to detect every ray possible. To the extent that no one is ever satisfied with the resolution of any system and always wishes for better, there is the need to be able to get spatial resolution better than the intrinsic 2 mm currently achievable. Some better collimation system, such as envisioned in a coincidence detection system like that used in PET, might make it possible to avoid stopping so many of the rays with the collimator.

We have now seen that energy resolution, sensitivity and resolving time of the detector are all bound up together to produce the spatial resolution of the instrument as well as the more obvious temporal resolution. The need to collimate to create an image rather than a blush greatly decreases the numbers of counts and makes Poisson statistics a major determinant of the appearance of nuclear medical images.

Table 4.4 shows a calculation for the NaI(TI)-based Anger camera showing 0.06% efficiency for the detection system. Thus the number of counts per second is not high and so is well within the temporal resolving capabilities of the detector system. The problem is the 0.06% efficiency, which is the effect of both the crystal thickness being optimized for imaging rather than for stopping all the gamma rays, and the lead collimation. Improvements in nuclear medicine imaging resolution can only come if both these factors are addressed.

4.1.3 Various Detector Configurations

The detectors in clinical nuclear medicine are NaI(TI) crystals. In research applications, other substances are employed, but the engineering considerations for the use of other detectors are more complex and have been less thoroughly explored (Table 4.3).

The possibilities for configuring the detectors have been increasing, although the older methods tend to be discarded as the new ones are exploited (see Table 4.2). This is in part because each laboratory cannot afford to have one of every kind of instrument, although there are tasks for which each one is ideally suited.

The first instruments possible for plane-projection imaging consisted of a moving single crystal probe, called a rectilinear scanner. The probe consisted of a detector (beginning with NaI(TI) but later incorporating small semiconductors) that was collimated by a focused lead collimator of appreciable thickness (often 2 in. of lead or more) with hole sizes and thicknesses of septa consonant with the intended energy and organ size and depth to be imaged. The collimated detector was caused to move across the patient at a constant speed; the pulses from the detector were converted to visible signals either by virtue of markings on a sheet or of light flashes exposing a film. This detector could see only one spot at a time, so only slow temporal changes in activity could be appreciated. A small organ such as the thyroid could be imaged in this fashion very satisfactorily. Bone imaging also could be done with the later versions of this instrument.

To enlarge the size of the detector, several probes, each with its own photomultiplier tube and collimator, could be used. Versions of this idea were used to create dual-probe instruments to image both sides of the patient simultaneously, bars of probes to sweep down the patient and create a combined image, etc.

To go further with the multiple crystals to create yet larger fields of view, the autofluoroscope (Figure 4.1) combined crystals in a rectangular array. For each to have its own photomultiplier tube required too many PMTs, so the instrument was designed with a light pipe to connect each crystal with a PMT to indicate its row and a second one to indicate its column. The crystals are separated by lead septa to prevent scattered photons from one crystal affecting the next. Because of the large number of crystals and PMTs, the instrument is very fast, but because of the size of the crystals, the resolution is coarse. To improve the resolution, the collimator is often jittered so that each crystal is made to see more than one field of view to create a better resolved image. For those dynamic examinations in which temporal resolution is more important than spatial resolution, the system has a clear advantage. It has not been very popular for general use, however. In its commercial realization, the field of view was not large enough to image either lungs or livers or bones in any single image fashion.

As large NaI(TI) crystals became a reality, new ways to use them were conceived. The Anger camera (Figure 4.2) is one of the older of these methods. The idea is to use a single crystal of diameter large enough to image a significant part of the human body and to back the crystal by an array of photomultiplier

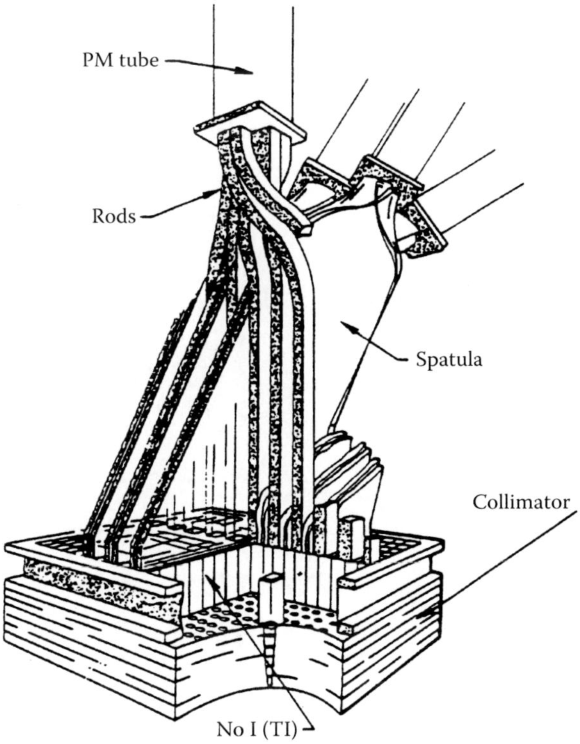

FIGURE 4.1 The Bender–Blau autofluoroscope is a multicrystal imager with a rectangular array of crystals connected to PMTs by plastic light guides. There is a PMT for each row of crystals and a PMT for each column of crystals, so an N by M array would have $(N + M)$ PMTs.

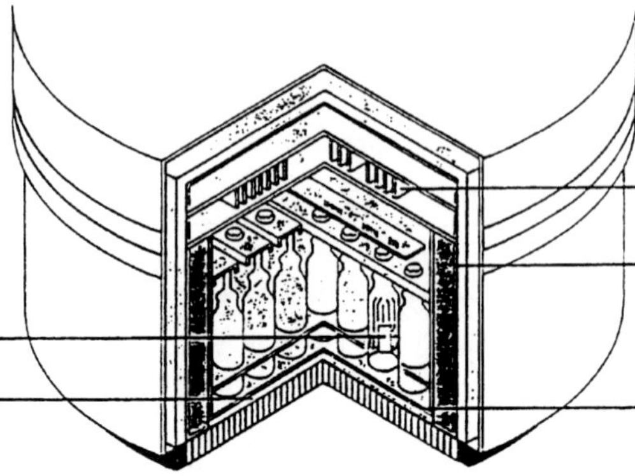

FIGURE 4.2 Anger camera detector design. This figure shows a cross section through the camera head. The active surface is pointed down. Shielding surrounds the assembly on the sides and top.

tubes to give positional sensitivity. Each PMT is assigned coordinates (Figure 4.3). When a photon is absorbed by the crystal, a number of PMTs receive light and therefore emit signals. The *X* and *Y* signal values for the emanation are determined by the strength of the signal from each of the tubes and its *x* and *y* position, and the energy of the emanation (which determines if it will be used to create the image) is the sum of all the signals (the *Z* pulse). If the discriminator passes the *Z* pulse, then the *X* and *Y* signals are sent to whatever device is recording the image, be it an oscilloscope and film recording system or the analog-to-digital (A/D) converters of a computer system. More recently, the A/D conversion is done earlier in the system so that the *X* and *Y* signals are themselves digital. The Anger camera is the major instrument in use in nuclear medicine today. It has been optimized for use with the 140-keV radiation from Tc-99m, although collimators have been designed for lower and higher energies, as well as optimized for higher sensitivity and higher resolution. The early systems used circular crystals, while the current configuration is likely to be rectangular or square.

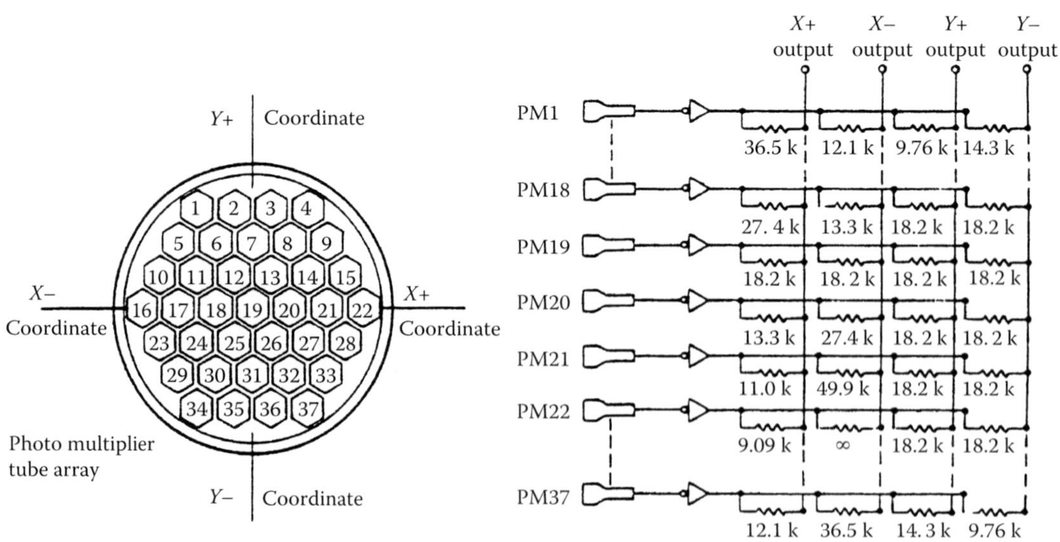

FIGURE 4.3 An array of PMTs in the Anger camera showing the geometric connection between the PMTs and the *X* and *Y* output.

A combination of the Anger positional logic and the focused collimator in a scanner produced the PhoCon instrument, which, because of the design of its collimators, had planar tomographic capabilities (the instrument could partially resolve activity in different planes, parallel to the direction of movement of the detector).

4.1.4 Ancillary Electronic Equipment for Detection

The detectors used in nuclear medicine are attached to preamplifiers, amplifiers, and pulse shapers to form a signal that can be examined for information about the energy of the detected photon (Figure 4.4). The energy discriminator has lower and upper windows that are set with reference radionuclides so that typically the particular nuclide in use can be dialed in along with the width of the energy window. A photon with an energy that falls in the selected range will cause the creation of a pulse of a voltage that falls in between the levels; all other photon energies will cause voltages either too high or too low. If only gross features are being recorded, any of the instruments may be used as probe detectors and the results recorded on strip-chart recordings of activity vs. time.

The PMT "multiplies" photons (Figure 4.5) because it has a quartz entrance window which is coated to release electrons when it absorbs a light photon and there is a voltage drop; the number of electrons released is proportional to the amount of light that hits the coating. The electrons are guided through a hole and caused to hit the first dynode, which is coated with a special substance to allow it to release electrons when it is hit by an electron. There are a series of dynodes each with a voltage that pulls the electrons from the last dynodes toward it. The surface coating not only releases electrons but also multiplies the electron shower. In a cascade through 10–12 dynodes, there is a multiplication of approximately 106, so that pulses of a few electrons become currents of the order of 10–12 A. The PMTs must be protected from other influences, such as stray radioactivity or strong magnetic fields, which might cause extraneous electron formation or curves in the electron path. Without the voltage drop from one dynode to the next, there is no cascade of electrons and no counting.

For imaging, the x and y positions of those photons in the correct energy range will be recorded in the image because they have a Z pulse. Once the pulse has been accepted and the position determined, that position may be recorded to make an image either in analog or digital fashion; a spot may be made on an oscilloscope screen and recorded on film or paper, or the position may be digitized and stored in a computer file for later imaging on an oscilloscope screen and/or for photography. In general, the computers required are very similar to those used for other imaging modalities, except for the hardware that allows the acceptance of the pulse. The software is usually specifically created for nuclear medicine because of the unique needs for determination of function.

The calibration of the systems follows a similar pattern, no matter how simple or complex the instrument. Most probe detectors must be "peaked," which means that the energy of the radioactivity must be connected with some setting of the instrument, often meant to read in kiloelectronvolts. This is accomplished by counting a sample with the instrument, using a reasonably narrow energy window, while varying the high voltage until the count rate reading is a maximum. The window is then widened for counting samples to encompass all the energy peak being counted. The detector is said to be liner if it can be set with one energy and another energy can be found where it should be on the kiloelectronvolt scale.

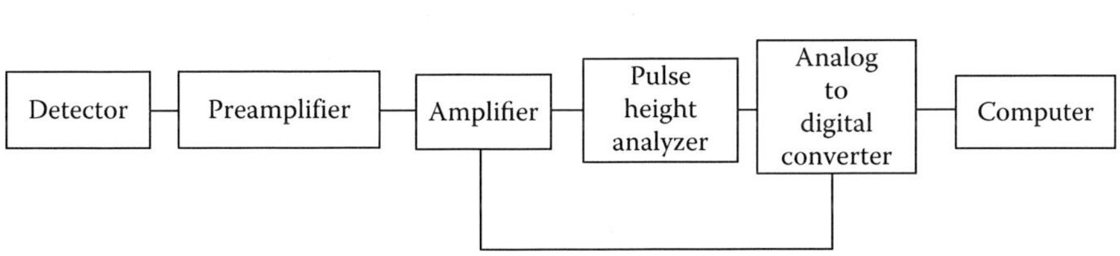

FIGURE 4.4 Schematic drawing of a generalized detector system. There would be a high-voltage power supply for the detector in an NaI(TI)-PMT detector system.

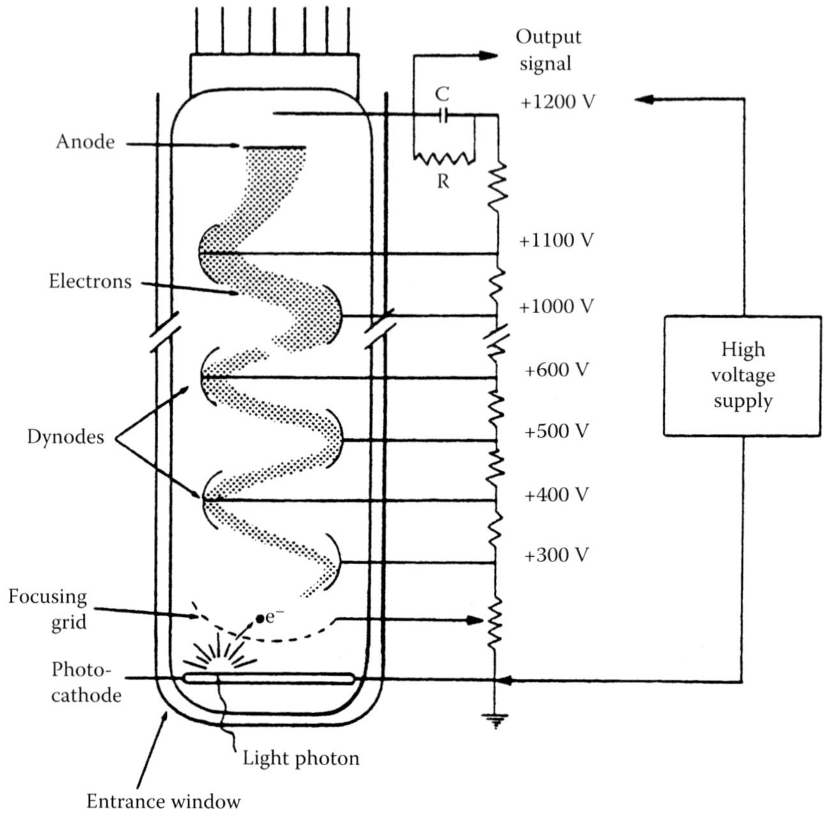

FIGURE 4.5 Schematic drawing of a photomultiplier tube (PMT). Each of the dynodes and the anode is connected to a separate pin in the tube socket. The inside of the tube is evacuated of all gas. Dynodes are typically copper with a special oxidized coating for electron multiplication.

To ensure that the images of the radioactivity accurately depict the distribution in the object being imaged, the system must be initialized correctly and tested at intervals. The several properties that must be calibrated and corrected are sensitivity, uniformity, energy pulse shape, and linearity.

These issues are addressed in several ways. The first is that all the PMTs used in an imaging system must be chosen to have matched sensitivities and energy spectra. Next, during manufacture, and at intervals during maintenance, the PMTs' response to voltage is matched so that the voltage from the power supply causes all the tubes to have maximum counts at the same voltage. The sensitivities of all the tubes are also matched during periodic maintenance. Prior to operation, usually at the start of each working day, the user will check the radioactive peak and then present the instrument with a source of activity to give an even exposure over the whole crystal. This uniform "flood" is recorded. The image may be used by the instrument for calibration; recalibration is usually performed at weekly intervals. The number of counts needed depends on the use the instrument is to be put to, but generally the instrument must be tested and calibrated with numbers of counts at least equal to those being emitted by the patients and other objects being imaged.

Because the PMT placement means that placement of the x and y locations is not perfect over the face of the crystal but has the effect of creating wiggly lines that may be closer together over the center of the PMT and farther apart at the interstices between tubes, the image may suffer from spatial nonlinearity. This can be corrected for by presenting the system with a lead pattern in straight-line bars or holes in rows and using a hard-wired or software method to position the X and Y signals correctly. This is called a linearity correction. In addition, there may be adjustments of the energy spectra of each tube to make them match each other so that variations in the number of kiloelectronvolts included in the window (created by varying the discriminator settings) will not create variations in sensitivity. This is called an energy correction.

4.1.5 Place of Planar Imaging in Nuclear Medicine Today: Applications and Economics

There are various ways of thinking about diagnostic imaging and nuclear medicine. If the reason for imaging the patient is to determine the presence of disease, then there are at least two possible strategies. One is to do the most complicated examination that will give a complete set of results on all patients. The other is to start with a simple examination and hope to categorize patients, perhaps into certain abnormals and all others, or even into abnormals, indeterminates, and normals. Then a subsequent, more complex examination is used to determine if there are more abnormals, or perhaps how abnormal they are. If the reason for imaging the patient is to collect a set of data that will be compared with results from that patient at a later time and with a range of normal results from all patients, then the least complex method possible for collecting the information should be used in a regular and routine fashion so that the comparison are possible.

In the former setting, where the complexity of the examination may have to be changed after the initial results are seen, in order to take full advantage of the dose of radioactive material that has been given to the patient, it is sensible to have the equipment available that will be able to perform the more complex examination and not to confine the equipment available to that capable of doing only the simple first examination. For this reason and because for some organs, such as the brain, the first examination is a SPECT examination, the new Anger cameras being sold today are mostly capable of doing rotating SPECT. The added necessities that SPECT brings to the instrument specifications are of degree: better stability, uniformity, and resolution. Thus they do not obviate the use of the equipment for planar imaging but rather enhance it. In the setting of performing only the examination necessary to define the disease, the Anger SPECT camera can be used for plane projection imaging and afterward for SPECT to further refine the examination. There are settings in which a planar camera will be purchased because of the simplicity of all the examinations (as in a very large laboratory that can have specialized instruments, a thyroid practice, or in the developing countries), but in the small- to medium-sized nuclear medicine practice, the new cameras being purchased are all SPECT-capable.

Nuclear medicine studies are generally less expensive than x-ray computed tomography or magnetic resonance imaging and more so than planar x-ray or ultrasound imaging. The general conduct of a nuclear medicine laboratory is more complex than these others because of the radioactive materials and the accompanying regulations. The specialty is practice both in clinics and in hospitals, but again, the complication imposed by the presence of the radioactive materials tips the balance of the practices toward the hospital. In that setting the practitioners may be imagers with a broad range of studies offered or cardiologists with a concentration on cardiac studies. Thus the setting also will determine what kind of instrument is most suitable.

Defining Terms

Energy resolution: Full width at half maximum of graph of detected counts vs. energy, expressed as a percentage of the energy.

Poisson statistics: Expresses probability in situations of equal probability for an event per unit of time, such as radioactive decay or cosmic-ray appearance. The standard deviation of a mean number of counts is the square root of the mean number of counts, which is a decreasing fraction of the number of counts when expressed as a fraction of the number of counts.

rad: The unit of radiation energy absorption (dose) in matter, defined as the absorption of 100 ergs per gram of irradiated material. The unit is being replaced by the gray, an SI unit, where 1 gray (Gy) = 100 rad.

Further Information

A good introduction to nuclear medicine, written as a text for technologists, is *Nuclear Medicine Technology and Techniques*, edited by D.R. Bernier, J.K. Langan, and L.D. Wells. A treatment of many of the nuclear medicine physics issues is given in L.E. Williams' *Nuclear Medicine Physics*, published in three volumes. Journals that publish nuclear medicine articles include the monthly *Journal of Nuclear Medicine, the European Journal of Nuclear Medicine, Clinical Nuclear Medicine, IEEE Transactions in Nuclear Science, IEEE Transactions in Medical Imaging, and Medical Physics*. Quarterly and annual publications include *Seminars in Nuclear Medicine, Yearbook of Nuclear Medicine, and Nuclear Medicine Annual.*

The Society of Nuclear Medicine holds an annual scientific meeting that includes scientific papers, continuing education which could give the novice a broad introduction, poster sessions, and a large equipment exhibition. Another large meeting, devoted to many radiologic specialties is the Radiologic Society of North America's annual meeting, held just after Thanksgiving.

4.2 SPECT (Single-Photon Emission Computed Tomography)

Benjamin M.W. Tsui

During the last three decades, there has been much excitement in the development of diagnostic radiology. The development is fueled by inventions and advances made in a number of exciting new medical imaging modalities, including ultrasound (US), x-ray CT (computed tomography), PET (positron emission tomography), SPECT (single-photon emission computed tomography), and MRI (magnetic resonance imaging). These new imaging modalities have revolutionized the practice of diagnostic radiology, resulting in substantial improvement in patient care.

Single-photon emission computed tomography (SPECT) is a medical imaging modality that combines conventional nuclear medicine (NM) imaging techniques and CT methods. Different from x-ray CT, SPECT uses radioactive-labeled pharmaceuticals, that is, radiopharmaceuticals, that distribute in different internal tissues or organs instead of an external x-ray source. The spatial and uptake distributions of the radiopharmaceuticals depend on the biokinetic properties of the pharmaceuticals and the normal or abnormal state of the patient. The gamma photons emitted from the radioactive source are detected by radiation detectors similar to those used in conventional nuclear medicine. The CT method requires projection (or planar) image data to be acquired from different views around the patient. These projection data are subsequently reconstructed using image reconstruction methods that generate cross-sectional images of the internally distributed radiopharmaceuticals. The SPECT images provide much improved contrast and detailed information about the radiopharmaceutical distribution as compared with the planar images obtained from conventional nuclear medicine methods.

As an emission computed tomographic (ECT) method, SPECT differs from PET in the types of radionuclides used. PET uses radionuclides such as C-11, N-13, O-15, and F-18 that emit positrons with subsequent emission of two coincident 511 keV annihilation photons. These radionuclides allow studies of biophysiologic functions that cannot be obtained from other means. However, they have very short half-lives, often requiring an on-site cyclotron for their production. Also, detection of the annihilation photons requires expensive imaging systems. SPECT uses standard radionuclides normally found in nuclear medicine clinics and which emit individual gamma-ray photons with energies that are much lower than 511 keV. Typical examples are the 140-keV photons from Tc-99m and the ~70-keV photons from TI-201. Subsequently, the costs of SPECT instrumentation and of performing SPECT are substantially less than PET.

Furthermore, substantial advances have been made in the development of new radiopharmaceuticals, instrumentation, and image processing and reconstruction methods for SPECT. The results are much improved quality and quantitative accuracy of SPECT images. These advances, combined with the relatively lower costs, have propelled SPECT to become an increasingly more important diagnostic tool in nuclear medicine clinics.

This section will present the basic principles of SPECT and the instrumentation and image processing and reconstruction methods that are necessary to reconstruct SPECT images. Finally, recent advances and future development that will continue to improve the diagnostic capability of SPECT will be discussed.

4.2.1 Basic Principles of SPECT

Single-photon emission computed tomography (SPECT) is a medical imaging technique that is based on the conventional nuclear medicine imaging technique and tomographic reconstruction methods. General review of the basic principles, instrumentation, and reconstruction technique for SPECT can be found in a few review articles (Jaszczak et al., 1980; Jaszczak and Coleman, 1985a; Barrett, 1986; Jaszczak and Tsui, 1994).

4.2.1.1 SPECT Imaging Process

The imaging process of SPECT can be simply depicted as in Figure 4.6. Gamma-ray photons emitted from the internal distributed radiopharmaceutical penetrate through the patient's body and are detected by a single or a set of collimated radiation detectors. The emitted photons experience interactions with the intervening tissues through basic interactions of radiation with matter (Evans, 1955). The photoelectric effect absorbs all the energy of the photons and stops their emergence from the patient's body. The other major interaction is Compton interaction, which transfers part of the photon energy to free electrons. The original photon is scattered into a new direction with reduced energy that is dependent on the scatter angle. Photons that escape from the patient's body include those that have not experienced any interactions and those which have experienced Compton scattering. For the primary photons from the commonly used radionuclides in SPECT, for example, 140-keV of TC-99m and ~70-keV of TI-201, the probability of pair production is zero.

Most of the radiation detectors used in current SPECT systems are based on a single or multiple NaI(TI) scintillation detectors. The most significant development in nuclear medicine is the scintillation camera (or Anger camera) that is based on a large-area (typically 40 cm in diameter) NaI(TI) crystal

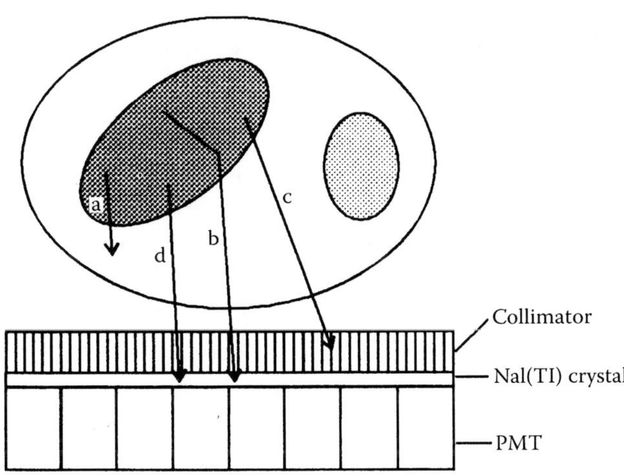

FIGURE 4.6 The conventional nuclear medicine imaging process. Gamma-ray photons emitted from the internally distributed radioactivity may experience photoelectric (a) or scatter (b) interactions. Photons that are not traveling in the direction within the acceptance analog of the collimator (c) will be intercepted by the lead collimator. Photons that experience no interaction and travel within the acceptance angle of the collimator will be detected (d).

(Anger, 1958, 1964). An array of photomultiplier tubes (PMTs) is placed at the back of the scintillation crystal. When a photon hits and interacts with the crystal, the scintillation generated will be detected by the array of PMTs. An electronic circuitry evaluates the relative signals from the PMTs and determines the location of interaction of the incident photon in the scintillation crystal. In addition, the scintillation cameras have built-in energy discrimination electronic circuitry with finite energy resolution that provides selection of the photons that have not been scattered or been scattered within a small scattered angle. The scintillation cameras are commonly used in commercial SPECT systems.

Analogous to the lens in an optical imaging system, a scintillation camera system consists of a collimator placed in front of the NaI(TI) crystal for the imaging purpose. The commonly used collimator is made of a large number of parallel holes separated by lead septa (Anger, 1964; Keller, 1968; Tsui, 1988). The geometric dimensions, that is, length, size, and shape of the collimator apertures, determine the directions of photons that will be detected by the scintillation crystals or the geometric response of the collimator. The width of the geometric response function increases (or the spatial resolution worsens) as the source distance from the collimator increases. Photons that do not pass through the collimator holes properly will be intercepted and absorbed by the lead septal walls of the collimator. In general, the detection efficiency is approximately proportional to the square of the width of the geometric response function of the collimator. This tradeoff between detection efficiency and spatial resolution is a fundamental property of a typical SPECT system using conventional collimators.

The amount of radioactivity that is used in SPECT is restricted by the allowable radiation dose to the patient. Combined with photon attenuation within the patient, the practical limit on imaging time, and the tradeoff between detection efficiency and spatial resolution of the collimator, the number of photons that are collected by a SPECT system is limited. These limitations resulted in SPECT images with relatively poor spatial resolution and high statistical noise fluctuations as compared with other medical imaging modalities. For example, currently a typical brain SPECT image has a total of about 500K counts per image slice and a spatial resolution in the order of approximately 8 mm. A typical myocardial SPECT study using TI-201 has about 150K total count per image slice and a spatial resolution of approximately 15 mm.

In SPECT, projection data are acquired from different views around the patient. Similar to x-ray CT, image processing and reconstruction methods are used to obtain transaxial or cross-sectional images from the multiple projection data. These methods consist of preprocessing and calibration procedures before further processing, mathematical algorithms for reconstruction from projections, and compensation methods for image degradation due to photon attenuation, scatter, and detector response.

The biokinetics of the radiopharmaceutical used, anatomy of the patient, instrumentation for data acquisition, preprocessing methods, image reconstruction techniques, and compensation methods have important effects on the quality and quantitative accuracy of the final SPECT images. A full understanding of SPECT cannot be accomplished without clear understanding of these factors. The biokinetics of radiopharmaceuticals and conventional radiation detectors have been described in the previous section on conventional nuclear medicine. The following subsections will present the major physical factors that affect SPECT and a summary review of the instrumentation, image reconstruction techniques, and compensation methods that are important technological and engineering aspects in the practice of SPECT.

4.2.1.2 Physical and Instrumentation Factors That Affect SPECT Images

There are several important physical and instrumentation factors that affect the measured data and subsequently the SPECT images. The characteristics and effects of these factors can be found in a few review articles (Jaszczak et al., 1981; Jaszczak and Tsui, 1994; Tsui et al., 1994a,b). As described earlier, gamma-ray photons that emit from an internal source may experience photoelectric absorption within the patient without contributing to the acquired data, Compton scattering with change in direction and loss of energy, or no interaction before exiting the patient's body. The exiting photons will be further selected by the geometric response of the collimator–detector. The photoelectric and Compton interactions and the characteristics of the collimator–detector have significant effects on both the quality and quantitative accuracy of SPECT image.

Photon attenuation is defined as the effect due to photoelectric and Compton interactions resulting in a reduced number of photons that would have been detected without them. The degree of attenuation is determined by the linear attenuation coefficient, which is a function of photon energy and the amount and types of materials contained in the attenuating medium. For example, the attenuation coefficient for the 140-keV photon emitted from the commonly used Tc-99m in water or soft tissue is 0.15 cm^{-1}. This gives rise to a half-valued-layer, the thickness of material that attenuates half the incident photons, or 4.5 cm H_2O for the 140-keV photon. Attenuation is the most important factor that affects the quantitative accuracy of SPECT images.

Attenuation effect is complicated by the fact that within the patient the attenuation coefficient can be quite different in various organs. The effect is most prominent in the thorax, where the attenuation coefficients range from as low as 0.05 cm^{-1} in the lung to as high as 0.18 cm^{-1} in the compact bone for the 140-keV photons. In x-ray CT, the attenuation coefficient distribution is the target for image reconstruction. In SPECT, however, the wide range of attenuation coefficient values and the variations of attenuation coefficient distributions among patients are major difficulties in obtaining quantitative accurate SPECT images. Therefore, compensation for attenuation is important to ensure good image quality and quantitatively high accuracy in SPECT. Review of different attenuation methods that have been used in SPECT is a subject of discussion later in this chapter.

Photons that have been scattered before reaching the radiation detector provide misplaced spatial information about the origin of the radioactive source. The results are inaccurate quantitative information and poor contrast in the SPECT images. For radiation detectors with perfect energy discrimination, scattered photons can be completely rejected. In a typical scintillation camera system, however, the energy resolution is in the order of 10% at 140 keV. With this energy resolution, the ratio of scattered to scattered total photons detected by a typical scintillation detector is about 20–30% in brain and about 30–40% in cardiac and body SPECT studies for 140-keV photons. Furthermore, the effect of scatter depends on the distribution of the radiopharmaceutical, the proximity of the source organ to the target organ, and the energy window used in addition to the photon energy and the energy resolution of the scintillation detector. The compensation of scatter is another important aspect of SPECT to ensure good image quality and quantitative accuracy.

The advances in SPECT can be attributed to simultaneous development of new radiopharmaceuticals, instrumentation, reconstruction methods, and clinical applications. Most radiopharmaceuticals that are developed for conventional nuclear medicine can readily be used in SPECT, and review of these developments is beyond the scope of this chapter. Recent advances include new agents that are labeled with iodine and technetium for blood perfusion for brain and cardiac studies. Also, the use of receptor agents and labeled antibiotics is being investigated. These developments have resulted in radiopharmaceuticals with improved uptake distribution, biokinetics properties, and potentially new clinical applications. The following subsections will concentrate on the development of instrumentation and image reconstruction methods that have made substantial impact on SPECT.

4.2.2 SPECT Instrumentation

Review of the advances in SPECT instrumentation can be found in several recent articles (Jaszczak et al., 1980; Rogers and Ackermann, 1992; Jaszczak and Tsui, 1994). A typical SPECT system consists of a single or multiple units of radiation detectors arranged in a specific geometric configuration and a mechanism for moving the radiation detector(s) or specially designed collimators to acquire data from different projection views. In general, SPECT instrumentation can be divided into three general categories (1) arrays of multiple scintillation detectors, (2) one or more scintillation cameras, and (3) hybrid scintillation detectors combining the first two approaches. In addition, special collimator designs have been proposed for SPECT for specific purposes and clinical applications. The following is a brief review of these SPECT systems and special collimators.

4.2.2.1 Multidetector SPECT System

The first fully functional SPECT imaging acquisition system was designed and constructed by Kuhl and Edwards (Kuhl and Edwards, 1963, 1964, 1968) in the 1960s, well before the conception of x-ray CT. As shown in Figure 4.7a, the MARK IV brain SPECT system consisted of four linear arrays of eight discrete NaI(TI) scintillation detectors assembled in a square arrangement. Projection data were obtained by rotating the square detector array around the patient's head. Although images from the pioneer MARK IV SPECT system were unimpressive without the use of proper reconstruction methods that were developed in later years, the multidetector design has been the theme of several other SPECT systems that were developed. An example is the Gammatom-1 developed by Cho et al. (1982). The design concept also was used in a dynamic SPECT system (Stokely et al., 1980) and commercial multidetector SPECT systems marketed by Medimatic, A/S (Tomomatic-32). Recently, the system design was extended to a multislice SPECT system with the Tomomatic-896, consisting of eight layers of 96 scintillation detectors. Also, the system allows both body and brain SPECT imaging by varying the aperture size.

Variations of the multiple-detectors arrangement have been proposed for SPECT system designs. Figure 4.7b shows the Headtome-II system by Shimadzu Corporation (Hirose et al., 1982), which consists of a stationary array of scintillation detectors arranged in a circular ring. Projection data are obtained by a set of collimator vanes that swings in front of the discrete detectors. A unique Cleon brain SPECT system (see Figure 4.7c), originally developed by Union Carbide Corporation in the 1970s, consists of 12 detectors that scan both radially and tangentially (Stoddart and Stoddart, 1979). Images from the original system were unimpressive due to inadequate sampling, poor axial resolution, and a reconstruction algorithm that did not take full advantage of the unique system design and data acquisition strategy. A much improved version of the system with a new reconstruction method (Moore et al., 1984) is currently marketed by Strichman Corporation.

The advantages of multidetector SPECT systems are their high sensitivity per image slice and high counting rate capability resulting from the array of multidetectors fully surrounding the patient. However, disadvantages of multidetector SPECT systems include their ability to provide only one or a few noncontiguous cross-sectional image slices. Also, these systems are relatively more expensive compared with camera-based SPECT systems described in the next section. With the advance of multicamera SPECT systems, the disadvantages of multidetector SPECT systems outweigh their advantages. As a result, they are less often found in nuclear medicine clinics.

4.2.2.2 Camera-Based SPECT Systems

The most popular SPECT systems are based on single or multiple scintillation cameras mounted on a rotating gantry. The successful design was developed almost simultaneously by three separate groups

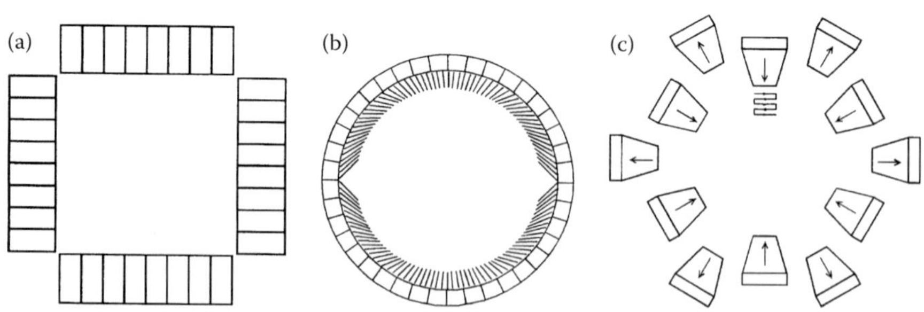

FIGURE 4.7 Examples of multidetector-based SPECT systems. (a) The MARK IV system consists of four arrays of eight individual NaI(TI) detectors arranged in a square configuration. (b) The Headtome-II system consists of a circular ring of detectors. A set of collimator vanes that swings in front of the discrete detector is used to collect projection data from different views. (c) A unique Cleon brain SPECT system consists of 12 detectors that scan both radially and tangentially.

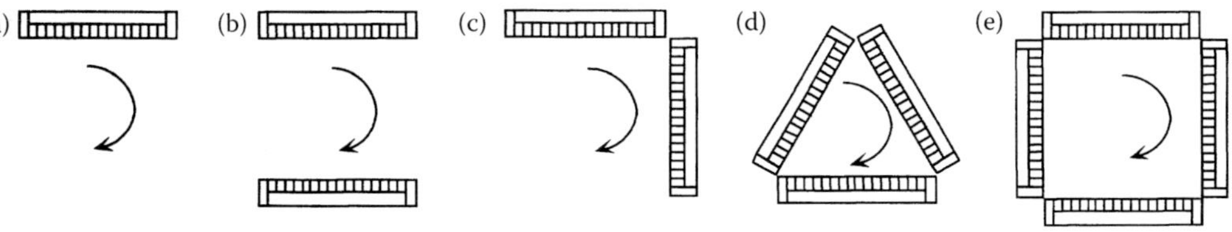

FIGURE 4.8 Examples of camera-based SPECT systems. (a) Single-camera system. (b) Dual-camera system with the two cameras placed at opposing sides of patient during rotation. (c) Dual-camera system with the two cameras placed at right angles. (d) Triple-camera system. (e) Quadruple-camera system.

(Budinger and Gullberg, 1977; Jaszczak et al., 1977; Keyes et al., 1977). In 1981, General Electric Medical Systems offered the first commercial SPECT system based on a single rotating camera and brought SPECT to clinical use. Today, there are over ten manufacturers (e.g., ADAC, Elscint, General Electric, Hitachi, Picker, Siemens, Sopha, Toshiba, and Trionix) offering an array of commercial SPECT systems in the marketplace.

An advantage of camera-based SPECT systems is their use of off-the-shelf scintillation cameras that have been widely used in conventional nuclear medicine. These systems usually can be used in both conventional planar and SPECT imaging. Also, camera-based SPECT systems allow truly three-dimensional (3D) imaging by providing a large set of contiguous transaxial images that cover the entire organ of interest. They are easily adaptable for SPECT imaging of the brain or body by simply changing the radius of rotation of the camera.

A disadvantage of a camera-based SPECT system is its relatively low counting rate capability. The dead time of a typical state-of-the-art scintillation camera gives rise to a loss of 20% of its true counts at about 80K counts per second. A few special high-count-rate systems give the same count rate loss at about 150K counts per second. For SPECT systems using a single scintillation camera, the sensitivity per image slice is relative low compared with a typical multidetector SPECT system.

Recently, SPECT systems based on multiple cameras became increasingly more popular. Systems with two (Jaszczak et al., 1979a), three (Lim et al., 1980, 1985), and four cameras provide increased sensitivity per image slice that is proportional to the number of cameras. Figure 4.8 shows the system configurations of these camera-based SPECT systems. The dual-camera systems with two opposing cameras (Figure 4.8b) can be used for both whole-body scanning and SPECT, and those with two right-angled cameras (Figure 4.8c) are especially useful for 180-degree acquisition in cardiac SPECT. The use of multicameras has virtually eliminated the disadvantages of camera-based SPECT systems as compared with multidetector SPECT systems. The detection efficiency of camera-based SPECT systems can be further increased by using converging-hole collimators such as fan, cone, and astigmatic collimators at the cost of a smaller field of view. The use of converging-hole collimators in SPECT will be described in a later section.

4.2.2.3 Novel SPECT System Designs

There are several special SPECT systems designs that do not fit into the preceding two general categories. The commercially available CERESPECT (formerly known as ASPECT) (Genna and Smith, 1988) is a dedicated brain SPECT system. As shown in Figure 4.9a, it consists of a single fixed-annular NaI(TI) crystal that completely surrounds the patient's head. Similar to a scintillation camera, an array of PMTs and electronics circuitry are placed behind the crystal to provide positional and energy information about photons that interact with the crystal. Projection data are obtained by rotating a segmented annular collimator with parallel holes that fits inside the stationary detector. A similar system is also being developed by Larsson et al. (1991) in Sweden.

Several unique SPECT systems are currently being developed in research laboratories. They consist of modules of small scintillation cameras that surround the patient. The hybrid designs combine

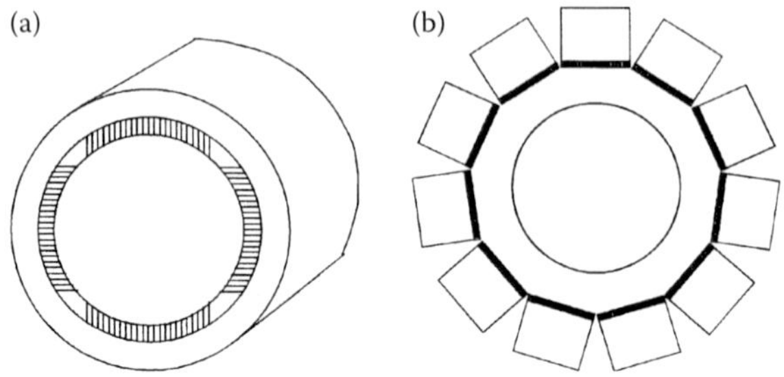

FIGURE 4.9 Examples of novel SPECT system designs. (a) The CERESPECT brain SPECT system consists of a single fixed annular NaI(TI) crystal and a rotating segmented annular collimator. (b) The SPRINT II brain SPECT system consists of 11 detector modules and a rotating lead ring with slit opening.

the advantage of multidetector and camera-based SPECT systems with added flexibility in system configuration. An example is the SPRINT II brain SPECT system developed at the University of Michigan (Rogers et al., 1988). As shown in Figure 4.9b, the system consists of 11 detector modules arranged in a circular ring around the patient's head. Each detector module consists of 44 one-dimensional (1D) bar NaI(TI) scintillation cameras. Projection data are required through a series of narrow slit openings on a rotating lead ring that fits inside the circular detector assemblies. A similar system was developed at the University of Iowa (Chang et al., 1990) with 22 detector modules, each consisting of four bar detectors. A set of rotating focused collimators is used to acquire projection data necessary for image reconstruction. At the University of Arizona, a novel SPECT system is being developed that consists of 20 small modular scintillation cameras (Milster et al., 1990) arranged in a hemispherical shell surrounding the patient's head (Rowe et al., 1992). Projection data are acquired through a stationary hemispherical array of pinholes that are fitted inside the camera array. Without moving parts, the system allows acquisition of dynamic 3D SPECT data.

4.2.2.4 Special Collimator Designs for SPECT Systems

Similar to conventional nuclear medicine imaging, parallel-hole collimators (Figure 4.10a) are commonly used in camera-based SPECT systems. As described earlier, the tradeoff between detection efficiency and spatial resolution of parallel-hole collimator is a limitating factor for SPECT. A means to improve SPECT system performance is to improve the tradeoff imposed by the parallel-hole collimation.

To achieve this goal, converging-hole collimator designs that increase the angle of acceptance of incoming photons without sacrificing spatial resolution have been developed. Examples are fan-beam (Jaszczak et al., 1979b; Tsui et al., 1986), cone-beam (Jaszczak et al., 1987), astigmatic (Hawman and Hsieh, 1986), and more recently varifocal collimators. As shown in Figure 4.10b, the collimator holes converge to a line that is oriented parallel to the axis of rotation for a fan-beam collimator, to a point for a cone-beam collimator, and to various points for a varifocal collimator, respectively. The gain in detection efficiency of a typical fan-beam and cone-beam collimator is about 1.5 and 2 times of that of a parallel-hole collimator with the same spatial resolution. The anticipated gain in detection efficiency and corresponding decrease in image noise are the main reasons for the interest in applying converging-hole collimators in SPECT.

Despite the advantage of increased detection efficiency, the use of converging-hole collimators in SPECT poses special problems. The tradeoff for increase in detection efficiency as compared with parallel-hole collimators is a decrease in field of view (see Figure 4.10). Consequently, converging-hole collimators are restricted to imaging small organs or body parts such as the head (Jaszczak et al., 1979b; Tsui et al., 1986) and heart (Gullberg et al., 1991). In addition, the use of converging-hole collimators requires special data-acquisition strategies and image reconstruction algorithms. For example, for cone-beam tomography using a conventional single planar orbit, the acquired projection data become

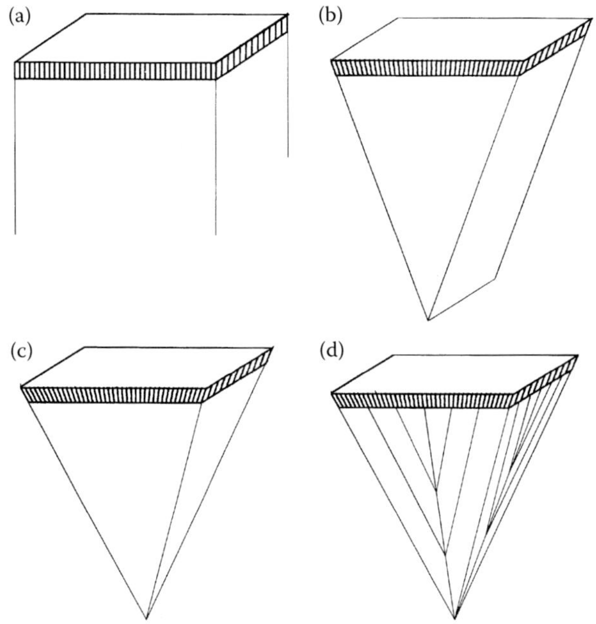

FIGURE 4.10 Collimator designs used in camera-based SPECT systems. (a) The commonly used parallel-hole collimator. (b) The fan-beam collimator, where the collimator holes are converged to a line that is parallel to the axis of rotation. (c) The cone-beam collimator, where the collimator holes are converged to a point. (d) A varifocal collimator, where the collimator holes are converged to various focal points.

increasingly insufficient for reconstructing transaxial image sections that are further away from the central plane of the cone-beam geometry. Active research is underway to study special rotational orbits for sufficient projection data acquisition and 3D image reconstruction methods specific for cone-beam SPECT.

4.2.3 Reconstruction Methods

As discussed earlier, SPECT combines conventional nuclear medicine image techniques and methods for image reconstruction from projections. Aside from radiopharmaceuticals and instrumentation, image reconstruction methods are another important engineering and technological aspect of the SPECT imaging technique.

In x-ray CT, accurate transaxial images can be obtained through the use of standard algorithms for image reconstruction from projections. The results are images of attenuation coefficient distribution of various organs within the patient's body. In SPECT, the goal of image reconstruction is to determine the distribution of administered radiopharmaceutical in the patient. However, the presence of photon attenuation affects the measured projection data. If conventional reconstruction algorithms are used without proper compensation for the attenuation effects, inaccurate reconstructed images will be obtained. Effects of scatter and the finite collimator–detector response impose additional difficulties on image reconstruction in SPECT.

In order to achieve quantitatively accurate images, special reconstruction methods are required for SPECT. Quantitatively accurate image reconstruction methods for SPECT consist of two major components. They are the standard algorithms for image reconstruction from projections and methods that compensate for the image-degrading effects described earlier. Often, image reconstruction algorithms are inseparable from the compensation methods, resulting in a new breed of reconstruction method not found in other tomographic medical imaging modalities. The following ssssssections will present the reconstruction problem and a brief review of conventional algorithms for image reconstruction from projections. Then quantitative SPECT reconstruction methods that include additional compensation methods will be described.

4.2.3.1 Image Reconstruction Problem

Figure 4.11 shows a schematic diagram of the two-dimensional (2D) image reconstruction problem. Let $f(x, y)$ represent a 2D object distribution that is to be determined. A 1D detector array is oriented at an angle q with respect to the x-axis of the laboratory coordinates system (x, y). The data collected into each detector element at location t, called the projection data $p(t, q)$, is equal to the sum of $f(x, y)$ along a ray that is perpendicular to the detector array and intersects the detector at position t; that is,

$$p(t,\theta) = c\int_{-\alpha}^{\alpha} f(x,y)\,ds \tag{4.1}$$

where (s, t) represents a coordinate system with s along the direction of the ray sum and t parallel to the 1D detector array, and c is the gain factor of the detection system. The angle between the s and x-axes is θ. The relationship between the source position (x, y), the projection angle θ, and the position of detection on the 1D detector array is given by

$$t = y\cos\theta - x\sin\theta \tag{4.2}$$

In 2D tomographic imaging, the 1D detector array rotates around the object distribution $f(x, y)$ and collects projection data from various projection data from various projection angles θ. The integral transform of the object distribution to its projections given by Equation 4.1 is called the Radon transform (Radon, 1917). The goal of image reconstruction is to solve the inverse Radon transform. The solution is the reconstructed image estimate $\hat{f}(x,y)$ of the object distribution $f(x, y)$.

In x-ray CT, the measured projection data is given by

$$p'(t,\theta) = c_t I_o \exp\left[-\int_{-\alpha}^{+\alpha} \mu(x,y)\,ds\right] \tag{4.3}$$

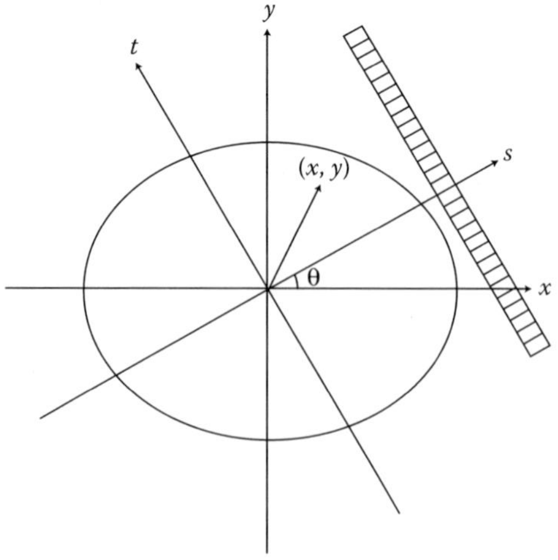

FIGURE 4.11 Schematic diagram of the two-dimensional image reconstruction problem. The projection data are line integrals of the object distribution along rays that are perpendicular to the detector. A source point (x, y) is projected onto a point $p(t, \theta)$, where t is a position along the projection and θ is the projection angle.

where I_o is the intensity of the incident x-ray, $\mu(x, y)$ is the 2D attenuation coefficient, and c_t is the gain factor which transforms x-ray intensity to detected signals. The reconstruction problem can be rewritten as

$$p(t,\theta) = \ln\left[\frac{I_o}{p'(t,\theta)}\right] = \int_{-\alpha}^{+\alpha} \mu(x,y)\,ds \tag{4.4}$$

with the goal to solve for the attenuation coefficient distribution $\mu(x, y)$. Also, in x-ray CT, if parallel rays are used, the projection data at opposing views are the same, that is, $p(t, \theta) = p(\theta + p)$, and projection data acquired over 180 degrees will be sufficient for reconstruction. The number of linear samples along the 1D projection array and angular samples, that is, the number of projection views, over 180 degrees must be chosen carefully to avoid aliasing error and resolution loss in the reconstructed images.

In SPECT, if the effects of attenuation, scatter, and collimator–detector response are ignored, the measured projection data can be written as the integral of radioactivity along the projection rays; that is,

$$p(t,\theta) = c_e \int_{-\alpha}^{+\alpha} \rho(x,y)\,ds \tag{4.5}$$

where $\rho(x,y)$ is the radioactivity concentration distribution of the object, and C_e is the gain factor which transforms radioactivity concentration to detected signals. Equation 4.5 fits in the form of the Radon transform, and similar to x-ray CT, the radioactivity distribution can be obtained by solving the inverse Radon transform problem.

If attenuation is taken into consideration, the attenuated Radon transform (Gullberg, 1979) can be written as

$$p(t,\theta_i) = c_e \int_{-\alpha}^{+\alpha} \rho(x,y)\exp\left[-\int_{(x,y)}^{+\alpha} \mu(u,v)\,ds'\right] ds \tag{4.6}$$

where $\mu(u,v)$ is the 2D attenuation coefficient distribution, and $\int_{x,y}^{+\alpha} \mu(u,v)\,ds$ is the attenuation factor for photons that originate from (x, y), travel along the direction perpendicular to the detector array, and are detected by the collimator–detector. A major difficulty in SPECT image reconstruction lies in the attenuation factor, which makes the inverse problem given by Equation 4.6 difficult to solve analytically. However, the solution is important in cardiac SPECT, where the widely different attenuation coefficients are found in various organs within the thorax. Also, due to the attenuation factor, the projection views at opposing angles are different. Hence full 360-degree projection data are usually necessary for image reconstruction in SPECT.

Different from x-ray CT, small differences in attenuation coefficient are not as important in SPECT. When the attenuation coefficient in the body region can be considered constant, the attenuated Radon transform given by Equation 4.6 can be written as (Tretiak and Metz, 1980)

$$p(t,\theta) = c_e \int_{-\alpha}^{+\alpha} \rho(x,y)\exp[-\mu l(x,y)]\,ds \tag{4.7}$$

where μ is the constant attenuation coefficient in the body region, and $l(x, y)$ is the path length between the point (x, y) and the edge of the attenuator (or patient's body) along the direction of the projection ray. The solution of the inverse problem with constant attenuator has been a subject of several investigations. It forms the basis for analytical methods for compensation of uniform attenuation described later in this chapter.

When scatter and collimator–detector response are taken into consideration, the assumption that the projection data can be represented by line integrals given by Equation 4.1 to Equation 4.7 will no longer be exactly correct. Instead, the integration will have to include a wider region covering the field of view of the collimator–detector (or the collimator–detector response function). The image reconstruction problem is further complicated by the nonstationary properties of the collimator–detector and scatter response functions and their dependence on the size and composition of the patient's body.

4.2.3.2 Algorithms for Image Reconstruction from Projections

The application of methods for image reconstruction from projections was a major component in the development of x-ray CT in the 1970s. The goal was to solve for the inverse Radon transform problem given in Equation 4.1. There is an extensive literature on these reconstruction algorithms. Reviews of the applications of these algorithms to SPECT can be found in several articles (Budinger and Gullberg, 1974; Brooks and Di Chiro, 1975, 1976; Barrett, 1986).

Simple backprojection: An intuitive image reconstruction method is simple backprojection. Here, the reconstructed image is formed simply by spreading the values of the measured projection data uniformly along the projection ray into the reconstructed image array. By backprojecting the measured projection data from all projection views, an estimate of the object distribution can be obtained. Mathematically, the simple backproject operation is given by

$$\hat{f}(x,y) = \sum_{j=1}^{m} p(y\cos\theta_j - x\sin\theta_j, \theta_j)\Delta\theta \tag{4.8}$$

where θ_j is the jth projection angle, m is the number of projection views, and $\Delta\theta$ is the angular spacing between adjacent projections. The simple backprojected image $\hat{f}(x,y)$ is a poor approximation of the true object distribution $f(x, y)$. It is equivalent to the true object distribution blurred by a blurring function in the form of $1/r$.

There are two approaches for accurate image reconstruction, and both have been applied to SPECT. The first approach is based on direct analytical methods and is widely used in commercial SPECT systems. The second approach is based on statistical criteria and iterative algorithms. They have been found useful in reconstruction methods that include compensation for the image-degrading effects.

Analytical reconstruction algorithms: filtered backprojection: The most widely used analytical image-reconstruction algorithm is the filtered backprojection (FBP) method, which involves backprojecting the filtered projections (Bracewell and Riddle, 1967; Ramachandran and Lakshminarayanan, 1971). The algorithm consists of two major steps:

1. Filter the measured projection data at different projection angles with a special function
2. Backproject the filtered projection data to form the reconstructed image

The first step of the filtered backprojection method can be implemented in two different ways. In the spatial domain, the filter operation is equivalent to convolving the measured projection data using a special convolving function $h(t)$; that is,

$$p'(t,\theta) = p(t,\theta) * h(t) \tag{4.8a}$$

where $*$ is the convolution operation. With the advance of FFT methods, the convolution operation can be replaced by a more efficient multiplication in the spatial frequency domain. The equivalent operation consists of three steps:

1. Fourier-transform the measured projection data into spatial frequency domain using the FFT method, that is, $P(v, \theta) = \text{FT}\{p(t, \theta)\}$, where FT is the Fourier transform operation.
2. Multiply the Fourier-transformed projection data with a special function that is equal to the Fourier transform of the special function used in the convolution operation described above, that is, $P'(v,\theta) = P(v,\theta) \cdot H(v)$, where $H(v) = \text{FT}\{h(x)\}$ is the Fourier transform of $h(x)$.
3. Inverse Fourier transform the product $P'(v,\theta)$ into spatial domain.

Again, the filtered projections from different projection angles are backprojected to form the reconstructed images.

The solution of the inverse Radon transform given in Equation 4.1 specifies the form of the special function. In the spatial domain, the special function h(x) used in the convolution operation in Equation 4.8 is given by

$$h(x) = \frac{1}{2(\Delta x)^2}\left\{\text{sinc}\left[\frac{x}{(\Delta x)}\right]\right\} - \frac{1}{4(\Delta x)^2}\left\{\text{sinc}^2\left[\frac{x}{2(\Delta x)}\right]\right\} \tag{4.9}$$

where Δx is the linear sampling interval, and $\text{sinc}(z) = [\sin(z)]/z$. The function $h(x)$ consists of a narrow central peak with high magnitude and small negative side lobes. It removes the blurring from the $1/r$ function found in the simple backprojected images.

In the frequency domain, the special function $H(v)$ is equivalent to the Fourier transform of $h(x)$ and is a truncated ramp function given by

$$H(\text{v}) = |\text{v}| \cdot \text{rect}(\text{v}) \tag{4.10}$$

where $|\text{v}|$ is the ramp function, and

$$\text{rect}(v) = \begin{cases} 1, & |v| \leq 0.5 \\ 0, & v < 0.5 \end{cases} \tag{4.11}$$

that is, the rectangular function rect (v) has a value of 1 when the absolute value of v is less than the Nyquist frequency at 0.5 cycles per pixel.

For noisy projection data, the ramp function tends to amplify the high-frequency noise. In these situations, an additional smoothing filter is often applied to smoothly roll off the high-frequency response of the ramp function. Examples are Hann and Butterworth filters (Huesman et al., 1977). Also, deconvolution filters have been used to provide partial compensation of spatial resolution loss due to the collimator-detection response and noise smoothing. Examples are the Metz and Wiener filters (see later).

Iterative reconstruction algorithms: Another approach to image reconstruction is based on statistical criteria and iterative algorithms. They were investigated for application in SPECT before the development of analytical image reconstruction methods (Kuhl and Edwards, 1968; Gordon et al., 1970; Gilbert, 1972; Goitein, 1972). The major drawbacks of iterative reconstruction algorithms are the extensive computations and long processing time required. For these reasons, the analytical reconstruction methods have gained widespread acceptance in clinical SPECT systems. In recent years, there has been renewed interest in the use of iterative reconstruction algorithms in SPECT to achieve accurate quantitation by compensating for the image-degrading effects.

A typical iterative reconstruction algorithm starts with an initial estimate of the object source distribution. A set of projection data is estimated from the initial estimate using a projector that models the imaging process. The estimated projection data are compared with the measured projection data at the same projection angles, and their differences are calculated. Using an algorithm derived from specific statistical criteria, the differences are used to update the initial image estimate. The updated image estimate is then used to recalculate a new set of estimated projection data that are again compared with the measured projection data. The procedure is repeated until the difference between the estimated and measured projection data are smaller than a preselected small value. Statistical criteria that have been used in formulating iterative reconstruction algorithms include the minimum mean squares error (MMSE) (Budinger and Gullberg, 1977), weighted least squares (WLS) (Huesman et al., 1977), maximum entropy (ME) (Minerbo, 1979), maximum likelihood (ML) (Shepp and Vardi, 1982), and maximum a posteriori approaches (Geman and McClure, 1985; Barrett, 1986; Levitan and Herman, 1987; Liang and Hart, 1987; Johnson et al., 1991). Iterative algorithms that have been used in estimating the reconstructed images include the conjugate gradient (CG) (Huesman et al., 1977) and expectation maximization (EM) (Lange and Carson, 1984).

Recently, interest in the application of iterative reconstruction algorithms in SPECT has been revitalized. The interest is sparked by the need to compensate for the spatially variant and/or nonstationary image-degrading factors in the SPECT imaging process. The compensation can be achieved by modeling the imaging process that includes the image-degrading factors in the projection and backprojection operations of the iterative steps. The development is aided by advances made in computer technology and custom-dedicated processors. The drawback of long processing time in using these algorithms is substantially reduced. Discussion of the application of iterative reconstruction algorithms in SPECT will be presented in a later section.

4.2.3.3 Compensation Methods

For a typical SPECT system, the measured projection data are severely affected by attenuation, scatter, and collimator–detector response. Direct reconstruction of the measured projection data without compensation of these effects produces images with artifacts, distortions, and inaccurate quantitation. In recent years, substantial efforts have been made to develop compensation methods for these image-degrading effects. This development has produced much improved quality and quantitatively accurate reconstructed images. The following sections will present a brief review of some of these compensation methods.

Compensation for attenuation: Methods for attenuation compensation can be grouped into two categories (1) methods that assume the attenuation coefficient is uniform over the body region, and (2) methods that address situations of nonuniform attenuation coefficient distribution. The assumption of uniform attenuation can be applied to SPECT imaging of the head and abdomen regions. The compensation methods seek to solve for the inverse of the attenuated Radon transform given in Equation 4.7. For cardiac and lung SPECT imaging, nonuniform attenuation compensation methods must be used due to the very different attenuation coefficient values in various organs in the thorax. Here, the goal is to solve the more complicated problem of the inverse of the attenuated Radon transform in Equation 4.6.

There are several approximate methods for compensating uniform attenuation. They include methods that preprocess the projection data or postprocess the reconstructed image. The typical preprocess methods are those which use the geometric or arithmetic mean (Sorenson, 1974) of projections from opposing views. These compensation methods are easy to implement and work well with a single, isolated source. However, they are relatively inaccurate for more complicated source configurations. Another method achieves uniform attenuation compensation by processing the Fourier transform of the sinogram (Bellini et al., 1979). The method provides accurate compensation even for complicated source configurations.

A popular compensation method for uniform attenuation is that proposed by Chang (1978). The method requires knowledge of the body contour. The information is used in calculating the average attenuation factor at each image point from all projection views. The array of attenuation factors is used to multiply the reconstructed image obtained without attenuation compensation. The result is the attenuation-compensated image. An iterative scheme also can be implemented for improved accuracy. In general, the Chang method performs well for uniform attenuation situations. However, the noise level in the reconstructed images increases with iteration number. Also, certain image features tend to fluctuate as a function of iteration. For these reasons, no more than one or two iterations are recommended.

Another class of methods for uniform attenuation compensation is based on analytical solution of the inverse of the attenuation Radon transform given in Equation 4.7 for a convex-shaped medium (Tretiak and Metz, 1980; Gullberg and Budinger, 1981). The resultant compensation method involves multiplying the projection data by an exponential function. Then the FBP algorithm is used in the image reconstruction except that the ramp filter is modified such that its value is zero in the frequency range between 0 and $\mu/2\pi$, where μ is the constant attenuation coefficient. The compensation method is easy to implement and provides good quantitative accuracy. However, it tends to amplify noise in the resulting image, and smoothing is required to obtain acceptable image quality (Gullberg and Budinger, 1981).

An analytical solution for the more complicated inverse attenuated Radon transform with nonuniform attenuation distribution (Equation 4.6) has been found difficult (Gullberg, 1979). Instead, iterative approaches have been used to estimate a solution of the problem. The application is especially important in cardiac and lung SPECT studies. The iterative methods model the attenuation distribution in the projection and backprojection operations (Manglos et al., 1987; Tsui et al., 1989). The ML criterion with the EM algorithm (Lange and Carson, 1984) has been used with success (Tsui et al., 1989). The compensation method requires information about the attenuation distribution of the region to be imaged. Recently, transmission CT methods are being developed using existing SPECT systems to obtain attenuation distribution from the patient. The accurate attenuation compensation of cardiac SPECT promises to provide much improved quality and quantitative accuracy in cardiac SPECT images (Tsui et al., 1989, 1994a).

Compensation for scatter: As described earlier in this chapter, scattered photons carry misplaced positional information about the source distribution resulting in lower image contrast and inaccurate quantitation in SPECT images. Compensation for scatter will improve image contrast for better image quality and images that will more accurately represent the true object distribution. Much research has been devoted to develop scatter compensation methods that can be grouped into two general approaches. In the first approach, various methods have been developed to estimate the scatter contribution in the measured data. The scatter component is then subtracted from the measured data or from the reconstructed images to obtain scatter-free reconstructed images. The compensation method based on this approach tends to increase noise level in the compensated images.

One method estimates the scatter contribution as a convolution of the measured projection data with an empirically derived function (Axelsson et al., 1984). Another method models the scatter component as the convolution of the primary (or unscattered) component of the projection data with an exponential function (Floyd et al., 1985). The convolution method is extended to 3D by estimating the 2D scatter component (Yanch et al., 1988). These convolution methods assume that the scatter response function is stationary, which is only an approximation.

The scatter component also has been estimated using two energy windows acquisition methods. One method estimates the scatter component in the primary energy window from the measured data obtained from a lower and adjacent energy window (Jaszczak et al., 1984, 1985b). In a dual photopeak window (DPW) method, two nonoverlapping windows spanning the primary photopeak window are used (King et al., 1992). This method provides more accurate estimation of the scatter response function.

Multiple energy windows also have been used to estimate the scatter component. One method uses two satellite energy windows that are placed directly above and below the photopeak window to estimate the scatter component in the center window (Ogawa et al., 1991). In another method, the energy spectrum detected at each image pixel is used to predict the scatter contribution (Koral et al., 1988). An energy-weighted acquisition (EWA) technique acquires data from multiple energy windows. The images reconstructed from these data are weighted with energy-dependent factors to minimize scatter contribution to the weighted image (DeVito et al., 1989; DeVito and Hamill, 1991). Finally, the holospectral imaging method (Gagnon et al., 1989) estimates the scatter contribution from a series of eigenimages derived from images reconstructed from data obtained from a series of multiple energy windows.

In the second approach, the scatter photons are utilized in estimating the true object distribution. Without subtracting the scatter component, the compensated images are less noisy than those obtained from the first approach. In one method, an average scatter response function can be combined with the geometric response of the collimator–detector to form the total response of the imaging system (Gilland et al., 1988; Tsui et al., 1994a). The total response function is then used to generate a restoration filter for an approximate geometric and scatter response compensation (see later).

Another class of methods, characterizes the exact scatter response function and incorporates it into iterative reconstruction algorithms for accurate compensation for scatter (Floyd et al., 1985; Frey and Tsui, 1992). Since the exact scatter response functions are nonstationary and are asymmetric in shape, implementation of the methods requires extensive computations. However, efforts are being made to parameterize the scatter response function and to optimize the algorithm for substantial reduction in processing time (Frey and Tsui, 1991; Frey et al., 1993).

Compensation for collimator–detector response: As described earlier, for a typical collimator–detector, the response function broadens as the distance from the collimator face increases. The effect of the collimator–detector response is loss of spatial resolution and blurring of fine detail in SPECT images. Also, the spatially variant detector response function will cause nonisotropic point response in SPECT images (Knesaurek et al., 1989; Maniawski et al., 1991). The spatially variant collimator–detector response is a major difficulty in its exact compensation.

By assuming an average and stationary collimator–detector response function, restoration filters can be used to provide partial and approximate compensation for the effects of the collimator–detector. Examples are the Metz (King et al., 1984, 1986) and Wiener (Penney et al., 1990) filters, where the inverse of the average collimator–detector response function is used in the design of the restoration filters. Two-dimensional compensation is achieved by applying the 1D restoration filters to the 1D projection data, and 3D compensation by applying the 2D filters to the 2D projection images (Tsui et al., 1994b).

Analytical methods have been developed for compensation of the spatially variant detector response. A spatially variant filtering method has been proposed which is based on the frequency distance principle (FDP) (Edholm et al., 1986; Lewitt et al., 1989). The method has been shown to provide an isotropic point response function in phantom SPECT images (Glick et al., 1993).

Iterative reconstruction methods also have been used to accurately compensate for both nonuniform attenuation and collimator–detector response by modeling the attenuation distribution and spatially variant detector response function in the projection and backprojection steps. The compensation methods have been applied in 2D reconstruction (Tsui et al., 1988; Formiconi et al., 1990), and more recently in 3D reconstruction (Zeng et al., 1991; Tsui et al., 1994b). It has been found that the iterative reconstruction methods provide better image quality and more accurate quantitation when compared with the conventional restoration filtering techniques. Furthermore, 3D compensation outperforms 2D compensation at the expense of more extensive computations (Tsui et al., 1994b).

4.2.4 Sample SPECT Images

This section presents sample SPECT images to demonstrate the performance of various reconstruction and compensation methods. Two data sets were used. The first set was acquired from a 3D physical phantom that mimics a human brain perfusion study. The phantom study provided knowledge of the true radioactivity distribution for evaluation purposes. The second data set was obtained from a patient myocardial SPECT study using thallium-201.

Figure 4.12a shows the radioactivity distribution from a selected slice of a 3D brain phantom manufactured by the Data Spectrum Corporation. The phantom design was based on PET images from a

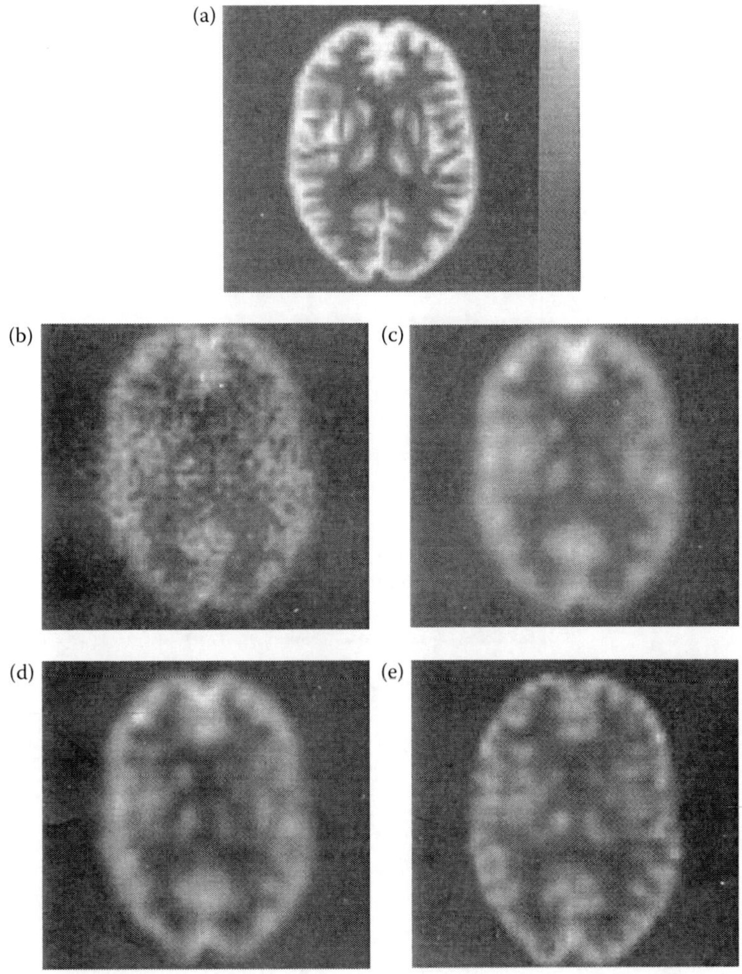

FIGURE 4.12 Sample images from a phantom SPECT study. (a) Radioactivity distribution from a selected slice of a 3D brain phantom. (b) Reconstructed image obtained from the FBP algorithm without any compensation. (c) Reconstructed image obtained with the application of noise-smoothing filter and compensation for uniform attenuation and scatter. (d) Similar to (c) except for an additional application of a Metz filter to partially compensate for the collimator–detector blurring. (e) Reconstructed image similar to that obtained from the iterative ML-EM algorithm that accurately models the attenuation and spatially variant detector response. (From Tsui B.M.W., Frey E.C., Zhao X.-D. et al. 1994. *Phys. Med. Biol.* 39: 509. Reprinted with permission.)

normal patient to simulate cerebral blood flow (Hoffman et al., 1990). The phantom was filled with water containing 74 mBq of Tc-99m. A single-camera-based GE 400AC/T SPECT system fitted with a high-resolution collimator was used for data collection. The projection data were acquired into 128 × 128 matrices at 128 views over 360 degrees. Figure 4.12b shows the reconstructed image obtained from the FBP algorithm without any compensation. The poor image quality is due to statistical noise fluctuations, effects of attenuation (especially at the central portion of the image), loss of spatial resolution due to the collimator–detector response, and loss of contrast due to scatter.

Figure 4.12c shows the reconstructed image obtained with the application of a noise-smoothing filter and compensation for the uniform attenuation and scatter. The resulting image has lower noise level, reduced attenuation effect, and higher contrast as compared with the image shown in Figure 4.12b. Figure 4.12d is similar to Figure 4.12c except for an additional application of a Metz filter to partially compensate for the collimator–detector blurring. Figure 4.12e shows the reconstructed image obtained from the iterative ML-EM algorithm that accurately modeled the attenuation and spatially variant detector response. The much superior image quality is apparent.

Figure 4.13a shows a selected FBP reconstructed transaxial image slice from a typical patient myocardial SPECT study using thallium-201. Figure 4.13b shows the reconstructed image obtained from the Chang algorithm for approximate nonuniform attenuation compensation and 2D processing using a Metz filter for approximate compensation for collimator–detector response. Figure 4.13c shows the reconstructed image obtained from the iterative ML-EM algorithm using a measured transmission

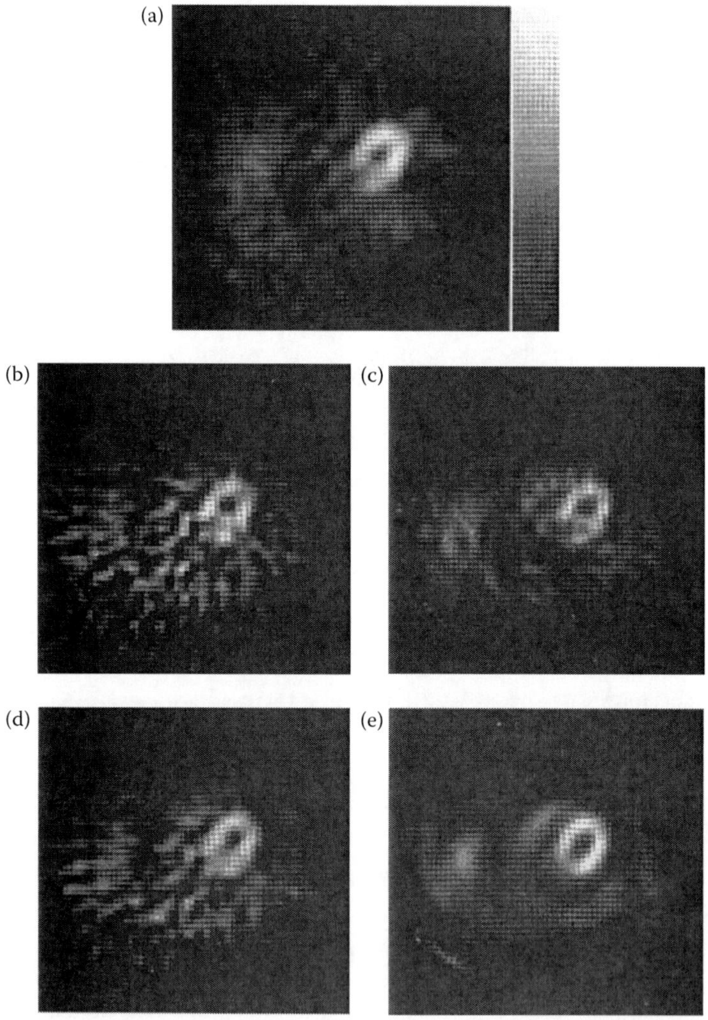

FIGURE 4.13 Sample images from a patient myocardial SPECT study using TI-201. (a) A selected transaxial image slice from a typical patient myocardial SPECT study using TI-201. The reconstructed image was obtained with the FBP algorithm without any compensation. (b) Reconstructed image obtained from the Chang algorithm for approximate nonuniform attenuation compensation and 2D processing using a Metz filter for approximate compensation for collimator–detector response. (c) Reconstructed image obtained from the iterative ML-EM algorithm using a measured transmission CT image for accurate attenuation compensation and 2D model of the collimator–detector response for accurate collimator–detector response compensation. (d) Similar to (b) except that the Metz filter was implemented in 3D. (e) Similar to (c) except that a 3D model of the collimator–detector response is used.

CT image for accurate attenuation compensation and a 2D model of the collimator–detector response for accurate collimator–detector response compensation. The reconstructed image in Figure 4.13d is similar to that in Figure 4.13b except that the Metz filter was implemented in 3D. Finally, the reconstructed image in Figure 4.13e is similar to that in Figure 4.13c except that a 3D model of the collimator–detector response is used. The superior image quality obtained from using an accurate 3D model of the imaging process is evident.

4.2.5 Discussion

The development of SPECT has been a combination of advances in radiopharmaceuticals, instrumentation, image processing and reconstruction methods, and clinical applications. Although substantial progress has been made during the last decade, there are many opportunities for contributions from biomedical engineering in the future.

The future SPECT instrumentation will consist of more detector area to fully surround the patient for high detection efficiency and multiple contiguous transaxial slice capability. Multicamera SPECT systems will continue to dominate the commercial market. The use of new radiation detector materials and detector systems with high spatial resolution will receive increased attention. Continued research is needed to investigate special converging-hole collimator design geometries, fully 3D reconstruction algorithms, and their clinical applications.

To improve image quality and to achieve quantitatively accurate SPECT images will continue to be the goals of image processing and image reconstruction methods for SPECT. An important direction of research in analytical reconstruction methods will involve solving the inverse Radon transform, which includes the effects of attenuation, the spatially variant collimator–detector response function, and scatter. The development of iterative reconstruction methods will require more accurate models of the complex SPECT imaging process, faster and more stable iterative algorithms, and more powerful computers and special computational hardware.

These improvements in SPECT instrumentation and image reconstruction methods, combined with newly developed radiopharmaceuticals, will bring SPECT images with increasingly higher quality and more accurate quantitation to nuclear medicine clinics for improved diagnosis and patient care.

References

Anger H.O. 1958. Scintillation camera. *Rev. Sci. Instrum.* 29: 27.

Anger H.O. 1964. Scintillation camera with multichannel collimators. *J. Nucl. Med.* 5: 515.

Axelsson B., Msaki P., and Israelsson A. 1984. Subtraction of Compton-scattered photons in single-photon emission computed tomography. *J. Nucl. Med.* 25: 490.

Barrett H.H. 1986. Perspectives on SPECT. *Proc. SPIE* 671: 178.

Bellini S., Piacentini M., Cafforio C. et al. 1979. Compensation of tissue absorption in emission tomography. *IEEE Trans. Acoust. Speech Signal Process. ASSP* 27: 213.

Bracewell R.N. and Riddle A.C. 1967. Inversion of fan-beam scans in radio astronomy. *Astrophys. J.* 150: 427.

Brooks R.A. and Di Chiro G. 1975. Theory of image reconstruction in computed tomography. *Radiology* 117: 561.

Brooks R.A. and Di Chiro G. 1976. Principles of computer assisted tomography (CAT) in radiographic and radioisotopic imaging. *Phys. Med. Biol.* 21: 689.

Budinger T.F. and Gullberg G.T. 1974. Three-dimensional reconstruction in nuclear medicine emission imaging. *IEEE Trans. Nucl. Sci. NS* 21: 2.

Budinger T.F. and Gullberg G.T. 1977. Transverse section reconstruction of gamma-ray emitting radionuclides in patients. In M.M. Ter-Pogossian, M.E. Phelps, G.L. Brownell et al. (Eds), *Reconstruction Tomography in Diagnostic Radiology and Nuclear Medicine*. Baltimore, University Park Press.

Chang L.T. 1978. A method for attenuation correction in radionuclide computed tomography. *IEEE Trans. Nucl. Sci. NS* 25: 638.

Chang W., Huang G., and Wang L. 1990. A multi-detector cylindrical SPECT system for phantom imaging. In *Conference Record of the 1990 Nuclear Science Symposium*, Vol. 2, pp. 1208–1211. Piscataway, NJ, IEEE.

Cho Z.H., Yi W., Jung K.J. et al. 1982. Performance of single photon tomography system-Gamma-tom-1. *IEEE Trans. Nucl. Sci. NS* 29: 484.

DeVito R.P. and Hamill J.J. 1991. Determination of weighting functions for energy-weighted acquisition. *J. Nucl. Med.* 32: 343.

DeVito R.P., Hamill J.J., Treffert J.D., and Stoub E.W. 1989. Energy-weighted acquisition of scintigraphic images using finite spatial filters. *J. Nucl. Med.* 30: 2029.

Edholm P.R., Lewitt R.M., and Lindholm B. 1986. Novel properties of the Fourier decomposition of the sinogram. *Proc. SPIE* 671: 8.

Evans R.D. 1955. *The Atomic Nucleus*. Malabar, FL, Robert E. Krieger.

Floyd C.E., Jaszczak R.J., Greer K.L., and Coleman R.E. 1985. Deconvolution of Compton scatter in SPECT. *J. Nucl. Med.* 26: 403.

Formiconi A.R., Pupi A., and Passeri A. 1990. Compensation of spatial system response in SPECT with conjugate gradient reconstruction technique. *Phys. Med. Biol.* 34: 69.

Frey E.C., Ju Z.-W., and Tsui B.M.W. 1993. A fast projector-backprojector pair modeling the asymmetric, spatially varying scatter response function for scatter compensation in SPECT imaging. *IEEE Trans. Nucl. Sci. NS* 40: 1192.

Frey E.C. and Tsui B.M.W. 1991. Spatial properties of the scatter response function in SPECT. *IEEE Trans. Nucl. Sci. NS* 38: 789.

Frey E.C. and Tsui B.M.W. 1992. A comparison of scatter compensation methods in SPECT: subtraction-based techniques versus iterative reconstruction with an accurate scatter model. In *Conference Record of the 1992 Nuclear Science Symposium and the Medical Imaging Conference*, October 27–31, Orlando, FL, pp. 1035–1037.

Gagnon D., Todd-Pokropek A., Arsenault A., and Dupros G. 1989. Introduction to holospectral imaging in nuclear medicine for scatter subtraction. *IEEE Trans. Med. Imag.* 8: 245.

Geman S. and McClure D.E. 1985. Bayesian image analysis: an application to single photon emission tomography. In *Proceedings of the Statistical Computing Section*. Washington, American Statistical Association.

Genna S. and Smith A. 1988. The development of ASPECT, an annular single crystal brain camera for high efficiency SPECT. *IEEE Trans. Nucl. Sci. NS* 35: 654.

Gilbert P. 1972. Iterative methods for the three-dimensional reconstruction of an object from projections. *J. Theor. Biol.* 36: 105.

Gilland D.R., Tsui B.M.W., Perry J.R. et al. 1988. Optimum filter function for SPECT imaging. *J. Nucl. Med.* 29: 643.

Glick S.J., Penney B.C., King M.A., and Byrne C.L. 1993. Non-iterative compensation for the distance-dependent detector response and photon attenuation in SPECT imaging. *IEEE Trans. Med. Imag.* 13: 363.

Goitein M. 1972. Three-dimensional density reconstruction from a series of two-dimensional projections. *Nucl. Instrum. Meth.* 101: 509.

Gordon R. 1974. A tutorial on ART (Algebraic reconstruction techniques). *IEEE Trans. Nucl. Sci.* 21: 78.

Gordon R., Bender R., and Herman G.T. 1970. Algebraic reconstruction techniques (ART) for three-dimensional electron microscopy and x-ray photography. *J. Theor. Biol.* 29: 471.

Gullberg G.T. 1979. The attenuated Radon transform: theory and application in medicine and biology. Ph.D. dissertation, University of California at Berkeley.

Gullberg G.T. and Budinger T.F. 1981. The use of filtering methods to compensate for constant attenuation in single-photon emission computed tomography. *IEEE Trans. Biomed. Eng. BME* 28: 142.

Gullberg G.T., Christian P.E., Zeng G.L. et al. 1991. Cone beam tomography of the heart using single-photon emission-computed tomography. *Invest. Radiol.* 26: 681.

Hawman E.G. and Hsieh J. 1986. An astigmatic collimator for high sensitivity SPECT of the brain. *J. Nucl. Med.* 27: 930.

Hirose Y., Ikeda Y., Higashi Y. et al. 1982. A hybrid emission CT-HEADTOME II. *IEEE Trans. Nucl. Sci. NS* 29: 520.

Hoffman E.J., Cutler P.D., Kigby W.M., and Mazziotta J.C. 1990. 3-D phantom to simulate cerebral blood flow and metabolic images for PET. *IEEE Trans. Nucl. Sci. NS* 37: 616.

Huesman R.H., Gullberg G.T., Greenberg W.L., and Budinger T.F. 1977. *RECLBL Library Users Manual, Donner Algorithms for Reconstruction Tomography*. Lawrence Berkeley Laboratory, University of California.

Jaszczak R.J., Chang L.T., and Murphy P.H. 1979. Single photon emission computed tomography using multi-slice fan beam collimators. *IEEE Trans. Nucl. Sci. NS* 26: 610.

Jaszczak R.J., Chang L.T., Stein N.A., and Moore F.E. 1979. Whole-body single-photon emission computed tomography using dual, large-field-of-view scintillation cameras. *Phys. Med. Biol.* 24: 1123.

Jaszczak R.J. and Coleman R.E. 1985. Single photon emission computed tomography (SPECT) principles and instrumentation. *Invest. Radiol.* 20: 897.

Jaszczak R.J., Coleman R.E., and Lim C.B. 1980. SPECT: single photon emission computed tomography. *IEEE Trans. Nucl. Sci. NS* 27: 1137.

Jaszczak R.J., Coleman R.E., and Whitehead F.R. 1981. Physical factors affecting quantitative measurements using camera-based single photon emission computed tomography (SPECT). *IEEE Trans. Nucl. Sci. NS* 28: 69.

Jaszczak R.J., Floyd C.E., and Coleman R.E. 1985. Scatter compensation techniques for SPECT. *IEEE Trans. Nucl. Sci. NS* 32: 786.

Jaszczak R.J., Floyd C.E., Manglos S.M. et al. 1987. Cone beam collimation for single photon emission computed tomography: Analysis, simulation, and image reconstruction using filtered backprojection. *Med. Phys.* 13: 484.

Jaszczak R.J., Greer K.L., Floyd C.E. et al. 1984. Improved SPECT quantification using compensation for scattered photons. *J. Nucl. Med.* 25: 893.

Jaszczak R.J., Murphy P.H., Huard D., and Burdine J.A. 1977. Radionuclide emission computed tomography of the head with 99 mTc and a scintillation camera. *J. Nucl. Med.* 18: 373.

Jaszczak R.J. and Tsui B.M.W. 1994. Single photon emission computed tomography. In H.N. Wagner and Z. Szabo (Eds), Principles of Nuclear Medicine, 2nd ed. Philadelphia, Saunders.

Johnson V.E., Wong W.H., Hu X., and Chen C.T. 1991. Image restoration using Gibbs priors: boundary modeling, treatment of blurring and selection of hyperparameters. *IEEE Trans. Pat.* 13: 413.

Keller E.L. 1968. Optimum dimensions of parallel-hole, multiaperture collimators for gamma-ray camera. *J. Nucl. Med.* 9: 233.

Keyes J.W., Jr, Orlandea N., Heetderks W.J. et al. 1977. The humogotron— a scintillation-camera transaxial tomography. *J. Nucl. Med.* 18: 381.

King M.A., Hademenos G., and Glick S.J. 1992. A dual photopeak window method for scatter correction. *J. Nucl. Med.* 33: 605.

King M.A., Schwinger R.B., Doherty P.W., and Penney B.C. 1984. Two-dimensional filtering of SPECT images using the Metz and Wiener filters. *J. Nucl. Med.* 25: 1234.

King M.A., Schwinger R.B., and Penney B.C. 1986. Variation of the count-dependent Metz filter with imaging system modulation transfer function. *Med. Phys.* 25: 139.

Knesaurek K., King M.A., Glick S.J. et al. 1989. Investigation of causes of geometric distortion in 180 degree and 360 degree angular sampling in SPECT. *J. Nucl. Med.* 30: 1666.

Koral K.F., Wang X., Rogers W.L., and Clinthorne N.H. 1988. SPECT Compton-scattering correction by analysis of energy spectra. *J. Nucl. Med.* 29: 195.

Kuhl D.E. and Edwards R.Q. 1963. Image separation radioisotope scanning. *Radiology* 80: 653.

Kuhl D.E. and Edwards R.Q. 1964. Cylindrical and section radioisotope scanning of the liver and brain. *Radiology* 83: 926.

Kuhl D.E. and Edwards R.Q. 1968. Reorganizing data from transverse section scans of the brain using digital processing. *Radiology* 91: 975.

Lange K. and Carson R. 1984. EM reconstruction algorithms for emission and transmission tomography. *J. Comput. Assist. Tomogr.* 8: 306.

Larsson S.A., Hohm C., Carnebrink T. et al. 1991. A new cylindrical SPECT Anger camera with a decentralized transputer based data acquisition system. *IEEE Trans. Nucl. Sci. NS* 38: 654.

Lassen N.A., Sveinsdottir E., Kanno I. et al. 1978. A fast moving single photon emission tomograph for regional cerebral blood flow studies in man. *J. Comput. Assist. Tomogr.* 2: 661.

Levitan E. and Herman G.T. 1987. A maximum a posteriori probability expectation maximization algorithm for image reconstruction in emission tomography. *IEEE Trans. Med. Imag. MI* 6: 185.

Lewitt R.M., Edholm P.R., and Xia W. 1989. Fourier method for correction of depth dependent collimator blurring. *Proc. SPIE* 1092: 232.

Liang Z. and Hart H. 1987. Bayesian image processing of data from constrained source distribution: I. Nonvalued, uncorrelated and correlated constraints. *Bull. Math. Biol.* 49: 51.

Lim C.B., Chang J.T., and Jaszczak R.J. 1980. Performance analysis of three camera configurations for single photon emission computed tomography. *IEEE Trans. Nucl. Sci. NS* 27: 559.

Lim C.B., Gottschalk S., Walker R. et al. 1985. Tri-angular SPECT system for 3-D total organ volume imaging: design concept and preliminary imaging results. *IEEE Trans. Nucl. Sci. NS* 32: 741.

Manglos S.H., Jaszczak R.J., Floyd C.E. et al. 1987. Nonisotropic attenuation in SPECT: phantom test of quantitative effects and compensation techniques. *J. Nucl. Med.* 28: 1584.

Maniawski P.J., Morgan H.T., and Wackers F.J.T. 1991. Orbit-related variations in spatial resolution as a source of artifactual defects in thallium-201 SPECT. *J. Nucl. Med.* 32: 871.

Milster T.D., Aarsvold J.N., Barrett H.H. et al. 1990. A full-field modular gamma camera. *J. Nucl. Med.* 31: 632.

Minerbo G. 1979. Maximum entropy reconstruction from cone-beam projection data. *Comput. Biol. Med.* 9: 29.

Moore S.C., Doherty M.D., Zimmerman R.E., and Holman B.L. 1984. Improved performance from modifications to the multidetector SPECT brain scanner. *J. Nucl. Med.* 25: 688.

Ogawa K., Harata Y., Ichihara T. et al. 1991. A practical method for position-dependent Compton scatter correction in SPECT. *IEEE Trans. Med. Imag.* 10: 408.

Penney B.C., Glick S.J., and King M.A. 1990. Relative importance of the errors sources in Wiener restoration of scintigrams. *IEEE Trans. Med. Imag.* 9: 60.

Radon J. 1917. Uber die bestimmung von funktionen durch ihre integral-werte langs gewisser mannigfaltigkeiten. *Ber. Verh. Sachs. Akad. Wiss.* 67: 26.

Ramachandran G.N. and Lakshminarayanan A.V. 1971. Three-dimensional reconstruction from radiographs and electron micrographs: application of convolutions instead of Fourier transforms. *Proc. Natl Acad. Sci. USA* 68: 2236.

Roger W.L. and Ackermann R.J. 1992. SPECT instrumentation. *Am. J. Physiol. Imag.* 314: 105.

Rogers W.L., Clinthorne N.H., Shao L. et al. 1988. SPRINT II: a second-generation single photon ring tomograph. *IEEE Trans. Med. Imag.* 7: 291.

Rowe R.K., Aarsvold J.N., Barrett H.H. et al. 1992. A stationary, hemispherical SPECT imager for 3D brain imaging. *J. Nucl. Med.* 34: 474.

Shepp L.A. and Vardi Y. 1982. Maximum likelihood reconstruction for emission tomography. *IEEE Trans. Med. Imag. MI* 1: 113.

Sorenson J.A. 1974. Quantitative measurement of radiation in vivo by whole body counting. In G.H. Hine and J.A. Sorenson (Eds), Instrumentation in Nuclear Medicine, Vol. 2, pp. 311–348. New York, Academic Press.

Stoddart H.F. and Stoddart H.A. 1979. A new development in single gamma transaxial tomography Union Carbide focused collimator scanner. *IEEE Trans. Nucl. Sci. NS* 26: 2710.

Stokely E.M., Sveinsdottir E., Lassen N.A., and Rommer P. 1980. A single photon dynamic computer assisted tomography (DCAT) for imaging brain function in multiple cross-sections. *J. Comput. Assist. Tomogr.* 4: 230.

Tretiak O.J. and Metz C.E. 1980. The exponential Radon transform. *SIAM J. Appl. Math.* 39: 341.

Tsui B.M.W. 1988. Collimator design, properties and characteristics. Chapter 2. In G.H. Simmons (Ed.), *The Scintillation Camera*, pp. 17–45. New York, The Society of Nuclear Medicine.

Tsui B.M.W., Frey E.C., Zhao X.-D. et al. 1994. The importance and implementation of accurate 3D compensation methods for quantitative SPECT. *Phys. Med. Biol.* 39: 509.

Tsui B.M.W., Gullberg G.T., Edgerton E.R. et al. 1986. The design and clinical utility of a fan beam collimator for a SPECT system. *J. Nucl. Med.* 247: 810.

Tsui B.M.W., Gullberg G.T., Edgerton E.R. et al. 1989. Correction of nonuniform attenuation in cardiac SPECT imaging. *J. Nucl. Med.* 30: 497.

Tsui B.M.W., Hu H.B., Gilland D.R., and Gullberg G.T. 1988. Implementation of simultaneous attenuation and detector response correction in SPECT. *IEEE Trans. Nucl. Sci. NS* 35: 778.

Tsui B.M.W., Zhao X.-D., Frey E.C., and McCartney W.H. 1994. Quantitative single-photon emission computed tomography: basics and clinical considerations. *Semin. Nucl. Med.* 24: 38.

Yanch J.C., Flower M.A., and Webb S. 1988. Comparison of deconvolution and windowed subtraction techniques for scatter compensation in SPECT. *IEEE Trans. Med. Imag.* 7: 13.

Zeng G.L., Gullberg G.T., Tsui B.M.W., and Terry J.A. 1991. Three-dimensional iterative reconstruction algorithms with attenuation and geometric point response correction. *IEEE Trans. Nucl. Sci. NS* 38: 693.

5

Ultrasound

Richard L. Goldberg
University of North Carolina

Stephen W. Smith
Duke University

Jack G. Mottley
University of Rochester

K. Whittaker Ferrara
Riverside Research Institute

5.1 Transducers

Richard L. Goldberg and Stephen W. Smith

An ultrasound transducer generates acoustic waves by converting magnetic, thermal, and electrical energy into mechanical energy. The most efficient technique for medical ultrasound uses the piezoelectric effect, which was first demonstrated in 1880 by Jacques and Pierre Curie (Curie and Curie, 1880). They applied a stress to a quartz crystal and detected an electrical potential across opposite faces of the material. The Curies also discovered the inverse piezoelectric effect by applying an electric field across the crystal to induce a mechanical deformation. In this manner, a piezoelectric transducer converts an oscillating electric signal into an acoustic wave, and vice versa.

Many significant advances in ultrasound imaging have resulted from innovation in transducer technology. One such instance was the development of linear-array transducers. Previously, ultrasound systems had made an image by manually moving the transducer across the region of interest. Even the faster scanners had required several seconds to generate an ultrasound image, and as a result, only static targets could be scanned. On the other hand, if the acoustic beam could be scanned rapidly, clinicians could visualize moving targets such as a beating heart. In addition, real-time imaging would provide instantaneous feedback to the clinician of the transducer position and system settings.

To implement real-time imaging, researchers developed new types of transducers that rapidly steer the acoustic beam. Piston-shaped transducers were designed to wobble or rotate about a fixed axis to mechanically steer the beam through a sector-shaped region. Linear sequential arrays were designed to electronically focus the beam in a rectangular image region. Linear phased-array transducers were designed to electronically steer and focus the beam at high speed in a sector image format.

This section describes the application of piezoelectric ceramics to transducer arrays for medical ultrasound. Background is presented on transducer materials and beam steering with phased arrays. Array performance is described, and the design of an idealized array is presented.

5.1.1 Transducer Materials

Ferroelectric materials strongly exhibit the piezoelectric effect, and they are ideal materials for medical ultrasound. For many years, the ferroelectric ceramic lead-zirconate-titanate (PZT) has been the standard transducer material for medical ultrasound, in part because of its high electromechanical conversion efficiency and low intrinsic losses. The properties of PZT can be adjusted by modifying the ratio of zirconium to titanium and introducing small amounts of other substances, such as lanthanum (Berlincourt, 1971). Table 5.1 shows the material properties of linear-array elements made from PZT-5H.

PZT has a high dielectric constant compared with many piezoelectric materials, resulting is favorable electrical characteristics. The ceramic is mechanically strong, and it can be machined to various shapes and sizes. PZT can operate at temperatures up to 100°C or higher, and it is stable over long periods of time.

The disadvantages of PZT include its high **acoustic impedance** ($Z = 30$ MRayls) compared with body tissue ($Z = 1.5$ MRayls) and the presence of **lateral modes** in array elements. One or more acoustic matching layers can largely compensate for the acoustic impedance mismatch. The effect of lateral modes can be diminished by choosing the appropriate element dimensions or by subdicing the elements.

Other piezoelectric materials are used for various applications. Composites are made from PZT interspersed in an epoxy matrix (Smith, 1992). Lateral modes are reduced in a composite because of its inhomo-geneous structure. By combining the PZT and epoxy in different ratios and spatial distributions, one can tailor the composite's properties for different applications. Polyvinylidene difluoride (PVDF) is a ferroelectric polymer that has been used effectively in high-frequency transducers (Sherar and Foster, 1989). The copolymer of PVDF with trifluoroethylene has an improved electromechanical conversion efficiency. Relaxor ferroelectric materials, such as lead-magnesium-niobate (PMN), become piezoelectric when a large direct-current (dc) bias voltage is applied (Takeuchi et al., 1990). They have a very large dielectric constant ($\varepsilon > 20{,}000\varepsilon_0$), resulting in higher transducer capacitance and a lower electrical impedance.

5.1.2 Scanning with Array Transducers

Array transducers use the same principles as acoustic lenses to focus an acoustic beam. In both cases, variable delays are applied across the transducer aperture. With a sequential or phased array, however,

TABLE 5.1 Material Properties of Linear-Array Elements Made of PZT-5H

Parameter	Symbol	Value	Units
Density	ρ	7500	kg/m^3
Speed of sound	c	3970	m/sec
Acoustic impedance	Z	29.75	MRayls
Relative dielectric constant	$\varepsilon/\varepsilon_0$	1475	None
Electromechanical coupling coefficient	k	0.698	None
Mechanical loss tangent	$\tan\delta_m$	0.015	None
Electrical loss tangent	$\tan\delta_e$	0.02	None

the delays are electronically controlled and can be changed instantaneously to focus the beam in different regions. Linear arrays were first developed for radar, sonar, and radio astronomy (Allen, 1964; Bobber, 1970), and they were implemented in a medical ultrasound system by Somer in 1968 (Somer, 1968).

Linear-array transducers have increased versatility over piston transducers. Electronic scanning involves no moving parts, and the focal point can be changed dynamically to any location in the scanning plane. The system can generate a wide variety of scan formats, and it can process the received echoes for other applications, such as dynamic receive focusing (von Ramm and Thurstone, 1976), correction for phase aberrations (Flax and O'Donnell, 1988; Trahey et al., 1990), and synthetic aperture imaging (Nock and Trahey, 1992).

The disadvantages of linear arrays are due to the increased complexity and higher cost of the transducers and scanners. For high-quality ultrasound images, many identical array elements are required (currently 128 and rising). The array elements are typically less than a millimeter on one side, and each has a separate connection to its own transmitter and receiver electronics.

The widespread use of array transducers for many applications indicates that the advantages often outweigh the disadvantages. In addition, improvement in transducer fabrication techniques and integrated circuit technology have led to more advanced array transducers and scanners.

5.1.2.1 Focusing and Steering with Phased Arrays

This section describes how a phased-array transducer can focus and steer an acoustic beam along a specific direction. An ultrasound image is formed by repeating this process over 100 times to interrogate a two-(2D) or three-dimensional (3D) region of the medium.

Figure 5.1a illustrates a simple example of a six-element linear array focusing the transmitted beam. One can assume that each array element is a point source that radiates a spherically shaped wavefront

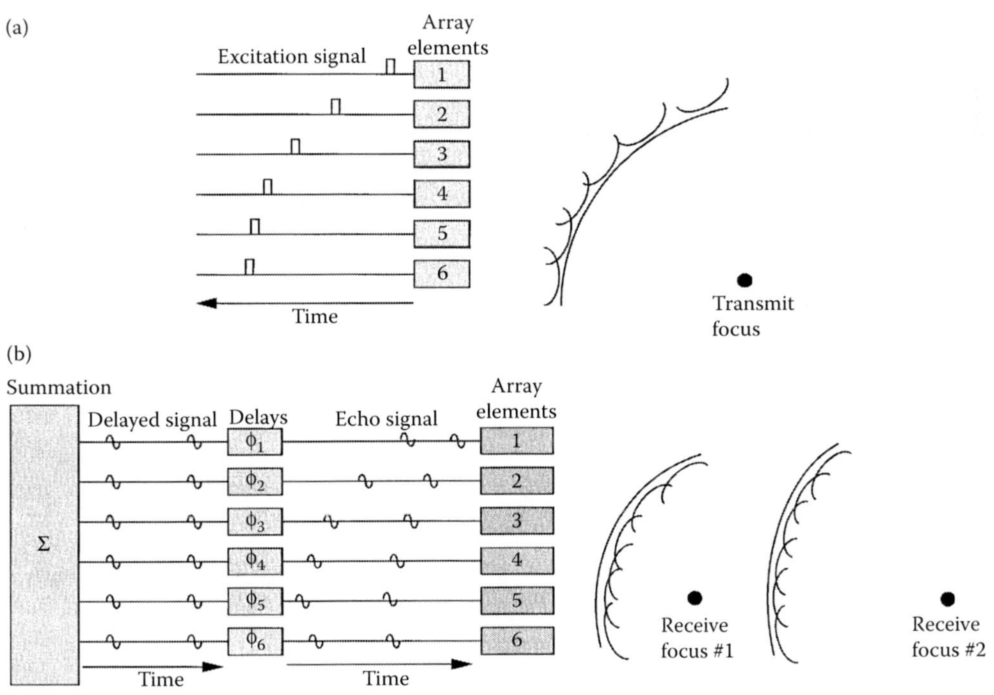

FIGURE 5.1 Focusing and steering an acoustic 1 Focusing beam using a phased array. A 6-element linear array is shown (a) in the transmit mode and (b) in the receive mode. Dynamic focusing in receive allows the scanner focus to track the range of returning echoes.

into the medium. Since the top element is farthest from the focus in this example, it is excited first. The remaining elements are excited at the appropriate time intervals so that the acoustic signals from all the elements reach the focal point at the same time. According to Huygens' principle, the net acoustic signal is the sum of the signals that have arrived from each source. At the focal point, the contributions from every element add in phase to produce a peak in the acoustic signal. Elsewhere, at least some of the contributions add out of phase, reducing the signal relative to the peak.

For receiving an ultrasound echo, the phased array works in reverse. Figure 5.1b shows an echo originating from focus 1. The echo is incident on each array element at a different time interval. The received signals are electronically delayed so that the delayed signals add in phase for an echo originating at the focal point. For echoes originating elsewhere, at least some of the delayed signals will add out of phase, reducing the receive signal relative to the peak at the focus.

In the receive mode, the focal point can be dynamically adjusted so that it coincides with the range of returning echoes. After transmission of an acoustic pulse, the initial echoes return from targets near the transducer. Therefore, the scanner focuses the phased array on these targets, located at focus 1 in Figure 5.1b. As echoes return from more distant targets, the scanner focuses at a greater depth (focus 2 in the figure). Focal zones are established with adequate depth of field so that the targets are always in focus in receive. This process is called dynamic receive focusing and was first implemented by von Ramm and Thurstone in 1976 (von Ramm and Thurstone, 1976).

5.1.2.2 Array-Element Configurations

An ultrasound image is formed by repeating the preceding process many times to scan a 2D or 3D region of tissue. For a 2D image, the scanning plane is the *azimuth dimension;* the *elevation dimension* is perpendicular to the azimuth scanning plane. The shape of the region scanned is determined by the array-element configuration, described in the paragraph below.

Linear Sequential Arrays: Sequential linear arrays have as many as 512 elements in current commercial scanners. A subaperture of up to 128 elements is selected to operate at a given time. As shown in Figure 5.2a, the scanning lines are directed perpendicular to the face of the transducer; the acoustic beam is focused but not steered. The advantage of this scheme is that the array elements have high sensitivity when the beam is directed straight ahead. The disadvantage is that the field of view is limited to the rectangular region directly in front of the transducer. Linear-array transducers have a large footprint to obtain an adequate field of view.

Curvilinear Arrays: Curvilinear or convex arrays have a different shape than sequential linear arrays, but they operate in the same manner. In both cases, the scan lines are directed perpendicular to the transducer face. A curvilinear array, however, scans a wider field of view because of its convex shape, as shown in Figure 5.2b.

Linear Phased Arrays: The more advanced linear phased arrays have 128 elements. All the elements are used to transmit and receive each line of data. As shown in Figure 5.2c, the scanner steers the ultrasound beam through a sector-shaped region in the azimuth plane. Phased arrays scan a region that is significantly wider than the footprint of the transducer, making them suitable for scanning through restricted acoustic windows. As a result, these transducers are ideal for cardiac imaging, where the transducer must scan through a small window to avoid the obstructions of the ribs (bone) and lungs (air).

1.5D Arrays: The so-called 1.5D array is similar to a 2D array in construction but a 1D array in operation. The 1.5D array contains elements along both the azimuth and elevation dimensions. Features such dynamic focusing and phase correction can be implemented in both dimensions to improve image quality. Since a 1.5D array contains a limited number of elements in elevation (e.g., 3 to 9 elements), steering is not possible in that direction. Figure 5.2d illustrates a B-scan made with a 1.5D phased array. Linear sequential scanning is also possible with 1.5D arrays.

2D Phased Arrays: A 2D phased-array has a large number of elements in both the azimuth and elevation dimensions. Therefore, 2D arrays can focus and steer the acoustic beam in both dimensions.

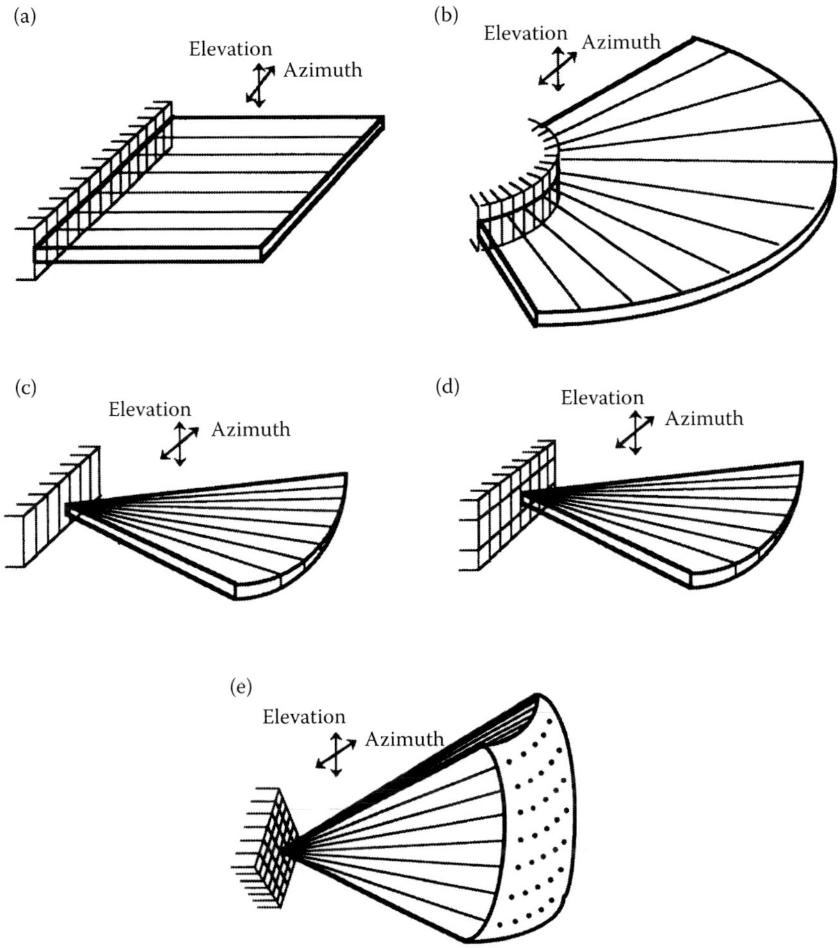

FIGURE 5.2 Array-element configurations and the region scanned by the acoustic beam. (a) A sequential linear array scans a rectangular region; (b) a curvilinear array scans a sector-shaped region; (c) a linear phased array scans a sector-shaped region; (d) a 1.5D array scans a sector-shaped region; and (e) a 2D array scans a pyramidal-shaped region.

Using parallel receive processing (Shattuck et al., 1984), a 2D array can scan a pyramidal region in real time to produce a volumetric image, as shown in Figure 5.2e (von Ramm and Smith, 1990).

5.1.3 Linear-Array Transducer Performance

Designing an ultrasound transducer array involves many compromises. Ideally, a transducer has high sensitivity or signal-to-noise ratio (SNR), good spatial resolution, and no artifacts. The individual array elements should have wide **angular response** in the steering dimensions, low cross-coupling, and an electrical impedance matched to the transmitter.

Figure 5.3a illustrates the connections to the transducer assembly. The transmitter and receiver circuits are located in the ultrasound scanner and are connected to the array elements through 1 to 2 m of coaxial cable. **Electrical matching networks** can be added to tune out the capacitance of the coaxial cable and/or the transducer element and increase the SNR.

A more detailed picture of six-transducer elements is shown in Figure 5.3b. Electrical leads connect to the ground and signal electrodes of the piezoelectric material. Acoustically, the array elements are loaded

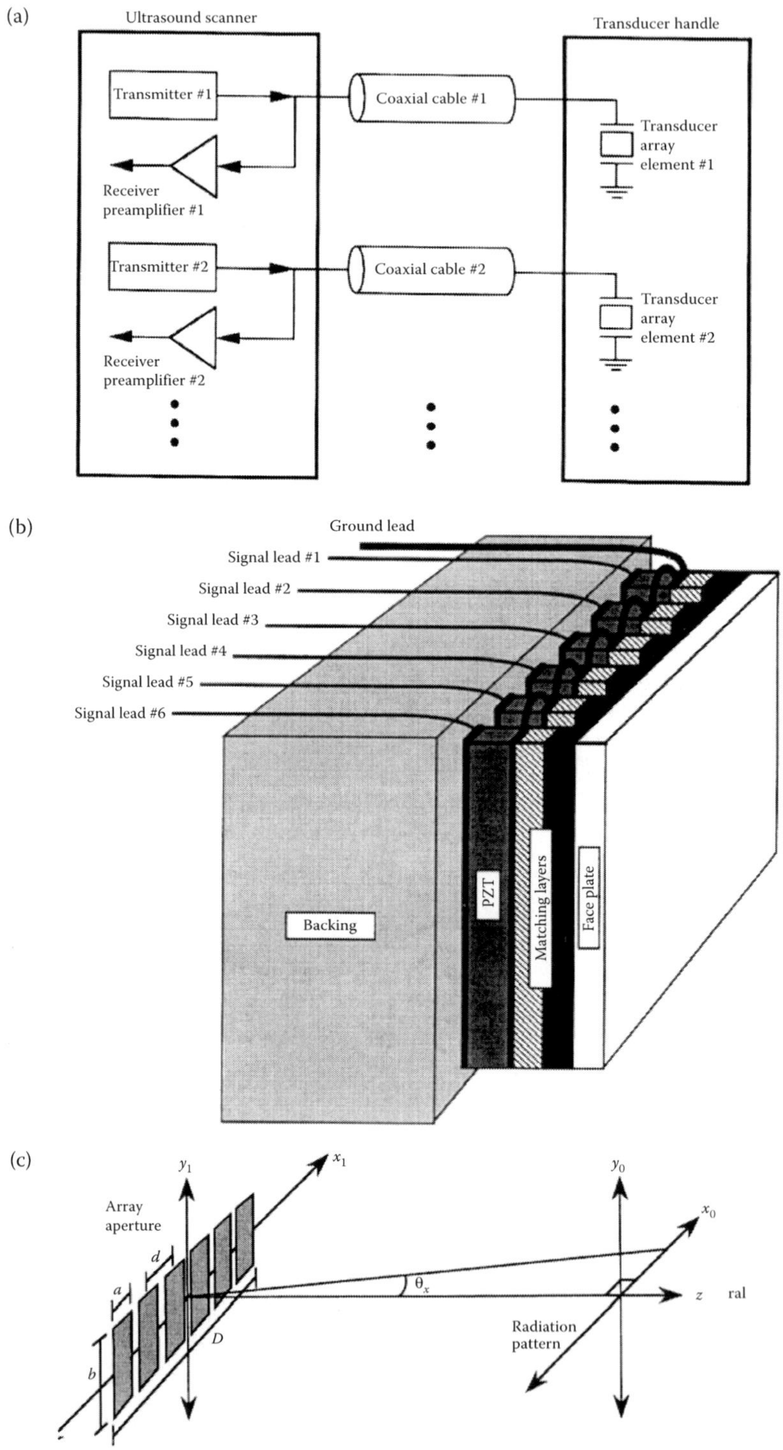

FIGURE 5.3 (a) The connections between the ultrasound scanner and the transducer assembly for two elements of an array. (b) A more detailed picture of the transducer assembly for six elements of an array. (c) Coordinate system and labeling used to describe an array transducer.

on the front side by one or two **quarter-wave matching layers** and the tissue medium. The matching layers may be made from glass or epoxy. A backing material, such as epoxy, loads the back side of the array elements. The faceplate protects the transducer assembly and also may act as an acoustic lens. Faceplates are often made from silicone or polyurethane.

The following sections describe several important characteristics of an array transducer. Figure 5.3c shows a six-element array and its dimensions. The element thickness, width, and length are labeled as *t*, *a*, and *b*, respectively. The interelement spacing is *d*, and the total aperture size is *D* in azimuth. The acoustic wavelength in the load medium, usually human tissue, is designated as l, while the wavelength in the transducer material is λ_t.

Examples are given below for a 128-element linear array operating at 5 MHz. The array is made of PZT-5H with element dimensions of $0.1 \times 5 \times 0.3$ mm³. The interelement spacing is $d = 0.15$ mm in azimuth, and the total aperture is $D = 128 \times 0.15$ mm $= 19.3$ mm (see Table 5.1 for the piezoelectric material characteristics). The elements have an epoxy backing of $Z = 3.25$ MRayls. For simplicity, the example array does not contain a $\lambda/4$ matching layer.

5.1.3.1 Axial Resolution

Axial resolution determines the ability to distinguish between targets aligned in the axial direction (the direction of acoustic propagation). In pulse-echo imaging, the echoes off of two targets separated by r/2 have a path length difference of *r*. If the acoustic pulse length is r, then echoes off the two targets are just distinguishable. As a result, the **axial resolution** is often defined as one-half the pulse length (Christensen, 1988). A transducer with a high resonant frequency and a broad bandwidth has a short acoustic pulse and good axial resolution.

5.1.3.2 Radiation Pattern

The radiation pattern of a transducer determines the insonified region of tissue. For **good lateral resolution** and sensitivity, the acoustic energy should be concentrated in a small region. The radiation pattern for a narrow-band or continuous-wave (CW) transducer is described by the Rayleigh-Sommerfeld diffraction formula (Goodman, 1986). For a pulse-echo imaging system, this diffraction formula is not exact due to the broadband acoustic waves used. Nevertheless, the Rayleigh-Sommerfeld formula is a reasonable first-order approximation to the actual radiation pattern.

The following analysis considers only the azimuth scanning dimension. Near the focal point or in the far field, the Fraunhofer approximation reduces the diffraction formula to a Fourier transform formula. For a circular or rectangular aperture, the far field is at a range of

$$z > \frac{D^2}{4\lambda} \tag{5.1}$$

Figure 5.3c shows the coordinate system used to label the array aperture and its radiation pattern. The array aperture is described by

$$\text{Array}(c_1) = \text{rect}\left(\frac{x_1}{a}\right) * \text{comb}\left(\frac{x_1}{d}\right) \cdot \text{rect}\left(\frac{x_1}{D}\right) \tag{5.2}$$

where the rect(*x*) function is a rectangular pulse of width *x*, and the comb(x) function is a delta function repeated at intervals of *x*. The diffraction pattern is evaluated in the x_0 plane at a distance *z* from the transducer, and θ_x is the angle of the point x_0 from the normal axis. With the Fraunhofer approximation, the normalized diffraction pattern is given by

$$\text{P}_x(\theta_x) = \text{sinc}\left(\frac{a\sin\theta_x}{\lambda}\right).\text{comp}\left(\frac{d\sin\theta_x}{\lambda}\right) * \text{sinc}\left(\frac{D\sin\theta_x}{\lambda}\right) \tag{5.3}$$

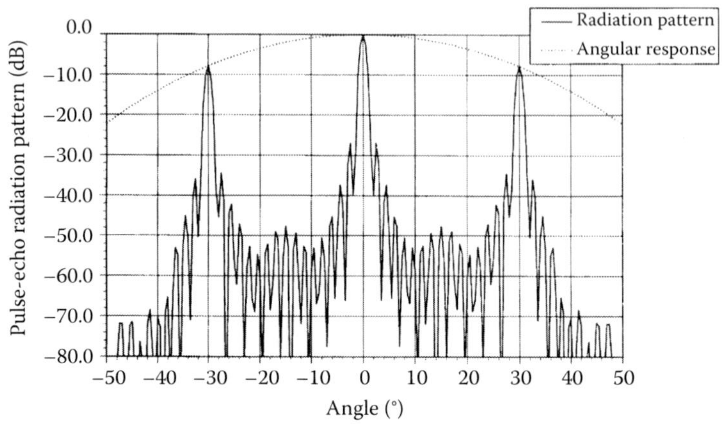

FIGURE 5.4 Radiation pattern of Equation 5.3 for a 16-element array with $a = \lambda$, $d = 2\lambda$, and $D = 32\lambda$. The angular response, the first term of Equation 5.3, is also shown as a dashed line.

in azimuth, where the Fourier transform of Equation 5.2 has been evaluated at the spatial frequency

$$f_x = \frac{x_0}{\lambda z} = \frac{\sin \theta_x}{\lambda} \tag{5.4}$$

Figure 5.4 shows a graph of Equation 5.3 for a 16-element array with $a = \lambda$, $d = 2\lambda$, and $D = 32\lambda$. In the graph, the significance of each term is easily distinguished. The first term determines the angular response weighting, the second term determines the location of **grating lobes** off-axis, and the third term determines the shape of the main lobe and the grating lobes. The significance of *lateral resolution*, angular response, and grating lobes is seen from the CW diffraction pattern.

Lateral resolution determines the ability to distinguish between targets in the azimuth and elevation dimensions. According to the Rayleigh criterion (Goodman, 1986), the lateral resolution can be defined by the first null in the main lobe, which is determined from the third term of Equation 5.3.

$$\theta_x = \sin^{-1} \frac{\lambda}{D} \tag{5.5}$$

in the **azimuth dimension**. A larger aperture results in a more narrow main lobe and better resolution.

A broad angular response is desired to maintain sensitivity while steering off-axis. The first term of Equation 5.3 determines the one-way angular response. The element is usually surrounded by a soft baffle, such as air, resulting in an additional cosine factor in the radiation pattern (Selfridge et al., 1980). Assuming transmit/receive reciprocity, the pulse-echo angular response for a single element is

$$P_x(\theta_x) = \frac{\sin^2(\pi a/\lambda \cdot \sin \theta_x)}{\pi a/\lambda \cdot \sin \theta_x} \cdot \cos^2 \theta_x \tag{5.6}$$

in the azimuth dimension. As the aperture size becomes smaller, the element more closely resembles a point source, and the angular response becomes more broad. Another useful indicator is the −6-dB angular response, defined as the full-width at half-maximum of the angular response graph.

Grating lobes are produced at a location where the path length difference to adjacent array elements is a multiple of a wavelength (the main lobe is located where the path length difference is zero). The acoustic contributions from the elements constructively interfere, producing off-axis peaks. The term grating

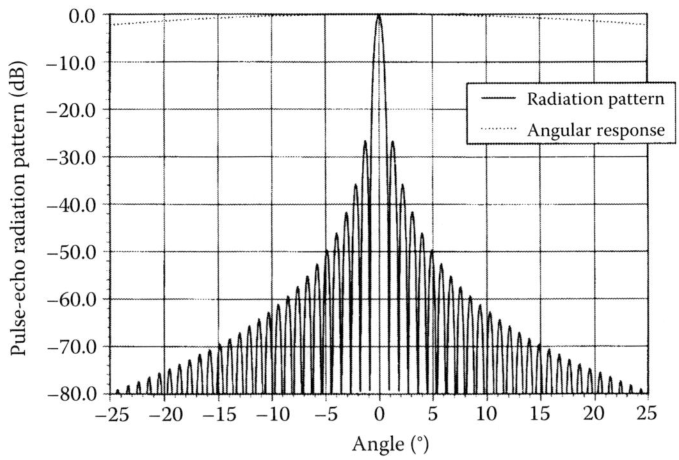

FIGURE 5.5 Radiation pattern of the example array element with $a = 0.1$ mm, $d = 0.15$ mm, $D = 19.2$ mm, and $\lambda = 0.3$ mm. The angular response of Equation 5.6 was substituted into Equation 5.3 for this graph.

lobe was originally used to describe the optical peaks produced by a diffraction grating. In ultrasound, grating lobes are undesirable because they represent acoustic energy steered away from the main lobe. From the comb function in Equation 5.3, the grating lobes are located at

$$\theta_x = \sin^{-1}\frac{i\lambda}{d} \quad i = 1,2,3,\ldots \tag{5.7}$$

in azimuth.

If d is a wavelength, then grating lobes are centered at ±90 degrees from the steering direction in that dimension. Grating lobes at such large angles are less significant because the array elements have poor angular response in those regions. If the main lobe is steered at a large angle, however, the grating lobes are brought toward the front of the array. In this case, the angular response weighting produces a relatively weak main lobe and a relatively strong grating lobe. To eliminate grating lobes at all steering angles, the interelement spacing is set to $\lambda/2$ or less (Steinberg, 1967).

Figure 5.5 shows the theoretical radiation pattern of the 128-element example. For this graph, the angular response weighting of Equation 5.6 was substituted into Equation 5.3. The lateral resolution, as defined by Equation 5.7, $\theta_x = 0.9$ degrees at the focal point. The −6-dB angular response is ±40 degrees from Equation 5.6.

5.1.3.3 Electrical Impedance

The electric impedance of an element relative to the electrical loads has a significant impact on transducer signal-to-noise ratio (SNR). At frequencies away from resonance, the transducer has electrical characteristics of a capacitor. The construction of the transducer is a parallel-plate capacitor with clamped capacitance of

$$C_0 = \varepsilon^s \frac{ab}{t} \tag{5.8}$$

where ε^s is the clamped dielectric constant.

Near resonance, equivalent circuits help to explain the impedance behavior of a transducer. The simplified circuit of Figure 5.6a is valid for transducers operating at series resonance without losses and

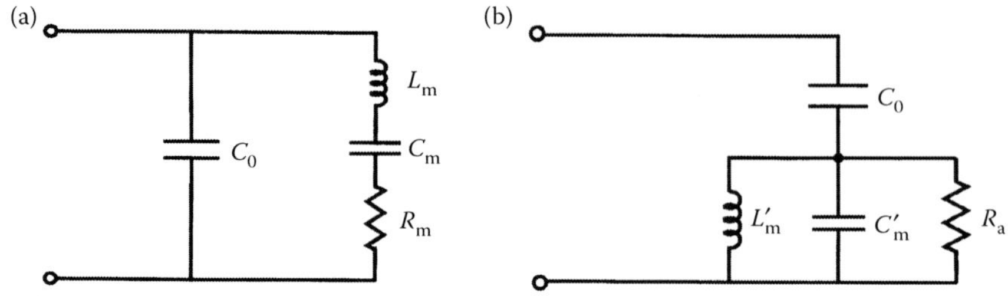

FIGURE 5.6 Simplified equivalent circuits for a piezoelectric transducer (a) near-series resonance and (b) near-parallel resonance.

with low acoustic impedance loads (Kino, 1987). The mechanical resistance R_m represents the acoustic loads as seen from the electrical terminals:

$$R_m = \frac{\pi}{4k^2\omega c_0} \cdot \frac{Z_1 + Z_2}{Z_C} \tag{5.9}$$

where k is the electromechanical coupling coefficient of the piezoelectric material, Z_C is the acoustic impedance of the piezoelectric material, Z_1 is the acoustic impedance of the transducer backing, and Z_2 is the acoustic impedance of the load medium (body tissue). The power dissipated through R_m corresponds to the acoustic output power from the transducer.

The mechanical inductance L_m and mechanical capacitance C_m are analogous to the inductance and capacitance of a mass-spring system. At the series resonant frequency of

$$f_s = \frac{1}{2\pi\sqrt{L_m C_m}} \tag{5.10}$$

the impedances of these components add to zero, resulting in a local impedance minimum.

The equivalent circuit of Figure 5.6a can be redrawn in the form shown in Figure 5.6b. In this circuit, Q is the same as before, but the mechanical impedances have values of L'_m, C'_m, and R_a. The resistive component R_a is

$$R_a = \frac{4k^2}{\pi\omega C_0} \cdot \frac{Z_C}{Z_1 - Z_2} \tag{5.11}$$

The inductor and capacitor combine to form an open circuit at the parallel resonant frequency of

$$f_p = \frac{1}{2\pi\sqrt{L'_m C'_m}} \tag{5.12}$$

The parallel resonance, which is at a slightly higher frequency than the series resonance, is indicated by a local impedance maximum.

Figure 5.7 shows a simulated plot of magnitude and phase versus frequency for the example array element described at the beginning of this subsection. The series resonance frequency is immediately identified at 5.0 MHz with an impedance minimum of $|Z| = 350\ \Omega$. Parallel resonance occurs at 6.7 MHz with an impedance maximum of $|Z| = 4000\ \Omega$. Note the capacitive behavior (approximately −90-degree phase) at frequenciess far from resonance.

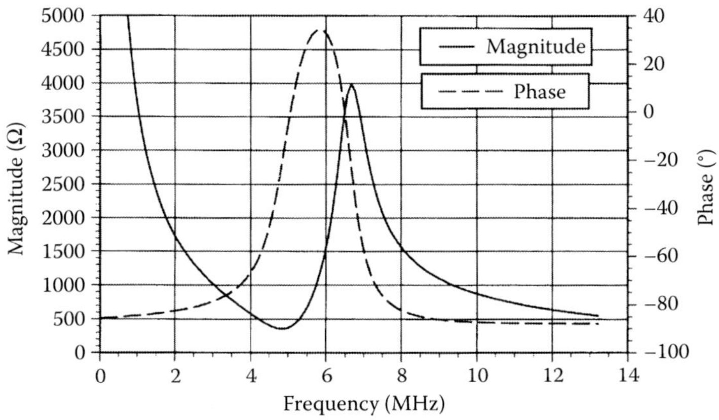

FIGURE 5.7 Complex electrical impedance of the example array element. Series resonance is located at 5.0 MHz, and parallel resonance is located at 6.7 MHz.

5.1.4 Designing a Phased-Array Transducer

In this section the design of an idealized phased-array transducer is considered in terms of the performance characteristics described above. Criteria are described for selecting array dimensions, acoustic backing and matching layers, and electrical matching networks.

5.1.4.1 Choosing Array Dimensions

The array element thickness is determined by the parallel resonant frequency. For $\lambda/2$ resonance, the thickness is

$$t = \frac{\lambda_t}{2}\frac{c_t}{2f_p} \tag{5.13}$$

where c_t is the longitudinal speed of sound in the transducer material.

There are three constraints for choosing the element width and length (1) a nearly square cross-section should be avoided so that lateral vibrations are not coupled to the thickness vibration; as a rule of thumb (Kino and DeSilets, 1979),

$$a/t = 0.6 \quad \text{or} \quad a/t = 10 \tag{5.14}$$

(2) a small width and length are also desirable for a wide angular response weighting function; and (3) an interelement spacing of $\lambda/2$ or less is necessary to eliminate grating lobes.

Fortunately, these requirements are consistent for PZT array elements. For all forms of PZT, $c_t > 2c$, where c is the speed of sound in body tissue (an average of 1540 m/sec). At a given frequency, then $\lambda_t > 2\lambda$. Also, Equation 5.13 states that $\lambda_t = 2t$ at a frequency of f_p. By combining these equations, $t > \lambda$ for PZT array elements operating at a frequency of f_p. If $d = \lambda/2$, then $a < \lambda/2$ because of the finite kerf width that separates the elements. Given this observation, then $a < t/2$. This is consistent with Equation 5.14 to reduce lateral modes.

An element having $d = \lambda/2$ also has adequate angular response. For illustrative purposes, one can assume a zero kerf width so that $a = \lambda/2$. In this case, the −6-dB angular response is $\theta_x = \pm 35$ degrees according to Equation 5.6.

The array dimensions determine the transducer's lateral resolution. In the azimuth dimension, if $d = \lambda/2$, then the transducer aperture is $D = n\lambda/2$, where n is the number of elements in a fully sampled array. From Equation 5.5, the lateral resolution in azimuth is

$$\theta_x = \sin^{-1}\frac{2}{n} \tag{5.15}$$

Therefore, the lateral resolution is independent of frequency in a fully sampled array with $d = \lambda/2$. For this configuration, the lateral resolution is improved by increasing the number of elements.

5.1.4.2 Acoustic Backing and Matching Layers

The backing and matching layers affect the transducer bandwidth and sensitivity. While a lossy, matched backing improves bandwidth, it also dissipates acoustic energy that could otherwise be transmitted into the tissue medium. Therefore, a low-impedance acoustic backing is preferred because it reflects the acoustic pulses toward the front side of the transducer. In this case, adequate bandwidth is maintained by acoustically matching the transducer to the tissue medium using matching layers.

Matching layers are designed with a thickness of $\lambda/4$ at the center frequency and an acoustic impedance between those of the transducer Z_T and the load medium Z_L. The ideal acoustic impedances can be determined from several different models (Hunt et al., 1983). Using the KLM equivalent circuit model (Desilets et al., 1978), the ideal acoustic impedance is

$$Z_1 = \sqrt[3]{Z_T Z_L^2} \tag{5.16}$$

for a single matching layer. For matching PZT-5H array elements (Z_T = 30 MRayls) to a water load (Z_L = 1.5 MRayls), a matching layer of Z_1 = 4.1 MRayls should be chosen. If two matching layers are used, they should have acoustic impedances of

$$Z_1 \sqrt[7]{Z_T^4 Z_L^3} \tag{5.17a}$$

$$Z_2 \sqrt[7]{Z_T Z_L^6} \tag{5.17b}$$

In this case, Z_1 = 8.3 MRayls and Z_2 = 2.3 MRayls for matching PZT-5H to a water load.

When constructing a transducer, a practical matching layer material is not always available, with the ideal acoustic impedance (Equation 5.16 or Equation 5.17). Adequate bandwidth is obtained by using materials that have an impedance close to the ideal value. With a single matching layer, for example, conductive epoxy can be used with Z = 5.1 MRayls.

5.1.4.3 Electrical Impedance Matching

Signal-to-noise ratio and bandwidth are also improved when electrical impedance of an array element is matched to that of the transmit circuitry. Consider the simplified circuit in Figure 5.8 with a transmitter of impedance R_0 and a transducer of real impedance R_t. The power output is proportional to the power dissipated in R_t, as expressed as

$$P_{out} = \frac{V_{out}^2}{R_t} \quad \text{where } V_{out} = \frac{R_t}{R_0 + R_t} V_{in} \tag{5.18}$$

The power available from the transmitter is

$$P_{in} = \frac{(V_{in}/2)^2}{R_0} \tag{5.19}$$

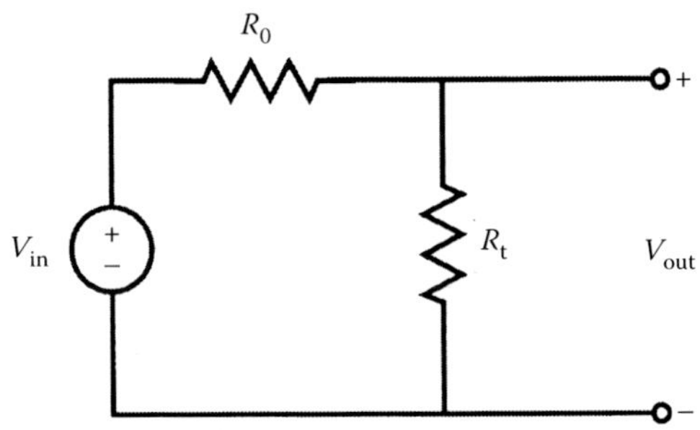

FIGURE 5.8 A transducer of real impedance R_t being excited by a transmitter with source impedance R_0 and source voltage V_{in}.

into a matched load. From the two previous equations, the power efficiency is

$$\frac{P_{out}}{P_{in}} = \frac{4R_0R_t}{(R_0 + R_t)^2} \tag{5.20}$$

For a fixed-source impedance, the maximum efficiency is obtained by taking the derivative of Equation 5.20 with respect to R_t and setting it to zero. Maximum efficiency occurs when the source impedance is matched to the transducer impedance, $R_0 = R_t$.

In practice, the transducer has a complex impedance of R_m in parallel with C_0 (see Figure 5.6), which is excited by a transmitter with a real impedance of 50 Ω. The transducer has a maximum efficiency when the imaginary component is tuned out and the real component is 50 Ω. This can be accomplished with electrical matching networks.

The capacitance C_0 is tuned out in the frequency range near Ω_0 using an inductor of

$$L_0 = \frac{1}{\omega_0^2 C_0} \tag{5.21}$$

for an inductor in shunt, or

$$L_1 \frac{1}{\omega_0^2 C_0 + 1/R_m^2 C_0} \tag{5.22}$$

for an inductor in series. The example array elements described in the preceding subsection have $C_0 = 22$ pF and $R_m = 340$ Ω at series resonance of 5.0 MHz. Therefore, tuning inductors of $L_0 = 46$ mH or $L_1 = 2.4$ mH should be used.

A shunt inductor also raises the impedance of the transducer, as seen from the scanner, while a series inductor lowers the terminal impedance (Hunt et al., 1983). For more significant changes in terminal impedance, transformers are used.

A transformer of turns ratio 1:N multiplies the terminal impedance by $1/N^2$. In the transmit mode, N can be adjusted so that the terminal impedance matches the transmitter impedance. In the receive mode, the open-circuit sensitivity varies as $1/N$ because of the step-down transformer. The lower terminal impedance of the array element, however, provides increased ability to drive an electrical load.

More complicated circuits can be used for better electrical matching across a wide bandwidth (Hunt et al., 1983). These circuits can be either passive, as above, or active. Inductors also can be used in the scanner to tune out the capacitance of the coaxial cable that loads the transducer on receive.

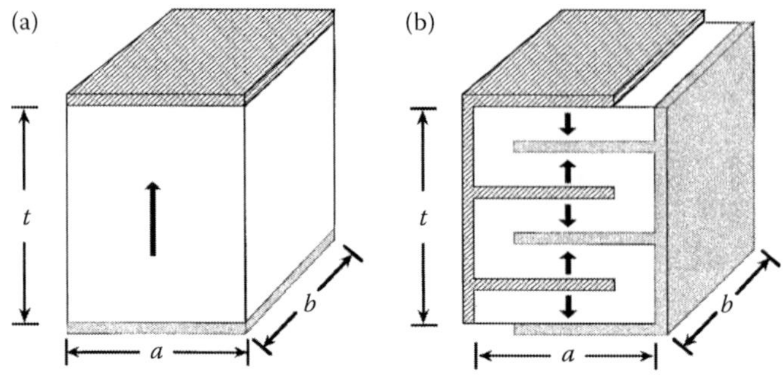

FIGURE 5.9 (a) Conventional single-layer ceramic; (b) five-layer ceramic of the same overall dimensions. The layers are electrically in parallel and acoustically in series. The arrows indicate the piezoelectric poling directions of each layer.

Another alternative for electrical matching is to use multilayer piezoelectric ceramics (Goldberg and Smith, 1994). Figure 5.9 shows an example of a single layer and a five-layer array element with the same overall dimensions of a, b, and t. Since the layers are connected electrically in parallel, the clamped capacitance of a multilayer ceramic (MLC) element is

$$C_0 = N \cdot \varepsilon^S \cdot \frac{ab}{t/N} = N^2 \cdot C_{\text{single}} \tag{5.23}$$

where C_{single} is the capacitance of the single-layer element (Equation 5.8). As a result, the MLC impedance is reduced by a factor of N^2. Acoustically, the layers of the MLC are in series so the $\lambda/2$ resonant thickness is t, the stack thickness.

To a first order, an N-layer ceramic has identical performance compared with a 1:N transformer, but the impedance is transformed within the ceramic. MLCs also can be fabricated in large quantities more easily than hand-wound transformers. While MLCs do not tune out the reactive impedance, they make it easier to tune a low capacitance array element. By lowering the terminal impedance of an array element, MLCs significantly improve transducer SNR.

5.1.5 Summary

The piezoelectric transducer is an important component in the ultrasound imaging system. The transducer often consists of a liner array that can electronically focus an acoustic beam. Depending on the configuration of array elements, the region scanned may be sector shaped or rectangular in two dimensions or pyramidal shaped in three dimensions.

The transducer performance large determines the resolution and the signal-to-noise ratio of the resulting ultrasound image. The design of an array involves many compromises in choosing operating frequency and array-element dimensions. Electrical matching networks and quarter-wave matching layers may be added to improve transducer performance.

Further improvements in transducer performance may result from several areas of research. Newer materials, such as composites, are gaining widespread use in medical ultrasound. In addition, 1.5D arrays or 2D arrays may be employed to control the acoustic beam in both azimuth and elevation. Problems in fabrication and electrical impedance matching must be overcome to implement these arrays in an ultrasound system.

Defining Terms

Acoustic impedance: In an analogy to transmission line impedance, the acoustic impedance is the ratio of pressure to particle velocity in a medium; more commonly, it is defined as $Z = \rho c$, where ρ = density and c = speed of sound in a medium [the units are kg/(m^2 sec) or Rayls].

Angular response: The radiation pattern versus angle for a single element of an array.

Axial resolution: The ability to distinguish between targets aligned in the axial direction (the direction of acoustic propagation).

Azimuth dimension: The lateral dimension that is along the scanning plane for an array transducer.

Electrical matching networks: Active or passive networks designed to tune out reactive components of the transducer and match the transducer impedance to the source and receiver impedance.

Elevation dimension: The lateral dimension that is perpendicular to the scanning plane for an array transducer.

Grating lobes: Undesirable artifacts in the radiation pattern of a transducer; they are produced at a location where the path length difference to adjacent array elements is a multiple of a wavelength.

Lateral modes: Transducer vibrations that occur in the lateral dimensions when the transducer is excited in the thickness dimension.

Lateral resolution: The ability to distinguish between targets in the azimuth and elevation dimensions (perpendicular to the axial dimension).

Quarter-wave matching layers: One or more layers of material placed between the transducer and the load medium (water or human tissue); they effectively match the acoustic impedance of the transducer to the load medium to improve the transducer bandwidth and signal-to-noise ratio.

References

Allen J.L. 1964. Array antennas: New applications for an old technique. *IEEE Spect.* 1: 115.

Berlincourt D. 1971. Piezoelectric crystals and ceramics. In O.E. Mattiat (Ed.), *Ultrasonic Transducer Materials*. New York, Plenum Press.

Bobber R.J. 1970. *UnderwaterElectro acoustic Measurements*. Washington, Naval Research Laboratory.

Christensen D.A. 1988. *Ultrasonic Bioinstrumentation*. New York, Wiley.

Curie P. and Curie J. 1980. Development par pression de l'electricite polaire dans les cristaux hemiedres a faces enclinees. *Comp. Rend.* 91: 383.

Desilets C.S., Fraser J.D., and Kino G.S. 1978. The design of efficient broad-band piezoelectric transducers. *IEEE Trans. Son. Ultrason.* SU-25: 115.

Flax S.W. and O'Donnell M. 1988. Phase aberration correction using signals from point reflectors and diffuse scatters: Basic principles. *IEEE Trans. Ultrason. Ferroelec. Freq. Contr.* 35: 758.

Goldberg R.L. and Smith S.W. 1994. Multi-layer piezoelectric ceramics for two-dimensional array transducers. *IEEE Trans. Ultrason. Ferroelec. Freq. Contr.*

Goodman W. 1986. *Introduction to Fourier Optics*. New York, McGraw-Hill.

Hunt J.W., Arditi M., and Foster F.S. 1983. Ultrasound transducers for pulse-echo medical imaging. *IEEE Trans. Biomed. Eng.* 30: 453.

Kino G.S. 1987. *Acoustic Waves*. Englewood Cliffs, NJ, Prentice-Hall.

Kino G.S. and DeSilets C.S. 1979. Design of slotted transducer arrays with matched backings. *Ultrason. Imag.* 1: 189.

Nock L.F. and Trahey G.E. 1992. Synthetic receive aperture imaging with phase correction for motion and for tissue inhomogeneities: I. Basic principles. *IEEE Trans. Ultrason. Ferroelec. Freq. Contr.* 39: 489.

Selfridge A.R., Kino G.S., and Khuri-Yahub B.T. 1980. A theory for the radiation pattern of a narrow strip acoustic transducer. *Appl. Phys. Lett.* 37: 35.

Shattuck D.P., Weinshenker M.D.,Smith S.W., and von Ramm O.T. 1984. Explososcan: A parallel processing technique for high speed ultrasound imaging with linear phased arrays. *J. Acoust. Soc. Am.* 75:1273.

Sherar M.D. and Foster F.S. 1989. The design and fabrication of high frequency poly(vinylidene fluoride) transducers. *Ultrason. Imag.* 11: 75.

Smith W.A. 1992. New opportunities in ultrasonic transducers emerging from innovations in piezoelectric materials. In F.L. Lizzi (Ed.), *New Developments in Ultrasonic Transducers and Transducer Systems*, pp. 3–26. New York, SPIE.

Somer J.C. 1968. Electronic sector scanning for ultrasonic diagnosis. *Ultrasonics* 153.

Steinberg B.D. 1976. *Principles of Aperture and Array System Design*. New York, Wiley.

Takeuchi H., Masuzawa H., Nakaya C., and Ito Y. 1990. Relaxor ferroelectric transducers. *Proceedings of IEEE Ultrasonics Symposium*, IEEE cat no 90CH2938-9, pp. 697–705.

Trahey G.E., Zhao D., Miglin J.A., and Smith S.W. 1990. Experimental results with a real-time adaptive ultrasonic imaging system for viewing through distorting media. *IEEE Trans. Ultrason. Ferroelec. Freq. Contr.* 37: 418.

von Ramm O.T. and Smith S.W. 1990. Real time volumetric ultrasound imaging system. In *SPIE Medical Imaging IV: Image Formation, Vol* 1231, pp. 15–22. New York, SPIE.

von Ramm O.T. and Thurstone F.L. 1976. Cardiac imaging using a phased array ultrasound system: I. System design. *Circulation* 53: 258.

Further Information

A good overview of linear array design and performance is contained in von O.T. Ramm and S.W. Smith (1983), Beam steering with linear arrays, *IEEE Trans. Biomed. Eng.* 30: 438. The same issue contains a more general article on transducer design and performance: J.W. Hunt, M.' Arditi, and F.S. Foster (1983), Ultrasound transducers for pulse-echo medical imaging, *IEEE Trans. Biomed. Eng.* 30: 453.

The journal *IEEE Transactions on Ultrasonics, Ferroelectrics, and Frequency Control* frequently contains articles on medical ultrasound transducers. For subscription information, contact IEEE Service Center, 445 Hoes Lane, P.O. Box 1331, Piscataway, NJ 08855-1331, phone (800) 678-IEEE.

Another good source is the proceedings of the IEEE Ultrasonics Symposium, published each year. Also, the proceedings from *New Developments in Ultrasonics Transducers and Transducer Systems*, edited by F.L. Lizzi, was published by SPIE, Vol. 1733, in 1992.

5.2 Ultrasonic Imaging

Jack G. Mottley

It was recognized long ago that the tissues of the body are inhomogeneous and that signals sent into them, like pulses of high-frequency sound, are reflected and scattered by those tissues. **Scattering**, or redirection of some of an incident energy signal to other directions by small particles, is why we see the beam of a spotlight in fog or smoke. That part of the scattered energy that returns to the transmitter is called the **backscatter**.

Ultrasonic imaging of the soft tissues of the body really began in the early 1970s. At that time, the technologies began to become available to capture and display the echoes backscattered by structures within the body as images, at first as static **compound images** and later as real-time moving images. The development followed much the same sequence (and borrowed much of the terminology) as did radar and sonar, from initial crude single-line-of-sight displays (**A-mode**) to recording these side by side to build up recordings over time to show motion (**M-mode**), to finally sweeping the transducer either mechanically or electronically over many directions and building up two-dimensional views (**B-mode or 2D**).

Since this technology was intended for civilian use, applications had to wait for the development of inexpensive data handling, storage, and display technologies. A-mode was usually shown on

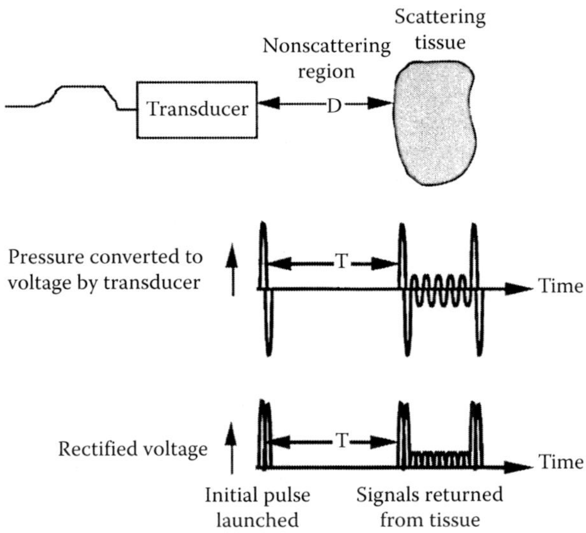

FIGURE 5.10 Schematic representation of the signal received from along a single line of sight in a tissue. The rectified voltage signals are displayed for A-mode.

oscilloscopes, M-modes were printed onto specially treated light-sensitive thermal paper, and B-mode was initially built up as a static image in analog scan converters and shown on television monitors. Now all modes are produced in real time in proprietary scan converters, shown on television monitors, and recorded either on commercially available videotape recorders (for organs or studies in which motion is a part of the diagnostic information) or as still frames on photographic film (for those cases in which organ dimensions and appearance are useful, but motion is not important).

Using commercial videotape reduces expenses and greatly simplifies the review of cases for quality control and training, since review stations can be set up in offices or conference rooms with commonly available monitors and videocassette recorders, and tapes from any imaging system can be played back. Also, the tapes are immediately available and do not have to be chemically processed.

Since the earliest systems were mostly capable of showing motion, the first applications were in studying the heart, which must move to carry out its function. A-mode and M-mode displays (see Figures 5.10 through 5.12) were able to demonstrate the motion of valves, thickening of heart chamber walls, relationships between heart motion and pressure, and other parameters that enabled diagnoses of heart problems that had been difficult or impossible before. For some valvular diseases, the preferred display format for diagnosis is still the M-mode, on which the speed of valve motions can be measured and the relations of valve motions to the electrocardiogram (ECG) are easily seen.

Later, as 2D displays became available, ultrasound was applied more and more to imaging of the soft abdominal organs and in obstetrics (Figure 5.13). In this format, organ dimensions and structural relations are seen more easily, and since the images are now made in real time, motions of organs such as the heart are still well appreciated. These images are used in a wide variety of areas from obstetrics and gynecology to ophthalmology to measure the dimensions of organs or tissue masses and have been widely accepted as a safe and convenient imaging modality.

5.2.1 Fundamentals

Strictly speaking, ultrasound is simply any sound wave whose frequency is above the limit of human hearing, which is usually taken to be 20 kHz. In the context of imaging of the human body, since frequency and wavelength (and therefore resolution) are inversely related, the lowest frequency of sound commonly used is around 1 MHz, with a constant trend toward higher frequencies in order to obtain

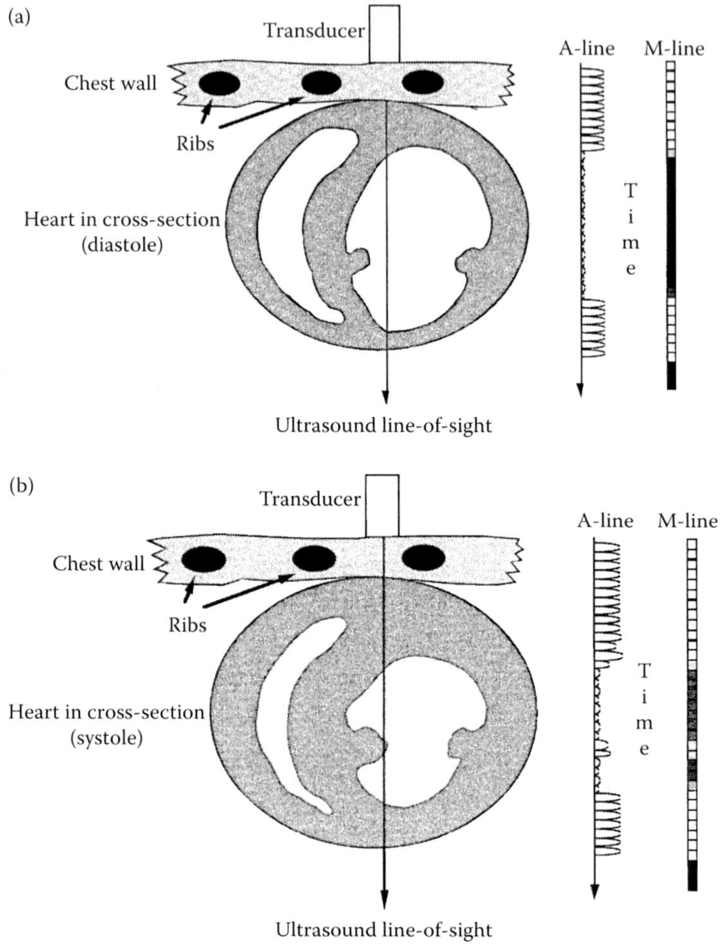

FIGURE 5.11 Example of M-mode imaging of a heart at two points during the cardiac cycle. (a) Upper panel shows heart during diastole (relaxation) with a line of sight through it and the corresponding A-line converted to an M-line. (b) The lower panel shows the same heart during systole (contraction) and the A- and M-lines. Note the thicker walls and smaller ventricular cross-section during systole.

better resolution. Axial resolution is approximately one wavelength, and at 1 MHz, the wavelength is 1.5 mm in most soft tissues, so one must go to 1.5 MHz to achieve 1-mm resolution.

Attenuation of ultrasonic signals increases with frequency in soft tissues, and so a tradeoff must be made between the depth of penetration that must be achieved for a particular application and the highest frequency that can be used. Applications that require deep penetration (e.g., cardiology, abdominal, obstetrics) typically use frequencies in the 2- to 5-MHz range, while those applications which only require shallow penetration but high resolution (e.g., ophthalmology, peripheral vascular, testicular) use frequencies up to around 20 MHz. Intra-arterial imaging systems, requiring submillimeter resolution, use even higher frequencies of 20 to 50 MHz, and laboratory applications of ultrasonic microscopy use frequencies up to 100 or even 200 MHz to examine structures within individual cells.

There are two basic equations used in ultrasonic imaging. One relates the (one-way) distance d of an object that caused an echo from the transducer to the (round-trip) time delay t and speed of sound in the medium c:

$$d = \frac{1}{2}tc \tag{5.24}$$

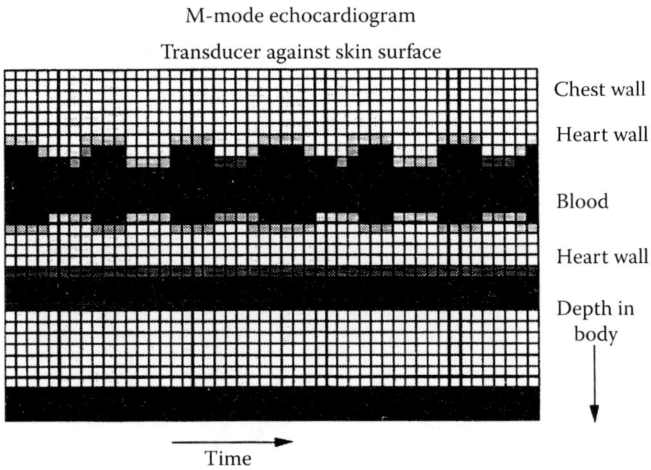

FIGURE 5.12 Completed M-mode display obtained by showing the M-lines of Figure 5.11 side by side. The motion of the heart walls and their thickening and thinning are well appreciated. Often the ECG or heart sounds are also shown in order to coordinate the motions of the heart with other physiologic markers.

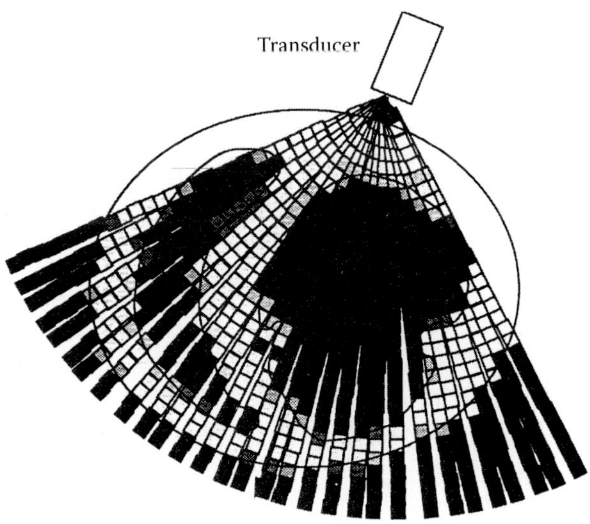

FIGURE 5.13 Schematic representation of a heart and how a 2D image is constructed by scanning the transducer.

The speed of sound in soft body tissues lies in a fairly narrow range from 1450 to 1520 m/sec. For rough estimates of time of flight, one often uses 1500 m/sec, which can be converted to 1.5 mm/μsec, a more convenient set of units. This leads to delay times for the longest-range measurements (20 cm) of 270 μsec.

To allow echoes and reverberations to die out, one needs to wait several of these periods before launching the next interrogating pulse, so pulse repetition frequencies of about a kilohertz are possible.

The other equation relates the received signal strength $S(t)$ to the transmitted signal $T(t)$, the transducer's properties $B(t)$, the attenuation of the signal path to and from the scatterer $A(t)$, and the strength of the scatterer $h(t)$:

$$S(t) = T(t) \otimes B(t) \otimes A(t) \otimes \eta(t) \tag{5.25}$$

where ⊗ denotes time-domain convolution. Using the property of Fourier transforms that a convolution in the time domain is a multiplication in the frequency domain, this is more often written in the frequency domain as

$$S(f) = T(f)B(f)A(f)\eta(f) \tag{5.26}$$

where each term is the Fourier transform of the corresponding term in the time-domain expression 5.25 and is written as a function of frequency f.

The goal of most imaging applications is to measure and produce an image based on the local values of the scattering strength, which requires some assumptions to be made concerning each of the other terms. The amplitude of the transmitted signal $T(f)$ is a user-adjustable parameter that simply adds a scale factor to the image values, unless it increases the returned signal to the point of saturating the receiver amplifier. Increasing the transmit power increases the strength of return from distant or faint echoes simply by increasing the power that illuminates them, like using a more powerful flashlight lets you see farther at night. Some care must be taken to not turn the transmit power up too high, since very high power levels are capable of causing acoustic cavitation or local heating of tissues, both of which can cause cellular damage. Advances in both electronics and transducers make it possible to transmit more and more power. For this reason, new ultrasonic imaging systems are required to display an index value that indicates the transmitted power. If the index exceeds established thresholds, it is possible that damage may occur, and the examiner should limit the time of exposure.

Most imaging systems are fairly narrow band, so the transducer properties $B(f)$ are constant and produce only a scale factor to the image values. On phased-array systems it is possible to change the depth of focus on both transmit and receive. This improves image quality and detection of lesions by matching the focusing characteristics of the transducer to best image the object in question, like focusing a pair of binoculars on a particular object.

As the ultrasonic energy travels along the path from transmitter to scatterer and back, attenuation causes the signal to decrease with distance. This builds up as a line integral from time 0 to time t as

$$A(f,t) = e^{-\int_0^t \alpha(f)c\,dt'}$$

An average value of attenuation can be corrected for electronically by increasing the gain of the imaging system as a function of time (variously called time gain compensation [TGC] or depth gain compensation [DGC]). In addition, some systems allow for lateral portions of the image region to have different attenuation by adding a lateral gain compensation in which the gain is increased to either side of the center region of the image.

Time gain compensation is usually set to give a uniform gray level to the scattering along the center of the image. Most operators develop a "usual" setting on each machine, and if it becomes necessary to change those settings to obtain acceptable images on a patient, then that indicates that the patient has a higher attenuation or that there is a problem with the electronics, transducer, or acoustic coupling.

5.2.2 Applications and Example Calculations

As an example of calculating the time of flight of an ultrasonic image, consider the following.

> *Example 5.1*: A tissue has a speed of sound $c = 1460$ m/sec, and a given feature is 10 cm deep within. Calculate the time it will take an ultrasonic signal to travel from the surface to the feature and back.
>
> *Answer:* $t = 2 \times (10\text{ cm})/(1460\text{ m/sec}) = 137\ \mu\text{sec}$, where the factor of 2 is to account for the round trip the signal has to make (i.e., go in and back out).

Example 5.2: Typical soft tissues attenuate ultrasonic signals at a rate of 0.5 dB/cm/MHz. How much attenuation would be suffered by a 3-MHz signal going through 5 cm of tissue and returning?
Answer: $a = 3\text{ MHz} \times (0.5\text{ dB/cm/MHz})/(8.686\text{ dB/neper}) = 0.173\text{ neper/cm}$, $A\ (3\text{ MHz}, 5\text{ cm}) = e^{(-0.173\text{ neper/cm})\times(5\text{ cm})\times 2} = 0.177$.

5.2.3 Economics

Ultrasonic imaging has many economic advantages over other imaging modalities. The imaging systems are typically much less expensive than those used for other modalities and do not require special preparations of facilities such as shielding for x-rays or uniformity of magnetic field for MRI. Most ultrasonic imaging systems can be rolled easily from one location to another, so one system can be shared among technicians or examining rooms or even taken to patients' rooms for critically ill patients.

There are minimal expendables used in ultrasonic examinations, mostly the coupling gel used to couple the transducer to the skin and videotape or film for recording. Transducers are reusable and amortized over many examinations. These low costs make ultrasonic imaging one of the least expensive modalities, far preferred over others when indicated. The low cost also means these systems can be a part of private practices and used only occasionally.

As an indication of the interest in ultrasonic imaging as an alternative to other modalities, in 1993, the *Wall Street Journal* reported that spending in the United States on MRI units was approximately \$520 million, on CT units \$800 million, and on ultrasonic imaging systems \$1000 million, and that sales of ultrasound systems was growing at 15% annually [1].

Defining Terms

A-mode: The original display of ultrasound measurements, in which the amplitude of the returned echoes along a single line is displayed on an oscilloscope.

Attenuation: The reduction is signal amplitude that occurs per unit distance traveled. Some attenuation occurs in homogeneous media such as water due to viscous heating and other phenomena, but that is very small and is usually taken to be negligible over the 10- to 20-cm distances typical of imaging systems. In inhomogeneous media such as soft tissues, the attenuation is much higher and increases with frequency. The values reported for most soft tissues lie around 0.5 dB/cm/MHz.

Backscatter: That part of a scattered signal that goes back toward the transmitter of the energy.

B-mode or 2D: The current display mode of choice. This is produced by sweeping the transducer from side to side and displaying the strength of the returned echoes as bright spots in their geometrically correct direction and distance.

Compound images: Images built up by adding, or compounding, data obtained from a single transducer or multiple transducers swept through arcs. Often these transducers were not fixed to a single point of rotation but could be swept over a surface of the body like the abdomen in order to build up a picture of the underlying organs such as the liver. This required an elaborate position-sensing apparatus attached to the patient's bed or the scanner and that the organ in question be held very still throughout the scanning process, or else the image was blurred.

M-mode: Followed A-mode by recording the strength of the echoes as dark spots on moving light-sensitive paper. Objects that move, such as the heart, caused standard patterns of motion to be displayed, and a lot of diagnostic information such as valve closure rates, whether valves opened or closed completely, and wall thickness could be obtained from M-mode recordings.

Real-time images: Images currently made on ultrasound imaging systems by rapidly sweeping the transducer through an arc either mechanically or electronically. Typical images might have 120 scan lines in each image, each 20 cm long. Since each line has a time of flight of 267 μsec, a single frame takes 120 × 267 μsec = 32 msec. It is therefore possible to produce images at standard video frame rates (30 frames/sec, or 33.3 msec/frame).

Reflection: Occurs at interfaces between large regions (much larger than a wavelength) of media with differing acoustic properties such as density or compressibility. This is similar to the reflection of light at interfaces and can be either total, like a mirror, or partial, like a half-silvered mirror or the ghostlike reflection seen in a sheet of glass.

Scattering: Occurs when there are irregularities or inhomogeneities in the acoustic properties of a medium over distances comparable with or smaller than the wavelength of the sound. Scattering from objects much smaller than a wavelength typically increases with frequency (the blue-sky law in optics), while that from an object comparable to a wavelength is constant with frequency (why clouds appear white).

Reference

1. Naj A.K. 1993. Industry focus: Big medical equipment makers try ultrasound market; cost-cutting pressures prompt shift away from more expensive devices. *Wall Street J.*, November 30, B-4.

Further Information

There are many textbooks that contain good introduction to ultrasonic imaging. *Physical Principles of Ultrasonic Diagnosis*, by P. N. Wells, is a classic, and there is a new edition of another classic, *Diagnostic Ultrasound: Principles, Instruments and Exercises*, 4th ed., by Frederick Kremkau. Books on medical imaging that contain introductions to ultrasonic imaging include *Medical Imaging Systems*, by Albert Macovski; *Principles of Medical Imaging*, by Kirk Shung, Michael Smith, and Benjamin Tsui; and *Foundations of Medical Imaging*, by Zang-Hee Cho, Joie P. Jones, and Manbir Singh.

The monthly journals *IEEE Transactions on Ultrasonics, Ferroelectrics, and Frequency Control* and *IEEE Transactions on Biomedical Engineering* often contain information and research reports on ultrasonic imaging. For subscription information, contact IEEE Service Center, 445 Hoes Lane, P.O. Box 1331, Piscataway, NJ 08855-1331, phone (800) 678-4333. Another journal that often contains articles on ultrasonic imaging is the *Journal of the Acoustical Society of America*. For subscription information, contact AIP Circulation and Fulfillment Division, 500 Sunnyside Blvd., Woodbury, NY 11797-2999, phone (800) 344-6908; e-mail: elecprod\@pinet.aip.org.

There are many journals that deal with medical ultrasonic imaging exclusively. These include *Ultrasonic Imaging the Journal of Ultrasound in Medicine*, American Institute of Ultrasound of Medicine (AIUM), 14750 Sweitzer Lane, Suite 100, Laurel, MD 20707-5906, and the Journal of Ultrasound in Medicine and Biology, Elsevier Science, Inc., 660 White Plains Road, Tarrytown, NY 10591-5153, e-mail: esuk.usa@elsevier.com.

There are also specialty journals for particular medical areas, for example, the *Journal of the American Society of Echocardiography*, that are available through medical libraries and are indexed in Index Medicus, Current Contents, Science Citation Index, and other databases.

5.3 Blood Flow Measurement Using Ultrasound

K. Whittaker Ferrara

In order to introduce the fundamental challenges of blood velocity estimation, a brief description of the unique operating environment produced by the ultrasonic system, intervening tissue, and the scattering of ultrasound by blood is provided. In providing an overview of the parameters that differentiate this problem from radar and sonar target estimation problems, an introduction to the fluid dynamics of the cardiovascular system is presented, and the requirements of specific clinical applications are summarized. An overview of blood flow estimation systems and their performance limitations is then presented. Next, an overview of the theory of moving target estimation, with its roots in radar and sonar signal processing, is provided. The application of this theory to blood velocity estimation is then reviewed, and a number of signal processing strategies that have been applied to this problem are considered. Areas of new research including three-dimensional (3D) velocity estimation and the use of ultrasonic contrast agents are described in the final section.

5.3.1 Fundamental Concepts

In blood velocity estimation, the goal is not simply to estimate the mean target position and mean target velocity. The goal instead is to measure the velocity profile over the smallest region possible and to repeat this measurement quickly and accurately over the entire target. Therefore, the joint optimization of spatial, velocity, and temporal resolution is critical. In addition to the mean velocity, diagnostically useful information is contained in the volume of blood flowing through various vessels, spatial variations in the velocity profile, and the presence of turbulence. While current methods have proven extremely valuable in the assessment of the velocity profile over an entire vessel, improved *spatial resolution* is required in several diagnostic situations. Improved *velocity resolution* is also desirable for a number of clinical applications. Blood velocity estimation algorithms implemented in current systems also suffer from a velocity ambiguity due to aliasing.

5.3.1.1 Unique Features of the Operating Environment

A number of features make blood flow estimation distinct from typical radar and sonar target estimation situations. The combination of factors associated with the beam formation system, properties of the intervening medium, and properties of the target medium lead to a difficult and unique operating environment. Figure 5.14 summarizes the operating environment of an ultrasonic blood velocity estimation system, and Table 5.2 summarizes the key parameters.

5.3.1.1.1 Beam Formation—Data Acquisition System

The transducer bandwidth is limited. Most current transducers are limited to a 50 to 75% fractional bandwidth due to their finite dimensions and a variety of electrical and mechanical properties. This limits the form of the transmitted signal. The transmitted pulse is typically a short pulse with a carrier frequency, which is the center **frequency** in the spectrum of the transmitted signal.

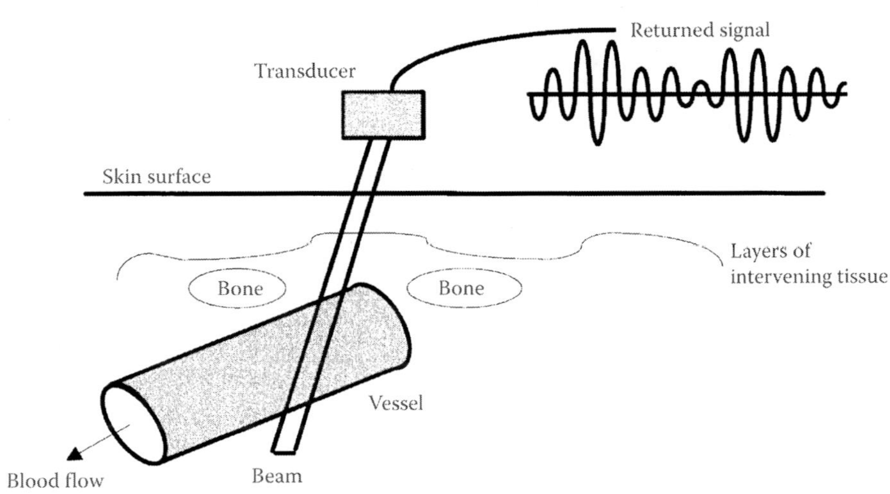

FIGURE 5.14 Operating environment for the estimation of blood velocity.

TABLE 5.2 Important Parameters

Typical transducer center frequency	2–10 MHz
Maximum transducer fractional bandwidth	50–75%
Speed of sound c	1500–1600 m/sec
Acoustic wavelength (c = 1540)	0.154–1.54 mm
Phased-array size	> 32·wavelength
Sample volume size	mm^3
Blood velocity	Normal; up to 1 m/sec
	Pathological: up to 8 m/sec
Vessel wall echo/blood echo	20–40 dB
Diameter of a red blood cell	8.5 μm
Thickness of a red blood cell	2.4 μm
Volume of a red blood cell	87 ± 6 mm^3
Volume concentration of cells (hematocrit)	45%
Maximum concentration without cell deformation	58%

Normal Adult Maximal Velocity (m/sec)		
Mitral	Flow	0.9
Tricuspid	Flow	0.5
Pulmonary	Artery	0.75
Left	Ventricle	0.9

Artery	Peak Systolic Velocity (cm/sec)	Standard Deviation
Proximal external iliac	99	22
Distal external iliac	96	13
Proximal common femoral	89	16
Distal common femoral	71	15
Proximal popliteal	53	9
Distal popliteal	53	24
Proximal peroneal	46	14
Distal peroneal	44	12

Federal agencies monitor four distinct intensity levels. The levels are TASA, TASP, TPSA, and TPSP, where T represents temporal, S represents spatial, A represents average, and P represents peak. Therefore, the use of long bursts requires a proportionate reduction in the transmitted peak power. This may limit the signal-to-noise ratio (SNR) obtained with a long transmitted burst due to the weak reflections from the complex set of targets within the body.

5.3.1.1.2 Intervening Medium

Acoustic windows, which are locations for placement of a transducer to successfully interrogate particular organs, are limited in number and size. Due to the presence of bone and air, the number of usable acoustic windows is extremely limited. The reflection of acoustic energy from bone is only 3 dB below that of a perfect reflector (Wells, 1977). Therefore, transducers cannot typically surround a desired imaging site. In many cases, it is difficult to find a single small access window. This limits the use of inverse techniques.

Intervening tissue produces acoustic refraction and reflection. Energy is reflected at unpredictable angles.

The **clutter**-to-signal ratio is very high. Clutter is the returned signal from stationary or slowly moving tissue, which can be 40 dB above the returned signal from blood. Movement of the vessel walls and valves during the cardiac cycle introduces a high-amplitude, low-frequency signal. This is typically considered to be unwanted noise, and a high-pass filter is used to eliminate the estimated wall frequencies.

The sampling rate is restricted. The speed of sound in tissue is low (~1540 m/sec), and each transmitted pulse must reach the target and return before the returned signal is recorded. Thus the sampling rate is restricted, and the aliasing limit is often exceeded.

The total observation time is limited (due to low acoustic velocity). In order to estimate the velocity of blood in all locations in a 2D field in real time, the estimate for each region must be based on the return from a limited number of pulses because of the low speed of sound.

Frequency-dependent attenuation affects the signal. Tissue acts as a low-pass transmission filter; the scattering functions as a high-pass filter. The received signal is therefore a distorted version of the transmitted signal. In order to estimate the effective filter function, the type and extent of each tissue type encountered by the wave must be known. Also, extension of the bandwidth of the transmitted signal to higher frequencies increases absorption, requiring higher power levels that can increase health concerns.

5.3.1.1.3 Target Scattering Medium (Red Blood Cells)

Multiple groups of scatterers are present. The target medium consists of multiple volumes of diffuse moving scatterers with velocity vectors that vary in magnitude and direction. The target medium is spread in space and velocity. The goal is to estimate the velocity over the smallest region possible.

There is a limited period of statistical stationarity. The underlying cardiac process can only be considered to be stationary for a limited time. This time was estimated to be 10 msec for the arterial system by Hatle and Angelsen (1985). If an observation interval greater than this period is used, the average scatterer velocity cannot be considered to be constant.

5.3.1.2 Overview of Ultrasonic Flow Estimation Systems

Current ultrasonic imaging systems operate in a pulse-echo (PE) or continuous-wave (CW) intensity mapping mode. In pulse-echo mode, a very short pulse is transmitted, and the reflected signal is analyzed. For a continuous-wave system, a lower-intensity signal is continuously transmitted into the body, and the reflected energy is analyzed. In both types of systems, an acoustic wave is launched along a specific path into the body, and the return from this wave is processed as a function of time. The return is due to reflected waves from structures along the line of sight, combined with unwanted noise. Spatial selectivity is provided by beam formation performed on burst transmission and reception. Steering of the beam to a particular angle and creating a narrow beam width at the depth of interest are accomplished by an effective lens applied to the ultrasonic transducer. This lens may be produced by a contoured material, or it may be simulated by phased pulses applied to a transducer array. The spatial weighting pattern will ultimately be the product of the effective lens on transmission and reception. The returned signal from the formed beam can be used to map the backscattered intensity into a two-dimensional gray-scale image, or to estimate target velocity. *We shall focus on the use of this information to estimate the velocity of red blood cells moving through the body.*

5.3.1.2.1 Single Sample Volume Doppler Instruments

One type of system uses the Doppler effect to estimate velocity in a single volume of blood, known as the sample volume, which is designated by the system operator. The Doppler shift frequency from a moving target can be shown to equal $2f_c v/c$, where f_c is the transducer center frequency in Hertz, c is the speed of sound within tissue, and v is the velocity component of the blood cells toward or away from the transducer. These "Doppler" systems transmit a train of long pulses with a well-defined carrier frequency and measure the Doppler shift in the returned signal. The spectrum of Doppler frequencies is proportional to the distribution of velocities present in the sample volume. The sample volume is on a cubic millimeter scale for typical pulse-echo systems operating in the frequency range of 2–10 MHz. Therefore, a thorough cardiac or peripheral vascular examination requires a long period. In these systems, 64–128 temporal samples are acquired for each estimate. The spectrum of these samples is typically computed using a fast Fourier transform (FFT) technique (Kay and Marple, 1981). The range of velocities present within the sample volume can then be estimated. The spectrum is scaled to represent velocity and plotted on the vertical axis. Subsequent spectral estimates are then calculated and plotted vertically adjacent to the first estimate.

5.3.1.2.2 Color Flow Mapping

In color flow mapping, a pseudo-color velocity display is overlaid on a 2D gray-scale image. Simultaneous amplitude and velocity information is thus available for a 2D sector area of the body. The clinical advantage is a reduction in the examination time and the ability to visualize the velocity profile as a 2D map. Figure 5.15 shows a typical color flow map of ovarian blood flow combined with the Doppler spectrum of the region indicated by the small graphic sample volume. The color flow map shows color-encoded velocities superimposed on the gray-scale image with the velocity magnitude indicated by the color bar on the side of the image. Motion toward the transducer is shown in yellow and red, and motion away from the transducer is shown in blue and green, with the range of colors representing a range of velocities to a maximum of 6 cm/sec in each direction. Velocities above this limit would produce aliasing for the parameters used in optimizing the instrument for the display of ovarian flow. A velocity of 0 m/sec would be indicated by black, as shown at the center of the color bar. Early discussions of the implementation of color flow mapping systems can be found in Curry and White (1978) and Nowicki and Reid (1981).

The lower portion of the image presents an intensity-modulated display of instantaneous Doppler components along the vertical axis. As time progresses, the display is translated along the horizontal axis to generate a Doppler time history for the selected region of interest (provided by Acuson Corporation, Mountain View, California).

Limitations of color flow instruments result in part from the transmission of a narrowband (long) pulse that is needed for velocity estimation but degrades spatial resolution and prevents mapping of the spatial-velocity profile. Due to the velocity gradient in each blood vessel, the transmission of a long pulse also degrades the velocity resolution. This is caused by the simultaneous examination of blood cells moving at different velocities and the resulting mixing of regions of the scattering medium, which can be distinctly resolved on a conventional B-mode image. Since the limited speed of acoustic propagation velocity limits the sampling rate, a second problem is aliasing of the Doppler frequency. Third, information regarding the presence of velocity gradients and turbulence is desired and is not currently available. Finally, estimation of blood velocity based on the Doppler shift provides only an estimate of the axial velocity, which is the movement toward or away from the transducer, and cannot be used to estimate movement across the transducer beam. It is the 3D velocity magnitude that is of clinical interest.

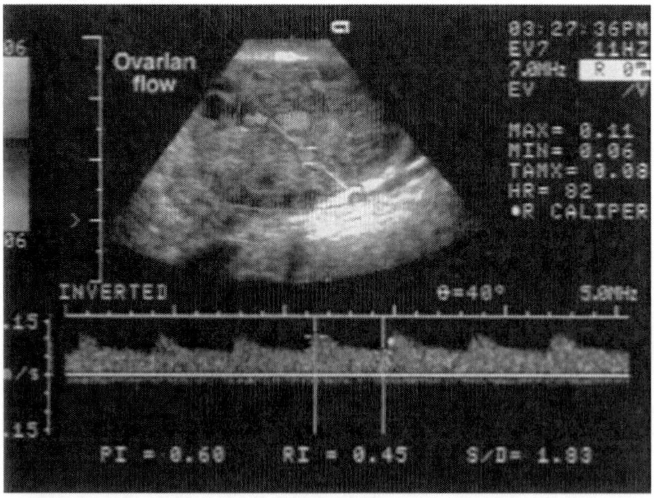

FIGURE 5.15 Flow map and Doppler spectrum for ovarian blood flow.

For a color flow map, the velocity estimation technique is based on estimation of the mean Doppler shift using signal-processing techniques optimized for rapid (real-time) estimation of velocity in each region of the image. The transmitted pulse is typically a burst of 4 to 8 cycles of the carrier frequency. Data acquisition for use in velocity estimation is interleaved with the acquisition of information for the gray-scale image. Each frame of acquired data samples is used to generate one update of the image display. An azimuthal line is a line that describes the direction of the beam from the transducer to the target. A typical 2D ultrasound scanner uses 128 azimuthal lines per frame and 30 frames per second to generate a gray-scale image. Data acquisition for the velocity estimator used in color flow imaging requires an additional 4 to 18 transducer firings per azimuthal line and therefore reduces both the number of azimuthal lines and the number of frames per second. If the number of lines per frame is decreased, spatial undersampling or a reduced examination area results. If the number of frames per second is decreased, temporal undersampling results, and the display becomes difficult to interpret.

The number of data samples available for each color flow velocity estimate is reduced to 4–18 in comparison with the 64–128 data samples available to estimate velocity in a single sample volume Doppler mode. This reduction, required to estimate velocity over the 2D image, produces a large increase in the estimator variance.

5.3.1.3 Fluid Dynamics and the Cardiovascular System

In order to predict and adequately assess blood flow profiles within the body, the fluid dynamics of the cardiovascular system will be briefly reviewed. The idealized case known as Poiseuille flow will be considered first, allowed by a summary of the factors that disturb Poiseuille flow.

A Poiseuille flow model is appropriate in a long rigid circular pipe at a large distance from the entrance. The velocity in this case is described by the equation $v/v_0 = 1 - (r/a)^2$, where v represents the velocity parallel to the wall, v_0 represents the center-line velocity, r is the radial distance variable, and a is the radius of the tube. In this case, the mean velocity is half the center-line velocity, and the volume flow rate is given by the mean velocity multiplied by the cross-sectional area of the vessel.

For the actual conditions within the arterial system, Poiseuille flow is only an approximation. The actual arterial geometry is tortuous and individualistic, and the resulting flow is perturbed by entrance effects and reflections. Reflections are produced by vascular branches and the geometric taper of the arterial diameter. In addition, spatial variations in vessel elasticity influence the amplitude and wave velocity of the arterial pulse. Several parameters can be used to characterize the velocity profile, including the Reynolds number, the Womersly number, the pulsatility index, and the resistive index. The pulsatility and resistive indices are frequently estimated during a clinical examination.

The Reynolds number is denoted Re and measures the ratio of fluid inertia to the viscous forces acting on the fluid. The Reynolds number is defined by $\mathrm{Re} = Dv'/\mu_k$, where v' is the average cross-sectional velocity μ_k is the kinematic viscosity, and D is the vessel diameter. *Kinematic viscosity* is defined as the fluid viscosity divided by the fluid density. When the Reynolds number is high, fluid inertia dominates. This is true in the aorta and larger arteries, and bursts of turbulence are possible. When the number is low, viscous effects dominate.

The Womersly number is used to describe the effect introduced by the unsteady, pulsatile nature of the flow. This parameter, defined by $a(\omega/\mu_k)^{1/2}$, where ω represents radian frequency of the wave, governs propagation along an elastic, fluid-filled tube. When the Womersly number is small, the instantaneous profile will be parabolic in shape, the flow is viscous dominated, and the profile is oscillatory and Poiseuille in nature. When the Womersly number is large, the flow will be blunt, inviscid, and have thin wall layers (Nicholas and O'Rourke, 1990).

The pulsatility index represents the ratio of the unsteady and steady velocity components of the flow. This shows the magnitude of the velocity changes that occur during acceleration and deceleration of blood constituents. Since the arterial pulse decreases in magnitude as it travels, this index is maximum in the aorta. The pulsatility index is given by the difference between the peak systolic and minimum diastolic values divided by the average value over one cardiac cycle. The Pourcelot, or resistance, index is the peak-to-peak swing in velocity from systole to diastole divided by the peak systolic value (Nichols and O'Rourke, 1990).

5.3.1.3.1 Blood Velocity Profiles

Specific factors that influence the blood velocity profile include the entrance effect, vessel curvature, skewing, stenosis, acceleration, secondary flows, and turbulence. These effects are briefly introduced in this section.

The entrance effect is a result of fluid flow passing from a large tube or chamber into a smaller tube. The velocity distribution at the entrance becomes blunt. At a distance known as the entry length, the fully developed parabolic profile is restored, where the entry length is given by 0.06Re·(2*a*) (Nerem, 1985). Distal to this point the profile is independent of distance.

If the vessel is curved, there will also be an entrance effect. The blunt profile in this case is skewed, with the peak velocity closer to the inner wall of curvature. When the fully developed profile occurs downstream, the distribution will again be skewed, with the maximal velocity toward the outer wall of curvature. Skewing also occurs at a bifurcation where proximal flow divides into daughter vessels. The higher-velocity components, which occurred at the center of the parent vessel, are then closer to the flow divider, and the velocity distribution in the daughter vessels is skewed toward the divider.

Stenosis, a localized narrowing of the vessel diameter, dampens the pulsatility of the flow and pressure waveforms. The downstream flow profile depends on the shape and degree of stenosis. Acceleration adds a flat component to the velocity profile. It is responsible for the flat profile during systole, as well as the negative flat component near the walls in the deceleration phase.

Secondary flows are swirling components which are superimposed on the main velocity profile. These occur at bends and branches, although regions of secondary flow can break away from the vessel wall and are then known as separated flow. These regions reattach to the wall at a point downstream.

One definition of turbulent flow is flow that demonstrates a random fluctuation in the magnitude and direction of velocity as a function of space and time. The intensity of turbulence is calculated using the magnitude of the fluctuating velocities. The relative intensity of turbulence is given by $I_t = u_{rms}/u_{mean}$, where u_{rms} represents the root-mean-square value of the fluctuating portion of the velocity, and u_{mean} represents the nonfluctuating mean velocity (Hinze, 1975).

5.3.1.4 Clinical Applications and Their Requirements

Blood flow measurement with ultrasound is used in estimating the velocity and volume of flow within the heart and peripheral arteries and veins. Normal blood vessels vary in diameter up to a maximum of 2 cm, although most vessels examined with ultrasound have a diameter of 1 to 10 mm. Motion of the vessel wall results in a diameter change of 5 to 10% during a cardiac cycle.

5.3.1.4.1 Carotid Arteries (Common, Internal, External)

The evaluation of flow in the carotid arteries is of great clinical interest due to their importance in supplying blood to the brain, their proximity to the skin, and the wealth of experience that has been developed in characterizing vascular pathology through an evaluation of flow. The size of the carotid arteries is moderate; they narrow quickly from a maximum diameter of 0.8 cm. The shape of carotid flow waveforms over the cardiac cycle can be related to the pathophysiology of the circulation. Numerous attempts have been made to characterize the parameters of carotid waveforms and to compare these parameters in normal and stenotic cases. A number of indices have been used to summarize the information contained in these waveforms. The normal range of the Pourcelot index is 0.55 to 0.75. Many researchers have shown that accurate detection of a minor stenosis requires accurate quantitation of the entire Doppler spectrum and remains very difficult with current technology. The presence of a stenosis causes spectral broadening with the introduction of lower frequency or velocity components.

5.3.1.4.2 *Cardiology*

Blood velocity measurement in cardiology requires analysis of information at depths up to 18 cm. A relatively low center frequency (e.g., 2.5 to 3.5 MHz) typically is used in order to reduce attenuation. Areas commonly studied and the maximum rate of flow include the following (Hatle, 1985):

5.3.1.4.3 *Aorta*

Aortic flow exhibits a blunt profile with entrance region characteristics. The entrance length is approximately 30 cm. The vessel diameter is approximately 2 cm. The mean Reynolds number is 2500 (Nerem, 1985), although the peak Reynolds number in the ascending aorta can range from 4300 to 8900, and the peak Reynolds number in the abdominal aorta is in the range of 400 to 1100 (Nichols and O'Rourke, 1990). The maximal velocity is on the order of 1.35 m/sec. The flow is skewed in the aortic arch with a higher velocity at the inner wall. The flow is unsteady and laminar with possible turbulent bursts at peak systole.

5.3.1.4.4 *Peripheral Arteries*

The peak systolic velocity (Hatsukami et al., 1992) in centimeters per second and standard deviation of the velocity measurement technique are provided below for selected arteries.

Nearly all the vessels above normally show some flow reversal during early diastole. A value of the pulsatility index of 5 or more in a limb artery is considered to be normal.

5.3.2 Velocity Estimation Techniques

Prior to the basic overview of theoretical approaches to target velocity estimation, it is necessary to understand a few basic features of the received signal from blood scatterers. It is the statistical correlation of the received signal in space and time that provides the opportunity to use a variety of velocity estimation strategies. Velocity estimation based on analysis of the frequency shift or the temporal correlation can be justified by these statistical properties.

Blood velocity mapping has unique features due to the substantial viscosity of blood and the spatial limitations imposed by the vessel walls. Because of these properties, groups of redblood cells can be tracked over a significant distance. Blood consists of a viscous incompressible fluid containing an average volume concentration of red blood cells of 45%, although this concentration varies randomly through the blood medium. The red blood cells are primarily responsible for producing the scattered wave, due to the difference in their acoustic properties in comparison with plasma. Recent research into the characteristics of blood has led to stochastic models for its properties as a function of time and space (Atkinson and Berry, 1974; Shung et al., 1976, 1992; Angelson, 1980; Mo and Cobbold, 1986). The scattered signal from an insonified spatial volume is a random process that varies with the fluctuations in the density of scatterers in the insonified area, the shear rate within the vessel, and the hematocrit (Atkinson and Berry, 1974; Mo and Cobbold, 1986; Ferrara and Algazi, 1994a,b).

Since the concentration of cells varies randomly through the vessel, the magnitude of the returned signal varies when the group of scatterers being insonified changes. The returned amplitude from one spatial region is independent of the amplitude of the signal from adjacent spatial areas. As blood flows through a vessel, it transports cells whose backscattered signals can be tracked to estimate flow velocities.

Between the transmission of one pulse and the next, the scatterers move a small distance within the vessel. As shown in Figure 5.16, a group of cells with a particular concentration which are originally located at depth D_1 at time T_1 move to depth D_2 at time T_2. The resulting change in axial depth produces a change in the delay of the signal returning to the transducer from each group of scatterers. This change in delay of the radiofrequency (RF) signal can be estimated in several ways. As shown in Figure 5.17, the returned signal from a set of sequential pulses then shows a random amplitude that can be used to estimate the velocity. Motion is detected using signal-processing techniques that estimate the shift of the signal between pulses.

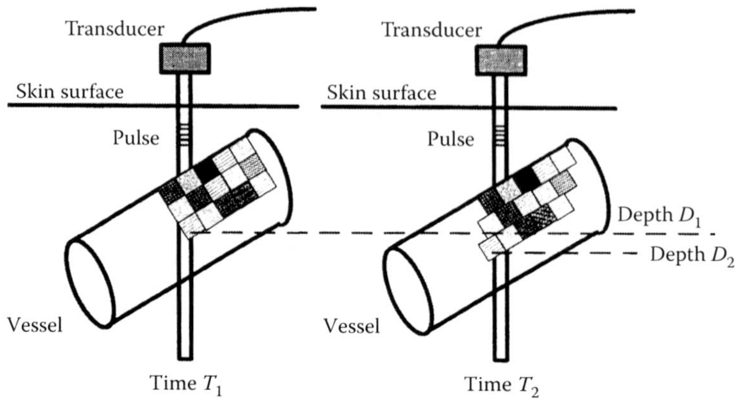

FIGURE 5.16 Random concentration of red blood cells within a vessel at times T_1 and T_2, where the change in depth from D_1 to D_2 would be used to estimate velocity.

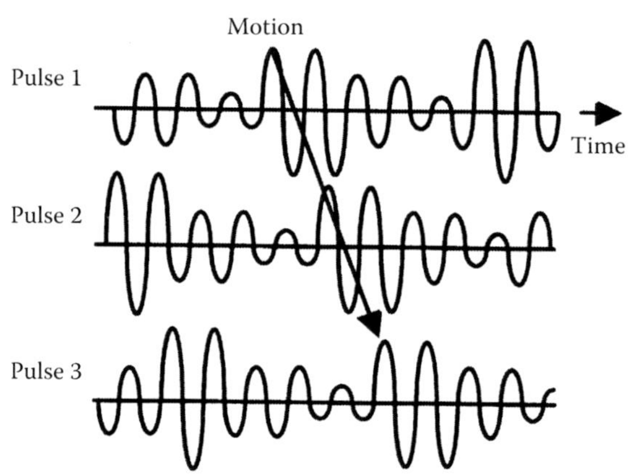

FIGURE 5.17 Received RF signal from three transmitted pulses, with a random amplitude which can be used to estimate the axial movement of blood between pulses. Motion is shown by the shift in the signal with a recognizable amplitude.

5.3.2.1 Clutter

In addition to the desired signal from the blood scatterers, the received signal contains clutter echoes returned from the surrounding tissue. An important component of this clutter signal arises from slowly moving vessel walls. The wall motion produces Doppler frequency shifts typically below 1 kHz, while the desired information from the blood cells exists in frequencies up to 15 kHz. Due to the smooth structure of the walls, energy is scattered coherently, and the clutter signal can be 40 dB above the scattered signal from blood. High-pass filters have been developed to remove the unwanted signal from the surrounding vessel walls.

5.3.2.2 Classic Theory of Velocity Estimation

Most current commercial ultrasound systems transmit a train of long pulses with a carrier frequency of 2 to 10 MHz and estimate velocity using the Doppler shift of the reflected signal. The transmission of a train of short pulses and new signal-processing strategies may improve the spatial resolution and quality of the resulting velocity estimate. In order to provide a basis for discussion and comparison of these techniques, the problem of blood velocity estimation is considered in this subsection from the view of classic velocity estimation theory typically applied to radar and sonar problems.

Important differences exist between classic detection and estimation for radar and sonar and the application of such techniques to medical ultrasound. The Van Trees (1971) approach is based on joint estimation of the Doppler shift and position over the entire target. In medical ultrasound, the velocity is estimated in small regions of a large target, where the target position is assumed to be known. While classic theories have been developed for estimation of all velocities within a large target by Van Trees and others, such techniques require a model for the velocity in each spatial region of interest. For the case of blood velocity estimation, the spatial variation in the velocity profile is complex, and it is difficult to postulate a model that can be used to derive a high-quality estimate. The theory of velocity estimation in the presence of spread targets is also discussed by Kennedy (1969) and Price (1968) as it applies to radar astronomy and dispersive communication channels.

It is the desire to improve the spatial and velocity resolution of the estimate of blood velocity that has motivated the evaluation of alternative wideband estimation techniques. Narrowband velocity estimation techniques use the Doppler frequency shift produced by the moving cells with a sample volume that is fixed in space. Wideband estimation techniques incorporate the change in delay of the returned pulse due to the motion of the moving cells. Within the classification of narrowband techniques are a number of estimation strategies to be detailed below. These include the fast Fourier transform (FFT), finite derivative estimation, the autocorrelator, and modern spectral estimation techniques, including autoregressive strategies. Within the classification of wideband techniques are cross-correlation strategies and the wideband maximum **likelihood** estimator (WMLE).

For improving the spatial mapping of blood velocity within the body, the transmission of short pulses is desirable. Therefore, it is of interest to assess the quality of velocity estimates made using narrowband and wideband estimators with transmitted signals of varying lengths. If $(2v/c)BT > 1$, where v represents the axial velocity of the target, c represents the speed of the wave in tissue, B represents the transmitted signal bandwidth, and T represents the total time interval used in estimating velocity within an individual region, then the change in delay produced by the motion of the red blood cells can be ignored (Van Trees, 1971).

This inequality is interpreted for the physical conditions of medical ultrasound in Figure 5.18. As shown in Figure 5.18, the value vT represents the axial distance traveled by the target while it is observed by the transducer beam, and $c/(2B)$ represents the effective length of the signal that is used to observe the moving cells. If $vT > c/(2B)$, the shift in the position of a group of red blood cells during their travel

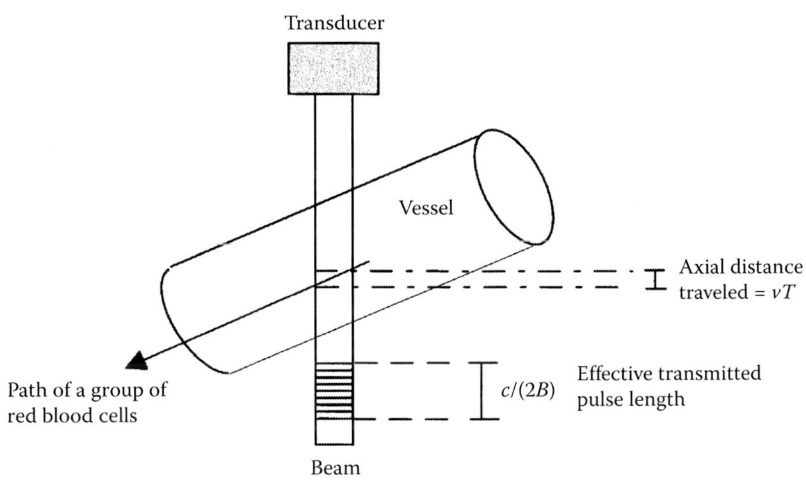

FIGURE 5.18 Comparison of the axial distance traveled and the effective length of the transmitted pulse.

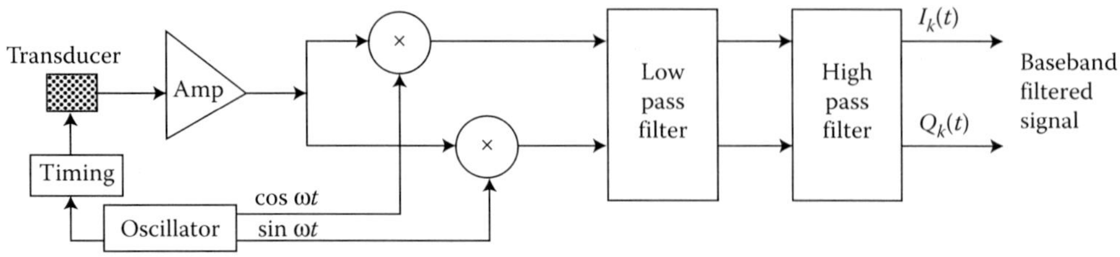

FIGURE 5.19 Block diagram of the system architecture required to generate the baseband signal used by several estimation techniques.

though the ultrasonic beam is not a detectable fraction of the signal length. This leads to two important restrictions on estimation techniques. First, under the "narrowband" condition of transmission of a long (narrowband) pulse, motion of a group of cells through the beam can only be estimated using the Doppler frequency shift. Second, if the inequality is not satisfied and therefore the transmitted signal is short (wideband), faster-moving red blood cells leave the region of interest, and the use of a narrowband estimation technique produces a biased velocity estimate. Thus two strategies can be used to estimate velocity. A long (narrowband) pulse can transmitted, and the signal from a fixed depth then can be used to estimate velocity. Alternatively, a short (wideband) signal can be transmitted in order to improve spatial resolution, and the estimator used to determine the velocity must move along with the red blood cells.

The inequality is now evaluated for typical parameters. When the angle between the axis of the beam and the axis of the vessel is 45 degrees, the axial distance traveled by the red blood cells while they cross the beam is equivalent to the lateral beam width. Using an axial distance vT of 0.75 mm, which is a reasonable lateral beam width, and an acoustic velocity of 1540 m/sec, the bandwidth of the transmitted pulse must be much less than 1.026 MHz for the narrowband approximation to be valid.

Due to practical advantages in the implementation of the smaller bandwidth required by **baseband signals**, the center frequency of the signal is often removed before velocity estimation. The processing required for the extraction of the baseband signal is shown in Figure 5.19. The returned signal from the transducer is amplified and coherently demodulated, through multiplication by the carrier frequency, and then a low-pass filter is applied to remove the signal sideband frequencies and noise. The remaining signal is the **complex envelope**. A high-pass filter is then applied to the signal from each fixed depth to remove the unwanted echoes from stationary tissue. The output of this processing is denoted as $I_k(t)$ for the in-phase signal from the kth pulse as a function of time and $Q_k(t)$ for the quadrature signal from the kth pulse.

5.3.2.3 Narrowband Estimation

Narrowband estimation techniques that estimate velocity for blood at a fixed depth are described in this subsection. Both the classic Doppler technique, which frequently is used in single-sample volume systems, and the autocorrelator, which frequently is used in color flow mapping systems, are included, as well as a finite derivative estimator and an autoregressive estimator, which have been the subject of previous research. The autocorrelator is used in real-time color flow mapping systems due to the ease of implementation and the relatively small bias and variance.

5.3.2.3.1 Classic Doppler Estimation

If the carrier frequency is removed by coherently demodulating the signal, the change in delay of the RF signal becomes a change in the phase of the baseband signal. The Doppler shift frequency from a moving target equals $2f_cv/c$. With a center frequency of 5 MHz, sound velocity of 1540 m/sec, and blood velocity of 1 m/sec, the resulting frequency shift is 6493.5 Hz. For the estimation of blood velocity, the Doppler shift is not detectable using a single short pulse, and therefore, the signal from a fixed depth and a train of pulses is acquired.

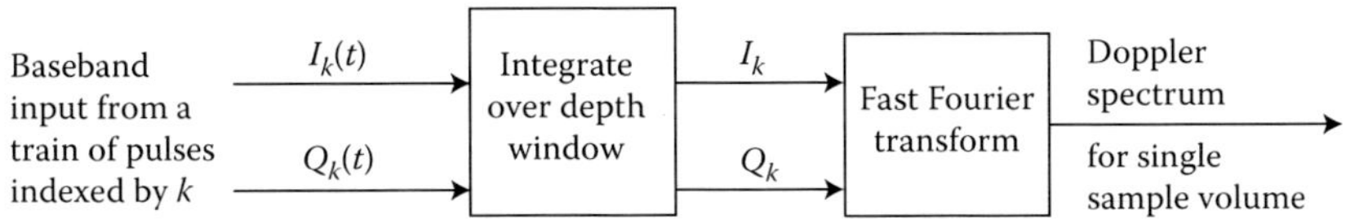

FIGURE 5.20 Block diagram of the system architecture required to estimate the Doppler spectrum from a set of baseband samples from a fixed depth.

A pulse-echo Doppler processing block diagram is shown in Figure 5.20. The baseband signal, from Figure 5.19, is shown as the input to this processing block. The received signal from each pulse is multiplied by a time window that is typically equal to the length of the transmitted pulse and integrated to produce a single data sample from each pulse. The set of data samples from a train of pulses is then Fourier-transformed, with the resulting frequency spectrum related to the axial velocity using the Doppler relationship.

Estimation of velocity using the Fourier transform of the signal from a fixed depth suffers from the limitations of all narrowband estimators, in that the variance of the estimate increases when a short pulse is transmitted. In addition, the velocity resolution produced using the Fourier transform is inversely proportional to the length of the data window. Therefore, if 64 pulses with a pulse repetition frequency of 5 kHz are used in the spectral estimate, the frequency resolution is on the order of 78.125 Hz (5000/64). The velocity resolution for a carrier frequency of 5 MHz and speed of sound of 1540 m/sec is then on the order of 1.2 cm/sec, determined from the Doppler relationship. Increasing the data window only improves the velocity resolution if the majority of the red blood cells have not left the sample volume and the flow conditions have not produced a decorrelation of the signal. It is this relationship between the data window and velocity resolution, a fundamental feature of Fourier transform techniques, that has motivated the use of autoregressive estimators. The frequency and velocity resolution are not fundamentally constrained by the data window using these modern spectral estimators introduced below.

5.3.2.3.2 *Autoregressive Estimation (AR)*

In addition to the classic techniques discussed previously, higher-order modern spectral estimation techniques have been used in an attempt to improve the velocity resolution of the estimate. These techniques are again narrowband estimation techniques, since the data samples used in computing the estimate are obtained from a fixed depth. The challenges encountered in applying such techniques to blood velocity estimation include the selection of an appropriate order which adequately models the data sequence while providing the opportunity for real-time velocity estimation and determination of the length of the data sequence to be used in the estimation process.

The goal in autogressive velocity estimation is to model the frequency content of the received signal by a set of coefficients which cold be used to reconstruct the signal spectrum. The coefficients $a(m)$ represent the AR parameters of the AR(p) process, where p is the number of poles in the model for the signal. Estimation of the AR parameters has been accomplished using the Burg and Levinson-Durban recursion methods. The spectrum $P(f)$ is then estimated using the following equation:

$$p(f) = k\left|1 + \sum_{m=1}^{p} a(m)\exp[-i2\pi mf]\right|^{-2}$$

The poles of the AR transfer function which lie within the unit circle can then be determined based on these parameters, and the velocity associated with each pole is determined by the Doppler equation.

Both autoregressive and autoregressive moving-average estimation techniques have been applied to single-sample-volume Doppler estimation. Order selection for single-sample-volume AR estimators is discussed in Kaluzinski (1989). Second-order autoregressive estimation has been applied to color flow mapping by Loupas and McDicken (1990) andAhnandPark (1991). Although two poles are not sufficient to model the data sequence, the parameters of a higher-order process cannot be estimated in real time. In addition, the estimation of parameters of a higher-order process using the limited number of data points available in color flow mapping produces a large variance. Loupas and McDicken have used the two poles to model the signal returned from blood. Ahn and Park have used one pole to model the received signal from blood and the second pole to model the stationary signal from the surrounding tissue.

While AR techniques are useful in modeling the stationary tissue and blood and in providing a high-resolution estimate of multiple velocity components, several problems have been encountered in the practical application to blood velocity estimation. First, the order required to adequately model any region of the vessel can change when stationary tissue is present in the sample volume or when the range of velocity components in the sample volume increases. In addition, the performance of an AR estimate degrades rapidly in the presence of white noise, particularly with a small number of data samples.

5.3.2.3.3 Autocorrelator

Kasai et al. (1985) and Barber et al. (1985) discussed a narrowband mean velocity estimation structure for use in color flow mapping. The phase of the signal correlation at a lag of one transmitted period is estimated and used in an inverse tangent calculation of the estimated mean Doppler shift f_{mean} of the returned signal. A block diagram of the autocorrelator is shown in Figure 5.21. The baseband signal is first integrated over a short depth window. The phase of the correlation at a lag of one pulse period is then estimated as the inverse tangent of the imaginary part of the correlation divided by the real part of the correlation. The estimated mean velocity v_{mean} of the scattering medium is then determined by scaling the estimated Doppler shift by several factors, including the expected center frequency of the returned signal.

The autocorrelator structure can be derived from the definition of instantaneous frequency, from the phase of the correlation at a lag of one period, or as the first-order autoregressive estimate of the mean frequency of a baseband signal. The contributions of uncorrelated noise should average to zero in both the numerator and denominator of the autocorrelator. This is an advantage because the autocorrelation estimate is unbiased when the input signal includes the desired flow signal and noise. Alternatively, in the absence of a moving target, the input to the autocorrelator may consist only of white noise. Under these conditions, both the numerator and denominator can average to values near zero, and the resulting output of the autocorrelator has a very large variance. This estimation structure must therefore be used with a power threshold that can determine the presence or absence of a signal from blood flow and set the output of the estimator to zero when this motion is absent.

The variance of the autocorrelation estimate increases with the transmitted bandwidth, and therefore, the performance is degraded by transmitting a short pulse.

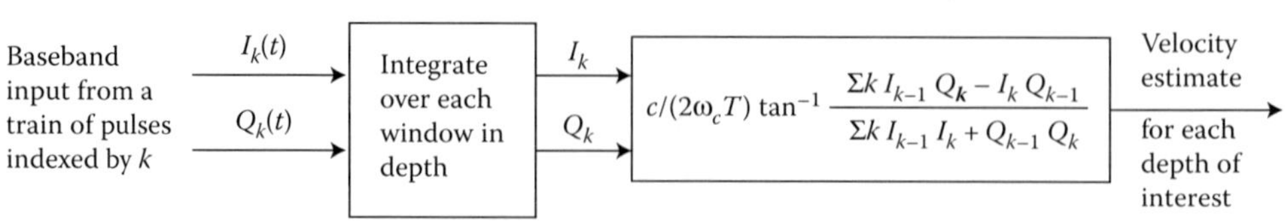

FIGURE 5.21 Block diagram of the system architecture required to estimate the mean Doppler shift for each depth location using the autocorrelator.

5.3.2.3.4 Finite Derivative Estimator (FDE)

A second approach to mean velocity or frequency estimation is based on a finite implementation of a derivative operator. The finite derivative estimator is derived based on the first and second moments of the spectrum. The basis for this estimator comes from the definition of the spectral centroid:

$$v_{\text{mean}} = \frac{\int \omega S(\omega)d\omega}{\int S(\omega)d\omega} \tag{5.27}$$

The mean velocity is given by v_{mean}, which is a scaled version of the mean frequency, where the scaling constant is given by k' and $S(\omega)$ represents the power spectral density. Letting $R_r(\cdot)$ represent the complex signal correlation and t represent the difference between the two times used in the correlation estimate, Equation 5.27 is equivalent to

$$v_{\text{mean}} = k' \frac{\left[(\partial/\partial\tau)R_r(\tau)\big|_{\tau=0}\right]}{R_r(0)} \tag{5.28}$$

Writing the baseband signal as the sum $I(t) + jQ(t)$ and letting E indicate the statistical expectation, Brody and Meindl (1974) have shown that the mean velocity estimate can be rewritten as

$$v_{\text{mean}} = \frac{k'E\{(\partial/\partial t)[I(t)]Q(t) - (\partial/\partial t)[Q(t)]I(t)\}}{E[I^2(t) + Q^2(t)]} \tag{5.29}$$

The estimate of this quantity requires estimation of the derivative of the in-phase portion $I(t)$ and quadrature portion $Q(t)$ of the signal. For an analog, continuous-time implementation, the bias and variance were evaluated by Brody and Meindl (1974). The discrete case has been studied by Kristoffersen (1986). The differentiation has been implemented in the discrete case as a finite difference or as a finite impulse response differentiation filter. The estimator is biased by noise, since the denominator represents power in the returned signal. Therefore, for nonzero noise power, the averaged noise power in the denominator will not be zero mean and will constitute a bias. The variance of the finite derivative estimator depends on the shape and bandwidth of the Doppler spectrum, as well as on the observation interval.

5.3.2.4 Wideband Estimation Techniques

It is desirable to transmit a short ultrasonic pulse in order to examine blood flow in small regions individually. For these short pulses, the narrowband approximation is not valid, and the estimation techniques used should track the motion of the red blood cells as they move to a new position over time. Estimation techniques that track the motion of the red blood cells are known as wideband estimation techniques and include cross-correlation techniques, the wideband maximum likelihood estimator and high time bandwidth estimation techniques. A thorough review of time-domain estimation techniques to estimate tissue motion is presented in Hein and O'Brien (1993).

5.3.2.4.1 Cross-Correlation Estimator

The use of time shift to estimate signal parameters has been studied extensively in radar. If the transmitted signal is known, a maximum likelihood (ML) solution for the estimation of delay has been discussed by Van Trees (1971) and others. If the signal shape is not known, the use of cross-correlation for delay estimation has been discussed by Helstrom (1968) and Knapp and Carter (1976)). If information regarding the statistics of the signal and noise are available, an MLE based on cross-correlation

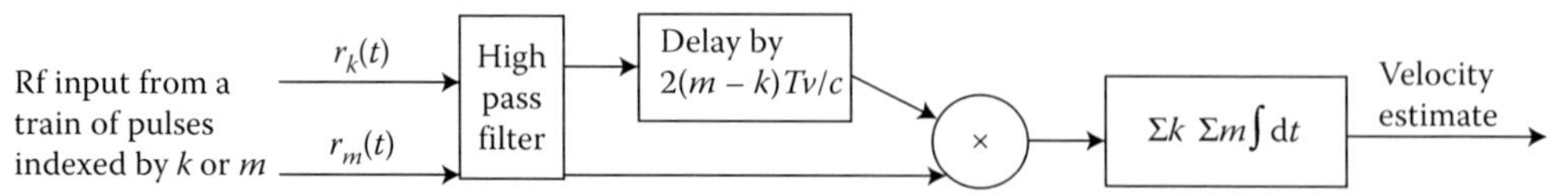

FIGURE 5.22 Block diagram of the system architecture required to estimate the velocity at each depth using a cross-correlation estimator.

has been proposed by Knapp and Carter (1976) known as the generalized correlation method for the estimation of time delay.

Several researchers have applied cross-correlation analysis to medical ultrasound. Bonnefous and Pesque (1986), Embree and O'Brien (1986), Foster et al. (1990), and Trahey et al. (1987) have studied the estimation of mean velocity based on the change in delay due to target movement. This analysis has assumed the shape of the transmitted signal to be unknown, and a cross-correlation technique has been used to estimate the difference in delay between successive pulses. This differential delay has then been used to estimate target velocity, where the velocity estimate is now based on the change in delay of the signal over an axial window, by maximizing the cross-correlation of the returned signal over all possible target velocities. Cross-correlation processing is typically performed on the radiofrequency (RF) signal, and a typical cross-correlation block diagram is shown in Figure 5.22. A high-pass filter is first applied to the signal from a fixed depth to remove the unwanted return from stationary tissue. One advantage of this strategy is that the variance is now inversely proportional to bandwidth of the transmitted signal rather than proportional.

5.3.2.4.2 Wideband Maximum Likelihood Estimator (WMLE)

Wideband maximum likelihood estimation is a baseband strategy with performance properties that are similar to cross-correlation. The estimate of the velocity of the blood cells is jointly based on the shift in the signal envelope and the shift in the carrier frequency of the returned signal. This estimator can be derived using a model for the signal that is expected to be reflected from the moving blood medium after the signal passes through intervening tissue. The processing of the signal can be interpreted as a filter matched to the expected signal. A diagram of the processing required for the wideband maximum likelihood estimator is shown in Figure 5.23 (Ferrara and Algazi, 1991). Assume that P pulses were transmitted. Required processing involves the delay of the signal from the $(P - k)$th pulse by an amount equal to $2v/ckT$, which corresponds to the movement of the cells between pulses for a specific v, followed by multiplication by a frequency which corresponds to the expected Doppler shift frequency of the baseband returned signal. The result of this multiplication is summed for all pulses, and the maximum likelihood velocity is then the velocity which produces the largest output from this estimator structure.

5.3.2.4.3 Estimation Using High-Time-Bandwidth Signals

Several researchers have also investigated the use of long wideband signals including "chirp" modulated signals and pseduo-random noise for the estimation of blood velocity. These signals are transmitted

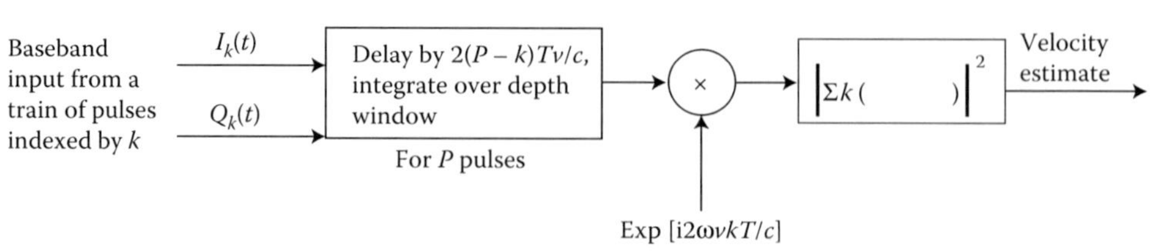

FIGURE 5.23 Block diagram of the system architecture required to estimate the velocity at each depth using the wideband MLE.

continuously (or with a short "flyback" time). Since these signals require continual transmission, the instantaneous power level must be reduced in order to achieve safe average power levels.

Bertram (1979) concluded that transmission of a "chirp" appears to give inferior precision for range measurement and inferior resolution of closely spaced multiple targets than a conventional pulse-echo system applied to a similar transducer. Multiple targets confuse the analysis. Using a simple sawtooth waveform, it is not possible to differentiate a stationary target at one range from a moving target at a different range. This problem could possibly be overcome with increasing and decreasing frequency intervals. Axial resolution is independent of the modulation rate, dependent only on the spectral frequency range (which is limited).

The limitations of systems that have transmitted a long pulse of random noise and correlated the return with the transmitted signal include reverberations from outside the sample volume which degrade the signal-to-noise ratio (the federally required reduction in peak transmitted power also reduces SNR), limited signal bandwidth due to frequency-dependent attenuation in tissue, and the finite transducer bandwidth (Cooper and McGillem, 1972; Bendick and Newhouse, 1974).

5.3.3 New Directions

Areas of research interest, including estimation of the 3D velocity magnitude, volume flow estimation, the use of high-frequency catheter-based transducers, mapping blood flow within malignant tumors, a new display mode known as color Doppler energy, and the use of contrast agents, are summarized in this subsection.

Estimation of the 3D velocity magnitude and beam vessel angle: Continued research designed to provide an estimate of the 3D magnitude of the flow velocity includes the use of crossed-beam Doppler systems (Wang and Yao, 1982; Overbeck et al., 1992) and tracking of speckle in two and three dimensions (Trahey et al., 1987). Mapping of the velocity estimate in two and three dimensions, resulting in a 3D color flow map has been described by Carson et al. (1992), Picot et al. (1993), and Cosgrove et al. (1990).

Volume flow estimation: Along with the peak velocity, instantaneous velocity profile, and velocity indices, a parameter of clinical interest is the volume of flow through vessels as a function of time. Estimation strategies for the determination of the volume of flow through a vessel have been described by Embree and O'Brien (1990), Gill (1979), Hottinger and Meindl (1979), and Uematsu (1981).

Intravascular ultrasound: It has been shown that intravascular ultrasonic imaging can provide information about the composition of healthy tissue and atheroma as well as anatomic data. A number of researchers have now shown that using frequencies of 30 MHz or above, individual layers and tissue types can be differentiated (de Kroon et al., 1991a,b; Lockwood et al., 1991). Although obvious changes in the vessel wall, such as dense fibrosis and calcification, have been identified with lower-frequency transducers, more subtle changes have been difficult to detect. Recent research has indicated that the character of plaque may be a more reliable predictor of subsequent cerebrovascular symptoms than the degree of vessel narrowing or the presence of ulceration (Merritt et al., 1992). Therefore, the recognition of subtle differences in tissue type may be extremely valuable. One signal-processing challenge in imaging the vascular wall at frequencies of 30 MHz or above is the removal of the unwanted echo from red blood cells, which is a strong interfering signal at high frequencies.

Vascular changes associated with tumors: Three-dimensional color flow mapping of the vascular structure is proposed to provide new information for the differentiation of benign and malignant masses. Judah Folkman and associates first recognized the importance of tumor vascularity in 1971 (Folkman et al., 1971). They hypothesized that the increased cell population required for the growth of a malignant tumor must be preceded by the production of new vessels. Subsequent work has shown that the walls of these vessels are deficient in muscular elements, and this deficiency results in a low impedance to flow (Gammill et al., 1976). This change can be detected by an increase in diastolic flow and a change in the resistive index.

More recently, Less et al. (1991) have shown that the vascular architecture of solid mammary tumors has several distinct differences from normal tissues, at least in the microvasculature. A type of network exists that exhibits fluctuations in both the diameter and length of the vessel with increasing branch order. Current color flow mapping systems with a center frequency of 5 MHz or above have been able to detect abnormal flow with varying degrees of clinical sensitivity from 40 to 82% (Belcaro et al., 1988; Balu-Maestro et al., 1991; Luska et al., 1992). Researchers using traditional Doppler systems have also reported a range of clinical sensitivity, with a general reporting of high sensitivity but moderate to low specificity. Burns et al., (1982) studied the signal from benign and malignant masses with 10-MHz CW Doppler. They hypothesized, and confirmed through angiography, that the tumors under study were fed by multiple small arteries, with a mean flow velocity below 10 cm/sec. Carson et al. (1992) compared 10-MHz CW Doppler to 5- and 7.5-MHz color flow mapping and concluded that while 3D reconstruction of the vasculature could provide significant additional information, color flow mapping systems must increase their ability to detect slow flow in small vessels in order to effectively map the vasculature.

Ultrasound contrast agents: The introduction of substances that enhance the ultrasonic echo signal from blood primarily through the production of microbubbles is of growing interest in ultrasonic flow measurement. The increased echo power may have a significant impact in contrast echocardiography, where acquisition of the signal from the coronary arteries has been difficult. In addition, such agents have been used to increase the backscattered signal from small vessels in masses that are suspected to be malignant. Contrast agents have been developed using sonicated albumen, saccharide microbubbles, and gelatin-encapsulated microbubbles.

Research to improve the sensitivity of flow measurement systems to low-velocity flow and small volumes of flow, with the goal of mapping the vasculature architecture, includes the use of ultrasonic contrast agents with conventional Doppler signal processing (Hartley et al., 1993), as well as the detection of the second harmonic of the transducer center frequency (Shrope and Newhouse, 1993).

Color Doppler energy: During 1993, a new format for the presentation of the returned signal from the blood scattering medium was introduced and termed color Doppler energy (CDE) or color power imaging (CPI). In this format, the backscattered signal is filtered to remove the signal from stationary tissue, and the remaining energy in the backscattered signal is color encoded and displayed as an overlay on the gray-scale image. The advantage of this signal-processing technique is the sensitivity to very low flow velocities.

Defining Terms

Baseband signal: The received signal after the center frequency component (carrier frequency) has been removed by demodulation.

Carrier frequency: The center frequency in the spectrum of the transmitted signal.

Clutter: An unwanted fixed signal component generated by stationary targets typically outside the region of interest (such as vessel walls).

Complex envelope: A signal expressed by the product of the carrier, a high-frequency component, and other lower-frequency components that comprise the envelope. The envelope is usually expressed in complex form.

Maximum likelihood: A statistical estimation technique that maximizes the probability of the occurrence of an event to estimate a parameter. ML estimate is the minimum variance, unbiased estimate.

References

Ahn Y. and Park S. 1991. Estimation of mean frequency and variance of ultrasonic Doppler signal by using second-order autoregressive model. *IEEE Trans. Ultrason. Ferroelec. Freq. Cont.* 38: 172.

Angelson B. 1980. Theoretical study of the scattering of ultrasound from blood. *IEEE Trans. Biomed. Eng.* 27: 61.

Atkinson P. and Berry M.V. 1974. Random noise in ultrasonic echoes diffracted by blood. *J. Phys. A: Math. Nucl. Gen.* 7: 1293.

Balu-Maestro C., Bruneton J.N., Giudicelli T. et al. 1991. Color Doppler in breast tumor pathology. *J. Radiol.* 72: 579.

Barber W., Eberhard J.W., and Karr S. 1985. A new time domain technique for velocity measurements using Doppler ultrasound. *IEEE Trans. Biomed. Eng.* 32: 213.

Belcaro G., Laurora G., Ricci A. et al. 1988. Evaluation of flow in nodular tumors of the breast by Doppler and duplex scanning. *Acta Chir. Belg.* 88: 323.

Bertram C.D. 1979. Distance resolution with the FM-CW ultrasonic echo-ranging system. *Ultrasound Med. Biol.* 61.

Bonnefous O. and Pesque P. 1986. Time domain formulation of pulse-Doppler ultrasound and blood velocity estimators by cross correlation. *Ultrason. Imag.* 8: 73.

Brody W. and Meindl J. 1974. Theoretical analysis of the CW Doppler ultrasonic flowmeter. *IEEE Trans. Biomed. Eng.* 21: 183.

Burns P.N., Halliwell M., Wells P.N.T., and Webb A.J. 1982. Ultrasonic Doppler studies of the breast. *Ultrasound Med. Biol.* 8: 127.

Carson P.L., Adler D.D., Fowlkes J.B. et al. 1992. Enhanced color flow imaging of breast cancer vasculature: Continuous wave Doppler and three-dimensional display. *J. Ultrasound Med.* 11: 77.

Cosgrove D.O., Bamber J.C., Davey J.B. et al. 1990. Color Doppler signals from breast tumors: Work in progress. *Radiology* 176: 175.

Curry G.R. and White D.N. 1978. Color coded ultrasonic differential velocity arterial scanner. *Ultrasound Med. Biol.* 4: 27.

de Kroon M.G.M., Slager C.J., Gussenhoven W.J. et al. 1991. Cyclic changes of blood echogenicity in high-frequency ultrasound. *Ultrasound Med. Biol.* 17: 723.

de Kroon M.G.M., van der Wal L.F., Gussenhoven W.J. et al. 1991. Backscatter directivity and integrated backscatter power of arterial tissue. *Int. J. Cardiac. Imag.* 6: 265.

Embree P.M. and O'Brien W.D. Jr. 1990. Volumetric blood flow via time-domain correlation: Experimental verification. *IEEE Trans. Ultrason. Ferroelec. Freq. Cont.* 37: 176.

Ferrara K.W and Algazi V.R. 1994a. A statistical analysis of the received signal from blood during laminar flow. *IEEE Trans. Ultrason. Ferroelec. Freq. Cont.* 41: 185.

Ferrara K.W. and Algazi V.R. 1994b. A theoretical and experimental analysis of the received signal from disturbed blood flow. *IEEE Trans. Ultrason. Ferroelec. Freq. Cont.* 41: 172.

Ferrara K.W. and Algazi V.R. 1991. A new wideband spread target maximum likelihood estimator for blood velocity estimation: I. Theory. *IEEE Trans. Ultrason. Ferroelec. Freq. Cont.* 38: 1.

Folkman J., Nerler E., Abernathy C., and Williams G. 1971. Isolation of a tumor factor responsible for angiogenesis. J. *Exp. Med.* 33: 275.

Foster S.G., Embree P.M., and O'Brien W.D. Jr. 1990. Flow velocity profile via time-domain correlation: Error analysis and computer simulation. *IEEE Trans. Ultrason. Ferroelec. Freq. Cont.* 37: 164.

Gammill S.L., Stapkey K.B., and Himmellarb E.H. 1976. Roenigenology—pathology correlative study of neovascularay. *AJR* 126: 376.

Gill R.W. 1979. Pulsed Doppler with B-mode imaging for quantitative blood flow measurement. *Ultrasound Med. Biol.* 5: 223.

Hartley C.J., Cheirif J., Collier K.R., and Bravenec J.S. 1993. Doppler quantification of echo-contrast injections *in vivo. Ultrasound Med. Biol.* 19: 269.

Hatle L. and Angelsen B. 1985. *Doppler Ultrasound in Cardiology*, 3rd ed. Philadelphia, Lea and Febiger.

Hatsukami T.S., Primozich J., Zierler R.E., and Strandness D.E. 1992. Color Doppler characteristics in normal lower extremity arteries. *Ultrasound Med. Biol.* 18(2): 167.

Hein I. and O'Brien W 1993. Current time domain methods for assessing tissue motion. *IEEE Trans. Ultrason. Ferroelec. Freq. Cont.* 40(2): 84.

Helstrom C.W. 1968. *Statistical Theory of Signal Detection*. London, Pergamon Press.

Hinze J.O. 1975. *Turbulence*. New York, McGraw-Hill.

Hottinger C.F. and Meindl J.D. 1979. Blood flow measurement using the attenuation compensated volume flowmeter. *Ultrason. Imag.* 1:1.

Kaluzinski K. 1989. Order selection in Doppler blood flow signal spectral analysis using autoregressive modelling. *Med. Biol. Eng. Com.* 27: 89.

Kasai C., Namekawa K., Koyano A., and Omoto R. 1985. Real-time two-dimensional blood flow imaging using an autocorrelation technique. *IEEE Trans. Sonics Ultrason.* 32.

Kay S. and Marple S.L. 1981. Spectrum analysis. A modern perspective. *Proc. IEEE* 69: 1380. Kennedy R.S. 1969. *Fading Dispersive Channel Theory*. New York, Wiley Interscience.

Knapp C.H. and Carter G.C. 1976. The generalized correlation method for estimation of time delay. *IEEE Trans. Acoust. Speech Signal Proc.* 24: 320.

Kristoffersen K. and Angelsen B.J. 1985. A comparison between mean frequency estimators for multigated Doppler systems with serial signal processing. *IEEE Trans. Biomed. Eng.* 32: 645.

Less J.R., Skalak T.C., Sevick E.M., and Jain R.K. 1991. Microvascular architecture in a mammary carcinoma: Branching patterns and vessel dimensions. *Cancer Res.* 51: 265.

Lockwood G.R., Ryan L.K., Hunt J.W., and Foster F.S. 1991. Measurement of the ultrasonic properties of vascular tissues and blood from 35-65 MHz. *Ultrasound Med. Biol.* 17: 653.

Loupas T. and McDicken W.N. 1990. Low-order AR models for mean and maximum frequency estimation in the context of Doppler color flow mapping. *IEEE Trans. Ultrason. Ferroelec. Freq. Cont.*37: 590.

Luska G., Lott D., Risch U., and von Boetticher H. 1992. The findings of color Doppler sonography in breast tumors. *Rofo Forts aufdem Gebiete derRontgens und derNeuen Bildg Verf 156:* 142.

Merritt C. and Bluth E. 1992. The future of carotid sonography. *AJR* 158: 37.

Mo L. and Cobbold R. 1986. A stochastic model of the backscattered Doppler ultrasound from blood. *IEEE Trans. Biomed. Eng.* 33: 20.

Nerem R.M. 1985. Fluid dynamic considerations in the application of ultrasound flowmetry. In S.A. Altobelli, W.F. Voyles, and E.R. Greene (Eds.), *Cardiovascular Ultrasonic Flowmetry*. New York, Elsevier.

Nichols W.W. and O'Rourke M.F. 1990. *McDonald's Blood Flow in Arteries: Theoretic, Experimental and Clinical principles*. Philadelphia, Lea and Febiger.

Nowicki A. and Reid J.M. 1981. An infinite gate pulse Doppler. *Ultrasound Med. Biol.* 7: 1.

Overbeck J.R., Beach K.W., and Strandness D.E. Jr. 1992. Vector Doppler: Accurate measurement of blood velocity in two dimensions. *Ultrasound Med. Biol.* 18: 19.

Picot P.A., Rickey D.W., and Mitchell R. et al. 1993. Three dimensional color Doppler mapping. *Ultrasound Med. Biol.* 19: 95.

Price R. 1968. Detectors for radar astronomy. In J. Evans and T. Hagfors (Eds.), *Radar Astronomy*. New York, McGraw-Hill.

Schrope B.A. and Newhouse V.L. 1993. Second harmonic ultrasound blood perfusion measurement. *Ultrasound Med. Biol.* 19: 567.

Shung K.K., Sigelman R.A., and Reid J.M. 1976. Scattering of ultrasound by blood. *IEEE Trans. Biomed. Eng.* 23: 460.

Shung K.K., Cloutier G., and Lim C.C. 1992. The effects of hematocrit, shear rate, and turbulence on ultrasonic Doppler spectrum from blood. *IEEE Trans. Biomed. Eng.* 39: 462.

Trahey G.E., Allison J.W., and Von Ramm O.T. 1987. Angle independent ultrasonic detection of blood flow. *IEEE Trans. Biomed. Eng.* 34: 964.

Uematsu S. 1981. Determination of volume of arterial blood flow by an ultrasonic device. *J. Clin. Ultrason.* 9: 209.

Van Trees H.L. 1971. *Detection, Estimation and Modulation Theory*, Part III. New York, Wiley.

Wang W. and Yao L. 1982. A double beam Doppler ultrasound method for quantitative blood flow velocity measurement. *Ultrasound Med. Biol.* 421.

Wells P.N.T. 1977. *Biomedical Ultrasonics*. London, Academic Press.

Further Information

The bimonthly journal *IEEE Transactions on Ultrasonics Ferroelectrics and Frequency* Control reports engineering advances in the area of ultrasonic flow measurement. For subscription information, contact IEEE Service Center, 445 Hoes Lane, P.O. Box 1331, Piscataway, NJ 08855-1331. Phone (800) 678-IEEE. The journal and the yearly conference proceedings of the IEEE Ultrasonic Symposium are published by the IEEE Ultrasonic Ferroelectrics and Frequency Control Society. Membership information can be obtained from the IEEE address above or from K. Ferrara, Riverside Research Institute, 330 West 42nd Street, New York, NY 10036.

The journal *Ultrasound in Medicine and Biology*, published 10 times per year, includes new developments in ultrasound signal processing and the clinical application of these developments. For subscription information, contact Pergamon Press, Inc., 660 White Plains Road, Tarrytown, NY 10591-5153. The American Institute of Ultrasound Medicine sponsors a yearly meeting which reviews new developments in ultrasound instrumentation and the clinical applications. For information, please contact American Institute of Ultrasound in Medicine, 11200 Rockville Pike, Suite 205, Rockville, MD 20852-3139; phone: (800) 638-5352.

6

Magnetic Resonance Microscopy

Xiaohong Zhou
Duke University Medical Center

G. Allan Johnson
Duke University Medical Center

Visualization of internal structures of opaque biologic objects is essential in many biomedical studies. Limited by the penetration depth of the probing sources (photons and electrons) and the lack of endogenous contrast, conventional forms of microscopy such as optical microscopy and electron microscopy require tissues to be sectioned into thin slices and stained with organic chemicals or heavy-metal compounds prior to examination. These invasive and destructive procedures, as well as the harmful radiation in the case of electron microscopy, make it difficult to obtain three-dimensional information and virtually impossible to study biologic tissues *in vivo*.

Magnetic resonance (MR) microscopy is a new form of microscopy that overcomes the aforementioned limitations. Operating in the radiofrequency (RF) range, MR microscopy allows biologic samples to be examined in the living state without bleaching or damage by ionizing radiation and in fresh and fixed specimens after minimal preparation. It also can use a number of endogenous contrast mechanisms that are directly related to tissue biochemistry, physiology, and pathology. Additionally, MR microscopy is digital and three-dimensional; internal structures of opaque tissues can be quantitatively mapped out in three dimensions to accurately reveal their histopathologic status. These unique properties provide new opportunities for biomedical scientists to attack problems that have been difficult to investigate using conventional techniques.

Conceptually, MR microscopy is an extension of magnetic resonance imaging (MRI) to the microscopic domain, generating images with spatial resolution better than 100 mm (Lauterbur, 1984). As such, MR microscopy is challenged by a new set of theoretical and technical problems (Johnson et al., 1992). For example, to improve isotropic resolution from 1 to 10 mm, signal-to-noise ratio (SNR) per voxel must be increased by a million times to maintain the same image quality. In order to do so, almost every component of hardware must be optimized to the fullest extent, pulse sequences have to be carefully designed to minimize any potential signal loss, and special software and dedicated computation facilities must be involved to handle large image arrays (e.g., 256^3). Over the past decade, development of MR microscopy has focused mainly on these issues. Persistent efforts by many researchers have recently lead to images with isotropic resolution of the order of ~10 μm (Cho et al., 1992; Johnson et al., 1992; Zhou and Lauterbur, 1992; Jacobs and Fraser, 1994). The significant resolution improvement opens up a broad range of applications, from histology to cancer biology and from toxicology to plant biology (Johnson et al., 1992). In this chapter we will first discuss the basic principles of MR microscopy, with special attention to such issues as resolution limits and sensitivity improvements. Then we will give an overview of the instrumentation. Finally, we will provide some examples to demonstrate the applications.

6.1 Basic Principles

6.1.1 Spatial Encoding and Decoding

Any digital imaging systems involve two processes. First, spatially resolved information must be encoded into a measurable signal, and second, the spatially encoded signal must be decoded to produce an image. In MR microscopy, the spatial encoding process is accomplished by acquiring nuclear magnetic resonance (NMR) signals under the influence of three orthogonal magnetic field gradients. There are many ways that a gradient can interact with a spin system. If the gradient is applied during a frequency-selective RF pulse, then the NMR signal arises only from a thin slab along the gradient direction. Thus a slice is selected from a three-dimensional (3D) object. If the gradient is applied during the acquisition of an NMR signal, the signal will consist of a range of spatially dependent frequencies given by

$$\omega(\vec{r}) = \gamma B_0 + \gamma \vec{G} \cdot \vec{r} \tag{6.1}$$

where γ is gyromagnetic ratio, B_0 is the static magnetic field, $\vec{G}$ is the magnetic field gradient, and $\vec{r}$ is the spatial variable. In this way, the spatial information along $\vec{G}$ direction is encoded into the signal as frequency variations. This method of encoding is called frequency encoding, and the gradient is referred to as a frequency-encoding gradient (or read-out gradient). If the gradient is applied for a fixed amount of time t_{pe} before the signal acquisition, then the phase of the signal, instead of the frequency, becomes spatially dependent, as given by

$$\phi(\vec{r}) = \int_0^{t_{pe}} \omega(\vec{r})\, dt = \phi_0 + \int_0^{t_{pe}} \gamma \vec{G} \cdot \vec{r}\, dt \tag{6.2}$$

where ϕ_0 is the phase originated from the static magnetic field. This encoding method is known as phase encoding, and the gradient is called a phase-encoding gradient.

Based on the three basic spatial encoding approaches, many imaging schemes can be synthesized. For two-dimensional (2D) imaging, a slice-selection gradient is first applied to confine the NMR signal in a slice. Spatial encoding within the slice is then accomplished by frequency encoding and by phase encoding. For 3D imaging, the slice-selection gradient is replaced by either a frequency-encoding or a phase-encoding gradient. If all spatial directions are frequency-encoded, the encoding scheme is called projection acquisition, and the corresponding decoding method is called projection reconstruction (Lauterbur, 1973; Lai and Lauterbur, 1981). If one of the spatial dimensions is frequency encoded while the rest are phase encoded, the method is known as Fourier imaging, and the image can be reconstructed simply by a multidimensional Fourier transform (Kumar et al., 1975; Edelstein et al., 1980). Although other methods do exist, projection reconstruction and Fourier imaging are the two most popular in MR microscopy.

Projection reconstruction is particularly useful for spin systems with short apparent T_2 values, such as protons in lung and liver. Since the T_2 of most tissues decreases as static magnetic field increases, the advantage of projection reconstruction is more obvious at high magnetic fields. Another advantage of projection reconstruction is its superior SNR to Fourier imaging. This advantage has been theoretically analyzed and experimentally demonstrated in a number of independent studies (Callaghan and Eccles, 1987; Zhou and Lauterbur, 1992; Gewalt et al., 1993). Recently, it also has been shown that projection reconstruction is less sensitive to motion and motion artifacts can be effectively reduced using sinograms (Glover and Pauly, 1992; Gmitro and Alexander, 1993; Glover and Noll, 1993). Unlike projection reconstruction, data acquisition in Fourier imaging generates Fourier coefficients of the image in a cartesian coordinate. Since multidimensional fast Fourier transform algorithms can be applied directly to the raw data, Fourier imaging is computationally more efficient than projection reconstruction. This advantage is most evident when reconstructing 3D images with large arrays (e.g., 2563). In addition, Fourier transform imaging is less prone to image artifacts arising from various off-resonance effects and is more robust in applications such as chemical shift imaging (Brown et al., 1982) and flow imaging (Moran, 1982).

6.1.2 Image Contrast

A variety of contrast mechanisms can be exploited in MR microscopy, including spin density (*r*), spin–spin relaxation time (T_1), spin-lattice relaxation time (T_2), apparent T_2 relaxation time (T_2), diffusion coefficient (*D*), flow and chemical shift (*d*). One of the contrasts can be highlighted by varying data-acquisition parameters or by choosing different pulse sequences. Table 6.1 summarizes the pulse sequences and data-acquisition parameters to obtain each of the preceding contrasts.

In high-field MR microscopy (>1.5*T*), T_2 and diffusion contrast are strongly coupled together. An increasing number of evidences indicate that the apparent T_2 contrast observed in high-field MR microscopy is largely due to microscopic magnetic susceptibility variations (Majumdar and Gore, 1988; Zhong and Gore, 1991). The magnetic susceptibility difference produces strong local magnetic field gradients. Molecular diffusion through the induced gradients causes significant signal loss. In addition, the large external magnetic field gradients required for spatial encoding further increase the diffusion-induced signal loss. Since the signal loss has similar dependence on echo time (TE) to T_2-related loss, the diffusion contrast mechanism is involved in virtually all T_2-weighted images. This unique contrast mechanism provides a direct means to probe the microscopic tissue heterogeneities and forms the basis for many histopathologic studies (Benveniste et al., 1992; Zhou et al., 1994).

Chemical shift is another unique contrast mechanism. Changes in chemical shift can directly reveal tissue metabolic and histopathologic stages. This mechanism exists in many spin systems such as ^1H, ^{31}P, and ^{13}C. Recently, Lean et al. (1993) showed that based on proton chemical shifts, MR microscopy can detect tissue pathologic changes with superior sensitivity to optical microscopy in a number of tumor models. A major limitation for chemical-shift MR microscopy is the rather poor spatial resolution, since most spin species other than water protons are of considerably low concentration and sensitivity.

TABLE 6.1 Choice of Acquisition Parameters for Different Image Contrasts

Contrast	TR[a]	TE[a]	Pulse Sequences[b]
r	3–5 $T_{1,max}$	$>T_{2,max}$	SE, GE
T_1	$-T_{1,avg}$	$>T_{2,min}$	SE, GE
T_2	3–5 $T_{1,max}$	$\sim T_{2,avg}$	SE, FSE
T_2^*	3–5 $T_{1,max}$	$\sim T^*_{s,avg}$	GE
D[c]	3–5 $T_{1,max}$	$>T_{2,min}$	Diffusion-weighted SE, GE, or FSE

[a] Subscripts min, max and avg stand for minimum, maximum and average values, respectively.

[b] SE: spin-echo; GE: gradient echo; FSE: fast spin echo.

[c] A pair of diffusion weighting gradients must be used.

In addition, the long data-acquisition time required to resolve both spatial and spectral information also appears as an obstacle.

6.2 Resolution Limits

6.2.1 Intrinsic Resolution Limit

Intrinsic resolution is defined as the width of the point-spread function originated from physics laws. In MR microscopy, the intrinsic resolution arises from two sources: natural linewidth broadening and diffusion (House, 1984; Callaghan and Eccles, 1988; Cho et al., 1988).

In most conventional pulse sequences, natural linewidth broadening affects the resolution limit only in the frequency-encoding direction. In some special cases, such as fast spin echo (Hennig et al., 1986) and echo planar imaging (Mansfield and Maudsley, 1977), natural linewidth broadening also imposes resolution limits in the phase-encoding direction (Zhou et al., 1993). The natural linewidth resolution limit, defined by

$$\Delta r_{\text{n.l.w.}} = \frac{2}{\gamma G T_2} \tag{6.3}$$

is determined by the T_2 relaxation time and can be improved using a stronger gradient G. To obtain 1 mm resolution from a specimen with $T_2 = 50$ msec, the gradient should be at least 14.9 G/cm. This gradient requirement is well within the range of most MR microscopes.

Molecular diffusion affects the spatial resolution in a number of ways. The bounded diffusion is responsible for many interesting phenomena known as edge enhancements (Hills et al., 1990; Hyslop and Lauterbur, 1991; Putz et al., 1991; Callaghan et al., 1993). They are observable only at the microscopic resolution and are potentially useful to detect microscopic boundaries. The unbounded diffusion, on the other hand, causes signal attenuation, line broadening, and phase misregistration. All these effects originate from an incoherent and irreversible phase-dispersion. The root-mean-square value of the phase dispersion is

$$\sigma = \gamma\left\{2D\int_0^t\left[\int_{t'}^{t} G(t'')\,dt''\right]^2 dt'\right\}^{1/2} \tag{6.4}$$

where t' and t are pulse-sequence-dependent time variables defined by Ahn and Cho (1989). Because of the phase uncertainty, an intrinsic resolution limit along the phase encoding direction arises:

$$\Delta r_{\text{pe}} = \frac{\sigma}{\gamma\int_0^t G_{\text{pe}}(t')\,dt'} \tag{6.5}$$

For a rectangularly shaped phase-encoding gradient, the preceding equation can be reduced to a very simple form:

$$\Delta r_{\text{pe}} = \sqrt{\frac{2}{3} D t_{\text{pe}}} \tag{6.6}$$

This simple result indicates that the diffusion resolution limit in the phase-encoded direction is determined only by the phase-encoding time t_{pe} (D is a constant for a chosen sample). This is so because the phase uncertainty is introduced only during the phase-encoding period. Once the spins are phase-encoded, they always carry the same spatial information no matter where they diffuse to. In the frequency-encoding direction, diffusion imposes resolution limits by broadening the point-spread function. Unlike natural linewidth broadening, broadening caused by diffusion is pulse-sequence-dependent. For the simplest pulse sequence, a 3D projection acquisition using free induction decays, the full width at half maximum (Callaghan and Eccles, 1988; McFarland, 1992; Zhou, 1992) is

$$\Delta r_{\text{fr}} = 8\left[\frac{D(\ln 2)^2}{3\gamma G_{\text{fr}}}\right]^{1/3} \tag{6.7}$$

Compared with the case of the natural linewidth broadening (Equation 6.3), the resolution limit caused by diffusion varies slowly with the frequency-encoding gradient G_{fr}. Therefore, to improve resolution by a same factor, a much larger gradient is required. With the currently achievable gradient strength, the diffusion resolution limit is estimated to be 5 to 10 μm.

6.2.2 Digital Resolution Limit

When the requirements imposed by intrinsic resolution limits are satisfied, image resolution is largely determined by the voxel size, provided that SNR is sufficient and the amplitude of physiologic motion is limited to a voxel. The voxel size, also known as digital resolution, can be calculated from the following equations:

Frequency-encoding direction:

$$\Delta x \equiv \frac{L_x}{N_x} = \frac{\Delta v}{2\pi\gamma G_x N_x} \tag{6.8}$$

Phase-encoding direction:

$$\Delta y \equiv \frac{L_y}{N_y} = \frac{\Delta\phi}{\gamma G_y t_{\text{pe}}} \tag{6.9}$$

where L is the field of view, N is the number of data points or the linear matrix size, G is the gradient strength, Δv is the receiver bandwidth, $\Delta\phi$ is the phase range of the phase-encoding data (e.g., if the data cover a phase range from $-\pi$ to $+\pi$, then $\Delta\phi = 2\pi$), and the subscripts x and y represent frequency- and phase-encoding directions, respectively. To obtain a high digital resolution, L should be kept minimal, while N maximal. In practice, the minimal field of view and the maximal data points are constrained by other experimental parameters. In the frequency-encoding direction, decreasing field of view results in an increase in gradient amplitude at a constant receiver bandwidth or a decrease in the bandwidth for a constant gradient (Equation 6.8). Since the receiver bandwidth must be large enough to keep the acquisition of NMR signals within a certain time window, the largest available gradient strength thus imposes the digital resolution limit. In the phase-encoding direction, $\Delta\phi$ is fixed at 2π in most experiments, and the maximum t_{pe} value is refrained by the echo time. Thus digital resolution is also determined by the maximum available gradient, as indicated by Equation 6.9. It has been estimated that in order to achieve 1 mm resolution with a phase-encoding time of 4 msec, the required gradient strength is as high as 587 G/cm. This gradient requirement is beyond the range of current MR microscopes. Fortunately, the requirement is fully relaxed in projection acquisition where no phase encoding is involved.

6.2.3 Practical Resolution Limit

The intrinsic resolution limits predict that MR microscopy can theoretically reach the micron regime. To realize the resolution, one must overcome several technical obstacles. These obstacles, or practical resolution limits, include insufficient SNR, long data-acquisition times, and physiologic motion. At the current stage of development, these practical limitations are considerably more important than other resolution limits discussed earlier and actually determine the true image resolution.

SNR is of paramount importance in MR microscopy. As resolution improves, the total number of spins per voxel decreases drastically, resulting in a cubic decrease in signal intensity. When the voxel signal intensity becomes comparable with noise level, structures become unresolvable even if the digital resolution and intrinsic resolution are adequate.

SNR in a voxel depends on many factors. The relationship between SNR and common experimental variables is given by

$$\mathrm{SNR} \propto \frac{B_1 B_0^2 \sqrt{n}}{\sqrt{4kT\Delta v(R_{\mathrm{coil}} + R_{\mathrm{sample}})}} \tag{6.10}$$

where B_1 is the RF magnetic field, B_0 is the static magnetic field, n is the number of average, T is the temperature, Δv is the bandwidth, k is the Boltzmann constant, and R_{coil} and R_{sample} are the coil and sample resistance, respectively. When small RF coils are used, R_{sample} is negligible. Since R_{coil} is proportional to due to skin effects, the overall SNR increases as $B_0^{7/4}$. This result strongly suggests that MR microscopy be performed at high magnetic field. Another way to improve SNR is to increase the B_1 field. This is accomplished by reducing the size of RF coils (Peck et al., 1990; McFarland and Mortara, 1992; Schoeniger et al., 1991; Zhou and Lauterbur, 1992). Although increasing B_0 and B_1 is the most common approach to attacking the SNR problem, other methods such as signal averaging, pulse-sequence optimization, and post data processing are also useful in MR microscopy. For example, diffusion-induced signal loss can be effectively minimized using diffusion-reduced-gradient (DRG) echo pulse sequences (Cho et al., 1992). Various forms of projection acquisition techniques (Hedges, 1984; McFarland and Mortara, 1992; Zhou and Lauterbur, 1992; Gewalt et al., 1993), as well as new k-space sampling schemes (Zhou et al., 1993), also have proved useful in SNR improvements. Recently, Black et al. (1993) used high-temperature superconducting materials for coil fabrication to simultaneously reduce coil resistance R_{coil} and coil temperature T. This novel approach can provide up to 70-fold SNR increase, equivalent to the SNR gain by increasing the magnetic field strength 11 times.

Long data-acquisition time is another practical limitation. Large image arrays, long repetition times (TR), and signal averaging all contribute to the overall acquisition time. For instance, a T_2-weighted image with a 256^3 image array requires a total acquisition time of more than 18 h (assuming TR = 500 msec and $n = 2$). Such a long acquisition time is unacceptable for most applications. To reduce the acquisition time while still maintaining the desired contrast, fast-imaging pulse sequences such as echo-planar imaging (EPI) (Mansfield and Maudsley, 1977), driven equilibrium Fourier transform (DEFT) (Maki et al., 1988), fast low angle shot (FLASH) (Haase et al., 1986), gradient refocused acquisition at steady state (GRASS) (Karis et al., 1987), and rapid acquisition with relaxation enhancement (RARE) (Hennig et al., 1986) have been developed and applied to MR microscopy. The RARE pulse sequence, or fast spin-echo (FSE) (Mulkern et al., 1990), is particularly useful in high-field MR microscopy because of its insensitivity to magnetic susceptibility effects as well as the reduced diffusion loss (Zhou et al., 1993). Using fast spin-echo techniques, a 256^3 image has been acquired in less than 2 h (Zhou et al., 1993).

For *in vivo* studies, the true image resolution is also limited by physiologic motion (Hedges, 1984; Wood and Henkelman, 1985). Techniques to minimize the motion effects have been largely focused on pulse sequences and post data processing algorithms, including navigator echoes (Ehman and Felmlee, 1989), motion compensation using even echo or moment nulling gradients, projection acquisition (Glover and

Noll, 1993), and various kinds of ghost-image decomposition techniques (Xiang and Henkelman, 1991). It should be noted, however, that by refining animal handling techniques and using synchronized data acquisition, physiologic motion effects can be effectively avoided and very high quality images can be obtained (Hedlund et al., 1986; Johnson et al., 1992).

6.3 Instrumentation

An MR microscope consists of a high-field magnet (>1.5 T), a set of gradient coils, an RF coil (or RF coils), and the associated RF systems, gradient power supplies, and computers. Among these components, RF coils and gradient coils are often customized for specific applications in order to achieve optimal performance. Some general guidelines to design customized RF coils and gradient coils are presented below.

6.3.1 Radiofrequency Coils

Many types of RF coils can be used in MR microscopy (Figure 6.1). The choice of a particular coil configuration is determined by specific task and specimen size. If possible, the smallest coil size should always be chosen in order to obtain the highest SNR. For *ex-vivo* studies of tissue specimens, solenoid coils are a common choice because of their superior B_1 field homogeneity, high sensitivity, as well as simplicity in fabrication. Using a 2.9 mm solenoid coil (5 turn), Zhou and Lauterbur (1992) have achieved the highest ever reported spatial resolution at 6.4 μm^3. Solenoid coil configurations are also used by others to obtain images with similar resolution (~10 μm) (Hedges, 1984; Cho et al., 1988; Schoeniger et al., 1991; McFarland and Mortara, 1992). Recently, several researchers began to develop microscopic solenoid coils with a size of a few hundred microns (Peck et al., 1990; McFarland and Mortara, 1992). Fabrication of these microcoils often requires special techniques, such as light lithography and electron-beam lithography.

(a) (b) (c)

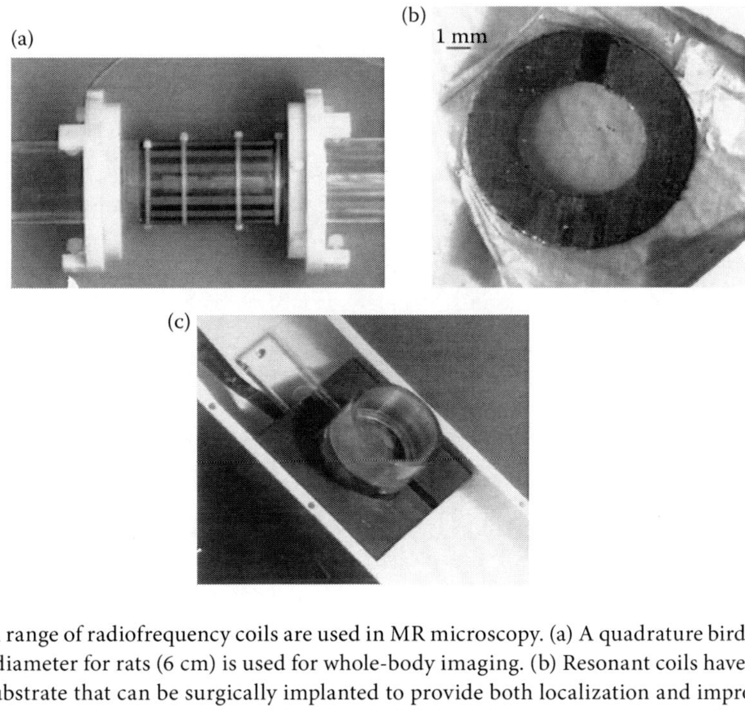

FIGURE 6.1 A range of radiofrequency coils are used in MR microscopy. (a) A quadrature birdcage coil scaled to the appropriate diameter for rats (6 cm) is used for whole-body imaging. (b) Resonant coils have been constructed on microwave substrate that can be surgically implanted to provide both localization and improved SNR. (c) MR microscopy of specimens is accomplished with a modified Helmholz coil providing good filling factors, high B homogeneity, and ease of access.

The direction of the B_1 field generated by a solenoid coil prevents the coil from being coaxially placed in the magnet. Thus accessing and positioning samples are difficult. To solve this problem, Banson et al. (1992) devised a unique Helmholtz coil that consists of two separate loops. Each loop is made from a microwave laminate with a dielectric material sandwiched between two copper foils. By making use of the distributed capacitance, the coil can be tuned to a desired frequency. Since the two loops of the coil are mechanically separated, samples can be easily slid into the gap between the loops without any obstruction. Under certain circumstances, the Helmholtz coil can outperform an optimally designed solenoid coil with similar dimensions.

For *in vivo* studies, although volume coils such as solenoid coils and birdcage coils can be employed, most high-resolution experiments are carried out using local RF coils, including surface coils (Rudin, 1987; Banson et al., 1992) and implanted coils (Hollett et al., 1987; Farmer et al., 1989; Zhou et al., 1994). Surface coils can effectively reduce coil size and simultaneously limit the field of view to a small region of interest. They can be easily adaptable to the shape of samples and provide high sensitivity in the surface region. The problem of inhomogeneous B_1 fields can be minimized using composite pulses (Hetherington et al., 1986) or adiabatic pulses (Ugurbil et al., 1987). To obtain high-resolution images from regions distant from the surface, surgically implantable coils become the method of choice. These coils not only give better SNR than optimized surface coils (Zhou et al., 1992) but also provide accurate and consistent localization. The latter advantage is particularly useful for time-course studies on dynamic processes such as the development of pathology and monitoring the effects of therapeutic drugs.

The recent advent of high-temperature superconducting (HTS) RF coils has brought new excitement to MR microscopy (Black et al., 1993). The substantial improvement, as discussed earlier, makes signal averaging unnecessary. Using these coils, the total imaging time will be solely determined by the efficiency to traverse the *k*-space. Although much research is yet to be done in this new area, combination of the HTS coils with fast-imaging algorithms will most likely provide a unique way to fully realize the potential of MR microscopy and eventually bring the technique into routine use.

6.3.2 Magnetic Field Gradient Coils

As discussed previously, high spatial resolution requires magnetic field gradients. The gradient strength increases proportional to the coil current and inversely proportional to the coil size. Since increasing current generates many undesirable effects (overheating, mechanical vibrations, eddy current, etc.), strong magnetic field gradient is almost exclusively achieved by reducing the coil diameter.

Design of magnetic field gradient coils for MR microscopy is a classic problem. Based on Maxwell equations, ideal surface current density can be calculated for a chosen geometry of the conducting surface. Two conducting surfaces are mostly used: a cylindrical surface parallel to the axis of the magnet and a cylindrical surface perpendicular to the axis. The ideal surface current density distributions for these two geometries are illustrated in Figure 6.2 (Suits and Wilken, 1989). After the ideal surface current density distribution is obtained, design of gradient coils is reduced to a problem of using discrete conductors with a finite length to approximate the continuous current distribution function. The error in the approximation determines the gradient linearity. Recent advancements in computer-based fabrication and etching techniques have made it feasible to produce complicated current density distributions. Using these techniques, nonlinear terms up to the eleventh order can be eliminated over a predefined cylindrical volume.

Another issue in gradient coil design involves minimizing the gradient rise time so that fast-imaging techniques can be implemented successfully and short echo times can be achieved to minimize signal loss for short T_2 specimens. The gradient rise time relies on three factors: the inductance over resistance ratio of the gradient coil, the time constant of the feedback circuit of the gradient power supply, and the decay rate of eddy current triggered by gradient switching. The time constant attributed to inductive resistance (L/R) is relatively short (<100 μsec) for most microscopy gradient coils, and the inductive resistance from the power supply can be easily adjusted to match the time constant of the coil. However, considering the high magnetic field gradient strength used in MR microscopy, eddy currents can be a serious problem. This problem is

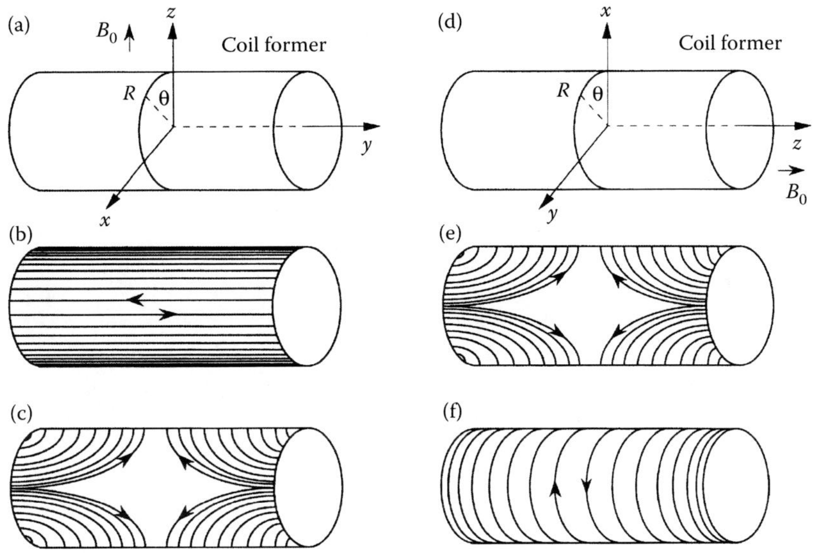

FIGURE 6.2 (a) Ideal surface current distributions to generate linear magnetic field gradients when the gradient coil former is parallel to the magnet bore: (b) for x gradient and (c) for z gradient. The analytical expressions for the current density distribution functions are $J_x = KG_x[R\cos\theta(\sin\theta i - \cos\theta j) - z\sin\theta k]$ and $J_z = KG_zZ[\sin\theta_i - \cos\theta_j]$. The y gradient can be obtained by rotating J_x 90°. (d) Ideal surface current distributions to generate linear magnetic field gradients when the gradient coil former is perpendicular to the magnet bore: (e) for x gradient and (f) for y gradient. The analytical expressions for the current density distribution functions are $J_x = KG_xR(\sin 2\theta_j)$ and $J_y = KG_y(-R\cos^2\theta_i + y\sin j + 0.5\,R\sin 2\theta_k)$. The z gradient can be obtained by rotating J_x 45°. (The graphs are adapted based on Suits, B.H. and Wilken, D.E. 1989. *J. Phys. E: Sci. Instrum.* 22: 565.)

even worsened when the gradient coils are closely placed in a narrow magnet bore. To minimize the eddy currents, modern design of gradient coils uses an extra set of coils so that the eddy currents can be actively cancelled (Mansfield and Chapman, 1986). Under close to optimal conditions, a rise time of <150 μsec can be achieved with a maximum gradient of 82 G/cm in a set of 8 cm coils (Johnson et al., 1992).

Although the majority of microscopic MR images are obtained using cylindrical gradient coils, surface gradient coils have been used recently in several studies (Cho et al., 1992). Similar to surface RF coils, surface gradient coils can be easily adapted to the shape of samples and are capable of producing strong magnetic field gradient in limited areas. The surface gradient coils also provide more free space in the magnet, allowing easy access to samples. A major problem with surface gradient coils is the gradient nonlinearity. But when the region of interest is small, high-quality images can still be obtained with negligible distortions.

6.4 Applications

MR microscopy has a broad range of applications. We include here several examples. Figure 6.3 illustrates an application of MR microscopy in *ex vivo* histology. In this study, a fixed sheep heart with experimentally-induced infarct is imaged at $2T$ using 3D fast-spin-echo techniques with T_1 (Figure 6.3a) and T_2 (Figure 6.3b) contrasts. The infarct region is clearly detected in both images. The region of infarct can be segmented from the rest of the tissue and its volume can be accurately measured. Since the image is three-dimensional, the tissue pathology can be examined in any arbitrary orientation. The nondestructive nature of MR microscopy also allows the same specimen to be restudied using other techniques as well as using other contrast mechanisms of MR microscopy. Obtaining this dimension of information from conventional histologic studies would be virtually impossible.

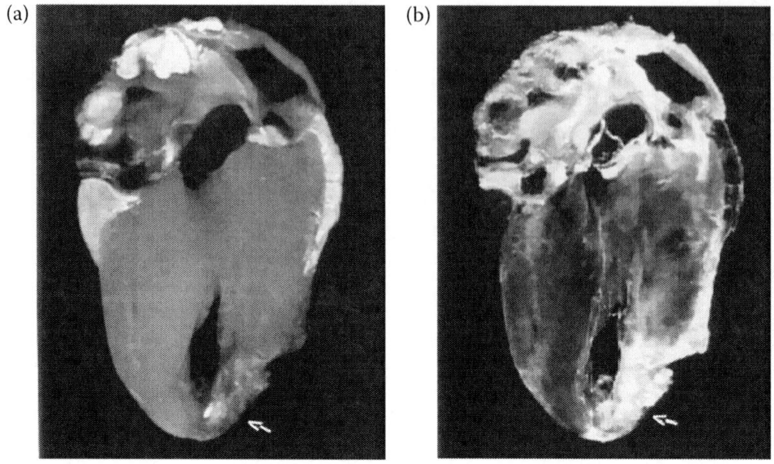

FIGURE 6.3 Selected sections of 3D isotropic images of a sheep heart with experimentally induced infarct show the utility of MRM in pathology studies. A 3D FSE sequence has been designed to allow rapid acquisition of either (a) T_1-weighted or (b) T_2-weighted images giving two separate "stains" for the pathology. Arrows indicate areas of necrosis.

The *in vivo* capability of MR microscopy for toxicologic studies is illustrated in Figure 6.4. In this study (Farmer et al., 1989), the effect of mercuric chloride in rat kidney is monitored in a single animal. Therefore, development of tissue pathology over a time period is directly observed without the unnecessary interference arising from interanimal variabilities. Since the kidney was the only region of interest in this study, a surgically implanted coil was chosen to optimize SNR and to obtain consistent localization. Figure 6.4 shows four images obtained from the same animal at different time points to show the

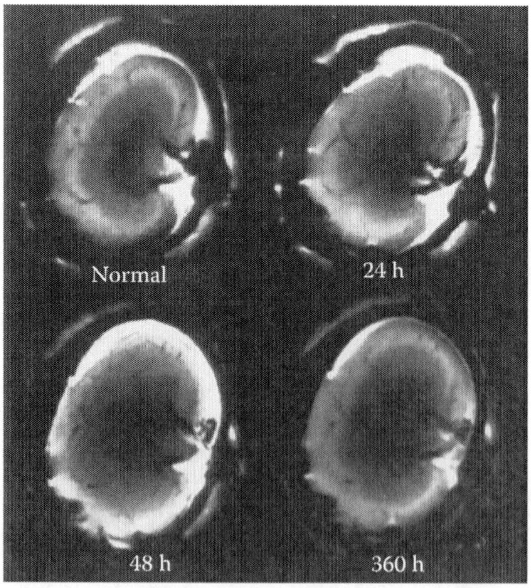

FIGURE 6.4 Implanted RF coils allow *in vivo* studies of deep structures with much higher spatial resolution by limiting the field of view during excitation and by increasing the SNR over volume coils. An added benefit is the ability to accurately localize the same region during a time course study. Shown here is the same region of a kidney at four different time points following exposure to mercuric chloride. Note the description of boundaries between the several zones of the kidney in the early part of the study followed by regeneration of the boundaries upon repair.

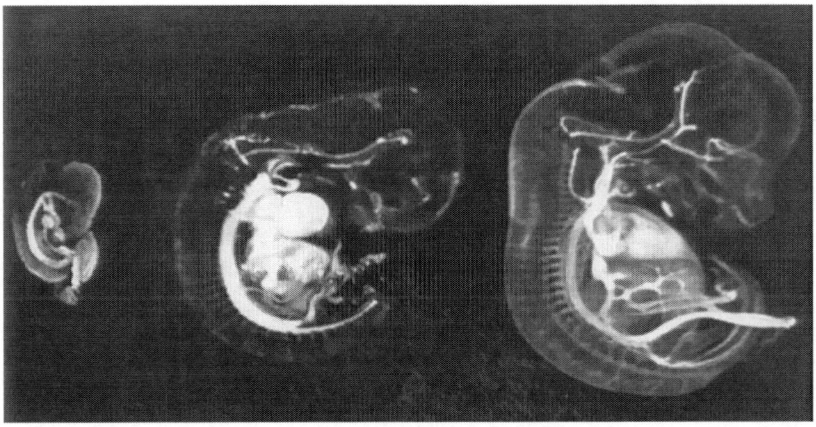

FIGURE 6.5 Isotropic 3D images of fixed mouse embryos at three stages of development have been volume rendered to allow visualization of the developing vascular anatomy.

progression and regression of the $HgCl_2$-induced renal pathology. Tissue damage is first observed 24 h after the animal was treated by the chemical, as evident by the blurring between the cortex and the outer medulla. The degree of damage is greater in the image obtained at 48 h. Finally, at 360 h, the blurred boundary between the two tissue regions completely disappeared, indicating full recovery of the organ. The capability to monitor tissue pathologic changes *in vivo*, as illustrated in this example, bodes well for a broad range of applications in pharmacology, toxicology, and pathology.

Development biology is another area where MR microscopy has found an increasing number of applications, Jacobs and Fraser (1994) used MR microscopy to follow cell movements and lineages in developing frog embryos. In their study, 3D images of the developing embryo were obtained on a time scale faster than the cell division time and analyzed forward and backward in time to reconstruct full cell divisions and cell movements. By labeling a 16-cell embryo with an exogenous contrast agent (Gd-DTPA), they successfully followed the progression from early cleavage and blastula stage through gastrulation, neurulation, and finally to tail bud stage. More important, they found that external ectodermal and internal mesodermal tissues extend at different rates during amphibian gastrulation and neurulation. This and many other key events in vertebrate embryogenesis would be very difficult to observe with optical microscopy. Another example in developmental biology is given in Figure 6.5. Using 3D high-field (9.4-*T*) MR microscopy with large image arrays (256^3), Smith et al. (1993) studied the early development of the circulatory system of mouse embryos. With the aid of a T_1 contrast agent made from bovine serum albumin and Gd-DTPA, vasculature such as ventricles, atria, aorta, cardinal sinuses, basilar arteries, and thoracic arteries are clearly identified in mouse embryos at between 9.5 and 12.5 days of gestation. The ability to study embryonic development in a noninvasive fashion provided great opportunities to explore many problems in transgenic studies, gene targeting, and *in situ* hybridization.

In less than 10 yr, MR microscopy has grown from a scientific curiosity to a tool with a wide range of applications. Although many theoretical and experimental problems still exist at the present time, there is no doubt that MR microscopy will soon make a significant impact in many areas of basic research and clinical diagnosis.

References

Ahn, C.B. and Cho, Z.H. 1989. A generalized formulation of diffusion effects in mm resolution nuclear magnetic resonance imaging. *Med. Phys.* 16: 22.

Banson, M.B., Cofer, G.P., Black, R.D., and Johnson, G.A. 1992a. A probe for specimen magnetic resonance microscopy. *Invest. Radiol.* 27: 157.

Banson, M.L., Cofer, G.P., Hedlund, L.W., and Johnson, G.A. 1992b. Surface coil imaging of rat spine at 7.0 T. *Magn. Reson. Imag.* 10: 929.

Benveniste, H., Hedlund, L.W., and Johnson, G.A. 1992. Mechanism of detection of acute cerebral ischemia in rats by diffusion-weighted magnetic resonance microscopy. *Stroke* 23: 746.

Black, R.D., Early, T.A., Roemer, P.B. et al. 1993. A high-temperature superconducting receiver for NMR microscopy. *Science* 259: 793.

Brown, T.R., Kincaid, B.M., and Ugurbil, K. 1982. NMR chemical shift imaging in three dimensions. *Proc. Natl Acad. Sci. USA* 79: 3523.

Callaghan, P.T. and Eccles, C.D. 1987. Sensitivity and resolution in NMR imaging. *J. Magn. Reson.* 71: 426.

Callaghan, P.T. and Eccles, C.D. 1988. Diffusion-limited resolution in nuclear magnetic resonance microscopy. *J. Magn. Reson.* 78: 1.

Callaghan, P.T., Coy, A., Forde, L.C., and Rofe, C.J. 1993. Diffusive relaxation and edge enhancement in NMR microscopy. *J. Magn. Reson. Ser. A* 101: 347.

Cho, Z.H., Ahn, C.B., Juh, S.C. et al. 1988. Nuclear magnetic resonance microscopy with 4 mm resolution: theoretical study and experimental results. *Med. Phys.* 15: 815.

Cho, Z.H., Yi, J.H., and Friedenberg, R.M. 1992. NMR microscopy and ultra-high resolution NMR imaging. *Rev. Magn. Reson. Med.* 4: 221.

Edelstein, W.A., Hutchison, J.M.S., Johnson, G., and Redpath, T. 1980. Spin warp NMR imaging and applications to human whole-body imaging. *Phys. Med. Biol.* 25: 751.

Ehman, R.L. and Felmlee, J.P. 1989. Adaptive technique for high-definition MR imaging of moving structures. *Radiology* 173: 255.

Farmer, T.H.R., Johnson, G.A., Cofer, G.P. et al. 1989. Implanted coil MR microscopy of renal pathology. *Magn. Reson. Med.* 10: 310.

Gewalt, S.L., Glover, G.H., MacFall, J.R. et al. 1993. MR microscopy of the rat lung using projection reconstruction. *Magn. Reson. Med.* 29: 99.

Glover, G.H. and Noll, D.C. 1993. Consistent projection reconstruction techniques for MRI. *Magn. Reson. Med.* 29: 345.

Glover, G.H. and Pauly, J.M. 1992. Projection reconstruction techniques for suppression of motion artifacts. *Magn. Reson. Med.* 28: 275.

Gmitro, A. and Alexander, A.L. 1993. Use of a projection reconstruction method to decrease motion sensitivity in diffusion-weighted MRI. *Magn. Reson. Med.* 29: 835.

Haase, A., Frahm, J., Matthaei, D. et al. 1986. FLASH imaging: rapid NMR imaging using low flip angle pulses. *J. Magn. Reson.* 67: 258.

Hedges, H.K. 1984. Nuclear magnetic resonance microscopy. Ph.D. dissertation, State University of New York at Stony Brook.

Hedlund, L.W., Dietz, J., Nassar, R. et al. 1986. A ventilator for magnetic resonance imaging. *Invest. Radiol.* 21: 18.

Hennig, J., Nauerth, A., and Friedburg, H. 1986. RARE imaging: a fast imaging method for clinical MR. *Magn. Reson. Med.* 3: 823.

Hetherington, H.P., Wishart, D., Fitzpatrick, S.M. et al. 1986. The application of composite pulses to surface coil NMR. *J. Magn. Reson.* 66: 313.

Hills, B.P., Wright, K.M., and Belton, P.S. 1990. The effects of restricted diffusion in nuclear magnetic resonance microscopy. *Magn. Reson. Imag.* 8: 755.

Hollett, M.D., Cofer, G.P., and Johnson, G.A. 1987. *In situ* magnetic resonance microscopy. *Invest. Radiol.* 22: 965.

House, W.V. 1984. NMR microscopy. *IEEE Trans. Nucl. Sci.* NS-31: 570.

Hyslop, W.B. and Lauterbur, P.C. 1991. Effects of restricted diffusion on microscopic NMR imaging. *J. Magn. Reson.* 94: 501.

Jacobs, R.E. and Fraser, S.E. 1994. Magnetic resonance microscopy of embryonic cell lineages and movements. *Science* 263: 681.

Johnson, G.A., Hedlund, L.W., Cofer, G.P., and Suddarth, S.A. 1992. Magnetic resonance microscopy in the life sciences. *Rev. Magn. Reson. Med.* 4: 187.

Karis, J.P., Johnson, G.A., and Glover, G.H. 1987. Signal to noise improvements in three dimensional NMR microscopy using limited angle excitation. *J. Magn. Reson.* 71: 24.

Kumar, A., Welti, D., and Ernst, R.R. 1975. NMR Fourier zeugmatography. *J. Magn. Reson.* 18: 69.

Lai, C.M. and Lauterbur, P.C. 1981. True three-dimensional image reconstruction by nuclear magnetic resonance zeugmatography. *Phys. Med. Biol.* 26: 851.

Lauterbur, P.C. 1973. Image formation by induced local interactions: examples employing nuclear magnetic resonance. *Nature* 242: 190.

Lauterbur, P.C. 1984. New direction in NMR imaging. *IEEE Trans. Nucl. Sci.* NS-31: 1010.

Lean, C.L., Russell, P., Delbridge, L. et al. 1993. Metastatic follicular thyroid diagnosed by 1H MRS. *Proc. Soc. Magn. Reson. Med.* 1: 71.

Majumdar, S. and Gore, J.C. 1988. Studies of diffusion in random fields produced by variations in susceptibility. *J. Magn. Reson.* 78: 41.

Maki, J.H., Johnson, G.A., Cofer, G.P., and MacFall, J.R. 1988. SNR improvement in NMR microscopy using DEFT. *J. Magn. Reson.* 80: 482.

Mansfield, P. and Chapman, B. 1986. Active magnetic screening of gradient coils in NMR imaging. *J. Magn. Reson.* 66: 573.

Mansfield, P. and Maudsley, A.A. 1977. Planar spin imaging by NMR. *J. Magn. Reson.* 27: 129.

McFarland, E.W. 1992. Time independent point-spread function for MR microscopy. *Magn. Reson. Imag.* 10: 269.

McFarland, E.W. and Mortara, A. 1992. Three-dimensional NMR microscopy: improving SNR with temperature and microcoils. *Magn. Reson. Imag.* 10: 279.

Moran, P.R. 1982. A flow velocity zeugmatographic interlace for NMR imaging in humans. *Magn. Reson. Imag.* 1: 197.

Mulkern, R.V., Wong, S.T.S., Winalski, C., and Jolesz, F.A. 1990. Contrast manipulation and artifact assessment of 2D and 3D RARE sequences. *Magn. Reson. Imag.* 8: 557.

Peck, T.L., Magin, R.L., and Lauterbur, P.C. 1990. Microdomain magnetic resonance imaging. *Proc. Soc. Magn. Reson. Med.* 1: 207.

Putz, B., Barsky, D., and Schulten, K. 1991. Edge enhancement by diffusion: microscopic magnetic resonance imaging of an ultrathin glass capillary. *Chem. Phys.* 183: 391.

Rudin, M. 1987. MR microscopy on rats *in vivo* at 4.7 T using surface coils. *Magn. Reson. Med.* 5: 443.

Schoeniger, J.S., Aiken, N.R., and Blackband, S.J. 1991. NMR microscopy of single neurons. *Proc. Soc. Magn. Reson. Med.* 2: 880.

Smith, B.R., Johnson, G.A., Groman, E.V., and Linney, E. 1993. Contrast enhancement of normal and abnormal mouse embryo vasculature. *Proc. Soc. Magn. Reson. Med.* 1: 303.

Suits, B.H. and Wilken, D.E. 1989. Improving magnetic field gradient coils for NMR imaging. *J. Phys. E: Sci. Instrum.* 22: 565.

Ugurbil, K., Garwood, M., and Bendall, R. 1987. Amplitude- and frequency-modulated pulses to achieve 90° plane rotation with inhomogeneous B_1 fields. *J. Magn. Reson.* 72: 177.

Wood, M.L. and Henkelman, R.M. 1985. NMR image artifacts from periodic motion. *Med. Phys.* 12: 143.

Xiang, Q.-S. and Henkelman, R.M. 1991. Motion artifact reduction with three-point ghost phase cancellation. *J. Magn. Reson. Imag.* 1: 633.

Zhong, J. and Gore, J.C. 1991. Studies of restricted diffusion in heterogeneous media containing variations in susceptibility. *Magn. Reson. Med.* 19: 276.

Zhou, X. and Lauterbur, P.C. 1992. NMR microscopy using projection reconstruction. In Blümich, B. and Kuhn, W. (eds.), *Magnetic Resonance Microscopy*, pp. 1–27. Weinheim, Germany, VCH.

Zhou, X., Cofer, G.P., Mills, G.I., and Johnson, G.A. 1992. An inductively coupled probe for MR microscopy at 7 T. *Proc. Soc. Magn. Reson. Med.* 1: 971.

Zhou, X., Cofer, G.P., Suddarth, S.A., and Johnson, G.A. 1993. High-field MR microscopy using fast spin-echoes. *Magn. Reson. Med.* 31: 60.

Zhou, X., Liang, Z.-P., Cofer, G.P. et al. 1993. An FSE pulse sequence with circular sampling for MR microscopy. *Proc. Soc. Magn. Reson. Med.* 1: 297.

Zhou, X., Maronpot, R.R., Mills, G.I. et al. 1994. Studies on bromobenzene-induced hepatotoxicity using *in vivo* MR microscopy. *Magn. Reson. Med.* 31: 619.

Zhou, Z. 1992. Nuclear magnetic resonance microscopy: new theoretical and technical developments. Ph.D. dissertation, University of Illinois at Urbana-Champaign.

Further Information

A detailed description of the physics of NMR and MRI can be found in Principles of Magnetic Resonance, by C.S., Slichter (3rd edition, Springer-Verlag, 1989), in NMR Imaging in Biology and Medicine, by P. Morris (Clarendon Press, 1986), and in Principles of Magnetic Resonance Microscopy, by P.T., Callaghan (Oxford Press, 1991). The latter two books also contain detailed discussions on instrumentation, data acquisition, and image reconstruction for conventional and microscopic magnetic resonance imaging.

Magnetic Resonance Microscopy, edited by Blümich and Kuhn (VCH, 1992), is particularly helpful to understand various aspects of MR microscopy, both methodology and applications. Each chapter of the book covers a specific topic and is written by experts in the field.

Proceedings of the Society of Magnetic Resonance (formerly Society of Magnetic Resonance in Medicine, Berkeley, California), published annually, documents the most recent developments in the field of MR microscopy. Magnetic Resonance in Medicine, Journal of Magnetic Resonance Imaging, and Magnetic Resonance Imaging, all monthly journals, contain original research articles and are good sources for up-to-date developments.

7 Positron-Emission Tomography

Thomas F. Budinger
University of California, Berkeley

Henry F. VanBrocklin
University of California, Berkeley

7.1 Radiopharmaceuticals

Thomas F. Budinger and Henry F. VanBrocklin

Since the discovery of artificial radioactivity a half century ago, radiotracers, radionuclides, and radionuclide compounds have played a vital role in biology and medicine. Common to all is radionuclide (radioactive isotope) production. This section describes the basic ideas involved in radionuclide production and gives examples of the applications of radionuclides. The field of radiopharmaceutical chemistry has fallen into subspecialties of positron-emission tomography (PET) chemistry and general radiopharmaceutical chemistry, including specialists in technetium chemistry, taking advantage of the imaging attributes of technetium-99m.

The two general methods of radionuclide production are neutron addition (activation) from neutron reactors to make neutron-rich radionuclides which decay to give off electrons and gamma rays and charged-particle accelerators (linacs and cyclotrons) which usually produce neutron-deficient isotopes that decay by electron capture and emission of x-rays, gamma rays, and positrons. The production of artificial radionuclides is governed by the number of neutrons or charged particles hitting an appropriate target per time, the cross section for the particular reaction, the number of atoms in the target, and the half-life of the artificial radionuclide:

$$A(t) = \frac{N\sigma\phi}{3.7 \times 10^{10}}\left(1 - e^{0.693t/T_{1/2}}\right) \tag{7.1}$$

where $A(t)$ is the produced activity in number of atoms per second, N is the number of target nuclei, σ is the cross section (probability that the neutron or charged particles will interact with the nucleus to form the artificial radioisotope) for the reaction, ϕ is the flux of charged particles, and $T_{1/2}$ is the half-life of the product. Note that N is the target mass divided by the atomic weight and multiplied by Avogadro's number (6.024×10^{23}) and σ is measured in cm^2. The usual flux is about 10^{14} neutrons per sec or, for charged particles, 10 to 100 mA, which is equivalent to 6.25×10^{13} to 6.25×10^{14} charged particles per second.

7.1.1 Nuclear Reactor-Produced Radionuclides

Thermal neutrons of the order of 10^{14} neutrons/sec/cm^2 are produced in a nuclear reactor usually during a controlled nuclear fission of uranium, though thorium or plutonium are also used. High specific activity neutron-rich radionuclides are produced usually through the (n, g), (n, p), or (n, a) reactions (Figure 7.1a). The product nuclides usually decay by b—followed by g. Most of the reactor-produced radionuclides are produced by the (n, g) reaction. The final step in the production of a radionuclide consists of the separation of the product nuclide from the target container by chemical or physical means.

An alternative method for producing isotopes from a reactor is to separate the fission fragments from the spent fuel rods. This is the leading source of ^{99}Mo for medical applications. The following two methods of ^{99}Mo production are examples of carrier-added and no-carrier-added radionuclide synthesis, respectively. In Figure 7.1a, the ^{99}Mo is production from ^{98}Mo. Only a small fraction of the ^{98}Mo nuclei will be converted to ^{99}Mo. Therefore, at the end of neutron bombardment, there is a mixture of both isotopes. These are inseparable by conventional chemical separation techniques, and both isotopes would participate equally well in chemical reactions. The ^{99}Mo from fission of ^{238}U would not contain any other isotopes of Mo and is considered carrier-free. Thus radioisotopes produced by any means having the same atomic number as the target material would be considered carrier-added. Medical tracer techniques obviate the need for carrier-free isotopes.

7.1.2 Accelerator-Produced Radionuclides

Cyclotrons and linear accelerators (linacs) are sources of beams of protons, deuterons, or helium ions that bombarded targets to produce neutron-deficient (proton-rich) radionuclides (Figure 7.1b). The neutron-deficient radionuclides produced through these reactions are shown in Table 7.1. These product nuclides (usually carrier-free) decay either by electron capture or by positron emission tomography or both, followed by g emission. In Table 7.2, most of the useful charged-particle reactions are listed.

The heat produced by the beam current on the target material can interfere with isotope production and requires efficient heat-removal strategies using extremely stable heat-conducting target materials such as metal foils, electroplates metals, metal powders, metal oxides, and salts melted on duralmin plate. All the modern targets use circulating cold deionized water and chilled helium gas to aid in

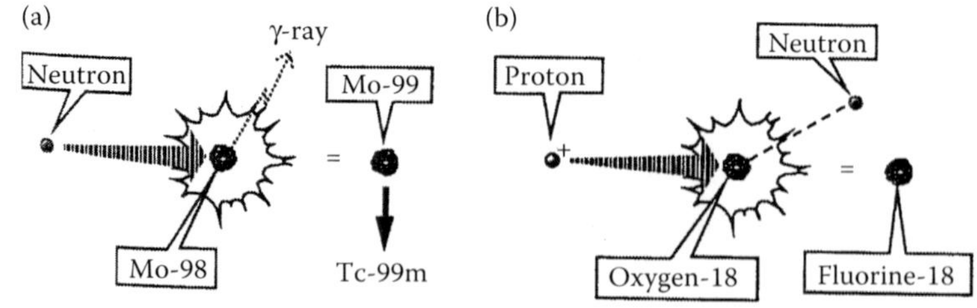

FIGURE 7.1 (a) High specific activity neutron-excess radionuclides are produced usually through the (n, g), (n, p) or (n, a) reactions. The product nuclides usually decay b—followed by g. Most of the reactor produced radionuclides are produced by the (n, g) reaction. (b) Cyclotrons and linear accelerators (linacs) are sources of beams of protons, deuterons, or helium ions which bombard targets to produce neutron-deficient radionuclides.

TABLE 7.1 Radionuclides Used in Biomedicine

Radionuclide	Half-Life	Application(s)
Arsenic-74[a]	17.9 d	A positron emitting chemical analog of phosphorus
Barium-128[a]	2.4 d	Parent in the generator system for producing the positron emitting ^{128}Cs, a potassium analog
Berylium-7[a]	53.37 d	Berylliosis studies
Bromine-77	57 h	Radioimmunotherapy
Bromin-82	35.3 h	Used in metabolic studies and studies of estrogen receptor content
Carbon-11[a]	20.3 min	Positron emitter for metabolism imaging
Cobalt-57[a]	270 d	Calibration of imaging instruments
Copper-62	9.8 min	Heart perfusion
Copper-64	12.8 h	Used as a clinical diagnostic agent for cancer and metabolic disorders
Copper-67	58.5 h	Radioimmunotherapy
Chromium-51	27.8 d	Used to assess red blood cell survival
Fluorine-18	109.7 min	Positron emitter used in glucose analogs uptake and neuroreceptor imaging
Gallium-68	68 min	Required in calibrating PET tomographs. Potential antibody level
Germanium-68[a]	287 d	Parent in the generator system for producing the positron emitting ^{68}Ga
Indium-111[a]	2.8 d	Radioimmunotherapy
Iodine-122	3.76 min	Positron emitter for blood flow studies
Iodine-123[a]	13.3 h	SPECT brain imaging agent
Iodine-124[a]	4.2 d	Radioimmunotherapy, positron emitter
Iodine-125	60.2 d	Used as a potential cancer therapeutic agent
Iodine-131	8.1 d	Used to diagnose and treat thyroid disorders including cancer
Iron-52[a]	8.2 h	Used as an iron tracer, positron emitter for bone-marrow imaging
Magnesium-28[a]	21.2 h	Magnesium tracer which decays to 2.3 in aluminum-28
Magnese-52m	5.6 d	Flow tracer for heart muscle
Mercury-195m[a]	40 h	Parent in the generator system for producing ^{195}mAu, which is used in cardiac blood pool studies
Molybdenum-99	67 h	Used to produce technetium-99m, the most commonly used radioisotope in clinical nuclear medicine
Nitrogen-13[a]	9.9 min	Positron emitter used as ^{13}NH for heart perfusion studies
Osmium-191	15 d	Decays to iridium-191 used for cardiac studies
Oxygen-15[a]	123 s	Positron emitter used for blood flow studies as H152O
Palladium-103	17 d	Used in the treatment of prostate cancer
Phosphorus-32	14.3 d	Used in cancer treatment, cell metabolism and kinetics, molecular biology, genetics research, biochemistry, microbiology, enzymology, and as a starter to make many basic chemicals and research products
Rhenium-188	17 h	Used for treatment of medullary thyroid carcinoma and alleviation of pain in bone metastases
Rubidium-82[a]	1.2 min	Positron emitter used for heart perfusion studies
Ruthemiun-97[a]	2.9 d	Hepatobiliary function, tumor and inflammation localization
Samarium-145	340 d	Treatment of ocular cancer
Samarium-153	46.8 h	Used to radiolabel various molecules as cancer therapeutic agents and to alleviate bone cancer pain
Scandium-47	3.4 d	Radioimmunotherapy
Scandium-47[a]	3.4 d	Used in the therapy of cancer

continued

TABLE 7.1 (continued) Radionuclides Used in Biomedicine

Radionuclide	Half-Life	Application(s)
Strontium-82[a]	64.0 d	Parent in the generator system for producing the positron emitting ^{82}Rb, a potassium analogue
Strontium-85	64 d	Used to study bone formation metabolism
Strontium-89	52 d	Used to alleviate metastatic bone pain
Sulfur-35	87.9 d	Used in studies of cell metabolism and kinetics, molecular biology, genetics research, biochemistry, microbiology, enzymology, and as a start to make many basic chemicals and research products
Technetium-99m	6 h	The most widely used radiopharmaceutical in nuclear medicine and produced from molybdenum-99
Thalium-201[a]	74 h	Cardiac imaging agent
Tin-117m	14.0 d	Palliative treatment of bone cancer pain
Tritium (hydrogen-3)	12.3 yr	Used to make tritiated water which is used as a starter for thousands of different research products and basic chemicals; used for life science and drug metabolism studies to ensure the safety of potential new drugs
Tungsten-178[a]	21.5 d	Parent in generator system for producing ^{178}Ta, short-lived scanning agent
Tungsten-188	69 d	Decays to rhenium-188 for treatment of cancer and rheumatoid arthritis
Vanadium-48[a]	16.0 d	Nutrition and environmental studies
Xenon-122[a]	20 h	Parent in the generator system for producing the positron emitting ^{122}I
Xenon-127[a]	36.4 d	Used in lung ventilation studies
Xenon-133	5.3 d	Used in lung ventilation and perfusion studies
Yttrium-88[a]	106.6 d	Radioimmunotherapy
Yttrium-90	64 h	Used to radiolabel various molecules as cancer therapeutic agents
Zinc-62[a]	9.13 h	Parent in the generator system for producing the positron emitting ^{62}Cu
Zirconium-89[a]	78.4 h	Radioimmunotherapy, positron emitter

[a] Produced by accelerated charged particles. Others are produced by neutron reactors.

cooling the target body and window foils. Cyclotrons used in medical studies have ^{11}C, ^{13}N, ^{15}O, and ^{18}F production capabilities that deliver the product nuclides on demand through computer-executed commands. The radionuclide is remotely transferred into a lead-shielded hot cell for processing. The resulting radionuclides are manipulated using microscale radiochemical techniques: small-scale synthetic methodology, ion-exchange chromatography, solvent extraction, electrochemical synthesis, distillation, simple filtration, paper chromatography, and isotopic carrier precipitation. Various relevant radiochemical techniques have been published in standard texts.

TABLE 7.2 Important Reactions for Cyclotron-Produced Radioisotopes

1. p, n	7. p, pn
2. $p, 2n$	8. $p, 2p$
3. d, n	9. d, p
4. $d, 2n$	10. $d, ^{4}$He
5. $d, ^{4}$He	11. p, d
6. $d, ^{4}$Hen	12. ^{4}He, n
	13. ^{3}He, p

7.1.3 Generator-Produced Radionuclides

If the reactor, cyclotron, or natural product radionuclide of long half-life decays to a daughter with nuclear characteristics appropriate for medical application, the system is called a medical radionuclide generator. There are several advantages afforded by generator-produced isotopes. These generators represent a convenient source of short-lived medical isotopes without the need for an on-site reactor or particle accelerator. Generators provide delivery of the radionuclide on demand at a site remote from the production facility. They are a source of both gamma- and positron-emitting isotopes.

The most common medical radionuclide generator is the ^{99}Mo Æ ^{99m}Tc system, the source of ^{99m}Tc, a gamma-emitting isotope currently used in 70% of the clinical nuclear medicine studies. The ^{99}Mo has a 67-h half-life, giving this generator a useful life of about a week. Another common generator is the ^{68}Ge Æ ^{68}Ga system. Germanium (half-life is 287 d) is accelerator-produced in high-energy accelerators (e.g., BLIP, LAMPF, TRIUMF) through the alpha-particle bombardment of ^{66}Zn. The 68Ge decays to ^{68}Ga, a positron emitter, which has a 68-min half-life. Gallium generators can last for several months.

The generator operation is fairly straightforward. In general, the parent isotope is bound to a solid chemical matrix (e.g., alumina column, anionic resin, Donux resin). As the parent decays, the daughter nuclide grows in. The column is then flushed ("milked") with a suitable solution (e.g., saline, hydrochloric acid) that elutes the daughter and leaves the remaining parent absorbed on the column. The eluent may be injected directly or processed into a radiopharmaceutical.

7.1.4 Radiopharmaceuticals

^{99m}Tc is removed from the generator in the form of TcO_4^- (pertechnetate). This species can be injected directly for imaging or incorporated into a variety of useful radiopharmaceuticals. The labeling of ^{99m}Tc usually involves reduction complexation/chelation. ^{99m}Tc-Sestamibi (Figure 7.2) is a radiopharmaceutical used to evaluate myocardial perfusion or to diagnose cancer. There are several reduction methods employed, including Sn(II) reduction in $NaHCO_3$ at pH of 8 and other reduction and complexation reactions such as $S_2O_3 = HCl$, $FeCl_3$ + ascorbic acid, $LiBH_4$, Zn + HCl, HCl, Fe(II), Sn(II)F_2, Sn(II) citrate, and Sn(II) tartrate reduction and complexation, electrolytic reduction, and *in vivo* labeling of red cells following Sn(II) pyrophosphate or Sn(II) DTPA administration. ^{131}I, ^{125}I, and ^{123}I labeling requires special reagents or conditions such as chloramine-T, widely used for protein labeling at 7.5 pH;

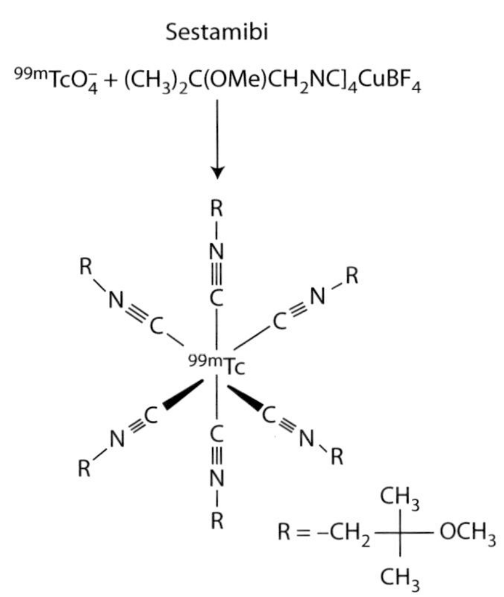

FIGURE 7.2 ^{99m}Tc-Sestamibi is a radiopharmaceutical used to evaluate myocardial perfusion or in the diagnosis of cancer using both computed tomography and scintigraphy techniques.

peroxidase + H_2O_2 widely used for radioassay tracers; isotopic exchange for imaging tracers; excitation labeling as in ^{123}Xe Æ ^{123}I diazotization plus iodination for primary amines; conjugation labeling with Bolton Hunter agent (N-succinimidyl 3-[4-hydroxy 5-(^{131,125,123}I)iodophenyl] propionate); hydroboration plus iodination; electrophilic destannylation; microdiffusion with fresh iodine vapor; and other methods. Radiopharmaceuticals in common use for brain perfusion studies are N-isopropyl-p [^{123}I] iodoamphetamine and ^{99m}Tc-labeled hexamethylpropyleneamine.

7.1.5 PET Radionuclides

For ^{11}C, ^{13}N, ^{15}O, and ^{18}F, the modes of production of short-lived positron emitters can dictate the chemical form of the product, as shown in Table 7.3. Online chemistry is used to make various PET agents and precursors. For example, ^{11}C cyanide, an important precursor for synthesis of other labeled compounds, is produced in the cyclotron target by first bombarding N_2 + 5% H_2 gas target with 20-MeV protons. The product is carbon-labeled methane, $^{11}CH_4$, which when combined with ammonia and passed over a platinum wool catalyst at 1000°C becomes $^{11}CN^-$, which is subsequently trapped in NaOH.

Molecular oxygen is produced by bombarding a gas target of $^{14}N_2$ + 2% O_2 with deuterons (6 to 8 MeV). A number of products (e.g., $^{15}O_2$, $C^{15}O_2$, $N^{15}O_2$, $^{15}O_3$, and $H_2{}^{15}O$) are trapped by soda lime followed by charcoal to give $^{15}O_2$ as the product. However, if an activated charcoal trap at 900°C is used before the soda lime trap, the $^{15}O_2$ will be converted to $C^{15}O$. The specific strategies for other online PET agents is given in Rayudu (1990).

Fluorine-18 (^{18}F) is a very versatile positron emitting isotope. With a 2-h half-life and two forms (F^+ and F^-), one can develop several synthetic methods for incorporating ^{18}F into medically useful compounds (Kilbourn, 1990). Additionally, fluorine forms strong bonds with carbon and is roughly the same size as a hydrogen atom, imparting metabolic and chemical stability of the molecules without drastically altering biologic activity. The most commonly produced ^{18}F radiopharmaceutical is 2-deoxy-2-[^{18}F]fluoroglucose (FDG). This radiotracer mimics part of the glucose metabolic pathway and has shown both hypo- and hypermetabolic abnormalities in cardiology, oncology, and neurology. The synthetic pathway for the production of FDG is shown in Figure 7.3.

The production of positron radiopharmaceuticals requires the rapid incorporation of the isotope into the desired molecule. Chemical techniques and synthetic strategies have been developed to facilitate these reactions. Many of these synthetic manipulations require hands-on operations by a highly trained chemist. Additionally, since positrons give off 2- to 511-keV gamma rays upon annihilation, proximity to the source can increase one's personal dose. The demand for the routine production of positron radiopharmaceuticals such as FDG has led to the development of remote synthetic devices. These devices can be human-controlled (i.e., flipping switches to open air-actuated valves), computer-controlled, or

TABLE 7.3 Major Positron-Emitting Radionuclides Produced by Accelerated Protons

Radionuclide	Half-Life	Reaction
Carbon-11	20 min	^{12}C (p, pn) ^{11}C
		^{14}N (p, α) ^{11}C
Nitrogen-12	10 min	^{16}O (p, α) ^{13}N
		^{13}C (p, n) ^{13}N
Oxygen-15	2 min	^{15}N (p, n) ^{15}C
		^{14}N (d, n) ^{15}O
Fluorine-18	110 min	^{18}O (p, n) ^{18}F
		^{20}Ne (d, α) ^{18}F

Note: $A(x,y)B$: A is target, x is the bombarding particle, y is the radiation product, and B is the isotope produced.

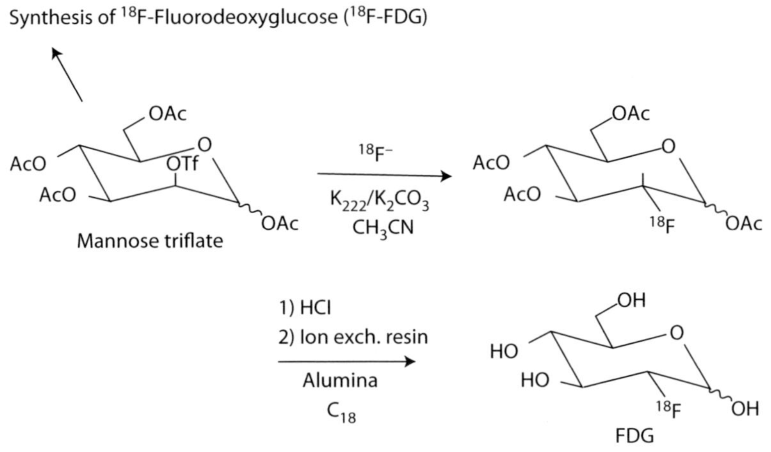

FIGURE 7.3 Schematic for the chemical production of deoxyglucose labeled with fluorine-18. Here K_{222} refers to Kryptofix and C_{18} denotes a reverse-phase high-pressure liquid chromatography column.

robotic. A sophisticated computer-controlled chemistry synthesis unit has been assembled for the fully automated production of ^{18}FDG (Padgett, 1989). A computer-controlled robot has been programmed to producc 6a-[^{18}F]fluoroestradiol for breast cancer imaging (Brodack et al., 1986; Mathias et al., 1987). Both these types of units increase the availability and reduce the cost of short-lived radiopharmaceutical production through greater reliability and reduced need for a highly trained staff. Additionally, these automated devices reduce personnel radiation exposure. These and other devices are being designed with greater versatility in mind to allow a variety of radiopharmaceuticals to be produced just by changing the programming and the required reagents.

These PET radionuclides have been incorporated into a wide variety of medically useful radiopharmaceuticals through a number of synthetic techniques. The development of PET scanner technology has added a new dimension to synthetic chemistry by challenging radiochemists to devise labeling and purification strategies that proceed on the order of minutes rather than hours or days as in conventional synthetic chemistry. To meet this challenge, radiochemists are developing new target systems to improve isotope production and sophisticated synthetic units to streamline routine production of commonly desired radiotracers as well as preparing short-lived radiopharmaceuticals for many applications.

Acknowledgments

This work was supported in part by the Director, Office of Energy Research, Office of Health and Environmental Research, Medical Applications and Biophysical Research Division of the U.S. Department of Energy, under contract No. DE-AC03-SF00098, and in part by NIH Grant HL25840.

References

Brodack, J.W., Dence, C.S., Kilbourn, M.R., and Welch, M.J. 1988. Robotic production of 2-deoxy-2-[^{18}F] fluoro-d-glucose: aroutine method of synthesis using tetrabutylammonium [^{18}F]fluoride. *Int. J. Radiat. Appl. Instrum. Part A: Appl. Radiat. Isotopes* 39: 699.

Brodack, J.W., Kilbourn, M.R., Welch, M.J., and Katzenellenbogen, J.A. 1986. Application of robotics to radiopharmaceutical preparation: controlled synthesis of fluorine-18 16 alpha-fluoroestradiol-17 beta. *J. Nucl. Med.* 27: 714.

Hupf, H.B. 1976. Production and purification of radionuclides. In *Radiopharmacy*. New York, Wiley.

Kilbourn, M.R. 1990. *Fluorine-18 Labeling of Radiopharmaceuticals.* Washington, National Academy Press.

Lamb, J. and Kramer, H.H. 1983. Commercial production of radioisotopes for nuclear medicine. In *Radiotracers for Medical Applications*, pp. 17–62. Boca Raton, FL, CRC Press.

Mathias, C.J., Welch, M.J., Katzenellenbogen, J.A. et al. 1987. Characterization of the uptake of 16 alpha-([^{18}F]fluoro)-17 beta-estradiol in DMBA-induced mammary tumors. *Int. J. Radiat. Appl. Instrum. Part B: Nucl. Med. Biol.* 14: 15.

Padgett, H.C., Schmidt, D.G., Luxen, A. et al. 1989. Computed-controlled radiochemical synthesis: a chemistry process control unit for the automated production of radiochemicals. *Int. J. Radiat. Appl. Instrum. Part A: Appl. Radiat. Isotopes* 40: 433.

Rayudu, G.V. 1990. Production of radionuclides for medicine. *Semin. Nucl. Med.* 20: 100.

Sorenson, J.A. and Phelps, M.E. 1987. *Physics in Nuclear Medicine.* New York, Grune & Stratton.

Steigman, J. and Eckerman, W.C. 1992. *The Chemistry of Technetium in Medicine.* Washington, National Academy Press.

Stocklin, G. 1992. Tracers for metabolic imaging of brain and heart: radiochemistry and radiopharmacology. *Eur. J. Nucl. Med.* 19: 527.

7.2 Instrumentation

Thomas F. Budinger

7.2.1 Background

The history of positron-emission tomography (PET) can be traced to the early 1950s, when workers in Boston first realized the medical imaging possibilities of a particular class of radioactive substances. It was recognized then that the high-energy photons produced by annihilation of the positron from positron-emitting isotopes could be used to describe, in three dimensions, the physiologic distribution of "tagged" chemical compounds. After two decades of moderate technological developments by a few research centers, widespread interest and broadly based research activity began in earnest following the development of sophisticated reconstruction algorithms and improvements in detector technology. By the mid-1980s, PET had become a tool for medical diagnosis and for dynamic studies of human metabolism.

Today, because of its million-fold sensitivity advantage over magnetic resonance imaging (MRI) in tracer studies and its chemical specificity, PET is used to study neuroreceptors in the brain and other body tissues. In contrast, MRI has exquisite resolution for anatomic (Figure 7.4) and flow studies as well as unique attributes of evaluating chemical composition of tissue but in the millimolar range rather

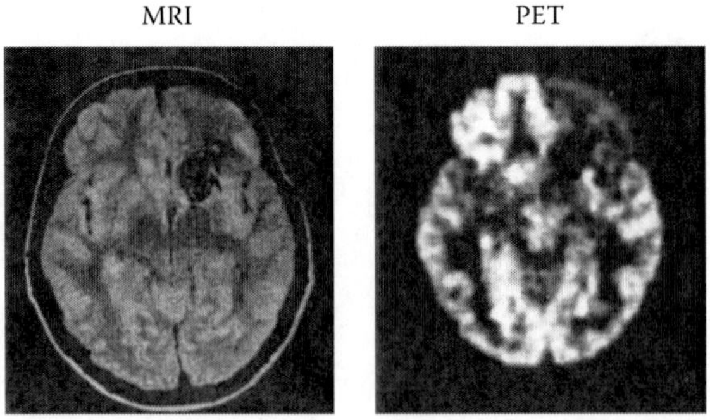

FIGURE 7.4 The MRI image shows the arteriovenous malformation (AVM) as an area of signal loss due to blood flow. The PET image shows the AVM as a region devoid of glucose metabolism and also shows decreased metabolism in the adjacent frontal cortex. This is a metabolic effect of the AVM on the brain and may explain some of the patient's symptoms.

than the nanomolar range of much of the receptor proteins in the body. Clinical studies include tumors of the brain, breast, lungs, lower gastrointestinal tract, and other sites. Additional clinical uses include Alzheimer's disease, Parkinson's disease, epilepsy, and coronary artery disease affecting heart muscle metabolism and flow. Its use has added immeasurably to our current understanding of flow, oxygen utilization, and the metabolic changes that accompany disease and that change during brain stimulation and cognitive activation.

7.2.2 PET Theory

PET imaging begins with the injection of a metabolically active tracer—a biologic molecule that carries with it a positron-emitting isotope (e.g., ^{11}C, ^{13}N, ^{15}O, or ^{18}F). Over a few minutes, the isotope accumulates in an area of the body for which the molecule has an affinity. As an example, glucose labeled with ^{11}C, or a glucose analogue labeled with ^{18}F, accumulates in the brain or tumors, where glucose is used as the primary source of energy. The radioactive nuclei then decay by positron emission. In positron (positive electron) emission, a nuclear proton changes into a positive electron and a neutron. The atom maintains its atomic mass but decreases its atomic number by 1. The ejected positron combines with an electron almost instantaneously, and these two particles undergo the process of annihilation. The energy associated with the masses of the positron and electron particles is 1.022 MeV in accordance with the energy E to mass m equivalence $E = mc^2$, where c is the velocity of light. This energy is divided equally between two photons that fly away from one another at a 180-degree angle. Each photon has an energy of 511 keV. These high-energy gamma rays emerge from the body in opposite directions, to be detected by an array of detectors that surround the patient (Figure 7.5). When two photons are recorded simultaneously by a pair of detectors, the annihilation event that gave rise to them must have occurred somewhere along the line connecting the detectors. Of course, if one of the photons is scattered, then the line of coincidence will be incorrect. After 100,000 or more annihilation events are detected, the distribution of the positron-emitting tracer is calculated by tomographic reconstruction procedures. PET reconstructs a two-dimensional (2D) image from the one-dimensional projections seen at different angles. Three-dimensional (3D) reconstructions also can be done using 2D projections from multiple angles.

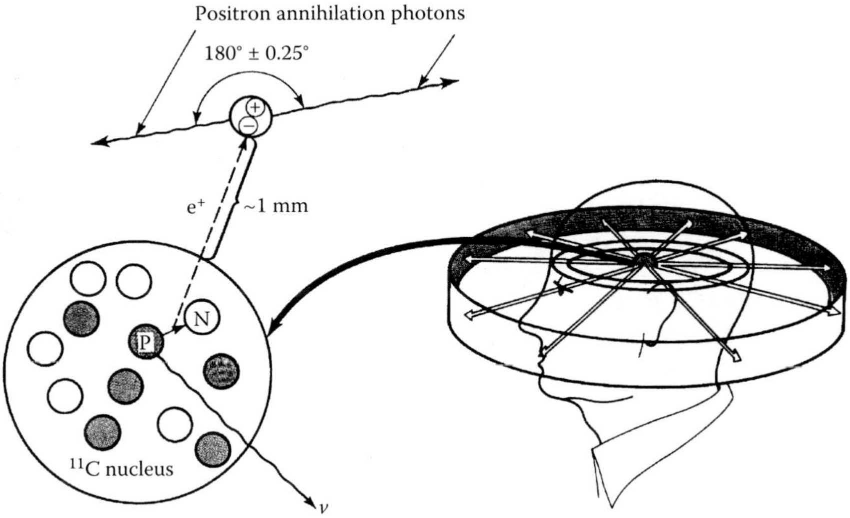

FIGURE 7.5 The physical basis of positron-emission tomography. Positrons emitted by "tagged" metabolically active molecules annihilate nearby electrons and give rise to a pair of high-energy photons. The photons fly off in nearly opposite directions and thus serve to pinpoint their source. The biologic activity of the tagged molecule can be used to investigate a number of physiologic functions, both normal and pathologic.

7.2.3 PET Detectors

Efficient detection of the annihilation photons from positron emitters is usually provided by the combination of a crystal, which converts the high-energy photons to visible-light photons, and a photomultiplier tube that produces an amplified electric current pulse proportional to the amount of light photons interacting with the photocathode. The fact that imaging system sensitivity is proportional to the square of the detector efficiency leads to a very important requirement that the detector be nearly 100% efficient. Thus other detector systems such as plastic scintillators or gas-filled wire chambers, with typical individual efficiencies of 20% or less, would result in a coincident efficiency of only 4% or less.

Most modern PET cameras are multilayered with 15 to 47 levels or transaxial layers to be reconstructed (Figure 7.6). The lead shields prevent activity from the patient from causing spurious counts in the tomograph ring, while the tungsten septa reject some of the events in which one (or both) of the 511-keV photons suffer a Compton scatter in the patient. The sensitivity of this design is improved by collection of data from cross-planes (Figure 7.6). The arrangement of scintillators and phototubes is shown in Figure 7.7.

The "individually coupled" design is capable of very high resolution, and because the design is very parallel (all the photomultiplier tubes and scintillator crystals operate independently), it is capable of very high data throughput. The disadvantages of this type of design are the requirement for many expensive photomultiplier tubes and, additionally, that connecting round photomultiplier tubes to rectangular scintillation crystals leads to problems of packing rectangular crystals and circular phototubes of sufficiently small diameter to form a solid ring.

The contemporary method of packing many scintillators for 511 keV around the patient is to use what is called a block detector design. A block detector couples several photomultiplier tubes to a bank of scintillator crystals and uses a coding scheme to determine the crystal of interaction. In the two-layer block (Figure 7.7), five photomultiplier tubes are coupled to eight scintillator crystals. Whenever one of the outside four photomultiplier tubes fires, a 511-keV photon has interacted in one of the two crystals attached to that photomultiplier tube, and the center photomultiplier tube is then used to determine whether it was the inner or outer crystal. This is known as a digital coding scheme, since each photomultiplier tube is either "hit" or "not hit" and the crystal of interaction is determined by a "digital" mapping of the hit pattern. Block detector designs are much less expensive and practical to form into a multilayer camera. However, errors in the decoding scheme reduce the spatial resolution, and since the entire block is "dead" whenever one of its member crystals is struck by a photon, the dead time is worse than with

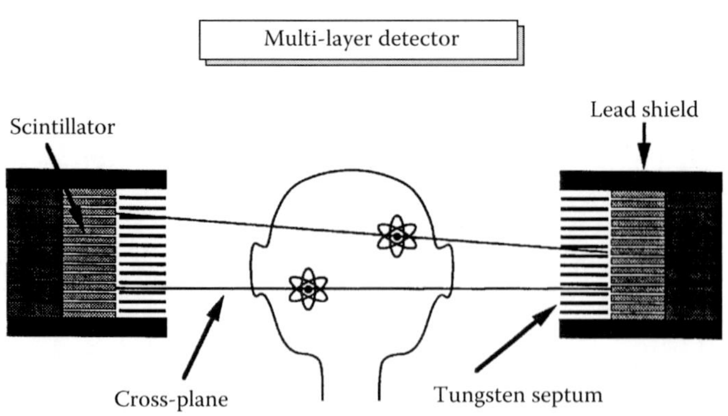

FIGURE 7.6 Most modern PET cameras are multilayered with 15 to 47 levels or transaxial layers to be reconstructed. The lead shields prevent activity from the patient from causing spurious counts in the tomograph ring, while the tungsten septa reject some of the events in which one (or both) of the 511-keV photons suffer a Compton scatter in the patient. The sensitivity of this design is improved by collection of data from cross-planes.

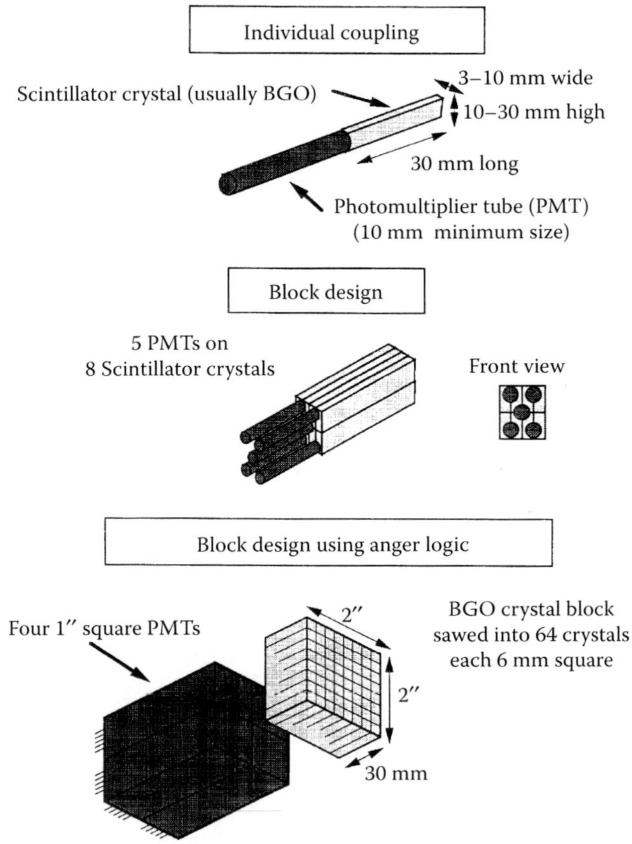

FIGURE 7.7 The arrangement of scintillators and phototubes is shown. The "individually coupled" design is capable of very high resolution, and because the design is very parallel (all the photomultiplier tubes and scintillator crystals operate independently), it is capable of very high data throughput. A block detector couples several photomultiplier tubes to a bank of scintillator crystals and uses a coding scheme to determine the crystal of interaction. In the two-layer block, five photomultiplier tubes are coupled to eight scintillator crystals.

individual coupling. The electronics necessary to decode the output of the block are straightforward but more complex than that needed for the individually coupled design.

Most block detector coding schemes use an analog coding scheme, where the ratio of light output is used to determine the crystal of interaction. In the example above, four photomultiplier tubes are coupled to a block of BGO that has been partially sawed through to form 64 "individual" crystals. The depth of the cuts are critical; that is, deep cuts tend to focus the scintillation light onto the face of a single photomultiplier tube, while shallow cuts tend to spread the light over all four photomultiplier tubes. This type of coding scheme is more difficult to implement than digital coding, since analog light ratios place more stringent requirements on the photomultiplier tube linearity and uniformity as well as scintillator crystal uniformity. However, most commercial PET cameras use an analog coding scheme because it is much less expensive due to the lower number of photomultiplier tubes required.

7.2.4 Physical Factors Affecting Resolution

The factors that affect the spatial resolution of PET tomographs are shown in Figure 7.8. The size of the detector is critical in determining the system's geometric resolution. If the block design is used, there is

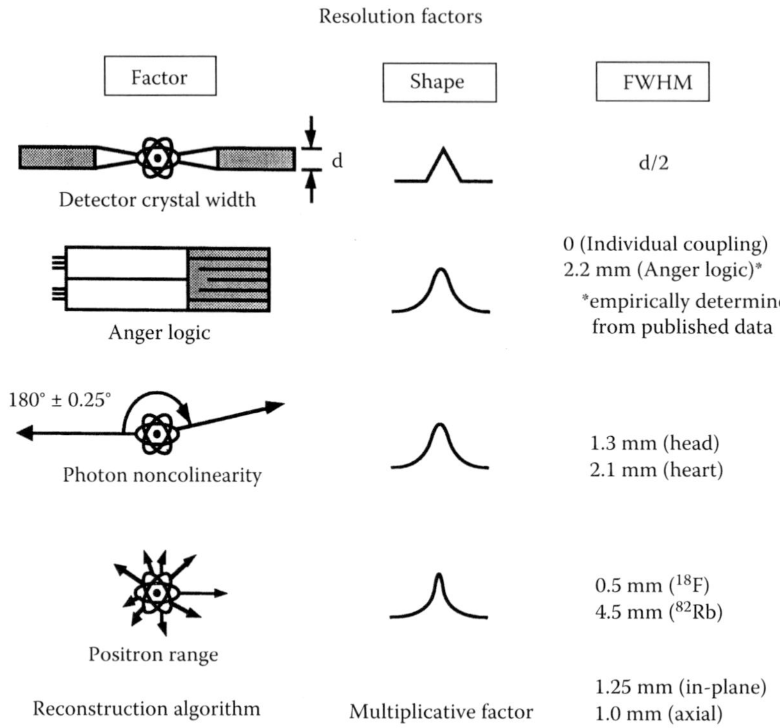

FIGURE 7.8 Factors contributing to the resolution of the PET tomograph. The contribution most accessible to further reduction is the size of the detector crystals.

a degradation in this geometric resolution by 2.2 mm for BGO. The degradation is probably due to the limited light output of BGO and the ratio of crystals (cuts) per phototube.

The angle between the paths of the annihilation photons can deviated from 180 degrees as a result of some residual kinetic motion (Fermi motion) at the time of annihilation. The effect on resolution of this deviation increases as the detector ring diameter increases so that eventually this factor can have a significant effect.

The distance the positron travels after being emitted from the nucleus and before annihilation causes a deterioration in spatial resolution. This distance depends on the particular nuclide. For example, the range of blurring for ^{18}F, the isotope used for many of the current PET studies, is quite small compared with that of the other isotopes. Combining values for these factors for the PET-600 tomograph, we can estimate a detector-pair spatial resolution of 2.0 mm and a reconstructed image resolution of 2.6 mm. The measured resolution of this system is 2.6 mm, but most commercially available tomographs use a block detector design (Figure 7.7), and the resolution of these systems is above 5 mm. The evolution of resolution improvement is shown in Figure 7.9.

The resolution evolutions discussed above pertain to results for the center or axis of the tomograph. The resolution at the edge of the object (e.g., patient) will be less by a significant amount due to two factors. First, the path of the photon from an "off-center" annihilation event typically traverses more than one detector crystal, as shown in Figure 7.10. This results in an elongation of the resolution spread function along the radius of the transaxial plane. The loss of resolution is dependent on the crystal density and the diameter of the tomograph detector ring. For a 60-cm diameter system, the resolution can deteriorate by a factor of 2 from the axis to 10 cm.

The coincidence circuitry must be able to determine coincident events with 10- to 20-nsec resolution for each crystal–crystal combination (i.e., chord). The timing requirement is set jointly by the time

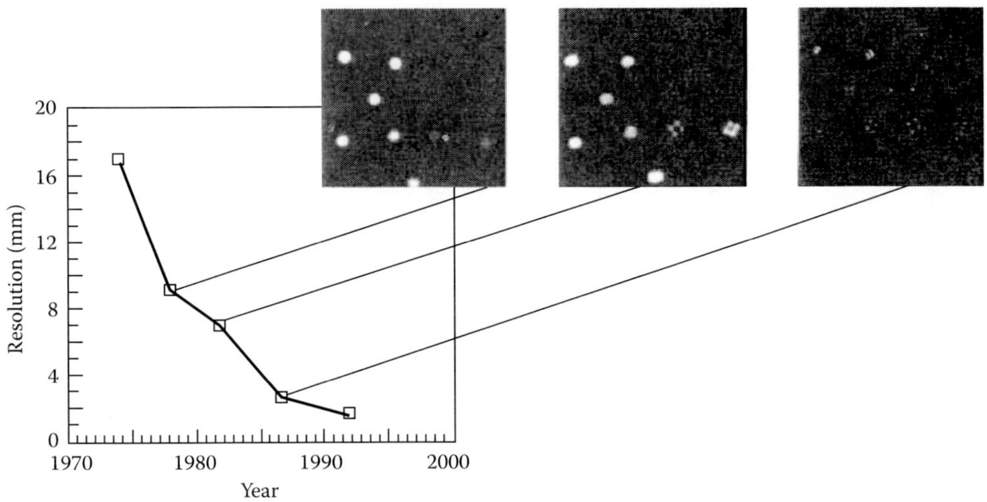

FIGURE 7.9 The evolution of resolution. Over the past decade, the resolving power of PET has improved from about 9 to 2.6 mm. This improvement is graphically illustrated by the increasing success with which one is able to resolve "hot spots" of an artificial sample that are detected and imaged by the tomographs.

of flight across the detector ring (4 nsec) and the crystal-to-crystal resolving time (typically 3 nsec). The most stringent requirement, however, is the vast number of chords in which coincidences must be determined (over 1.5 million in a 24-layer camera with septa in place and 18 million with the septa removed).

It is obviously impractical to have an individual coincidence circuit for each chord, so tomograph builders use parallel organization to solve this problem. A typical method is to use a high-speed clock (typically 200 MHz) to mark the arrival time of each 511-keV photon and a digital coincidence processor to search for coincident pairs of detected photons based on this time marker. This search can be done extremely quickly by having multiple sorters working in parallel.

The maximum event rate is also quite important, especially in septaless systems. The maximum rate in a single detector crystals is limited by the dead time due to the scintillator fluorescent lifetime

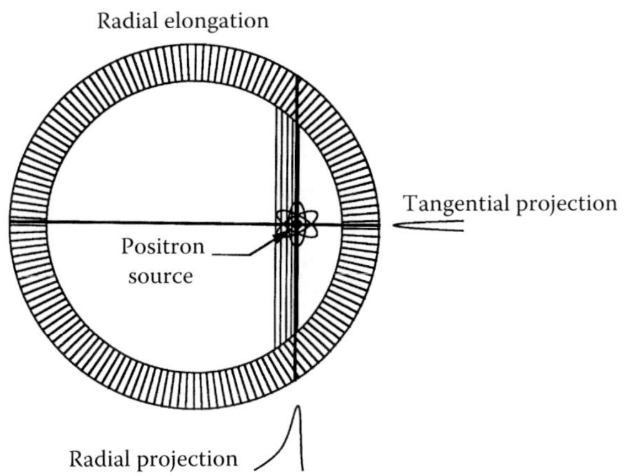

FIGURE 7.10 Resolution astigmatism in detecting off-center events. Because annihilation photons can penetrate crystals to different depths, the resolution is not equal in all directions, particularly at the edge of the imaging field. This problem of astigmatism will be taken into account in future PET instrumentation.

(typically 1 msec per event), but as the remainder of the scintillator crystals are available, the instrument has much higher event rates (e.g., number of crystals ¥ 1 msec). Combining crystals together to form contiguous blocks reduces the maximum event rate because the fluorescent lifetime applies to the entire block and a fraction of the tomograph is dead after each event.

7.2.5 Random Coincidences

If two annihilation events occurs within the time resolution of the tomograph (e.g., 10 nsec), then random coincident "events" add erroneous background activity to the tomograph and are significant at high event rates. These can be corrected for on a chord-by-chord basis. The noncoincidence event rate of each crystal pair is measured by observing the rate of events beyond the coincident timing window. The random rate for the particular chord R_{ij} corresponding to a crystal pair is

$$R_{ij} = r_i \times r_j \times 2\tau \quad R_{ij} = R_{ij} \tag{7.2}$$

where r_i and r_j are the event rates of crystal i and crystal j, and τ is the coincidence window width. As the activity in the subject increases, the event rate in each detector increases. Thus the random event rate will increase as the square of the activity.

7.2.6 Tomographic Reconstruction

Before reconstruction, each projection ray or chord receives three corrections: crystal efficiency, attenuation, and random efficiency. The efficiency for each chord is computed by dividing the observed count rate for that chord by the average court rate for chords with a similar geometry (i.e., length). This is typically done daily using a transmission source without the patient or object in place. Once the patient is in position in the camera, a transmission scan is taken, and the attenuation factor for each chord is computed by dividing its transmission count rate by its efficiency count rate. The patient is then injected with the isotope, and an emission scan is taken, during which time the random count rate is also measured. For each chord, the random event rate is subtracted from the emission rate, and the difference is divided by the attenuation factor and the chord efficiency. (The detector efficiency is divided twice because two separate detection's measurements are made—transmission and emission.) The resulting value is reconstructed, usually with the filtered backprojection algorithm. This is the same algorithm used in x-ray computed tomography (CT) and in projection MRI. The corrected projection data are formatted onto parallel- or fan-beam data sets for each angle. These are modified by a high-pass filter and backprojected.

The process of PET reconstruction is linear and shown by operators successively operating on the projection P:

$$A = \sum_{\theta} BPF^{-1}RF(P) \tag{7.3}$$

where A is the image, F is the Fourier transform, R is the ramp-shaped high-pass filter, F^{-1} is the inverse Fourier transform, BP is the backprojection operation, and θ denotes the superposition operation.

The alternative class of reconstruction algorithms involves iterative solutions to the classic inverse problem:

$$P = FA \tag{7.4}$$

where P is the projection matrix, A is the matrix of true data being sought, and F is the projection operation. The inverse is

$$A = F^{-1}P \tag{7.4a}$$

which is computed by iteratively estimating the data A' and modifying the estimate by comparison of the calculated projection set P' with the true observed projections P. The expectation-maximization algorithm solves the inverse problem by updating each pixel value ai in accord with

$$a_1^{k+1} = \sum p_j \frac{a_j^k f_{ij}}{\sum_i a_i^k f_{ij}} \tag{7.5}$$

where P is the measured projection, f_{ij} is the probability a source at pixel i will be detected in projection detector j, and k is the iteration.

7.2.7 Sensitivity

The sensitivity is a measure of how efficiently the tomograph detects coincident events and has units of count rate per unit activity concentration. It is measured by placing a known concentration of radionuclide in a water-filled 20-cm-diameter cylinder in the field of view. This cylinder, known as a phantom, is placed in the tomograph, and the coincidence event rate is measured. High sensitivity is important because emission imaging involves counting each event, and the resulting data are as much as 1000 times less than experienced in x-ray CT. Most tomographs have high individual detection efficiency for 511-keV photons impinging on the detector (>90%), so the sensitivity is mostly determined by geometric factors, that is, the solid angle subtended by the tomograph:

$$S = \frac{A\varepsilon^2\gamma \times 3.7 \times 10^4}{4\pi r^2} \text{(events/sec)/(mCi/cc)} \tag{7.6}$$

where

r = radius of tomograph
A = area of detector material seen by each point in the object ($2\pi r$ ¥ axial aperture)
ε = efficiency of scintillator
γ = attenuation factor

For a single layer, the sensitivity of a tomograph of 90 cm diameter (2-cm axial crystals) will be 15,000 events/sec/mCi/mL for a disk of activity 20 cm in diameter and 1 cm thick. For a 20-cm-diameter cylinder, the sensitivity will be the same for a single layer with shields or septa that limit the off-slice activity from entering the collimators. However, modern multislice instruments use septa that allow activity from adjacent planes to be detected, thus increasing the solid angle and therefore the sensitivity. This increase comes at some cost due to increase in scatter. The improvement in sensitivity is by a factor of 7, but after correction for the noise, the improvement is 4. The noise equivalent sensitivity S_{NE} is given by

$$S_{NE} = \frac{(\text{true events})^2}{\text{true} \times \text{scatter} \times \text{random}} \tag{7.7}$$

7.2.8 Statistical Properties of PET

The ability to map quantitatively the spatial distribution of a positron-emitting isotope depends on adequate spatial resolution to avoid blurring. In addition, sufficient data must be acquired to allow a statistically reliable estimation of the tracer concentration. The amount of available data depends on the biomedical accumulation, the imaging system sensitivity, and the dose of injected radioactivity. The propagation of errors due to the reconstruction process results in an increase in the noise over that expected for an independent signal, for example, by a factor proportional to the square root of the number of resolution elements (true pixels) across the image. The formula that deals with the general case of emission reconstruction (PET or SPECT) is

$$\%\ \text{uncertainty} = \frac{1.2 \times 100(\text{total no. of events})^{3/4}}{(\text{total no. of events})^{1/2}} \tag{7.8}$$

The statistical requirements are closely related to the spatial resolution, as shown in Figure 7.11.

For a given accuracy or a signal-to-noise ratio for a uniform distribution, the ratio of the number of events needed in a high-resolution system to that needed in a low-resolution system is proportional to

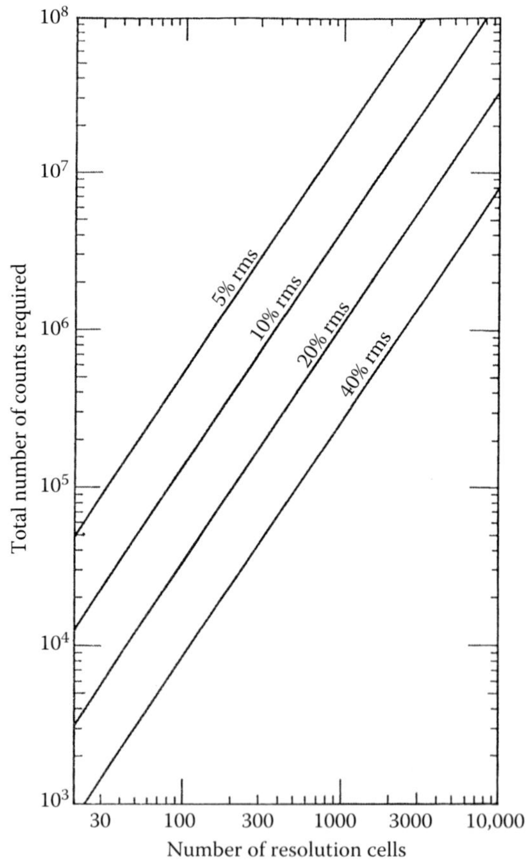

FIGURE 7.11 Statistical requirements and spatial resolution. The general relationship between the detected number of events and the number of resolution elements in an image is graphed for various levels of precision. These are relations for planes of constant thickness.

the 3/2 power of the ratio of the number of effective resolution elements in the two systems. Equation 7.8 and Figure 7.11 should be used not with the total pixels in the image but with the effective resolution cells. The number of effective resolution cells is the sum of the occupied resolution elements weighted by the activity within each element. Suppose, however, that the activity is mainly in a few resolution cells (e.g., 100 events per cell) and the remainder of the 10,000 cells have a background of 1 event per cell. The curves of Figure 7.11 would suggest unacceptable statistics; however, in this case, the effective number of resolution cells is below 100. The relevant equation for this situation is

$$\%\ \text{uncertainty} = \frac{1.2 \times 100(\text{no. of resolutiion cells})^{3/4}}{(\text{avg. no. of events per resolution cell in target})^{3/4}} \tag{7.9}$$

The better resolution gives improved results without the requirement for a drastic increase in the number of detected events is that the improved resolution increases contrast. (It is well known that the number of events needed to detect an object is inversely related to the square of the contrast.)

Acknowledgments

This work was supported in part by the Director, Office of Energy Research, Office of Health and Environmental Research, Medical Applications and Biophysical Research Division of the U.S. Department of Energy under contract No. DE-AC03-SF00098 and in part by NIH Grant HL25840. I wish to thank Drs. Stephen Derenzo and William Moses, who contributed material to this presentation.

References

1. Anger, H.O. 1963. Gamma-ray and positron scintillator camera. *Nucleonics* 21: 56.
2. Bailey, D.L. 1992. 3D acquisition and reconstruction in positron emission tomography. *Ann. Nucl. Med.* 6: 123.
3. Brownell, G.L. and Sweet, W.H. 1953. Localization of brain tumors with positron emitters. *Nucleonics* 11: 40.
4. Budinger, T.F., Greenberg, W.L., Derenzo, S.E. et al. 1978. Quantitative potentials of dynamic emission computed tomography. *J. Nucl. Med.* 19: 309.
5. Budinger, T.F., Gullberg, G.T., and Huesman, R.H. 1979. Emission computed tomography. In G.T. Herman (ed.), *Topics in Applied Physics: Image Reconstruction from Projections: Implementation and Applications*, pp. 147–246. Berlin, Springer-Verlag.
6. Cherry, S.R., Dahlbom, M., and Hoffman, E.J. 1991. 3D PET using a conventional multislice tomograph without septa. *J. Comput. Assist. Tomogr.* 15: 655.
7. Daube-Witherspoon, M.E. and Muehllehner, G. 1987. Treatment of axial data in three-dimensional PET. *J. Nucl. Med.* 28: 1717.
8. Derenzo, S.E., Huesman, R.H., Cahoon, J.L. et al. 1988. A positron tomograph with 600 BGO crystals and 2.6 mm resolution. *IEEE Trans. Nucl. Sci.* 35: 659.
9. Kinahan, P.E. and Rogers, J.G. 1989. Analytic 3D image reconstruction using all detected events. *IEEE Trans. Nucl. Sci.* 36: 964–968.
10. Shepp, L.A. and Vardi, Y. 1982. Maximum likelihood reconstruction for emission tomography. *IEEE Trans. Med. Imaging* 1: 113.
11. Ter-Pogossian, M.M., Phelps, M.E., Hoffman, E.J. et al. 1975. A positron-emission transaxial tomograph for nuclear imaging (PETT). *Radiology* 114: 89.

8

Electrical Impedance Tomography

N. Huber
University of Sussex

W. Wang
University of Sussex

D.C. Barber
Sheffield University

8.1 Electrical Impedance of Tissue

Human tissue is electrically conductive, with tissue conductivity varying significantly between different tissues. Electrical impedance tomography (EIT) is a relatively new imaging modality, first developed in Sheffield, UK in 1982 (Brown and Barber), whose purpose is to produce images from the internal conductivity distribution of an object, based on measurements collected from the surface of the object (Kohn and Vogelius, 1984a,b; Sylvester and Uhlmann, 1986). The original assumptions made were of the near-isotropic distribution of conductivity of the imaged object.

Conductivity images produced through EIT have been mostly cross-sectional, hence the association with tomography, even though this term is not entirely accurate. Moreover, it has been shown that tissue incorporates a reactive component, making it not merely conductive. Hence, it is more fitting to be referring to the *impedance* (or its inverse, the *admittance*) of tissue, rather than the conductivity or resistivity of tissue. Hence, the term *impedance* in EIT.

8.2 Conduction in Human Tissues

In short, tissue can be considered as a collection of conducting fluids (electrolytes) bound by insulating membranes, these membranes themselves existing in a conductive suspension. This, to all intents and purposes, resembles a leaky dielectric. The first model developed (Fricke, 1925) portrayed the cell as a multitude of resistances representing every minute section of the extracellular and intracellular media, and capacitances representing every minute section of the membrane. With the use of circuit theory, these elements can be combined such that an equivalent circuit is produced, as seen from the perspective

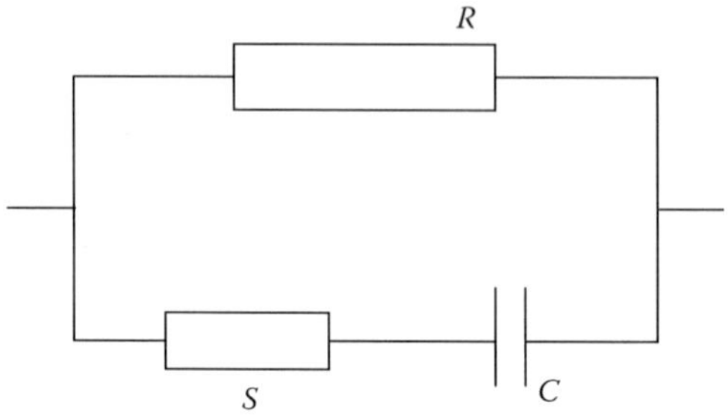

FIGURE 8.1 The Cole–Cole model of tissue impedance.

of the electrodes. It is interesting to note that, apart from a change in actual component values, both a single cell in suspension and tissue can be represented with the same electrical model.

The developed model for bulk tissue most widely in use today (Cole and Cole, 1941) consists of a resistor R in parallel with a combination of a capacitor C and a resistor S, as shown in Figure 8.1.

It is worth noting that S has also been portrayed as being in series with both R and C (for a more detailed review, refer to McAdams and Jossinet, 1995). This oversimplified model has been shown empirically to fit data well, when a term is added to represent the frequency dependence of the component values. When expressed in terms of impedance, tissue behavior with respect to frequency can be shown to be

$$Z = Z_\infty + \frac{(Z_0 - Z_\infty)}{1 + \left(j(f/f_c)\right)^{(1-\alpha)}} \tag{8.1}$$

where Z_0 is tissue impedance at very low frequency, Z_∞ the impedance at high frequency, both complex, and f_c is a characteristic frequency. Alpha (α) allows for the frequency dependency of the model components, and is found experimentally to match the particular tissue type.

Diseased tissue is known to exhibit altered R, S, and C due to structural change. By extracting the component values through complex measurements of impedance of tissue can allow for the differentiation of tissues and of different disease conditions. It is worth noting that although maximum accuracy in the determination of the model components can be obtained if both real and imaginary components are available, in principle, knowledge of the resistive component alone should enable the values to be determined, provided an adequate range of frequencies is used. This can have practical consequences for data collection, since accurate measurement of the capacitive component can prove difficult.

The prevailing assumption in EIT is that tissue is isotropic. However, this is not true in the macroscopic scale for various types of tissue due to their anistropic physical structure. The inherent anisotropy can be shown to prevent the existence of unique solutions to the conductivity distributions, since there are sets of different anisotropic conductivity distributions that give the same surface voltages, rendering them indistinguishable from each other. It is an open issue as to the degree to which tissue anisotropy affects the generation of useful images through EIT, which is actively being explored (e.g., Lionheart and Paridis, 2010).

8.3 Determination of Impedance Distribution

The distribution of electrical potential within an isotropic conducting object through which a low-frequency current is flowing is given by

$$\nabla(\sigma \nabla \varphi) = 0 \tag{8.2}$$

where φ is the potential distribution within the object and σ is the distribution of conductivity (generally admittivity) within the object (Brown and Barber, 1982). If the conductivity is uniform, this reduces to Laplace's equation. Strictly speaking, this equation is only correct for direct current, but for the frequencies of alternating current used in EIT (up to 1 MHz) and the sizes of objects being imaged, it can be assumed that this equation continues to describe the instantaneous distribution of potential within the conducting object. If this equation is solved for a given conductivity distribution and current distribution through the surface of the object, the potential distribution developed on the surface of the object may be determined. The distribution of potential will depend on several things. It will depend on the pattern of current applied and the shape of the object. It will also depend on the internal conductivity of the object, and it is this that needs to be determined. In theory, the current may be applied in a continuous and nonuniform pattern at every point across the surface. In practice, current is applied to an object through electrodes attached to the surface of the object. Theoretically, potential may be measured at every point on the surface of the object. Again, voltage on the surface of the object is measured in practice using electrodes (possibly different from those used to apply current) attached to the surface of the object. There will be a relationship, the forward solution, between an *applied current pattern* j_i, the conductivity distribution σ, and the surface potential distribution φ_i which can be formally represented as

$$\varphi_i = R(j_i, \sigma) \tag{8.3}$$

If σ and j_i are known, φ_i can be computed. For one current pattern j_i, knowledge of φ_i is not in general sufficient to uniquely determine σ. However, by applying a complete set of independent current patterns, it becomes possible to obtain sufficient information to determine σ, at least in the isotropic case. This is the inverse solution. In practice, measurements of surface potential or voltage can only be made at a finite number of positions, corresponding to electrodes placed on the surface of the object. This also means that only a finite number of independent current patterns can be applied. For N electrodes, $N - 1$ independent current patterns can be defined and $N(N - 1)/2$ independent measurements made. This latter number determines the limit of image resolution achievable with N electrodes. In practice, it may not be possible to collect all possible independent measurements. Since only a finite number of current patterns and measurements is available, the set of equations represented by Equation 8.3 can be rewritten as

$$v = A_c c \tag{8.4}$$

where v is now a concatenated vector of all voltage values for all current patterns, c is a vector of conductivity values, representing the conductivity distribution divided into uniform image *pixels*, and A_c a matrix representing the transformation of this conductivity vector into the voltage vector. Since A_c depends on the conductivity distribution, this equation is nonlinear. Although formally the preceding equation can be solved for c by inverting A_c, the nonlinear nature of this equation means that this cannot be done in a single step. An iterative procedure will therefore be needed to obtain c.

Examination of the physics of current flow shows that current tends to take the easiest path possible in its passage through the object. If the conductivity at some point is changed, the current path redistributes in such a way that the effects of this change are minimized. The practical effect of this is that it is possible to have fairly large changes in conductivity within the object which only produce relatively small changes in voltage at the surface of the object. The converse of this is that when reconstructing the conductivity distribution, small errors on the measured voltage data, both random and systematic, can translate into large errors in the estimate of the conductivity distribution. This effect forms, and

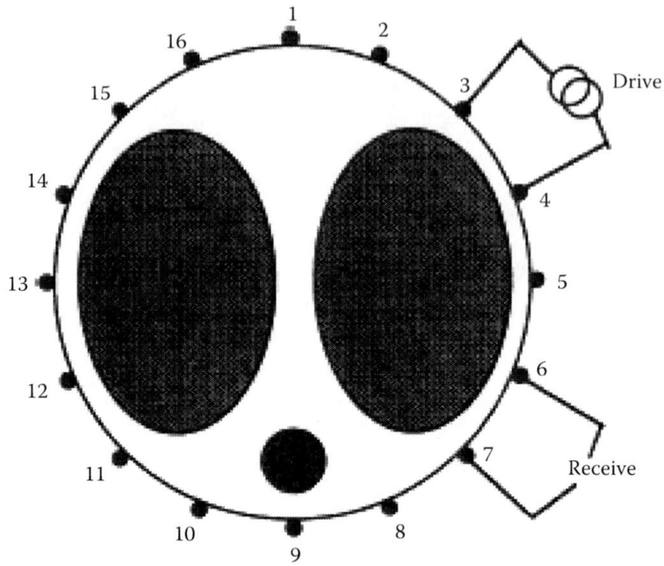

FIGURE 8.2 Idealized electrode positions around a conducting object with typical drive and measurement electrode pairs indicated.

will continue to form, a limit to the quality of reconstructed conductivity images in terms of resolution, accuracy, and sensitivity.

Any measurement of voltage must always be referred to a reference point. Usually this is one of the electrodes, which is given the nominal value of 0 V. The voltage on all other electrodes is determined by measuring the voltage difference between each electrode and the reference electrode. Alternatively, voltage differences may be measured between pairs of electrodes. A common approach is to measure the voltage between adjacent pairs of electrodes (Figure 8.2). Clearly, the measurement scheme affects the form of A_c. Choice of the pattern of applied currents and the voltage measurement scheme used can affect the accuracy with which images of conductivity can be reconstructed.

EIT is not a mature technology. However, it has been the subject of intensive research over the past few years, and this work is still continuing. Nearly all the research effort has been devoted to exploring the different possible ways of collecting data and producing images of tissue resistivity, with the aim of optimizing image reconstruction in terms of image accuracy, spatial resolution, and sensitivity.

Although the most current interest is in the use of EIT for medical imaging, there is also interest in its use in whole-body composition monitoring (Rebeyrol et al., 2010). Other applications include geophysical measurements (e.g., Marescot et al., 2003), and a comprehensive review of the applications of electrical tomography for industrial applications is given in York (2005).

8.3.1 Data Collection

8.3.1.1 Basic Requirements

Data are collected by applying a current to the object through electrodes connected to the surface of the object and then making measurements of the voltage on the object surface through the same or other electrodes. Although conceptually simple, technically this can be difficult. Great attention must be paid to the reduction of noise and the elimination of any voltage offsets on the measurements. The currents applied are alternating currents usually in the range 10 kHz to 1 MHz. Since tissue has a complex impedance, the voltage signals will contain in-phase and out-of-phase components. In principle, both of these can be measured. In practice, measurement of the out-of-phase (the capacitive) component is significantly more difficult because of the presence of unwanted (stray) capacitances between various parts of the voltage measurement system, including the leads from the data-collection apparatus to the electrodes. These stray capacitances can lead to appreciable leakage currents, especially at the higher frequencies, which translate into systematic errors on the voltage measurements. McEwan et al. (2007) offers an in-depth review of the errors affecting EIT systems, and a simplified, practical approach to measuring errors is given in Hahn et al. (2008).

The signal measured on an electrode, or between a pair of electrodes, oscillates at the same frequency as the applied current. The magnitude of this signal (usually separated into real and imaginary components) is determined, typically by demodulation and integration. The frequency of the demodulated signal is much less than the frequency of the applied signal, and the effects of stray capacitances on this signal are generally negligible. This realization has led some workers to propose that the signal demodulation and detection system be mounted as close to the electrodes as possible, ideally at the electrode site itself, and some systems have been developed that use this approach, although none with sufficient miniaturization of the electronics to be practical in a clinical setting. This solution is not in itself free of problems, but this approach is likely to be of increasing importance if the frequency range of applied currents is to be extended beyond 1 MHz, necessary if the value of the complex impedance is to be adequately explored as a function of frequency.

Various data-collection schemes have been proposed. Most data are collected from a two-dimensional (2D) configuration of electrodes, either from 2D objects or around the border of a plane normal to the principal axis of a cylindrical (in the general sense) object where that plane intersects the object surface. The simplest data-collection protocol is to apply a current between a pair of electrodes (often an adjacent pair) and measure the voltage difference between other adjacent pairs (see Figure 8.2). Although in principle voltage could be measured on electrodes through which current is simultaneously flowing, the presence of an electrode impedance, generally unknown, between the electrode and the body surface means that the voltage measured is not actually that on the body surface. Various means have been suggested for either measuring the electrode impedance in some way or including it as an unknown in the image-reconstruction process. However, in many systems, measurements from electrodes through which current is flowing are simply ignored. Electrode impedance is generally not considered to be a problem when making voltage measurements on electrodes through which current is not flowing, provided a voltmeter with sufficiently high input impedance is used, although, since the input impedance is always finite, every attempt should be made to keep the electrode impedance as low as possible. Using the same electrode for driving current and making voltage measurements, even at different times in the data-collection cycle, means that at some point in the data-collection apparatus, wires carrying current and wires carrying voltage signals will be brought close together in a switching system, leading to the possibility of leakage currents. There is a good argument for using separate sets of electrodes for driving and measuring to reduce this problem. Paulson et al. (1992) have also proposed this approach and have noted that it can aid in the modeling of the forward solution (see Section 8.3.2). Brown et al. (1994) have used this approach in making multifrequency measurements.

Clearly, the magnitude of the voltage measured will depend on the magnitude of the current applied. If a constant-current drive is used, this must be able to deliver a known current to a variety of input impedances with a stability of better than 0.1%. This is technically demanding. The best approach to this problem is to measure the current being applied, which can easily be done to this accuracy. These measurements are then used to normalize the voltage data.

The current application and data-collection regime will depend on the reconstruction algorithm used. Several EIT systems apply current in a distributed manner, with currents of various magnitudes being applied to several or all of the electrodes. These optimal currents (see Section 8.3.2) must be specified accurately, and again, it is technically difficult to ensure that the correct current is applied at each electrode. Although there are significant theoretical advantages to using distributed current patterns, the increased technical problems associated with this approach, and the higher noise levels associated with the increase in electronic complexity, may outweigh these advantages.

Although most EIT at present is 2D in the sense given above, it is intrinsically a three-dimensional (3D) imaging procedure, since current cannot be constrained to flow in a plane through a 3D object. 3D data collection does not pose any further problems apart from increased complexity due to the need for more electrodes. Whereas most data-collection systems to date have been based on 16 or 32 electrodes, 3D systems will require four times or more electrodes distributed over the surface of the object if adequate resolution is to be maintained. Technically, this will require "belts" or "vests" of electrodes that can be rapidly applied (McAdams et al., 1994). Some of these are already available, and the application of an adequate number of electrodes should not prove insuperable provided electrode-mounted electronics are not required. Metherall et al. (1996) describe a three-dimensional data-collection system and reconstruction algorithm and note the improved accuracy of three-dimensional images compared to two-dimensional images constructed using data collected from three-dimensional objects.

8.3.1.2 EIT Hardware Design Principles

The vast majority of EIT systems depend on the application of current patterns and the measurement of voltages. The architecture of systems is rather common, and at first glance straightforward (Figure 8.3).

The signal to be applied to the subject (commonly a sinusoid of known frequency) is first generated, and is fed to a voltage-to-current converter circuit. This converter can be either single- or multiple-output, providing a means of classifying EIT systems by the number of current sources used.

The number of electrodes of the system is, generally, larger than the number of sources, and hence a switching network to serve all electrodes with current is required, ordinarily implemented through multiplexers. In most cases, cables (with driven shields in most cases) supply the current to the electrodes. Measurements are taken from the electrodes in either single-ended or differential mode, demultiplexed appropriately. A differential amplifier is more frequently used to take a measurement between pairs of electrodes, both to achieve high common mode rejection and to reduce the dynamic range compared to measurements made with respect to ground. As the measurement has to be taken during the time of

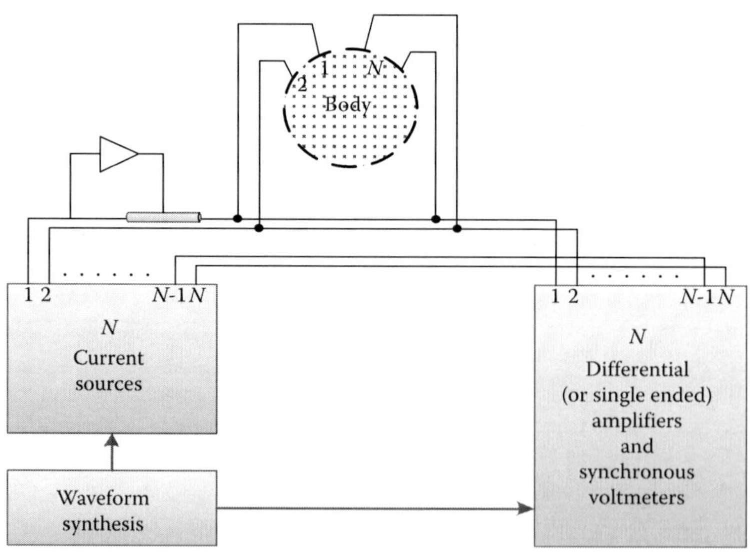

FIGURE 8.3 A typical multichannel EIT system. The number of current sources is commonly smaller than the number of electrodes, requiring multiplexing at the source and demultiplexing at the voltmeters. (Reproduced from Saulnier, G.J. 2005. *Electrical Impedance Tomography: Methods, History and Applications*, Institute of Physics Publishing, Bristol, pp. 67–104.)

application of the signal into the body under investigation, there is commonly digital synchronization between the signal generation and voltage measurement.

The challenges of EIT electronics design are not directly apparent. In simulations, current sources behave ideally, supplying constant current for a constant voltage input irrespective of load over a frequency bandwidth well beyond the medical requirements of EIT (i.e., >10 MHz). The output impedance of the current source is assumed infinite, the full current passed directly to the load. In practice, however, the output impedance of current sources is better modeled as a resistance and a capacitance in parallel. The output response of the source is, then, dependent on the load, with current actually shunted directly to ground rather than through the load, especially at higher frequencies. Stray capacitances external to the source introduced either by the PCB design/implementation or any cables used. A very thorough analysis of the techniques used in EIT instrumentation design can be found in Saulnier (2005), the author looking into every aspect of the hardware from signal generation to sampling, source design, and performance compensation.

8.3.2 Image Reconstruction

8.3.2.1 Basics of Reconstruction

Several different approaches to image reconstruction have been tried and established since 1982 (Brown and Barber, 1982; Wexler et al., 1985; Yorkey, 1986; Barber and Seagar, 1987; Kim and Woo, 1987; Hua et al., 1988; Breckon, 1990; Cheney et al., 1990; Zadehkoochak et al., 1991; Koire et al., 1992; Bayford, 1994; Morucci et al., 1994). A detailed review of reconstruction efforts can be found in work specifically on this topic by Lionheart (2004).

The most accurate approaches are based broadly on the following algorithm. For a given set of current patterns, a forward transform is set up for determining the voltages v produced from the conductivity distribution c (Equation 8.4). A_c is dependent on c, so it is necessary to assume an initial starting conductivity distribution c_0. This is usually taken to be uniform. Using A_c, the expected voltages v_0 are calculated and compared with the actual measured voltages v_m. Unless c_0 is correct (which it will not be initially), v_0 and v_m will differ. It can be shown that an improved estimate of c is given by

$$\Delta c = \left(S_c^t S_c\right)^{-1} S_c^t (v_0 - v_m) \tag{8.5}$$

$$c_1 = c_0 + \Delta c \tag{8.6}$$

where S_c is the differential of A_c with respect to c, the sensitivity matrix and S_c^t is the transpose of S_c. The improved value of c is then used in the next iteration to compute an improved estimate of v_m, that is, v_1. This iterative process is continued until some appropriate endpoint is reached. Although convergence is not guaranteed, in practice, convergence to the correct c in the absence of noise can be expected, provided a good starting value is chosen. Uniform conductivity seems to be a reasonable choice. In the presence of noise on the measurements, iteration is stopped when the difference between v and v_m is within the margin of error set by the known noise on the data.

There are some practical difficulties associated with this approach. One is that large changes in c may only produce small changes in v, and this will be reflected in the structure of S_c, making $S_c^t S_c$ very difficult to invert reliably. Various methods of regularization have been used, with varying degrees of success, to achieve stable inversion of this matrix although the greater the regularization applied, the poorer the resolution that can be achieved. A more difficult practical problem is that for convergence to be possible the computed voltages v must be equal to the measured voltages v_m when the correct conductivity values are used in the forward calculation. Although in a few idealized cases analytical solutions of the forward problem are possible, in general, numerical solutions must be used. Techniques such as the finite-element method (FEM) have been developed to solve problems of this type numerically. However, the accuracy of these methods has to be carefully examined (Paulson et al., 1992) and, while they are adequate for many applications, they may not be adequate for the EIT reconstruction problem, especially in the case of 3D objects. Accuracies of rather better than 1% appear to be required if image artifacts are to be minimized. Consider a situation in which the actual distortion of conductivity is uniform. Then the initial v should be equal to the v_m to an accuracy less than the magnitude of the noise. If this is not the case, then the algorithm will alter the conductivity distribution from uniform, which will clearly result in error. While the required accuracies have been approached under ideal conditions, there is only a limited amount of evidence at present to suggest that they can be achieved with data taken from human subjects.

8.3.2.2 Optimal Current Patterns

So far little has been said about the form of the current patterns applied to the object except that a set of independent patterns is needed. The simplest current patterns to use are those given by passing current into the object through one electrode and extracting current through a second electrode (a bipolar pattern). This pattern has the virtue of simplicity and ease of application. However, other current patterns are possible. Current can be passed simultaneously through many electrodes, with different amounts passing through each electrode. Indeed, an infinite number of patterns are possible, the only limiting condition being that the magnitude of the current flowing into the conducting object equals the magnitude of the current flowing out of the object. Isaacson (1986) has shown that for any conducting object there is a set of optimal current patterns and has provided an algorithm to compute them even if the conductivity distribution is initially unknown. Isaacson showed that by using optimal patterns, significant improvements in sensitivity could be obtained compared with simpler two-electrode current patterns. However, the additional computation and hardware required to use optimal current patterns compared with fixed, nonoptimal patterns are considerable.

Use of suboptimal patterns close to optimum will also produce significant gains. In general, the optimal patterns are very different from the patterns produced in the simple two-electrode case. The optimal patterns are often cosine-like patterns of current amplitude distributed around the object boundary rather than being localized at a pair of points, as in the two-electrode case. Since the currents are passed simultaneously through many electrodes, it is tempting to try and use the same electrodes for voltage measurements. This produces two problems. As noted above, measurement of voltage on an electrode through which an electric current is passing is compromised by the presence of electrode resistance, which causes a generally unknown voltage drop across the electrode, whereas voltage can be accurately measured on an electrode through which current is not flowing using a voltmeter of high input impedance. In addition, it has proved difficult to model current flow around an electrode through which current is flowing with sufficient accuracy to allow the reliable calculation of voltage on that electrode, which is needed for accurate reconstruction. It seems that separate electrodes should be used for voltage measurements with distributed current systems.

Theoretically, distributed (near-) optimal current systems have some advantages. As each of the optimal current patterns is applied, it is possible to determine if the voltage patterns produced contain any useful information or if they are simply noise. Since the patterns can be generated and applied in order of decreasing significance, it is possible to terminate application of further current patterns when no further information can be obtained. A consequence of this is that signal-to-noise ratios (SNR) can be maximized for a given total data-collection time. With bipolar current patterns, this option is not available. All patterns must be applied. Provided the SNR in the data is sufficiently good and only a limited number of electrodes are used, this may not be too important, and the extra effort involved in generating the optimal or near-optimal patterns may not be justified. However, as the number of electrodes is increased, the use of optimal patterns becomes more significant. It also has been suggested that the distributed nature of the optimal patterns makes the forward problem less sensitive to modeling errors. Although there is currently no firm evidence for this, this seems a reasonable assertion.

8.3.2.3 Three-Dimensional Imaging

Most published work so far on image reconstruction has concentrated on solving the 2D problem. However, real medical objects, that is, patients, are three-dimensional. Theoretically, as the dimensionality of the object increases, reconstruction should become better conditioned. However, unlike 3D x-ray images, which can be constructed from a set of independent 2D images, EIT data from 3D objects cannot be so decomposed and data from over the whole surface of the object is required for 3D reconstruction. The principles of reconstruction in 3D are identical to the 2D situation although practically the problem is quite formidable, principally because of the need to solve the forward problem in three dimensions. Some early work on 3D imaging was presented by Goble and Isaacson (1990). Metherall et al. (1996) have shown images using data collected from human subjects. Cherepenin et al. (2001) started using 3D imaging for impedance mammography, with other systems being reported since (e.g., Sze et al., 2011). With further theoretical proof of the necessity for 3D imaging over conventional 2D (Halter et al., 2007a), and lately it has been shown that the creation of patient-specific forward models through accurate knowledge of electrode positioning enhances performance of reconstruction algorithms used, and conforms more realistically with the breast geometry (Tizzard et al., 2010). Indeed, it has been stressed in recent publications that the use of accurate shape information is paramount to achieving much higher sensitivity, and works have been carried out to use *a priori* knowledge from multiple other modalities to this end (e.g., CT (Grychtol et al., 2011), MRI (Davidson et al., 2011), ultrasound (Borsic et al., 2011)), most performed for 3D EIT.

8.3.2.4 Single-Step Reconstruction

The complete reconstruction problem is nonlinear and requires iteration. However, each step in the iterative process is linear. Images reconstructed using only the first step of iteration effectively treat image formation as a linear process, an assumption approximately justified for small changes in conductivity from uniform. In the case, the functions A_c and S_c often can be precomputed with reasonable accuracy because they usually are computed for the case of uniform conductivity. Although the solution cannot be correct, since the nonlinearity is not taken into account, it may be useful, and first-step linear approximations have gained some popularity. Cheney et al. (1990) have published some results from a first-step process using optimal currents. Most, if not all, of the clinical images produced to date have used a single-step reconstruction algorithm (Barber and Seagar, 1987; Barber and Brown, 1990). The use of linearity for each reconstruction iteration has more recently been incorporated into the concerted effort of standardizing 2D pulmonary monitoring (Adler et al., 2009), evidently showing that this is the method of choice for a large number of researchers worldwide.

8.3.2.5 Differential Imaging

Ideally, the aim of EIT is to reconstruct images of the absolute distribution of conductivity (or admittivity). These images are known as absolute (or static) images. However, this requires that the forward problem can be solved to a high degree of accuracy, and this can be difficult. The magnitude of the voltage signal measured on an electrode or between electrodes will depend on the body shape, the electrode shape and position, and the internal conductivity distribution. The signal magnitude is in fact dominated by the first two effects rather than by conductivity. However, if a change in conductivity occurs within the object, then it can often be assumed that the change in surface voltage is dominated by this conductivity change. In differential (or dynamic) imaging, the aim is to image changes in conductivity rather than absolute values. If the voltage difference between a pair of (usually adjacent) electrodes before a conductivity change occurs is g_1 and the value after change occurs is g_2, then a normalized data value is defined as

$$\Delta g_n = 2\frac{g_1 - g_2}{g_1 + g_2} = \frac{\Delta g}{g_{\text{mean}}} \tag{8.7}$$

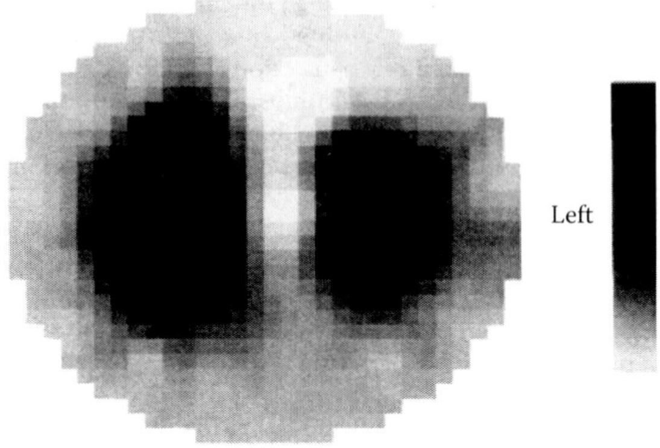

FIGURE 8.4 A differential conductivity image representing the changes in conductivity in going from maximum inspiration (breathing in) to maximum expiration (breathing out). Increasing blackness represents increasing conductivity.

Many of the effects of body shape (including interpreting the data as coming from a 2D object when in fact it is from a 3D object) and electrode placing at least partially cancel out in this definition. The values of the normalized data are determined largely by the conductivity changes. It can be argued that the relationship between the (normalized) changes in conductivity c_n and the normalized changes in boundary data g_n is given by

$$\Delta g_n = F\Delta c_n \tag{8.8}$$

where F is a sensitivity matrix which can be shown to be much less sensitive to object shape end electrode positions than the sensitivity matrix of Equation 8.5. Although images produced using this algorithm are not completely free of artifact, that is the only algorithm which has reliably produced images using data taken from human subjects. Metherall et al. (1996) used a version of this algorithm for 3D imaging and Brown et al. (1994) for multifrequency imaging, in this case imaging changes in conductivity with frequency. The principle disadvantage of this algorithm is that it can only image changes in conductivity, which must be either natural or induced. Figure 8.4 shows a typical differential image. This represents the changes in the conductivity of the lungs between expiration and inspiration.

8.3.3 Multifrequency Measurements

Differential algorithms can only image changes in conductivity. Absolute distributions of conductivity cannot be produced using these methods. In addition, any gross movement of the electrodes, either because they have to be removed and replaced or even because of significant patient movement, makes the use of this technique difficult for long-term measurements of changes. As an alternative to changes in time, differential algorithms can image changes in conductivity with frequency. Brown et al. (1994) have shown that if measurements are made over a range of frequencies and differential images produced using data from the lowest frequency and the other frequencies in turn, these images can be used to compute parametric images representing the distribution of combinations of the circuit values in Figure 8.1. For example, images representing the ratio of S to R, a measure of the ratio of intracellular to extracellular volume, can be produced, as well as images of $f_o = 1/(2p(RC + SC))$, the tissue characteristic frequency. Although not images of the absolute distribution of conductivity, they are images of absolute tissue properties. Some cancer research has also been performed on combining these parameters in ways considered related to pathological conditions for diagnosis of an abnormality (Wang et al., 2007a), but which are also useful for enhancing the imaging performance. Since these properties are related to tissue structure, they should produce images with useful contrast. Data sufficient to reconstruct an image can be collected in a time short enough to preclude significant patient movement, which means these images are robust against movement artifacts. Changes of these parameters with time can still be observed.

8.4 Areas of Clinical Application

EIT is yet to be widely accepted in the clinical environment. The strengths and weaknesses of EIT have to be weighed so that the most appropriate applications can be identified. This modality cannot compete in spatial resolution or resemblance to real-life with other anatomic imaging methods (MRI, CT, x-ray, etc.), but is safe, relatively cheap, does not involve any ionizing radiation, and has high temporal resolution, as well as the unique capability of providing images that are directly assistive to condition diagnosis.

The advantages of EIT have spurred a large amount of research in the field, of which excellent coverage is given in Holder (2005). Some applications are highlighted here for completeness, with particular emphasis on applications that have been tested clinically.

8.4.1 Cancer Detection in Soft Tissue

8.4.1.1 Tissue Impedance Spectroscopy

Tissue impedance (often referred to as electrical impedance spectroscopy, EIS) measurements *in vitro* not only have been used as the reference point for most clinical *in vivo* studies (Surowiec et al., 1988; Tunstall et al., 1997; Jossinet and Schmitt, 1999), but also have other practical applications. In an example, impedance spectroscopy was employed for identification of cancerous mouse pancreas by taking measurements of the electrical properties of the mouse pancreas (Qiao et al., 2008). In the context of cell biology, the results of other experiments are not disturbed by the electrical impedance measurements. Through the measurement of the impedance of tissue, the time taken to locate a malignant islet could be dramatically reduced.

The standard method for bioimpedance measurements is the "four-electrode chamber," a proven stable system for such applications (Ferris and Rose, 1972; Wang et al., 2006; Grimnes and Martinsen, 2008), but recent studies have also shown that cell culture imaging is possible for monitoring of cell growth through the use of impedance tomography at the micro-scale level (Linderholm et al., 2008). There has also been effort on imaging at the cellular level, such as the system discussed in Liu et al. (2010).

With the advent of nanoparticles, the EIT group of Middlesex University have suggested the use of gold nanoparticles as contrast enhancement media for *in vivo* applications (Callaghan et al., 2010). Their research recognizes that the impedance change caused by these particles is on the almost beyond the reach of current EIT systems, but the idea of EIT contrast enhancement is a very appealing one.

8.4.1.2 Breast Cancer

Early breast cancer detection has been linked directly with a reduction in mortality rate. The golden standard of detection, x-ray mammography, is limited by low sensitivity, more pronouncedly so in dense breasts, as well as lack of diagnostic ability, requiring needle aspirations for a positive diagnosis.

Various early *in vitro* studies have shown that the electrical properties of malignant tissue of the breast differ significantly to both benign and healthy tissue. Indicatively, fat tissue is at 0.5 mS/cm conductivity, connective tissue at 0.6–0.8 mS/cm, carcinoma at 1.5–2.2 mS/cm, mastopathy tissue at 2.7–3.5 mS/cm,

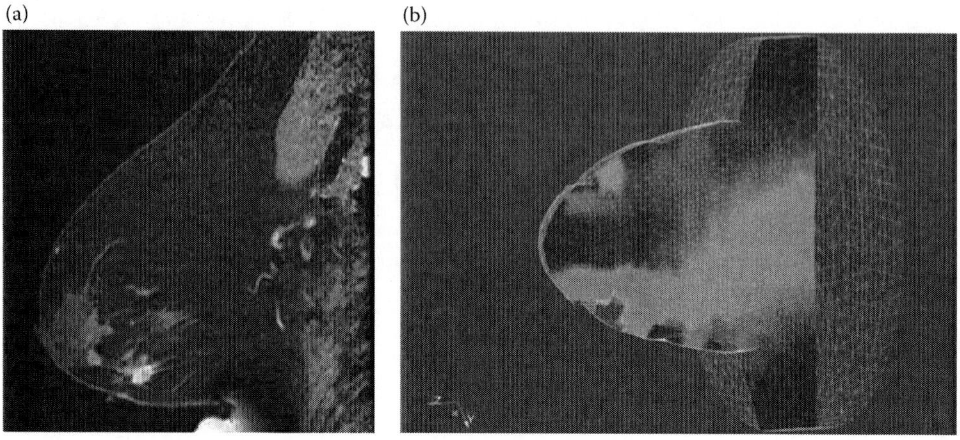

FIGURE 8.5 **(See color insert.)** (a) Sagittal contrast-enhanced breast MRI images. 1.3 cm cancerous mass visible at 5 cm from nipple. (b) 3D EIT reconstruction of breast. (Adapted from Tizzard, A. et al. 2010. *J. Phys. Conf. Ser.* 224(1): 012034.)

and fibroadenoma at 3.3–4.0 mS/cm, depending on the source (e.g., Surowiec et al., 1988; Tunstall et al., 1997; Jossinet and Schmitt, 1999). Consequently, EIT has been suggested and pursued by a large number of researchers for early breast cancer detection due to its noninvasive or ionizing nature, with the application sometimes referred to as electrical impedance mammography (EIM) (Wang et al., 1996). Indeed, one of the first electrical impedance-based devices to reach the market was for EIM (Assenheimer et al., 2001), even though not strictly tomographic. A number of clinical trials have shown that EIT is a promising modality, or at least an adjunctive diagnostic method to the more established techniques (Malich et al., 2001), and has also been suggested as a real-time, noninvasive monitoring method of breast tissue thermoablation (e.g., Lukaszewicz et al., 2010). There has also been suggestion of guiding noninvasive breast cancer treatment based on high-intensity focused ultrasound (HIFU) through the combined use of EIM and ultrasound (Lobstein-Adams et al., 2011).

The two main directions followed by researchers are those of ring-electrode configurations (Wheeler et al., 2002; Halter et al., 2005; Oh et al., 2007), or of planar-array electrodes (Cherepenin et al., 2002; Tang et al., 2002), with 16 to 256, and up to thousands of electrodes (Zhao et al., 2010), hand-held or fixed in a set patient interface, with the use or not of contact media, such as through a saline bath, with physical contact (dry) with the electrodes, or with recessed (wet) electrodes (Wang et al., 2007b). Reported results vary significantly due to the different nature of each device and the reconstruction approach followed. As an example, Figure 8.5 shows the 3D model generated using the system in Halter et al. (2005). However, the same sources of errors and hurdles seem to be common between all systems: low spatial resolution, especially at a distance from the electrodes, variable contact impedance and optimization of current driving, and voltage measurement methods.

8.4.1.3 Cervical Cancer

Cervical cancer accounts for approximately 10% of cancers diagnosed in women worldwide, with much higher incidence rates found mostly in Southern Africa and Central America. Currently, an impedance-based device is on an EU multicenter trial, having shown that the device can be used as an adjunct to the gold standard, colposcopy, for the diagnosis of high-grade cervical intraepithelial neoplasia (Balasubramani et al., 2009), based on the original work by Brown et al. (1998). Further effort has been carried out by Korjenevsky et al. (2010), following a different approach borrowed by EIM, also showing promising clinical results. The main difference between the two systems is that the latter is designed to produce impedance images of the cervix, while the former returns localized impedance measurements.

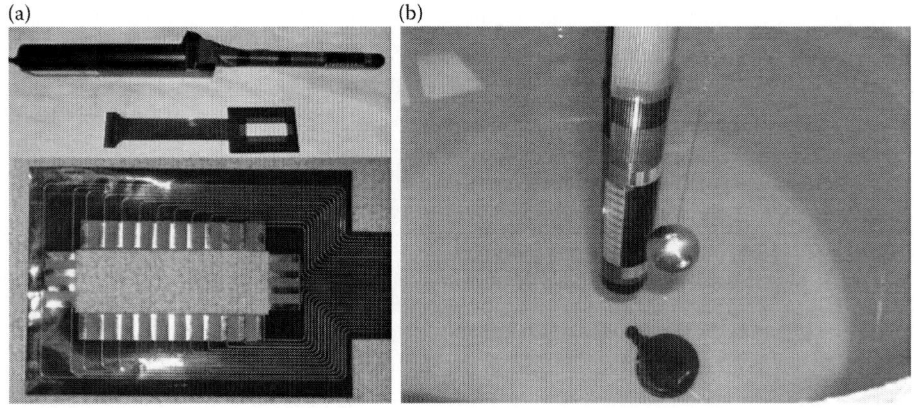

FIGURE 8.6 **(See color insert.)** (a) Top: flex circuit of electrodes wrapped around the ultrasound probe. (a) Bottom: magnified view of flex circuit of electrodes laid flat. (b) Imaging configuration; an inclusion (steel ball) is placed immediately in front of the acoustic window. (Adapted from Wan, Y. et al. 2010. *J. Phys. Conf. Ser.* 224(1): 012067.)

8.4.1.4 Skin Cancer

Melanoma is rapidly becoming the fastest-growing type of cancer, with a reported 120,000 new cases per year in the United States alone. EIT was first employed in studies related to skin conditions in the work by Ollmar and Emtestam (1992), where it was deemed useful as a tool for the noninvasive detection of skin irritation. More importantly, they later demonstrated the relationship between impedance spectral changes due to irritation caused by a variety of surfactants, and the related histopathology (Nicander et al., 1996). Their recent work (Aberg et al., 2004) has shown that multifrequency EIT can be used as a screening tool for the diagnosis of skin malignancies, with very high specificity and sensitivity, leading to the development of the latest commercial device to this end. Modeling of the skin is still under investigation with some very interesting findings (Grimnes and Martinsen, 2010; Martinsen et al., 2010).

8.4.1.5 Prostate Cancer

Prostate cancer is one of the most prevalent forms of the disease to afflict men, mostly of older age. Figures provided by Cancer Research UK state that approximately 913,000 men were diagnosed with prostate cancer in 2008. By the original work done at Dartmouth College, it was shown that the conductivity of cancerous versus healthy prostatic tissue removed from patients differed significantly over the entire frequency range (1 kHz to 1 MHz) (Halter et al., 2008). The work has taken an interesting direction, as there has been study into the incorporation of an EIT system into a conventional transrectal, ultrasonic (US) probe for prostate cancer detection *in vivo* (Borsic et al., 2009). The multimodality of this approach will provide better boundary detection of the prostate, increasing the accuracy of the segmentation of the organ into EIT voxels. The probe itself has recently been presented (Wan et al., 2010), consisting of a standard transrectal probe with a set of flexible electrodes positioned around the acoustic window (Figure 8.6), and clinical trials are under way.

8.4.2 Lung Imaging

Lung pathology can be imaged by conventional radiography, x-ray computed tomography, or magnetic resonance imaging (MRI), but there are clinical situations where it would be desirable to have a portable means of imaging regional lung ventilation which could, if necessary, generate repeated images over time. EIT, being inexpensive, relatively portable, and of high temporal resolution, is ideally suited for this application, but significant challenges, of course, remain.

Lung imaging is widely recognized as one of the most viable areas of research for EIT (Barber, 2005). Over the years, EIT has been applied to the imaging of a large number of medical conditions in the thoracic region, and the techniques, algorithms, and devices developed are too varied to cover fully here. Adequate coverage is given in Smit et al. (2005), with some comparisons against other, more established techniques. From the topical review by Frerichs (2000), it is evident that the body of work is extensive and has occupied EIT experts since the technique was introduced in 1985. The studies conducted have almost exclusively used ring electrode configurations, but the conditions of the tests have varied considerably. Animal (Hahn et al., 1995; Newell et al., 1996) and human studies have taken place, the latter with subjects being either healthy or with such conditions as pulmonary embolism (PE) (Leathard et al., 1994), emphysema (Harris et al., 1987; Eyüboglu et al., 1995) bronchial carcinoma (Holder and Temple, 1993; Morice et al., 1993; Shinkarenko et al., 1997), and chronic obstructive pulmonary disease (COPD) (Eyüboglu et al., 1995; Von Noordergraaf et al., 1998), among others. Subjects in these studies were under spontaneous or mechanical ventilation, depending on the study.

Specifically for the case of patients under mechanical ventilation, pioneering work performed by a close collaboration between clinicians and EIT experts showed the viability of the use of EIT for establishing the best ventilator regime to be used in critically ill patients by the bedside. In the work by Luepschen et al. (2007), it was conclusively shown that thoracic EIT can be used to dynamically image lung ventilation, hence, leading to the individualization of protective ventilation strategies and lung recruitment maneuvers. This research, then, successfully led to the development of the first commercially available, critical-care tool for such applications (for a very interesting read, visit Dräger, website).

Neonatal lung formation has been one of the first areas of research for EIT lung imaging, owing to the fact that noninvasiveness and lack of radiation are preferred in such applications. Early on, Hampshire et al. (1995) looked into the development of the neonatal lung, drawing conclusions on their differentiation to adult lungs. Later on, similar work showed that the definition of absolute conductivity of the lung in neonates could be used to determine lung density and air volume (Brown et al., 2002), and the later work in Brown et al. (2006).

Blood volume in the thorax has also been a topic of intense study, as it could allow for the detection of PE and COPD, as the conductivity of blood is about 6.7 mS/cm, which is approximately three times that of most intrathoracic tissues. The imaged quantity will be the change in conductivity due to replacement of tissue by blood (or vice versa) as a result of the pulsatile flow through the thorax. This may be relatively large in the cardiac ventricles but will be smaller in the peripheral lung fields. If a blood clot is present in the lung, the lung beyond the clot will not be perfused, and under favorable circumstances, this may be visualized using a gated blood volume image of the lung, and PE can be diagnosed if a ventilation image of the same region shows normal ventilation. This was experimentally shown by the work of Newell et al. (1996), whereby PE was simulated in a dog in each lung separately through occlusion of a major branch of the pulmonary artery, observing that the nonperfused lung had different electrical properties.

Research is ongoing in this very promising field. It is, hence, fitting that lung imaging is possibly the only application of EIT to enjoy a linear reconstruction algorithm developed from the consensus of most of the EIT groups in the world (Adler et al., 2009), which can be modified to suit specific devices and applications (example shown in Figure 8.7). Moreover, lung EIT is used as the testing ground of new reconstruction algorithms, such as those proposed in Borsic et al. (2010), where the efficacy of a novel reconstruction algorithm in preserving sharp conductivity boundaries was proven via the use of *in vivo* thoracic measurements. EIT of the lungs is, obviously, close to becoming a clinical reality.

8.4.3 Gastrointestinal System

A priori, it seems likely that EIT could be applied usefully to the measurement of motor activity in the gut. Electrodes can be applied with ease around the abdomen, and there are no large bony structures likely to seriously violate the assumption of constant initial conductivity. During motor activity, such as gastric emptying or peristalsis, there are relatively large movements of the conducting fluids within

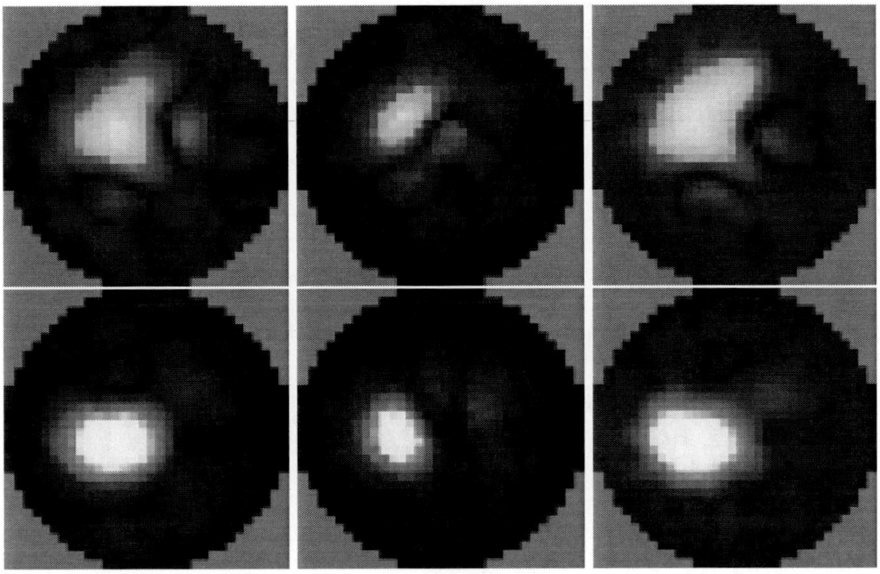

FIGURE 8.7 (**See color insert.**) Images of localized (top row) pneumothorax and (bottom row) pleural effusion in an anesthetized, ventilated pig. Left: GREIT (R_{GN}), center: Gauss–Newton (R_{GN}), right: Sheffield backprojection (R_{SBP}). (Adapted from Adler, A. et al. 2009. *Physiol. Meas.* 30: S35–S55.)

the bowel. The quantity of interest is the timing of activity, for example, the rate at which the stomach empties, and the absolute impedance change and its exact location in space area of secondary importance. The principal limitations of EIT of poor spatial resolution and amplitude measurement are largely circumvented in this application (Avill et al., 1987).

Most research has concentrated on the correlation of conductivity and consistency of meals, as imaged through EIT, and at this, the Sheffield Mark I system has been used almost exclusively (Soulsby et al., 2005). The accuracy of EIT has been validated in parallel, mostly by the use of scintigraphy, and the results are generally in agreement. Wright (1995), for example, investigated gastric emptying of liquid and semisolid meals in healthy volunteers, both male and female, showing that gastric emptying was quicker for males than females on semisolids, but there was no significant difference between the groups on liquids. There has also been some interest in the measurement of other aspects of gastrointestinal function such as esophageal activity (Erol et al., 1995).

The lack of clinical acceptance of EIT for gastrointestinal studies has been largely attributed to the lack of standardization of test meals, the acceptance of EIT by clinicians, and the lack of commercially available EIT systems (Soulsby et al., 2005). This, unfortunately, seems to have limited more extensive research in this particular application of EIT.

8.4.4 Brain Imaging

Brain imaging through EIT has been the effort mostly of the University College London (UCL) group, and their work is well documented in Holder et al. (2005), where the devices, method, and reconstruction algorithms developed by the group are summarized. The applications they envisage for EIT in this field are twofold: accurate localization of epileptic foci in subjects suffering from intractable epilepsy to assist surgical intervention, and detection of cell swelling in cerebral energy failure as caused by pathological conditions such as stroke and ischemia, among others.

For the former, the subjects are monitored through EEG for a period of a week before surgery, during which time their epilepsy suppressant medication is suspended, but this technique is beneficial for superficial foci only. For deeper foci, implantable electrodes are used, with obvious dangers, and this is where EIT may prove to be of great importance as a noninvasive alternative, even though a combination of both types of electrodes has been proposed (Manwaring et al., 2010). Holder et al. (1996) performed animal studies to this effect, reporting that induced epileptic seizures as measured invasively induce conductivity changes of about 10%. However, this signal is expected to be attenuated by a factor of 10 when measured at the scalp, due to the diffusion caused by the scalp, skull, and cerebral matter. Recent work (Fabrizi et al., 2009) showed that, in phantom studies, the EIT device in use (UCH Mk2.5) produced comparable localization results to standard 31-electrode EEG. The authors, however, were cautious to state that the SNR expected in measurements of human subjects would be 50% lower, and baseline noise would be of the order of 1% (as observed clinically in Fabrizi et al. (2006, 2007)). This would imply that localization in human subjects is still a challenge. This fact is further corroborated by simulations performed on a FEM of the skull for intracerebral hemorrhage (ICH) (Tehrani et al., 2010), establishing that the required noise level for detection of a 1 mL blood clot has to be $<0.005\%$, inclusive of all physiological and systematic error sources. This performance is at the limit of detectability of current EIT systems.

A very interesting and important application of EIT in brain monitoring would be of the imaging of neuronal depolarization, the high temporal resolution of EIT rendering it ideal for this. Early work has been performed, not only for the design of electrodes for this application (Gilad et al., 2007), but also on human subjects (Gilad and Holder, 2009). Unfortunately, the results obtained so far have not been of adequate quality for reliable imaging, due to the formidable nature of the challenge, but research is still ongoing in this exciting field.

8.5 New Directions

8.5.1 Multimodal Applications

EIT has been combined with other, more established modalities, which can benefit from its advantages. The use of EIT with ultrasound specifically for prostate cancer has already been discussed earlier. However, other applications exist where other modalities in conjunction with EIT are used, the most prominent one being that of breast cancer.

The Rensselaer group, New York, has devised a method to simultaneously scan both modalities with the use of a handheld probe (Kulkarni et al., 2010). A 6×6 array of electrodes, with the central two columns constructed by gold deposition on sonolucent material forms the front of a specially made housing for a standard ultrasound probe. The coupling to the patient, both electrically and acoustically, is provided through standard ultrasound gel. The impedance measurements are performed using the fourth generation of the Adaptive Current Tomograph (ACT4), developed by the same group (Saulnier et al., 2007).

The same group is also active in the combination of the golden standard for breast cancer detection, x-ray mammography, and EIT (Saulnier et al., 2007). Employing electron beam deposition, they applied thin layers of metal on Kapton to create radiolucent electrode arrays, which are subsequently adapted to the compression plates of an x-ray mammography system, performing simultaneous, dual-modality scans. Impedance measurements are again provided by the ACT4 system, with a frequency range of 3.33 kHz to 1 MHz.

8.5.2 Magnetic Resonance EIT

One of the most promising areas of research in EIT is that of magnetic resonance EIT (MREIT). During current injection into an electrically conductive medium, both a magnetic and an electric field are produced. The concept behind MREIT is to utilize the additional information derived from the magnetic field to assist in turning the ill-posed EIT reconstruction problem into a well-posed one, by effectively deducing the internal current pathways due to the resistivity distribution. The magnetic field can be detected using a conventional MRI scanner, but then the challenge remains in using this internal information in resistivity image reconstructions (Woo et al., 2005). A very thorough review of MREIT can be found in Woo and Seo (2008) and Seo and Woo (2011).

Ampère's law states that the internal current density $\mathbf{J} = (J_x, J_y, J_z)$ is related to the magnetic flux density $\mathbf{B} = (B_x, B_y, B_z)$ by $\mathbf{J} = \nabla \times (\mathbf{B}/\mu_0)$, where μ_0 is the magnetic permeability of free space. The novelty of MREIT lies in the use of an MRI scanner to measure B while actively injecting a current. However, measuring all orthogonal components of the flux density would require the subject to be rotated, a difficult task to achieve, causing problems such as pixel misalignment, which deteriorate the image quality (Woo et al., 2005). One of the most important breakthroughs in this respect was the development of the so-called harmonic B_z algorithm, developed by the Kyung Hee University, Korea, the driving force behind MREIT since its inception. The algorithm proposed (Seo et al., 2003; Oh et al., 2003) enables the reconstruction of conductivity images from a single magnetic flux component parallel to the main magnetic axis of the MRI scanner, this generally being B_z, requiring at least two data sets to be measured. However, the algorithm was found to perform poorly at boundaries of such regions with low conductivity as bones, lungs and the subject boundary, and the solution investigated currently is that of localized algorithm (Seo et al., 2008), which, interestingly, can provide scaled, regional conductivity images without knowing the conductivity distribution outside of the region. A software package (called CoReHA (Jeon et al., 2009)) now incorporates all the above algorithms for MREIT image reconstruction.

The research was limited to phantom studies (Figure 8.8) and postmortem animal tissue studies (e.g., Oh et al., 2005), but has recently been applied to *in vivo* animal and human subjects. The first experiment on a human subject was reported in Kim et al. (2009), where the human leg was imaged (Figure 8.9), and more recently, the human knee (Jeong et al., 2010). These recent images produced are encouraging, but significant challenges have been surpassed, and some are still under investigation. For example, carbon-hydrogel electrodes have been selected to minimize near-electrode artifacts, and a multiecho pulse sequence was developed to increase the SNR of MR magnitude images at lower injected current amplitudes, to be kept at 90% of subject sensation (Han et al., 2010). This latter problem has been the biggest stumbling block for human subject experiments to date. Other issues include the anisotropic nature of some tissues, and an improved electrode configuration and data-collection method (Woo et al., 2005).

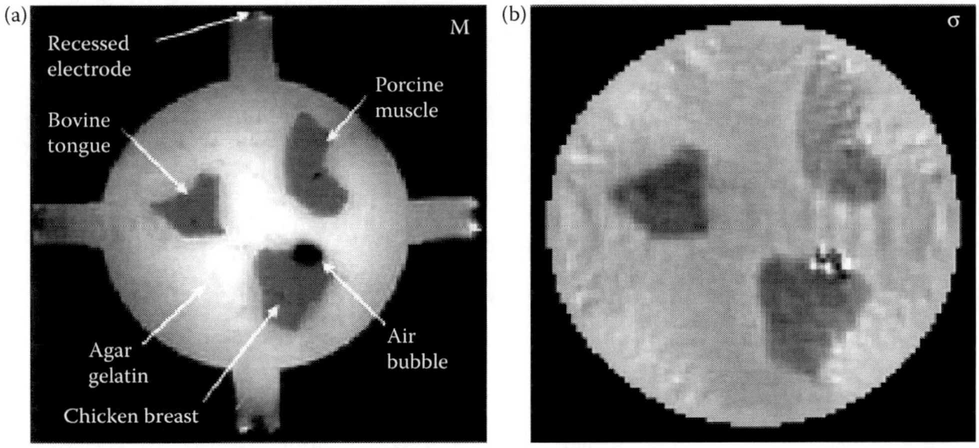

FIGURE 8.8 (a) MREIT biological test phantom, consisting of agar tissue mimicking material, with implanted animal tissue samples. (b) Results recorded from the test phantom using MREIT.

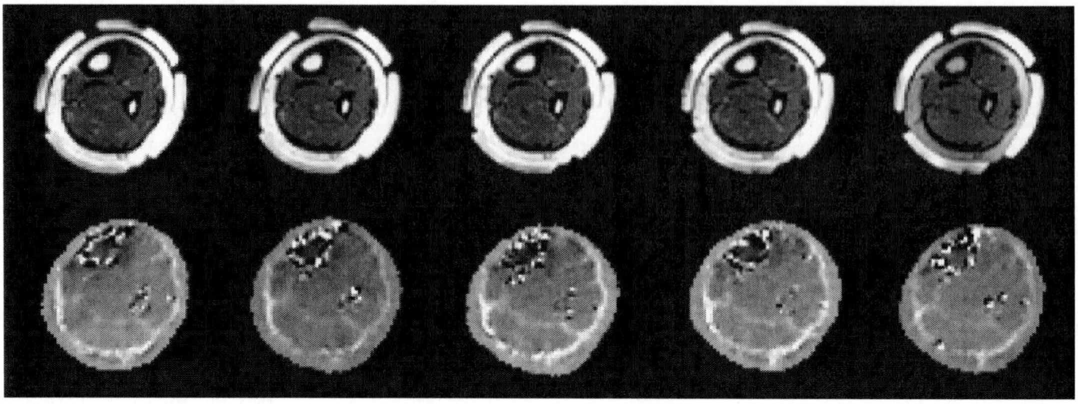

FIGURE 8.9 *In vivo* MREIT imaging experiment of a human leg using a 3 T MRI scanner. Multislice images and reconstructed equivalent isotropic conductivity images of a human leg are shown in the top and bottom rows, respectively. (Adapted from Woo, E.J. and Seo, J.K. 2008. *Physiol. Meas.* 29: R1–R26.)

MREIT is a medical imaging technique based on electrical impedance, but offering much higher resolution than conventional EIT. This advantage comes at the expense of nonportability and much higher expense due to the use of an MRI scanner, but its advantages, if put into routine clinical use, may prove it to be a new standard in medical imaging.

8.5.3 Magnetic Induction Tomography

Another newly developed technique based on the pioneering work from Sheffield Group in the 1980s, which is gaining wide interest and has seen significant advances in recent years, is magnetic induction tomography (MIT). This modality has been, somewhat arbitrarily, called "electromagnetic tomography" (EMT), "mutual inductance tomography" (again MIT), and "eddy current tomography," among others. For clarity, by MIT we are referring here to the field of study in which eddy currents are induced through the magnetic field from an excitation coil, and the magnetic fields subsequently caused by these eddy currents are then measured externally.

As described in depth in Griffiths (2005), the key advantage of MIT, compared with other electrical impedance-based modalities, is that there is no contact with the subject under investigation. This, of course, means that imaging can be performed without the need for time-consuming electrode placement. Further, contact artifacts will not affect imaging, and monitoring can be carried out over long periods of time without adverse effects to the subject.

In the sensing coil, the signal detected is a combination of the primary signal, $\mathbf{B}$, which is directly induced by the field from the excitation coil, and a much smaller, secondary signal, $\Delta\mathbf{B}$, caused by the magnetic field induced by the eddy currents in the subject. It can be shown that $\Delta\mathbf{B}$ is proportional to frequency and conductivity, and lags the primary signal by 90°, with the total detected field ($\mathbf{B} + \Delta\mathbf{B}$) lagging the primary field by an angle φ. In biological tissues, $\Delta\mathbf{B}$ is much smaller in magnitude than $\mathbf{B}$, and thus φ can be expressed as

$$\varphi \simeq \left|\frac{\Delta\mathbf{B}}{\mathbf{B}}\right| \propto \omega\sigma \tag{8.9}$$

showing that the phase angle between the total detected signal and the primary signal is proportional to the angular frequency, ω, of the primary signal, and σ is the conductivity of the sample. Hence, the electrical properties of the measured biological material can be extracted.

MIT is a promising modality, but is faced with significant challenges that so far have not allowed the production of a system with high-enough spatial resolution for clinical applications. Careful screening is required to reduce the effects of external to the system stimuli, and to minimize capacitive coupling that can swamp the wanted, inductive coupling-induced signal. Furthermore, as the primary signal is orders of magnitude larger than the secondary one, there has been interest in the elimination of the primary signal from the measurements, with varying degree of success. Other technical challenges include, among others, the limiting of the spatial resolution by the number of excitor/sensor combinations, high inherent noise figures compared to the measured signal and significant difficulties in the reconstruction algorithms akin to those of EIT (Griffiths, 2005). Further research is being carried out in confronting all the above challenges, and improved results of studies in this field are eagerly awaited.

8.6 Summary and Future Developments

EIT is still an emerging technology. In its development, several novel and difficult measurement and image-reconstruction problems have had to be addressed. Most of these have been satisfactorily solved. This is evident from the move toward standardization, especially with respect to a particular application. Furthermore, the adoption of information from other modalities into the EIT reconstruction problem has been hailed as a critical step forward that may finally see the adoption of EIT widely in clinical use.

The current generation of EIT imaging systems are multifrequency, with some capable of 3D imaging. These should be capable of greater quantitative accuracy and be less prone to image artifact and are likely to find a practical role in clinical diagnosis. Although there are still many technical problems to be answered and many clinical applications to be addressed, the technology may be close to coming of age.

What may possibly be the reason this promising modality has been hindered in its wide recognition is the need for the collaboration of individuals from different, and sometimes disparate, disciplines. Clinicians, hardware and software engineers, medical physicists, and inverse problem experts are required to collaborate with a single clinical application in mind, for any success to ever be gleaned. As EIT groups are maturing, and expertise in this field is now more widespread, it is reasonable to hope that EIT will find its way to widespread, routine clinical use in the near future.

Defining Terms

Absolute imaging: Imaging the actual distribution of conductivity.

Admittivity: The specific admittance of an electrically conducting material. For simple biomedical materials such as saline with no reactive component of resistance, this is, the same as conductivity.

Anisotropic conductor: A material in which the conductivity is dependent on the direction in which it is measured through the material.

Applied current pattern: In EIT, the electric current is applied to the surface of the conducting object via electrodes placed on the surface of the object. The spatial distribution of current flow through the surface of the object is the applied current pattern.

Bipolar current pattern: A current pattern applied between a single pair of electrodes.

Conductivity: The specific conductance of an electrically conducting material. The inverse of resistivity.

Differential imaging: An EIT imaging technique that specifically images changes in conductivity.

Distributed current: A current pattern applied through more than two electrodes.

Dynamic imaging: The same as differential imaging.

EIT: Electrical impedance tomography.

Forward transform or problem or solution: The operation, real or computational, that maps or transforms the conductivity distribution to surface voltages.

Impedivity: The specific impedance of an electrically conducting material. The inverse of admittivity. For simple biomedical materials such as saline with no reactive component of resistance, this is the same as resistivity.

Inverse transform or problem or solution: The computational operation that maps voltage measurements on the surface of the object to the conductivity distribution.

Optimal current: One of a set of current patterns computed for a particular conductivity distribution that produce data with maximum possible SNR.

Pixel: The conductivity distribution is usually represented as a set of connected piecewise uniform patches. Each of these patches is a pixel. The pixel may take any shape, but square or triangular shapes are most common.

Resistivity: The specific electrical resistance of an electrical conducting material. The inverse of conductivity.

Static imaging: The same as absolute imaging.

References

Aberg, P., Nicander, I., Hansson, J., Geladi, P., Holmgren, U., and Ollmar, S. 2004. Skin cancer identification using multifrequency electrical Impedance—A potential screening tool, *IEEE Trans. Biomed. Eng.* 51(12): 2097–2102.

Adler, A., Arnold, J.H., Bayford, R., Borsic, A., Brown, B., Dixon, P., Faes, T.J.C. et al. 2009. GREIT: A unified approach to 2D linear EIT reconstruction of lung images. *Physiol. Meas.* 30: S35–S55.

Assenheimer, M., Laver-Moskovitz, O., Malonek, D., Manor, D., Nahaliel, U., Nitzan, R., and Saad, A. 2001. The T-SCAN technology: Electrical impedance as a diagnostic tool for breast cancer detection. *Physiol. Meas.* 22(1): 1–8.

Avill, R.F., Mangnall, Y.F., Bird, N.C., Brown, B.H., Barber, D.C., Seagar, A.D., Johnson, A.G., and Read, N.W. 1987. Applied potential tomography: A new non-invasive technique for measuring gastric emptying. *Gastroenterology* 92: 1019.

Balasubramani, L., Brown, B.H., Healey, J., and Tidy, J.A. 2009. The detection of cervical intraepithelial neoplasia by electrical impedance spectroscopy: The effects of acetic acid and tissue homogeneity. *Gynecol. Oncol.* 115(2): 267–271.

Barber, D.C. 2005. EIT: The view from Sheffield. In D.S. Holder (Ed.), *Electrical Impedance Tomography: Methods, History and Applications*, Institute of Physics Publishing, Bristol, pp. 348–372.

Barber, D.C. and Brown, B.H. 1990. Progress in electrical impedance tomography. In D. Colton, R. Ewing and W. Rundell (Eds), *Inverse Problems in Partial Differential Equations*, pp. 149–162. New York: SIAM.

Barber, D.C. and Seagar, A.D. 1987. Fast reconstruction of resistive images. *Clin. Phys. Physiol. Meas.* 8: 47.

Bayford, R. 1994. Application of constrained optimization techniques in electrical impedance tomography, PhD thesis. Middlesex University, U.K.

Borsic, A., Graham, B.M., Adler, A., and Lionheart, W.R.B. 2010. *In vivo* impedance imaging with total variation regularization. *IEEE Trans. Med. Imag.* 29(1): 44–54.

Borsic, A., Halter, R., Wan, Y., Hartov, A., and Paulsen, K.D. 2009. Sensitivity study and optimization of a 3D electric impedance tomography prostate probe. *Physiol. Meas.* 30: S1–S18.

Borsic, A., Syed, H., Halter, R.J., and Hartov, A. 2011. Using ultrasound information in EIT reconstruction of the electrical properties of the prostate. In *Proceedings of the 12th International Conference in Electrical Impedance Tomography*, Bath, UK.

Breckon, W.R. 1990. Image Reconstruction in Electrical Impedance Tomography. Ph.D. thesis, School of Computing and Mathematical Sciences, Oxford Polytechnic, Oxford, UK.

Brown, B.H. and Barber, D.C. 1982. Applied potential tomography—A new in vivo medical imaging technique. In *Proceedings of Hospital Physicists' Association, Annual Conference*, Shefield.

Brown, B.H., Barber, D.C., Wang, W., Lu, L., Leathard, A.D., Smallwood, R.H., Hampshire, A.R., Mackay, R., and Hatzigalanis, K. 1994. Multifrequency imaging and modelling of respiratory related electrical impedance changes. *Physiol. Meas.* 15: A1.

Brown, B.H., Milnes, P., and Mills, G.H. 2006. Indirect measurement of lung density and air volume from electrical impedance tomography (EIT) data. In *Proceedings of 7th Conference on Biomed. Appl. Electrical Impedance Tomography*, Seoul, pp. 85–89.

Brown, D.C. and Seagar, A.D. 1987. The Sheffield data collection system. *Clin. Phys. Physiol. Meas.* 8(suppl A): 91–98.

Brown, B.H., Tidy, J., Boston, K., Blackett, A.D., and Sharp, F. 1998. Tetrapolar measurement of cervical tissue structure using impedance spectroscopy. In *IEEE/EMBS International Conference*, pp. 2886–2889.

Brown, B.H., Primhak, R.A., Smallwood, R.H., Milnes, P., Narracott, A.J., and Jackson, M.J. 2002. Neonatal lungs—Can absolute lung resistivity be determined non-invasively? *Med. Biol. Eng.* 40: 388–394.

Callaghan, M.F., Lund, T., Hashemzadeh, P., Roitt, I.M., and Bayford, R.H. 2010. An investigation of the impedance properties of gold nanoparticles. *J. Phys. Conf. Ser.* 224(1): 012058.

Cheney, M.D., Isaacson, D., Newell, J.C., Simske, S., and Goble, J. 1990. Noser: An algorithm for solving the inverse conductivity problem. *Int. J. Imag. Syst. Tech.* 2: 66–75.

Cherepenin, V.A., Karpov, A.Y., Korjenevsky, A., Kornienko, V., Mazaletskaya, A., Mazourov, D., and Meister, D. 2001. A 3D electrical impedance tomography (EIT) system for breast cancer detection. *Physiol. Meas.* 22(1): 9–18.

Cherepenin, V.A., Karpov, A.Y., Korjenevsky, A.V., Kornienko, V.N., Kultiasov, Y.S., Ochapkin, M.B., and Trochanova, O.V. 2002. Three-dimensional EIT imaging of breast tissues: System design and clinical testing. *IEEE Trans. Med. Imag.* 21(6): 662–667.

Cole, K.S. and Cole, R.H. 1941. Dispersion and absorption in dielectrics: I. Alternating current characteristics. *J. Chem. Phys.* 9: 431.

Davidson, J.L., Little, R.A., Wright, P., Naish, J., Kikinis, R., Parker, G.J.M., and McCann, H. 2011. Co-registration and fusion of EIT images with MRI scan data. In *Proceedings of the 12th International Conference in Electrical Impedance Tomography*, Bath, UK.

Dräger. Electrical impedance tomography: The realization of regional ventilation monitoring. http://www.draeger.com/media/10/08/98/10089883/rsp_eit_booklet_9066788_en.pdf.

Erol, R.A., Smallwood, R.H., Brown, B.H., Cherian, P., and Bardham, K.D. 1995. Detecting oesophageal-related changes using electrical impedance tomography. *Physiol. Meas.* 16(suppl 3A): 143–152.

Eyüboglu, B.M., Oner, A.F., Baysal, U., Biber, C., Keyf, A.I., Yilmaz, U., and Erdogan, Y. 1995. Application of electrical impedance tomography in diagnosis of emphysema—A clinical study. *Physiol. Meas.* 16: A191–A211.

Fabrizi, L., McEwan, A., Oh, T., Woo, E.J., and Holder, D.S. 2009. A comparison of two EIT systems suitable for imaging changes in epilepsy, *Physiol. Meas.*, 30: S103–S120.

Fabrizi, L., McEwan, A., Woo, E., and Holder, D.S. 2007. Analysis of resting noise characteristics of three EIT systems in order to compare suitability for time difference imaging with scalp electrodes during epileptic seizures. *Physiol. Meas.* 28: S217–S236.

Fabrizi, L., Sparkes, M., Horesh, L., Perez-JusteAbascal, J.F. , McEwan, A., Bayford, R.H., Elwes, R., Binnie, C.D., and Holder, D.S. 2006. Factors limiting the application of electrical impedance tomography for identification of regional conductivity changes using scalp electrodes during epileptic seizures in humans. *Physiol. Meas.* 27: 163–174.

Ferris, C.D. and Rose, D.R. 1972. An operational amplifier 4-electrode impedance bridge for electrolyte measurement. *Med. Biol. Eng.* 10: 647–654.

Frerichs, I. 2000. Electrical impedance tomography (EIT) in applications related to lung and ventilation: A review of experimental and clinical activities, *Physiol. Meas.* 21: R1–R21.

Fricke, H. 1925. The electrical capacity of suspensions with special reference to blood. *J. Gen. Physiol.* 9: 137–152.

Gilad, O. and Holder, D.S. 2009. Impedance changes recorded with scalp electrodes during visual evoked responses: Implications for electric impedance tomography of fast neural activity. *NeuroImage* 47: 412–522.

Gilad, O., Horesh, L., and Holder, D.S. 2007. Design of electrodes and current limits for low frequency electrical impedance tomography of the brain. *Med. Biol. Eng. Comput.* 45: 612–633.

Goble, J. and Isaacson, D. 1990. Fast reconstruction algorithms for three-dimensional electrical tomography. In *IEEE EMBS Proceedings of the 12th Annual International Conference*, Philadelphia, pp. 285–286.

Griffiths, H. 2005. Magnetic induction tomography. In D.S. Holder (Ed.), *Electrical Impedance Tomography: Methods, History and Applications*, Institute of Physics Publishing, Bristol, pp. 213–238.

Grimnes, S. and Martinsen, O.G. 2008. *Bioimpedance & Bioelectricity Basics*, 2nd Ed., Elsevier, Academic Press, Oxford, UK.

Grimnes, S. and Martinsen, O.G. 2010. Alpha dispersion in human tissue. *J. Phys. Conf. Ser.* 224(1): 012073.

Grychtol, B., Lionheart, W.R.B., Wolf, K.G., Bodenstein, M., and Adler, A. 2011. The importance of shape: Thorax models for GREIT. In *Proceedings of the 12th International Conference in EIT*, Bath, UK.

Hahn, G., Sipinková, I., Baisch, F., and Hellige, G. 1995. Changes in the thoracic impedance distribution under different ventilator conditions. *Physiol. Meas.* 16: A161–A173.

Hahn, G., Just, A., Dittmar, J., and Hellige, G. 2008. Systematic errors of EIT systems determined by easily-scalable resistive phantoms. *Physiol. Meas.* 29: S163–S172.

Halter, R., Hartov, A., and Paulsen, K.D. 2005. High frequency EIT for breast imaging. In *Conf. Biomed. Appl. Elec. Impedance Tomography, 6th*, London, UK.

Halter, R.J., Hartov, A., and Paulsen, K.D. 2007a. Experimental justification for using 3D conductivity reconstructions in electrical impedance tomography. *Physiol. Meas.* 28: S115–S127.

Halter, R., Schned, A., Heaney, J., Hartov, A., Schutz, S., and Paulson, K.D. 2007b. Electrical impedance spectroscopy of benign and malignant prostatic tissues. *J. Urol.* 4: 1580–1596.

Halter, R.J., Schned, A., Heaney, J., Hartov, A., Schutz, S., and Paulsen, K.D. 2008. Electrical impedance spectroscopy of benign and malignant prostatic tissue, *J. Urol.* 179(4): 1580–1586.

Hampshire, A.R., Smallwood, R.H., Brown, B.H., and Primhak, R.A. 1995. Multifrequency and parametric EIT images of neonatal lungs. *Physiol. Meas.* 16(suppl 3A): 175–189.

Han, Y.Q., Meng, Z.J., Jeong, W.C., Kim, Y.T., Minhas, A.S., Kim, H.J., Nam, H.S., Kwon, O., and Woo, E.J. 2010. MREIT conductivity imaging of canine head using multi-echo pulse sequence. *J. Phys. Conf. Ser.* 224(1): 012078.

Harris, N.D., Sugget, A.J., Barber, D.C., and Brown, B.H. 1987. Applications of applied potential tomography (APT) in respiratory medicine. *Clin. Phys. Physiol. Meas.* 8: 155–165.

Holder, D.S., Rao, A., and Hanquan, Y. 1996. Imaging of physiologically evoked responses by EIT tomography with cortical electrodes in the anaesthisised rabbit. *Physiol. Meas.* 17(A): 179–186.

Holder, D.S. and Temple, A.J. 1993. Effectiveness of the Sheffield EIT system in distinguishing patients with pulmonary pathology from a series of normal subjects. In D.S. Holder (Ed.), *Clinical and Physiological Applications of Electrical Impedance Tomography*, UCL Press, London, pp. 227–298.

Holder, D.S. (Ed.). 2005. *Electrical Impedance Tomography: Methods, History and Applications*, Institute of Physics Publishing, Bristol.

Hua, P., Webster, J.G., and Tompkins, W.J. 1988. A regularized electrical impedance tomography reconstruction algorithm. *Clin. Phys. Physiol. Meas.* 9(suppl A): 137–141.

Isaacson, D. 1986. Distinguishability of conductivities by electric current computed tomography. *IEEE Trans. Med. Imag.* 5: 91.

Jeon, K., Kim, H.J., Lee, C.O., Woo, E.J., and Seo, J.K. 2009. CoReHA: Conductivity reconstructor using harmonic algorithms for magnetic resonance electrical impedance tomography (MREIT). *J. Biomed. Eng. Res.* 30: 279–287.

Jeong, W.C., Kim, Y.T., Minhas, A.S., Lee, T.H., Kim, H.J., Nam, H.S., Kwon, O., and Woo, E.J. 2010. *In-vivo* conductivity imaging of human knee using 3 mA injection current in MREIT. *J. Phys. Conf. Ser.* 224(1): 012148.

Jossinet, J. and Schmitt, M. 1999. A review of parameters for the bioelectrical characterization of breast tissue. *Ann. NY Acad. Sci.* 20: 30–41.

Kim, H.J., Kim, Y.T., Minhas, A.S., Jeong, W.C., Woo, E.J., Seo, J.K., and Kwon, O.J. 2009. *In vivo* high-resolution conductivity imaging of the human leg using MREIT: The first human experiment. *IEEE. Trans. Med. Imag.* 28(11): 1681–1687.

Kim, H. and Woo, H.W. 1987. A prototype system and reconstruction algorithms for electrical impedance technique in medical imaging. *Clin. Phys. Physiol. Meas.* 8: 63.

Kohn, R.V. and Vogelius, M. 1984a. Determining the conductivity by boundary measurement. *Commun. Pure Appl. Math.* 37: 289.

Kohn, R.V. and Vogelius, M. 1984b. Identification of an unknown conductivity by means of the boundary. *SIAM-AMS Proc.* 14: 113.

Koire, C.J. 1992. EIT image reconstruction using sensitivity coefficient weighted backprojection. *Physiol. Meas.* 15(suppl 2A): 125–136.

Korjenevsky, A., Cherepenin, V., Trokhanova, O., and Tuykin, T. 2010. Gynecologic electrical impedance tomography. *J. Phys. Conf. Ser.* 224(1): 012070.

Kulkarni, R., Kao, T-J., Boverman, G., Saulnier, G.J., Isaacson, D., Szabo, T.L., and Newell J.C. 2010. A hand-held probe for combined ultrasound and electrical impedance tomography. *J. Phys. Conf. Ser.* 224(1): 012043.

Leathard, A.D., Brown, B.H., Campbell, J., Zhang, F., Morice, A.H., and Tayler, D. 1994. A comparison of ventilatory and cardiac related changes in EIT images of normal human lungs and of lungs with pulmonary embolism. *Physiol. Meas.* 15: A137.

Lindherholm, P., Marescot, L., Loke, M. H., and Renaud, P. 2008. Cell culture imaging using microimpedance tomography. IEEE. *Trans. Biomed. Eng.* 55(1): 138–145.

Lionheart, W.R.B. 2004. Review: EIT reconstruction algorithms: Pitfalls, challenges and recent developments. *Physiol. Meas.* 25(1): 125.

Lionheart, W.R.B. and Paridis, K. 2010. Finite elements and anisotropic EIT reconstruction. *J. Phys. Conf. Ser.* 224(1): 012022.

Liu, Q., Wi, H., Oh, T.I., Woo, E.J., and Seo, J.K. 2010. Development of a prototype micro-EIT system using three sets of 15 × 8 array electrodes. *J. Phys. Conf. Ser.* 224(1): 012161.

Lobstein-Adams, C., Beqo, N., and Wang, W. 2011. Electrical impedance assisted ultrasound guided focused ultrasound surgery: A comprehensive out-patient procedure? In *Proceedings of the 12th International Conference in Electrical Impedance Tomography*, Bath, UK.

Luepschen, H., Meier, T., Grossherr, M., Leibecke, T., Karsten, J., and Leonhardt, S. 2007. Protective ventilation using electrical impedance tomography, *Physiol. Meas.* 28: S247–S260.

Lukaszewicz, K., Wtorek, J., Bujnowski, A., and Skokowski, J. 2010. Monitoring of breast tissue thermoablation by means of impedance measurements. *J. Phys. Conf. Ser.* 224(1): 012136.

Malich, A., Boehm, T., Facius, M., Freesmeyer, M.G., Fleck, M., Anderson, R., and Kaiser, W.A. 2001. Differentiation of mammographically suspicious lesions: Evaluation of breast ultrasound, MRI mammography and electrical impedance scanning as adjunctive technologies in breast cancer detection. *Clin. Radiol.* 56: 278–283.

Manwaring, P.K., Halter, R.J., Borsic, A., and Hartov, A. 2010. A modified electrode configuration for brain EIT. *J. Phys. Conf. Ser.* 224(1): 012062.

Marescot, L., Loke, M.H., Chapelier, D., Delaloye, R., Lambiel, C., and Reynard, E. 2003. Assessing reliability of 2D resistivity imaging in mountain permafrost studies using the depth of investigation index method. *Near Surface Geophys.* 1: 57–68.

Martinsen, Ø.G., Grimnes, S., Lütken, C.A., and Johnsen, G.K. 2010. Memristance in human skin. *J. Phys. Conf. Ser.* 224(1): 012071.

McAdams, E.T. and Jossinet, J. 1995. Tissue impedance: A historical overview. *Physiol. Meas.* 16: A1–A13.

McAdams, E.T., McLaughlin, J.A., and Anderson, J.Mc.C. 1994. Multielectrode systems for electrical impedance tomography. *Physiol. Meas.* 15: A101.

McEwan, A., Cusick, G., and Holder, D.S., 2007. A review of errors in multi-frequency EIT instrumentation. *Physiol. Meas.* 28: S197–S215.

Metherall, P., Barber, D.C., Smallwood, R.H., and Brown, B.H. 1996. Three-dimensional electrical impedance tomography. *Physiol. Meas.* 15: A101.

Morice, A.H., Harris, N., Campbell, J., Zhang, F., and Brown, B. 1993. EIT in the investigation of chest disease. In *Clinical and Physiological Applications of Electrical Impedance Tomography*, UCL Press, London, pp. 655–677.

Morucci, J.P., Marsili, P.M., Granie, M., Dai, W.W., and Shi, Y. 1994. Direct sensitivity matrix approach for fast reconstruction in electrical impedance tomography. *Physiol. Meas.* 15(suppl 2A): 107–114.

Newell, J.C., Edic, P.M., Ren, X., Larson-Wiseman, J.L., and Danylecko, M.D. 1996. Assessment of acute pulmonary edema in dogs by electrical impedance imaging. *IEEE Trans. Biomed. Eng.* 43: 133–139.

Nicander, I., Ollmar, S., Eek, A., LundhRozell, B., and Emtestam, L. 1996. Correlation of impedance response patterns to histological findings in irritant skin reactions induced by various surfactants. *Br. J. Dermatol.* 134: 221–228.

Oh, S.H., Lee, B.I., Woo, E.J., Lee, S.Y., Kim, T-S., Kwon, O., and Seo, J.K. 2005. Electrical conductivity images of biological tissue phantoms in MREIT. *Physiol. Meas.* 26: S279–S288.

Oh, S.H., Lee, B.I., Woo, E.J., Lee, S.Y., Cho, M.H., Kwon, O., and Seo, J.K. 2003. Conductivity and current density image reconstruction using harmonic Bz algorithm in magnetic resonance electrical impedance tomography. *Phys. Med. Biol.* 48: 3101–3116.

Oh, T.I., Woo, E.J., and Holder, D. 2007. Multi-frequency EIT system with radially symmetric architecture: KHU Mark1. *Physiol. Meas.* 28: S183–S196.

Ollmar, S. and Emtestam, L. 1992. Electrical impedance applied to non-invasive detection of irritation in skin. *Contact Dermatitis* 27: 278–282.

Paulson, K., Breckon, W., and Pidcock, M. 1992. Optimal measurements in electrical impedance tomography, In *14th IEEE EMBS, Annual International Conference*, Paris.

Qiao, G., Wang, W., Sze, G., Hussain, W., Wang, L., and Al-Akaidi, M. 2008. Investigation of in-vitro bioimpedance test system for mouse pancreas. In *9th Electrical Impedance Tomography Conference*, Hanover, USA.

Rebeyrol, J., Moreno, M-V., Ribbe, E., and Vannicatte, A. 2010. Case study: Using monitoring of body composition data obtained by bioimpedance, in training of an elite male runner. In *ISEA, Proceedings of 8th*, 2, pp. 3059–3064.

Saulnier, G.J. 2005. EIT instrumentation. In D.S. Holder (Ed.), *Electrical Impedance Tomography: Methods, History and Applications*, Institute of Physics Publishing, Bristol pp. 67–104.

Saulnier, G.J., Liu, N., Tamma, C., Hongjun, X., Tzu-Jen Kao, Newell, J.C., and Isaacson, D. 2007. An electrical impedance spectroscopy system for breast cancer detection. *Proceedings of the 29th Annual International Conference on IEEE EMBS.* pp. 4154–4157.

Seo, J.K., Kim, S.W., Kim, S., Liu, J., Woo, E.J., Jeon, K., and Lee, C-O. 2008. Local harmonic B_z algorithm with domain decomposition in MREIT: Computer simulation study. *IEEE Trans. Med. Imag.* 27: 1754–1761.

Seo, J.K. and Woo, E.J. 2011. Magnetic resonance electrical impedance tomography (MREIT). *SIAM Rev.* 53(1): 40–68.

Seo, J.K., Yoon, J.R., Woo, E.J., and Kwon, O. 2003. Reconstruction of conductivity and current density images using only one component of magnetic field measurements. *IEEE Trans. Biomed. Eng.* 50: 1121–1124.

Shinkarenko, V.S., Chuchalin, A.G., Kostromina, E., Aïsanov, Z.R., Pashkova, T.L., Voloshina, N.A., and Iag'ia, T.N. 1997. Electrical impedance tomography in pulmonology. *Ter. Arkh.* 69: 48–51.

Smit, H.J., VonkNoordegraaf, A., van Genderingen, H.R., and Kunst, P.W.A. 2005. Imaging of the thorax by EIT. In D.S. Holder (Ed.), *Electrical Impedance Tomography: Methods, History and Applications*, Institute of Physics Publishing, Bristol, pp. 107–126.

Soulsby, C., Yazaki, E., and Evans, D.F. 2005. Applications of electrical impedance tomography in the gastrointestinal tract. In D.S. Holder (Ed.), *Electrical Impedance Tomography: Methods, History and Applications*, Institute of Physics Publishing, Bristol, pp. 186–206.

Surowiec, A.J., Stuchly, S.S., Barr, J.R., and Swarup, A. 1988. Dielectric properties of breast carcinoma and the surrounding tissues. *IEEE Trans. Biomed. Eng.* 35: 257–263.

Sylvester, J. and Uhlmann, G. 1986. A uniqueness theorem for an inverse boundary value problem in electrical prospection. *Commun. Pure Appl. Math.* 39: 91.

Sze, G., Wang, W., Barber, D.C., and Huber, N. 2011. Preliminary study of the sensitivity of the Sussex MK4 Electrical Impedance Mammography planar electrode system. In *Proceedings of the 12th International Conference in Electrical Impedance Tomography*, Bath, UK.

Tang, M., Wang, W., Wheeler, J., McCormick, M., and Dong, X., 2002. Effects of incompatible boundary information in EIT on the convergence behaviour of an iterative algorithm. *IEEE Trans. Biomed. Eng.* 21(6): 620–628.

Tehrani, J.N., Anderson, C., Jin, C., Van Schaik, A., Holder, D.S., and McEwan, A. 2010. Feasibility of electrical impedance tomography in heamorrhagic stroke treatment using adaptive mesh. *J. Phys. Conf. Ser.* 224(1): 012065.

Tizzard, A., Borsic, A., Halter, R., and Bayford, R. 2010. Generation and performance of patient-specific forward models for breast imaging with EIT. *J. Phys. Conf. Ser.* 224(1): 012034.

Tunstall, B., Wang, W., McCormick, M., Walker, R., and Rew, D. 1997. Preliminary *in vitro* studies of electrical impedance mammography (EIM): A future technique for non-invasive breast tissue imaging? *Breast* 6(4): 253.

Von Noordergraaf, A. A., Kunst, P.W., Janse, A., Marcus, J.T., Postmus, P.E., Faes, T.J., and deVries, P.M. 1998. Pulmonary perfusion measured by means of electrical impedance tomography. *Physiol. Meas.* 19(2): 263–273.

Wan, Y., Halter, R., Borsic, A., Manwaring, P., Hartov, A., and Paulsen, K.D. 2010. Sensitivity study of an ultrasound coupled transrectal electrical impedance tomography system for prostate imaging. *J. Phys. Conf. Ser.* 224(1): 012067.

Wang, L., Wang, W., Qiao, G., Brien, M., and Al-Akaidi, M. 2006. Preliminary report of optimisation of in-vitro studies with an in-vitro specimen measuring system. In *Proceedings of the 7th Conference on Biomedical Applications of Electrical Impedance Tomography, WC2006*, Korea.

Wang, W., Cheng, Z., and McCormick, M. 1996. Design of programmable wide bandwidth current source for an impedance tomography system. In *IEEE, EMBS 18th Annual International Conference*, Amsterdam, the Netherlands.

Wang, W., Wang, L., Huang, T., Tunstall, B., Gu, D-W., and Sze, G. 2007a. Parametric image: A step forward for virtual biopsy by EIT? In *IFMBE Proceedings 2007*. 17(1): 432–435.

Wang, W., Wang, L., Qiao, G., Prickett, P., Bramer, B., Tunstall, B., and Al-Akaidi, M. 2007b. Study into the repeatability of the electrode-skin interface utilizing electrodes commonly used in electrical impedance tomography. *In IFMBE Proceedings* 2007, 17(10): 336–339.

Wexler, A., Fry, B., and Neuman, M.R. 1985. Impedance-computed tomography: Algorithm and system. *Appl. Opt.* 24: 3985.

Wheeler, J., Wang, W., and Tang, M. 2002. A comparison of methods for measurement of spatial resolution in two-dimensional circular EIT images. *Physiol. Meas.* 23(1): 169–176.

Woo, E.J. and Seo, J.K. 2008. Magnetic resonance electrical impedance tomography (MREIT) for high-resolution conductivity imaging. *Physiol. Meas.* 29: R1–R26.

Woo, E.J., Seo, J.K., and Lee, S.Y. 2005. Magnetic resonance electrical impedance tomography (MREIT). In D.S. Holder (Ed.), *Electrical Impedance Tomography: Methods, History and Applications*, Institute of Physics Publishing, Bristol, pp. 239–294.

Wright, J.W. 1994. The effect of intraluminal content on gastrointestinal motility in man, PhD thesis, Nottingham, UK.

York, T. 2005. Electrical tomography for industrial applications. In D.S. Holder (Ed.), *Electrical Impedance Tomography: Methods, History and Applications*, Institute of Physics Publishing, Bristol, pp. 105–209.

Yorkey, T.J. 1986. Comparing Reconstruction Algorithms for Electrical Impedance Imaging. Ph.D. thesis, University of Wisconsin, Madison, WI.

Zadehkoochak, M., Blott, B.H., Hames, T.K., and George, R.E. 1991. Spectral expansion analysis in electrical impedance tomography. *J. Phys. D: Appl. Phys.* 24: 1911–1916.

Zhao, M., Liu, Q., Oh, T.I., Woo, E.J., and Seo, J.K. 2010. Development of a trans-admittance mammography (TAM) using 60 × 60 electrode array. *J. Phys. Conf. Ser.* 224(1): 012045.

9

Magnetic Resonance Imaging of Atherosclerosis

Chun Yuan
University of Washington

William S. Kerwin
University of Washington

Gador Canton
University of Washington

Jinnan Wang
Philips Research North America

Huijun Chen
University of Washington

Niranjan Balu
University of Washington

9.1 Introduction

Despite new drug therapies and increasing research, deaths from heart disease and stroke remain the leading causes of death in the United States, the developed world, and are increasingly prevalent in the developing world. Clinically, the majority of these deaths are attributed to atherosclerosis, literally meaning "hardening of the arteries" (Ross et al. 1999). Although the outcomes of the disease are clear, the exact process and biological mechanisms that lead to death or a clinical event by atherosclerosis remain an intensively studied area.

It has been thought that heart attack, stroke, and other clinical events from atherosclerosis are caused by vessel occlusion; a buildup of plaque which eventually blocks the artery, removing the blood supply. Thus, extensive medical imaging techniques have been developed that focus on identifying narrowing of blood vessels (stenosis). In medical terminology, these techniques are referred to as angiographic methods. Angiographic methods, with ultrasound, x-ray, computed tomography (CT), and magnetic resonance imaging (MRI), remain the standard methodology for clinical assessment of the severity of atherosclerotic disease, and are used as criteria for surgical decision making (Achenbach et al. 2010).

However, as indicated by the continued high rates of mortality, assessment of luminal narrowing provides an incomplete picture of the severity of atherosclerosis. In part, this is due to the phenomenon of arterial remodeling (Glagov et al. 1987)—the vessel wall is a responsive organ and atherosclerotic plaques may develop without infringing on the vessel lumen by pushing the vessel wall outward. It is the disease within the vessel wall that contributes to the progression and clinical complication of atherosclerosis. Recently, the concept of "vulnerable plaque" (VP) was introduced defining a series of different types of atherosclerotic plaques with higher risk of causing clinical events such as heart attack and stroke (Naghavi et al. 2003). Most of these plaque types involve plaque tissue compositions and distributions which lead to plaque rupture and subsequent clinical events. Vessel wall imaging (VWI) has also undergone rapid development, in parallel with the realization of the importance of VP.

VWI is defined as using imaging means to characterize atherosclerotic plaques of the vessel wall *in vivo*. Figure 9.1 shows a typical atherosclerotic plaque excised intact from the human carotid artery during a surgical procedure called carotid endarterectomy. It highlights the complexity of an advanced plaque and outlines the need for VWI, including ways to measure the size of an atherosclerotic plaque (volume, cross-sectional area, thickness), identify and quantify plaque tissue composition, determine inflammatory activity, and plaque–lumen interface condition such as ulcers and surface disruption.

This chapter provides an overview of MRI-based techniques for VWI and the basic hardware requirements and acquisition techniques, image processing tools which enable qualitative and quantitative data analysis, and advanced and developing techniques such as dynamic contrast imaging (DCE), and biomechanics.

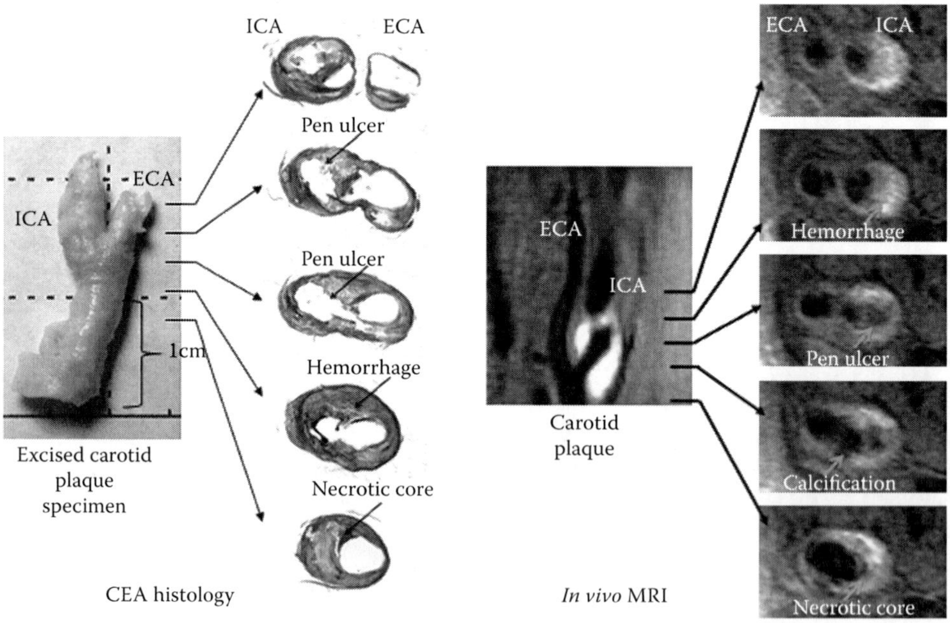

FIGURE 9.1 Two examples of carotid atherosclerotic plaque visualization. On the left is a photograph of a carotid endarterectomy (CEA) specimen measuring 2 cm in length. Serial sections of the specimen show the components of the plaque: intraplaque hemorrhage; necrotic core; calcification; and penetrating ulcer. The blue stain is a Mallory's trichrome. On the right are *in vivo* MRI images showing the common carotid bifurcation which separates into the external carotid (ECA) and internal carotid arteries (ICA). The longitudinal scan reveals a large bright plaque that almost occludes the lumen in the internal carotid. The cross-sectional images show the complexity of the lesion. Hemorrhage, calcification, necrotic core, and a penetrating ulcer are clearly visible.

9.2 Vessel Wall MRI Acquisition Techniques

VWI using MRI within a clinical feasible time frame is a demanding task that presents a number of technical challenges. Atherosclerotic lesions in the vessel wall with small but complicated plaque structures require high signal-to-noise ratio (SNR) imaging with high spatial resolution and good lumen-to-wall contrast. A high spatial resolution of 0.5–0.8 mm in the wall thickness direction is generally needed to image major arteries in the human body. Even higher resolutions may be needed when smaller or more peripheral arteries are to be imaged. To achieve this resolution, specially designed radiofrequency (RF) coils are needed to improve the SNR of the acquired images. Signal due to flowing blood and motion of surrounding soft tissues must also be considered in vessel wall MRI. Effective blood suppression is required for lumen boundary identification. Additionally, to detect and accurately quantify the various tissue components found in the atherosclerotic plaques, imaging techniques also need to be specifically optimized for certain types of tissue contrast. In addition to the above considerations, blood vessels are often surrounded by fat, and fat suppression is required to better define the margins of the outer wall. Techniques for motion artifact suppression can also help improve vessel wall boundary identification on images.

Recent developments in vessel wall MR imaging have successfully addressed many of the above-mentioned challenges, allowing VWI in clinical environments. These technical advancements will be discussed in detail throughout the section. The carotid arteries are frequent sites of atherosclerosis and several MR imaging techniques have been developed for carotid imaging. Since histological confirmation of image findings is best performed in the carotid artery using surgical endarterectomy specimens, many imaging techniques have been first developed in carotid arteries and then migrated to other arterial beds. In this section, we focus on carotid imaging techniques and their extension to other arterial beds.

9.2.1 Hardware

9.2.1.1 Field Strength

A whole-body clinical scanner can be used for VWI if it is equipped with dedicated RF coils to provide high SNR while maintaining the high spatial resolution required for discerning small plaque components. While field strengths from 1.5 T have been used for vessel wall MRI, higher main magnetic field strengths provide increased SNR and thereby higher increase contrast-to-noise ratio (CNR) in VWI (Figure 9.2). Compared to 1.5 T, black-blood carotid MRI at 3 T provides a 1.4–2.4 times higher CNR (Yarnykh et al. 2006). While plaque morphological measurements and identification of plaque components are similar at 1.5 and 3 T field strengths, appearance of some tissue components may differ at 3 T and higher field strengths due to increased susceptibility.

Moving to even higher field strengths such as 7 T can theoretically improve carotid VWI further, and a 3 times SNR gain at 7 T over 3 T has been demonstrated (Wiggins et al. 2009). However, the potential increase in SNR from high field imaging is outweighed by other issues at high field strengths, such as higher susceptibility, and B_0 and B_1 inhomogeneities. Furthermore, MR techniques based on balanced steady-state gradient echo imaging cannot be applied at 7 T fields due to higher field inhomogeneities. For these reasons, VWI is currently best performed at 3 T. 3 T VWI is capable of visualizing all major arterial beds, including carotid, aorta, femoral, and coronary. 3 T MR scanners are also widely available and used in clinical practice, making it possible to transition even advanced 3 T techniques to clinical practice. In the future, as MR techniques to mitigate effects of susceptibility and field inhomogeneities become available, VWI may become practical at higher field strengths.

9.2.1.2 Coils

Body coil transmission is used for VWI and is useful for uniform excitation and large coverage needed for spatial blood suppression sequences. However, body coil reception does not provide high-enough

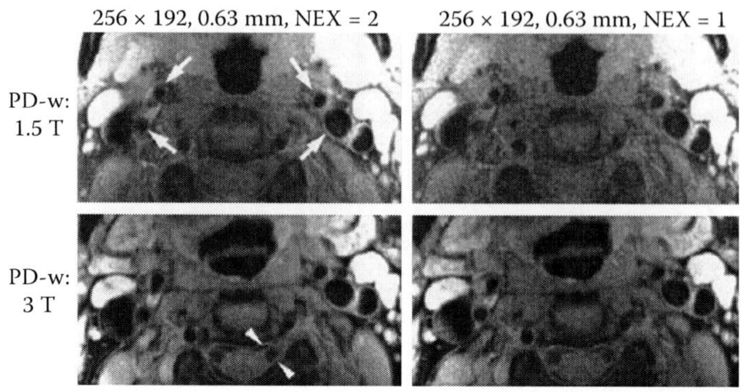

FIGURE 9.2 Axial carotid MRI sequences obtained at 1.5 and 3 T using the same sequence parameters. Note the improved SNR at 3 T with NEX = 1 showing improved vessel wall delineation (arrow) and appearance of small structures as nerve roots (arrowheads). With NEX = 2 at 1.5 T, some of the SNR loss can be offset but comes at the expense of increased scan time. (Adapted from Yarnykh VL et al. 2006. *J Magn Reson Imaging.* 5:691–698.)

SNR for visualizing the vessel wall. Surface coils can be applied close to the area of interest and pick up signal from a limited region of interest with high sensitivity. For example, the carotid bifurcation is a frequent location for atherosclerosis and is located close to the surface. Phased-array surface coils designed for carotid artery imaging provide improved SNR at the carotid bifurcation compared to body coil reception. Increasing the number of coil elements can provide improved coverage coupled with the inherent high SNR of surface coil imaging. Four- and eight-element bilateral phased-array carotid coils are now commercially available. Figure 9.3 shows an 8-element coil design and its commercial available design. The 8-element phased-array provides 1.7-fold SNR increase and increased longitudinal coverage compared to its 4-channel predecessor (Balu et al. 2009b). While these coils can be used for parallel imaging, VWI is often SNR limited. Therefore, parallel imaging in VWI is not preferred since it can decrease SNR. Thoracic vessels can be imaged with cardiac coils and peripheral vessels can be imaged with general-purpose surface coils.

9.2.2 Black-Blood Imaging Techniques

Effective suppression of the blood signal in the vessel wall is critical for accurate atherosclerotic plaque burden measurements because residual blood signal, commonly referred to as plaque-mimicking artifacts, can often be confused as part of the vessel wall, thus leading to inaccurate lumen/wall boundary segmentation. In addition, the effective suppression of the blood signal also allows for more prominent vessel wall visualization. *Black-blood (BB) imaging* indicates special MRI techniques in which the blood signal is suppressed. The suppression of blood signal generally takes the advantage of one or multiple of the following properties of the blood: the *inflow effect*, the unique magnetic relaxation values (T_1 and T_2 times) of the blood, and the diamagnetic nature of the blood. These properties are fairly unique to blood and therefore could be used to differentiate blood from surrounding tissues like the vessel wall.

9.2.2.1 Inflow-Suppression Technique

The *inflow-suppression* (IS) technique (Edelman et al. 1990) uses only the inflow effect for blood suppression. As shown in Figure 9.4, the essence of the IS technique is very straightforward: suppressing the upstream blood signal before it enters the imaging volume and acquiring the image once the imaging volume is filled with nullified blood.

In the IS technique, the upstream blood is usually suppressed by applying an excitation pulse (usually 90°) to flip all the spins and then the signal is suppressed with a spoiler gradient. With a certain

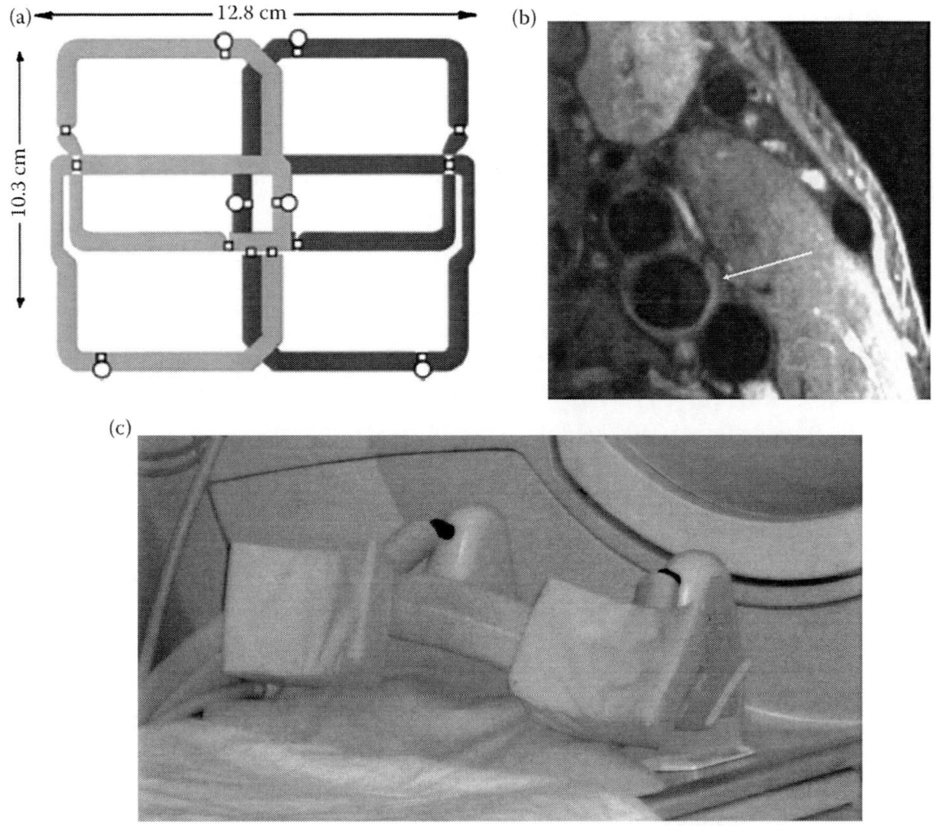

FIGURE 9.3 The 8-element carotid coil is made of bilateral coils of four elements each. The four elements of one side are shown in (a) with resistors and capacitors. High resolution (0.27 mm^2 in-plane resolution T_1w MRI) (b) obtained with the coil shows excellent delineation of the thin carotid walls. (c) A commercial version of the coil design in (a) showing the bilateral carotid "paddles" and integrated head rest. (Adapted from Balu N et al. 2009a. *ISMRM*, April, pp. 18–24, Honolulu, Hawaii.)

amount of delay time, after all prepared blood in the imaging volume is completely replaced by the nullified blood, a black-blood image can be acquired. The module itself usually takes no more than tens of milliseconds and can usually be combined with any kind of sequence (spin echo, gradient echo, etc.) for data acquisition. The delay time between the IS and acquisition modules is usually very short to avoid the blood signal increase due to the T_1 relaxation.

Compared to the other BB techniques, the advantage of the IS technique is the ease of implementation and directional selectivity. It is typically available in clinical MR scanners and no special optimization

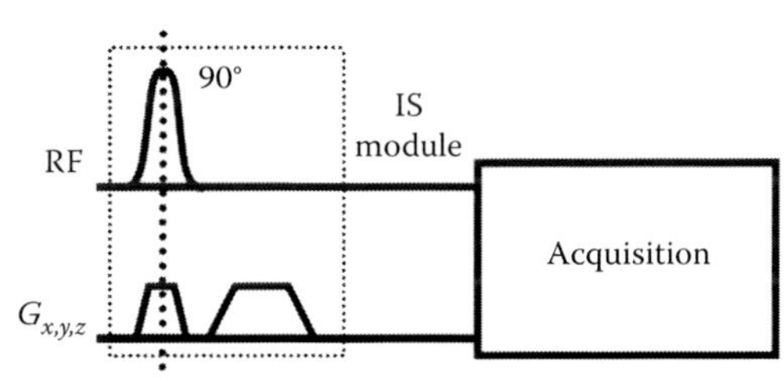

FIGURE 9.4 Pulse sequence diagram of the inflow-suppression technique.

is needed. Benefiting from the flexibility of placing the suppression bands, the blood from different directions can be selectively suppressed as long as the suppression bands are properly placed. The IS technique is therefore also frequently used to suppress only the arterial (or venous) flows without affecting flow in the other direction.

A major disadvantage of the IS technique is insufficient blood suppression. Owing to the short delay between the IS and acquisition modules, slow-flowing blood and/or blood flow without a through-plane velocity usually cannot be effectively replenished by the nullified blood, leading to an insufficient BB effect. The IS technique is therefore not frequently used for quantitative vessel wall MR applications; rather, it is more frequently used as an auxiliary tool to suppress blood signal when other imaging goals are explored, for example, to help maintain a black-blood contrast in perfusion MRI.

9.2.2.2 Double Inversion Recovery Technique and Its Variations

The *Double Inversion Recovery (DIR) technique* (Edelman et al. 1991) suppresses blood signal by using both the inflow effect and the T_1 relaxation time of the blood. Compared to the IS technique, the DIR technique can suppress blood more efficiently, especially in regions where slow and/or recirculating flow exists.

The pulse sequence of the DIR sequence is shown in Figure 9.5. It comprises two inversion pulses, where the first one is a nonselective RF pulse and the second is a selective RF pulse that covers only the imaging volume. The two RF pulses are usually placed as close as possible to minimize the magnetization perturbation for the spins inside the imaging volume. The net effect after the application of both pulses is that the magnetizations of all spins outside the imaging volume will be inverted while the magnetizations of the spins inside the imaging volume remain untouched. Since all blood magnetizations outside the imaging volume have been inverted, the images will be acquired after a delay time TI, at which time the blood signal reaches zero magnetization through T_1 relaxation.

The selection of the TI time is determined by both the T_1 relaxation time of the blood and the particular sequence used in acquisition. For example, if a standard spin echo sequence is used, the optimal TI is determined by

$$\mathrm{TI} = -T_1 \ln \frac{1 + e^{-\mathrm{TR}/T_1}}{2} \tag{9.1}$$

where T_1 is the blood relaxation time and TR is the repetition time of the sequence. Due to the relatively long T_1 time of the blood, the optimal TI time is usually a few hundred milliseconds.

Similar to the IS technique, the DIR technique also requires the blood in the imaging volume to be completely replenished by the inverted blood to achieve the best black-blood imaging effect. Due to the longer delay (TI) time, the DIR sequence allows for even slow flow in the imaging volume to be replenished. It therefore greatly relaxes the flow velocity requirement to achieve satisfactory BB effect.

However, DIR is still limited in a number of ways: (1) it still depends on the complete outflow of the blood to achieve satisfactory BB effect, which could be an issue for large-coverage three-dimensional (3D) imaging and in-plane blood flow suppression; (2) it is essentially a single-slice imaging scheme

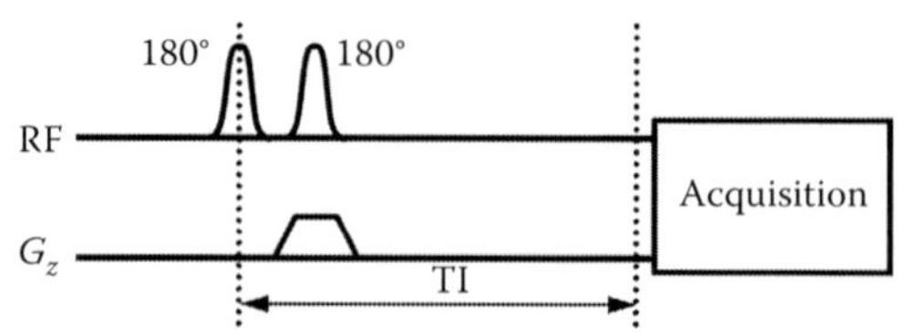

FIGURE 9.5 Pulse sequence diagram of the double inversion recovery technique.

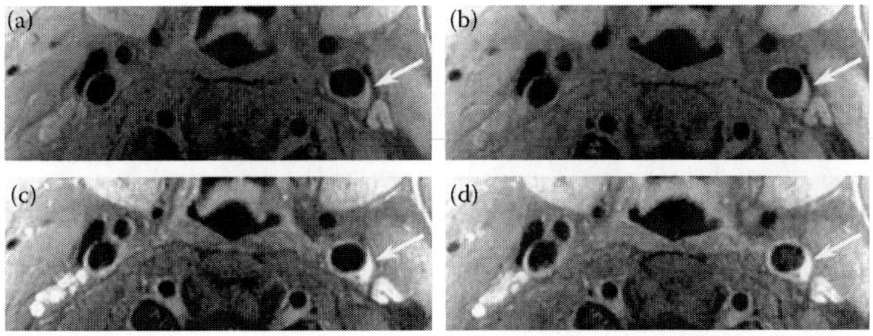

FIGURE 9.6 (a) and (b) Pre- and (c) and (d) postcontrast T_1-weighted black-blood images of carotid arteries: (a) and (c) images acquired with the QIR preparation; (b) and (d) images acquired with the DIR sequence. Note marked enhancement in the lesion and incomplete flow suppression in the left common carotid artery on (d) the postcontrast DIR image. Blood suppression is markedly improved postcontrast by use of the QIR preparation. (Adapted from Yarnykh VL and Yuan C. 2002. *Magn Reson Med.* 5:899–905.)

so it is usually very time consuming if a large volume of tissue is to be covered; and (3) it relies on the blood T_1 value for parameter optimization so the BB effect will become suboptimal or completely ineffective after the contrast agent is administered. To address these issues, a number of techniques have been proposed to improve the acquisition efficiency (Parker et al. 2002; Yarnykh and Yuan 2003; Itskovich et al. 2004), to improve lumen/wall contrast (Abd-Elmoniem et al. 2010), and to remove T_1 sensitivity (Yarnykh and Yuan 2002). As shown in Figure 9.6, the quadruple inversion recovery (QIR) technique has been shown to be very insensitive to the blood T_1 change and therefore maintains very good blood suppression efficiency even for postcontrast BB imaging when blood T_1 times are much lower than precontrast values. These solutions, however, usually come at a price of reduced blood suppression efficiency.

9.2.2.3 Motion-Sensitized-Driven Equilibrium Technique

Unlike IS and DIR, the *motion-sensitized-driven equilibrium (MSDE) technique* suppresses blood with no direct dependence on the inflow effect from the blood; rather, it suppresses blood based on phase dispersion of magnetization among moving spins (Koktzoglou and Li 2007; Wang et al. 2007). As shown in Figure 9.7, the MSDE sequence utilizes the motion-sensitizing gradient pulses to induce phase dispersions among spins with different velocities. For static tissues, no phase dispersion will be introduced by the gradient pairs. For flowing blood, however, different phases will be introduced for spins with different velocity vectors. If the phase variation is significant enough within a pixel, the signal will be dephased and black-blood contrast will be achieved.

Because the MSDE sequence does not directly rely on the flow replenishing rate for blood suppression, it provides more effective blood suppression compared to the IS and DIR techniques. As shown in Figure 9.8A, when different BB techniques were applied to image the carotid bulb, where stagnant flow tends to be found, only the MSDE technique completely removed the flow artifacts. For the same reason, the MSDE technique is also more suitable for tortuous arteries, in-plane flow suppression and

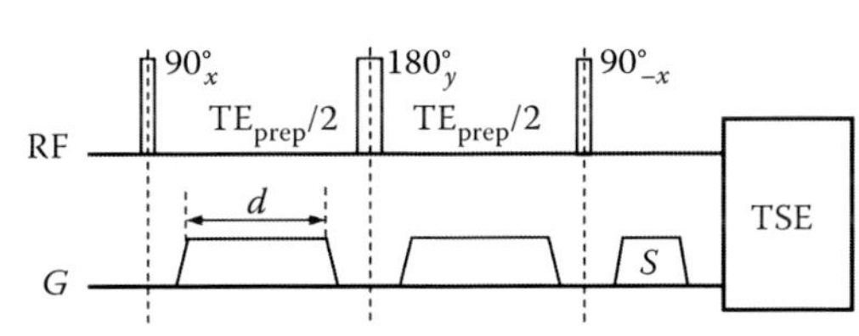

FIGURE 9.7 Pulse sequence diagram of the motion-sensitized-driven equilibrium (MSDE) sequence.

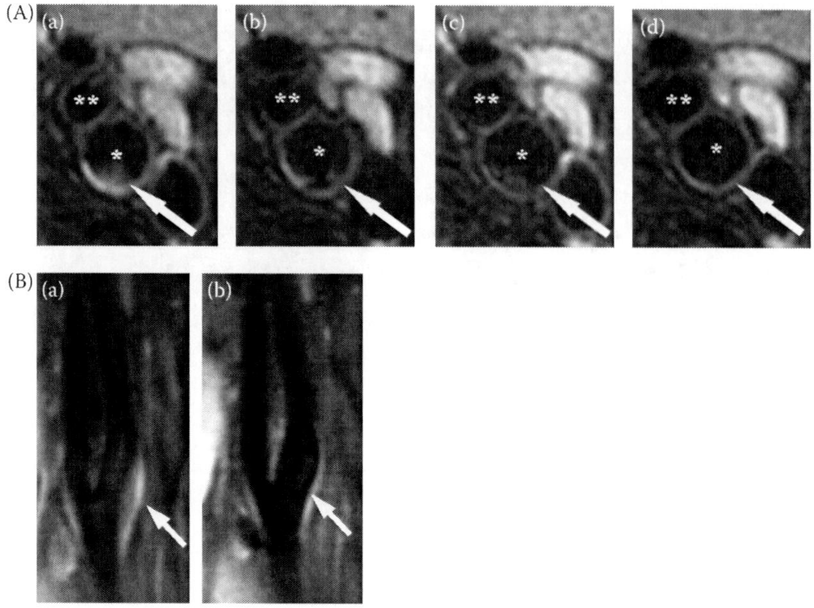

FIGURE 9.8 (A) Sample images illustrating the flow suppression capability of different BB imaging techniques: (a) IS, (b) DIR, (c) multislice DIR, and (d) MSDE. The residual flow signal is eliminated by the MSDE sequence but is visible for other techniques (arrows). All images are presented using the identical window settings. *, internal carotid lumen; **, external carotid lumen. (Adapted from Wang J et al. 2007. *Magn Reson Med.* 58(5):973–981.) (B) Sample sagittal carotid bifurcation images illustrating the improved flow suppression capability of MSDE for in-plane flows. Plaque-mimicking artifacts (arrows) can be observed on the multislice DIR images (a) but were completely removed on the MSDE images (b).

large-coverage 3D imaging, as shown in Figure 9.8B. The MSDE technique can also be used for postcontrast imaging since it does not rely on the blood T_1 time for blood suppression.

One limitation of the MSDE sequence is the signal drop caused by the inherent T_2/diffusion decay and/or system imperfection like magnetic field inhomogeneity and eddy current effects. In a typical clinical environment, this may cause ~12% of signal loss. An improved MSDE sequence (Wang et al. 2010b) has also been proposed to compensate for the signal loss caused by the RF transmission field inhomogeneity and eddy current effects. In the improved MSDE version, a second refocusing pulse is added reducing effects of field inhomogeneity.

9.2.2.4 Other Black-Blood Imaging Techniques

In addition to the BB techniques mentioned above, other black-blood imaging techniques based on different mechanisms have also been proposed. The T_2-IR technique achieves black-blood contrast by taking advantage of the longer T_1 and T_2 values of the blood compared to other tissues (Brown et al. 2010). In this technique, when a T_2 preparation module is combined with an inversion RF pulse, the black-blood effect can be achieved if the TI value is properly selected. Compared to the other techniques, the advantage of the T_2-IR technique is that it poses no reliance on the blood flow to achieve the black-blood effect. It can therefore usually achieve satisfactory blood suppression in even the most challenging regions. The T_2-IR technique, however, is still limited by the fact that it heavily relies on the blood T_1 and T_2 values for BB effect and therefore may need to be individually optimized for postcontrast imaging. Besides, it is also limited by the fact that it is a single slice imaging technique and therefore may have to be acquired at reduced time efficiency. Besides the inflow effect and blood relaxation properties, the difference in *diamagnetism* between blood and vessel wall has been used for VWI. The *susceptibility weighted imaging (SWI)* technique has been shown to demonstrate a distinct lumen/wall separation when used for peripheral artery plaque burden imaging (Yang et al. 2009). Since no flow velocity requirement is needed for this technique, the SWI image is also expected to achieve satisfactory lumen/wall separation in most circumstances. A current limitation of the technique is that no uniform phase enhancement maybe achieved for complicated plaque components where different tissues with various diamagnetic effect may present. This inhomogeneous phase shift may cause impaired lumen and/or outer wall boundaries segmentation (Yang et al. 2009).

9.2.3 Plaque Composition and Multicontrast MRI

The atherosclerotic plaque can contain several components such as calcification, lipid-rich necrotic core (LRNC), and intraplaque hemorrhage (IPH). Relaxation properties of tissues comprising each plaque component provide an indication of the type of plaque component. By varying pulse sequence parameters, multicontrast images of each component can be obtained and analyzed to identify the particular plaque component. For example, Figure 9.9 shows the appearance of IPH using a multicontrast MRI protocol. The area identified as IPH on MRI can be seen to agree well with IPH on histology. Appearance of LRNC and calcification on multicontrast MRI is shown in Figure 9.10A and B, respectively. Knowledge of the histological and chemical constituents of plaque components are important in choosing optimal pulse sequence parameters for plaque imaging and are considered next.

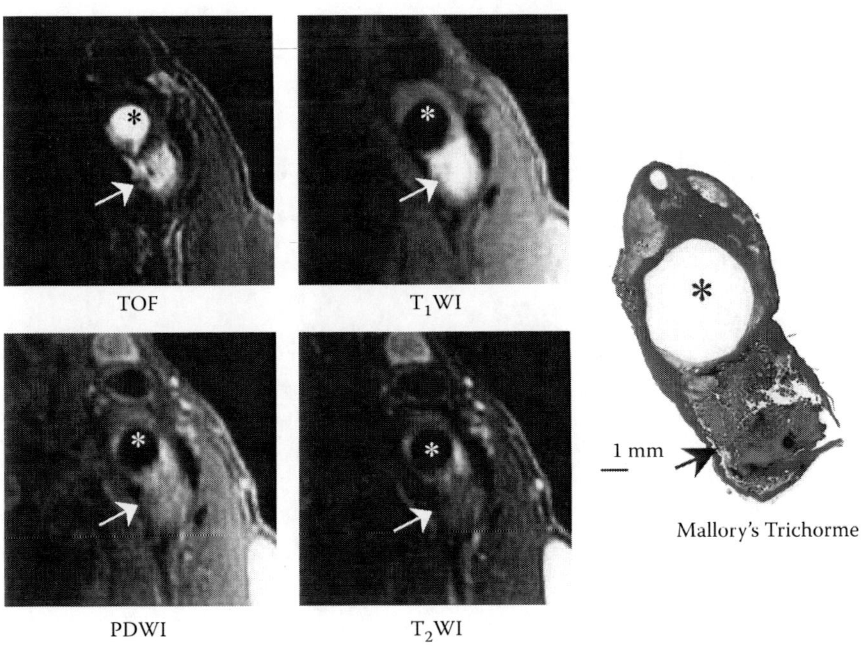

FIGURE 9.9 Multicontrast carotid MRI protocol with 3D-TOF, T_1-weighted, PD-weighted, and T_2-weighted images compared to histology (Mallory's trichrome stain). IPH is bright on T_1-weighted sequences such as TOF and T_1w. Corresponding area on histology verifies IPH (arrow). (Adapted from Cai J et al. 2002. *Circulation.* 106:1368–1373.)

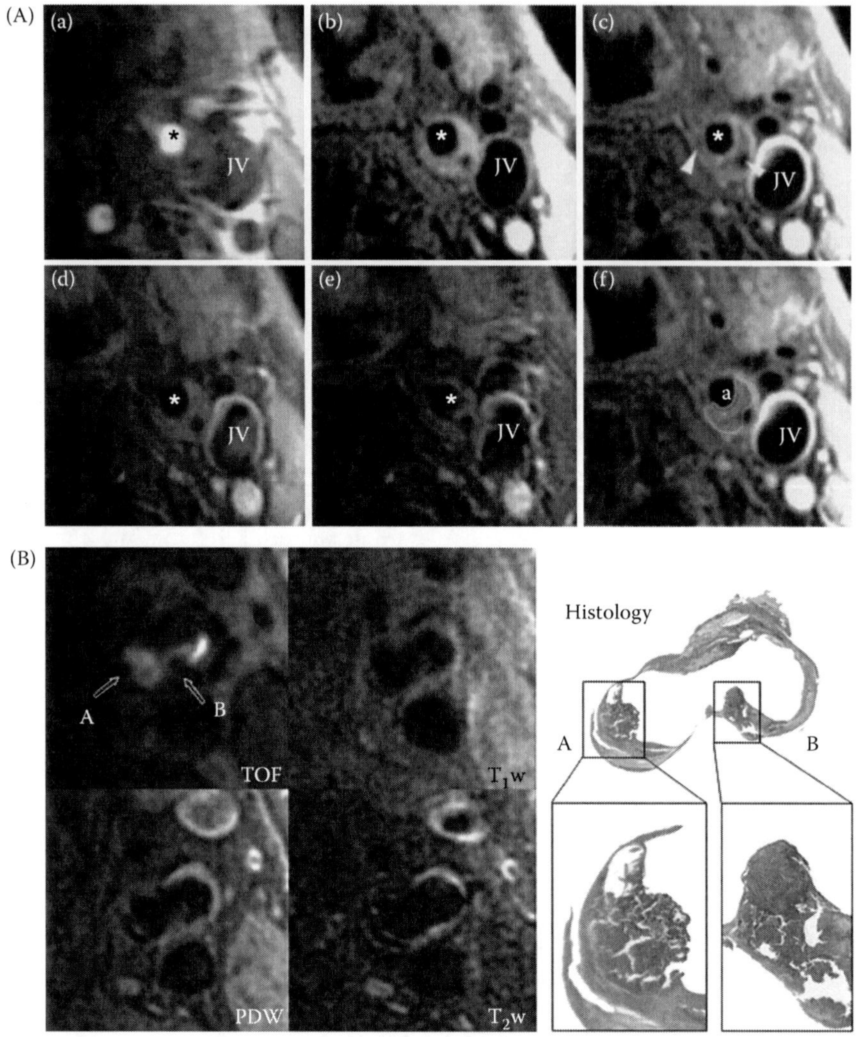

FIGURE 9.10 (A) Multicontrast MRI showing identification of LRNC. Comparing postcontrast T_1w (CE-T_1w) to precontrast T_1w (T_1w), nonenhancement of LRNC postcontrast (arrow) is clearly visible. (Adapted from Cai J et al. 2005. *Circulation*. 22:3437–3444.) (B) Calcification (arrows on TOF) appears hypointense on all contrast weightings. Boxes A and B on histology correspond to arrows on TOF. The juxtaluminal location of calcification marked by arrow B on TOF makes interpretation on dark blood images difficult but is well appreciated on the bright-blood TOF. (From Chu B et al. 2009. *J Am Coll Cardiol Img*. 2009(2):883–896.)

9.2.3.1 Intraplaque Hemorrhage and Thrombus

In advanced stages of atherosclerosis, there may be bleeding into the plaque from either the lumen after the surface is disrupted or the *vasa vasorum* which supplies blood to both the normal and abnormal parts of the vessel wall. Such bleeding into the plaque is termed intraplaque hemorrhage (IPH). Evolution of IPH signal over time is not yet fully understood but findings have been extrapolated to IPH from known characteristics of brain hemorrhage. Immediately after hemorrhage into the brain, intracellular hemoglobin from the red blood cells gets deoxygenated to deoxyhemoglobin and then oxidized to methemoglobin. The five unpaired electrons in methemoglobin render it paramagnetic with short T_1 relaxation times. After cell lysis, the paramagnetic methemoglobin becomes extracellular until it is reabsorbed by reparative processes. The high signal from methemoglobin on T_1-weighted images is also seen in IPH. Both acute IPH (intracellular methemoglobin) and older IPH (extracellular methemoglobin) are seen as hyperintense signals on T_1-weighted sequences such as black-blood T_1 turbo spin echo or bright-blood 3D time-of-flight sequences. Both sequences are used for IPH identification in a multicontrast review. However, highly T_1-weighted sequences such as magnetization-prepared rapid gradient echo (MP-RAGE) are more sensitive and specific for IPH detection (Ota et al. 2010).

In the MP-RAGE technique, additional T_1 weighting is imparted by a nonselective inversion pulse applied before a slice-selective gradient echo readout. Optimization of the inversion time is required for good blood suppression while at the same time providing high contrast for methemoglobin (Zhu et al. 2008). A sequence diagram of MP-RAGE is given in Figure 9.11. At 3 T, the T_1 of blood is around 1500 ms and of vessel wall is around 1000 ms. However, the T_1 relaxation times of methemoglobin in plaque is not fully characterized. If T_1 of methemoglobin is assumed to be 500 ms, Zhu et al. (2008) showed that the optimal sequence parameters that minimized blood signal while maximizing IPH contrast depended on TI. Short T_1 times provide the best contrast between vessel wall and intraluminal blood while at the same time providing good contrast for methemoglobin in IPH. However, longer TI times may be needed to improve overall signal. A range of TI times varying from 300 to 600 ms are used for IPH imaging at 3 T depending on scan time, coverage, and SNR constraints. Imaging parameters guided by Bloch simulation and adjusted by scanning subjects with IPH are usually used. Optimized MP-RAGE parameters show IPH as hyperintense regions while most tissues including vessel wall are not visible (Figure 9.11). Increased TI times can be used to show vessel wall and surrounding muscle/fibrous tissue but this comes at the expense of lowered contrast between IPH and the fibrous tissue of plaque. Long TI times can reduce the sensitivity to detect small IPH. Alternatively, increasing the number of averages can be used to visualize the vessel wall in addition to IPH but it comes at the expense of increased scan time or lowered resolution.

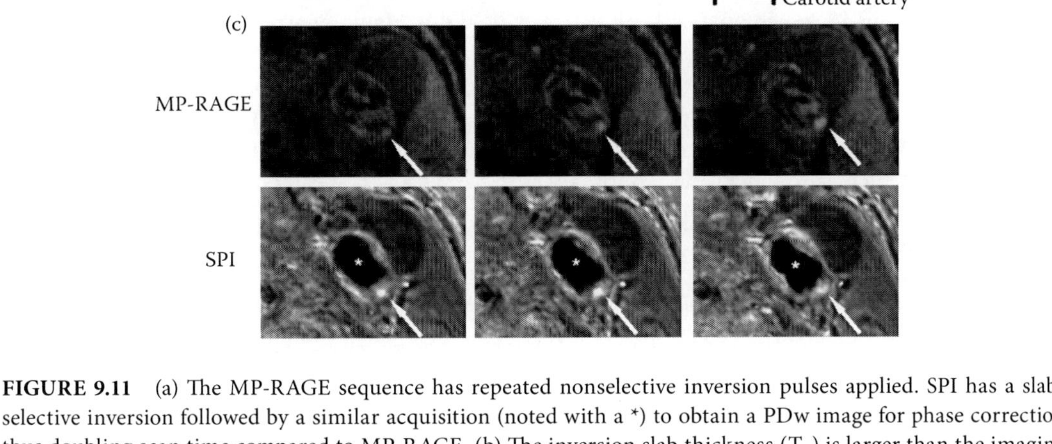

FIGURE 9.11 (a) The MP-RAGE sequence has repeated nonselective inversion pulses applied. SPI has a slab-selective inversion followed by a similar acquisition (noted with a *) to obtain a PDw image for phase correction thus doubling scan time compared to MP-RAGE. (b) The inversion slab thickness (T_{ir}) is larger than the imaging slab (T_{sl}). IPH signal is bright on both MP-RAGE and SPI (arrow in c). However note the better lumen delineation (*) by SPI. Corresponding MP-RAGE images show flow artifacts that can interfere with detection of *juxtaluminal* IPH. (Adapted from Wang J et al. 2010a. *Magn Reson Med.* 64(5):1332–1340.)

Optimization of MP-RAGE TI is difficult since both blood signal suppression and methemoglobin signal enhancement have to be balanced by choosing TI. Phase-sensitive reconstruction combined with slab-selective inversion instead of the nonselective inversion of MP-RAGE can be used to suppress blood signal relative to the vessel wall. This allows more room in the adjustment of TI for improving contrast of IPH relative the vessel wall. This technique called slab-selective phase-sensitive inversion recovery (SPI) (Wang et al. 2010a) is compared to MP-RAGE in Figure 9.11. Phase-sensitive correction requires an additional PD-weighted image for phase correction of the SPI image. Thus, phase correction doubles scan time over MP-RAGE but provides improved CNR for IPH and blood suppression. In MP-RAGE, blood inflow into the imaging slab is reinverted multiple times by the nonselective inversion pulse. However, in SPI, the slab thickness of the inversion pulse (T_{ir}) is optimized such that inflowing blood is only inverted once before image acquisition. The "inverted" blood in SPI provides "negative" blood signal compared to the vessel wall signal, whereas in MP-RAGE, blood signal is saturated by repetitive reinversion thereby reducing the vessel wall-blood CNR. The T_{ir} in SPI is governed by the following two equations:

$$\frac{T_{ir} - T_{sl}}{2} \rangle TI \times v_{max}$$

$$\frac{T_{ir} - T_{sl}}{2} \langle (TI + IRTR) \times v_{min}$$

where IRTR is the time interval between two consecutive IR pulses and v_{min} and v_{max} are the minimum and maximum velocities of inflowing blood, respectively. TI of SPI is adjusted to maximize IPH-vessel wall CNR. Figure 9.11 shows the improved IPH contrast and blood suppression of SPI compared to MP-RAGE.

Thrombus is a blood clot usually covering the rupture site. Thrombi can cause transient ischemic symptoms or stroke through emboli. The intracellular methemoglobin within a thrombus can be seen on T_1-weighted sequences especially MP-RAGE, which is also referred to as direct thrombus imaging (Moody et al. 2003). Use of SPI over MP-RAGE is also advantageous for thrombus imaging because of better blood suppression.

9.2.3.2 Lipid-Rich Necrotic Core

The hallmark of atherosclerosis is the accumulation of lipid within the vessel wall. It has been suggested that lipid may exist in several forms within the atherosclerotic plaque contributing to its varied appearance on T_1- and T_2-weighted MRI. Lipid rarely exists as a pure lipid pool in advanced atherosclerotic plaques. Instead, it is usually mixed with cells and cellular debris and presents a heterogeneous appearance on MRI. Such advanced lipid collections are called LRNC to emphasize their necrotic and varied contents. Fat suppression used in VWI does not suppress LRNC signal in contrast to subcutaneous fat. LRNC appears isotense to surrounding muscle on T_1-weighted sequences such as T_1w TSE and TOF images but are hypointense on T_2-weighted images. The shorter T_2 of LRNC compared to the vessel wall makes T_2-weighted images more reliable for identification of LRNC. Administration of gadolinium contrast agents such as Magnevist can improve the sensitivity and specificity for detection of LRNC. Compared to the normal vessel wall, uptake of contrast by LRNC is low. Thus, LRNC can be identified by its hypointensity on postcontrast T_1w compared to the precontrast T_1w (Figure 9.10A). Delineation of the LRNC boundaries also helps define the thickness of the overlying fibrous cap. Contrast-enhanced T_1w images have been shown to improve measurement reproducibility of LRNC (Takaya et al. 2006) and allow fibrous cap length measurements (Cai et al. 2005). Use of the QIR black-blood preparation allows the same sequence parameters to be used between pre- and postcontrast T_1w images, thereby simplifying both the image acquisition and image review. The low diffusivity of LRNC has also been used for its detection using *diffusion weighted imaging* (DWI). DWI images using reduced field-of-view echo planar imaging show lipid core as dark regions on the apparent diffusion coefficient (ADC) maps (Kim et al. 2009). However, due to challenging requirements of DWI, it is difficult to obtain ADC maps in the carotids with high SNR. Hence unlike IPH, LRNC is still best detected using a combination of weightings rather than a single sequence. Use of contrast-enhanced T_1w MRI provides the most reliable measurement of LRNC.

9.2.3.3 Loose Matrix

The fibrous tissue of plaque consists primarily of organized collagen fibers in the intact plaque. When there are reparative processes secondary to rupture, reabsorption of lipid, and so on, a loose fibrous tissue is laid down. Several different proteins, including hyaluronan, proteoglycans, and so on, can form a loose framework collectively referred to as "loose matrix." Although loose matrix in carotid plaque has not been shown to have any risk associated with it, loose matrix is often found in advanced plaques and may be a sign of tissue repair. Due to the high water content of loose matrix, it appears hyperintense on T_2-weighted images. Loose matrix has high permeability and therefore has high uptake of gadolinium contrast agents and is hyperintense on postcontrast T_1-weighted images compared to precontrast T_1-weighted images, where it is, in general, similar to other fibrous tissues.

9.2.3.4 Calcification

Calcification is a frequent finding in atherosclerotic plaques but its implications are not well understood. Calcification has been implicated in both plaque rupture and plaque stability. Calcification can serve as a measure of disease burden, especially in arterial beds that are difficult to image such as coronary arteries. Calcification has a very short T_2 relaxation time and appears hypointense on all traditional MRI contrast weightings such as T_1w, T_2w, and PDw. Size measurements of calcification using gradient echo techniques tend to overestimate the size due to effect of magnetic field inhomogeneity. This effect however is advantageous for visualizing small plaque components. Thus, spin echo images may be preferable for calcification size estimation but gradient echo sequences for detection of calcification. Identification of calcium may be difficult due to its negative contrast in two situations: (1) in SNR limited images and (2) when calcium is juxtaluminal and black-blood images are acquired. For the latter case, 3D-time-of-flight is often preferred since dark juxtaluminal calcification is easily identified against the bright-blood lumen. A better approach is to acquire images where calcium has positive contrast. Ultrashort echo time (UTE) MRI, a gradient echo technique, provides "positive contrast" for calcium by rapid image readout immediately following the excitation RF pulse. In UTE imaging, an ultrashort echo and a regular echo are acquired for each excitation RF pulse. Subtracting the regular echo image from the ultrashort echo image highlights short T_2 components such as calcifications (Herzka et al. 2008). Short echo times also reduce susceptibility issues and may provide a better size measurement of calcification. Since the readout must immediately follow the excitation RF, radial or spiral readouts are preferred over Cartesian phase encoding. Applications of UTE in plaque imaging are still under investigation.

9.2.4 Cardiac Gating and Motion Compensation

While all major blood vessels are subject to pulsation with each cardiac cycle, the effects on vessel wall images differ depending on the arterial beds. Use of motion compensation techniques such as cardiac or respiratory gating can prolong image acquisition time. Therefore, use of such techniques should consider both the necessity for motion compensation and image acquisition time to obtain high-quality images with short scan times. Coronary arteries, thoracic aorta, and abdominal aorta require vector cardiographic gating for image acquisition fully synchronized with the heartbeat. They also require diaphragmatic respiratory navigator to be applied every TR to reduce the incidence of motion artifacts from the abdominal wall and reduce motion artifacts of the heart. Alternatively, respiratory bellows may also be used for thoracic or abdominal aorta MRI. Carotid VWI has been known to be affected by cardiac, respiratory, and swallowing motion (Boussel et al. 2006). Swallowing motion had the largest effect with respiratory and cardiac motion having lesser effect. Swallowing motion is unique to carotid artery imaging and is difficult to correct for owing to its nonperiodic nature. Navigators positioned on the posterior pharyngeal wall can be used for detecting the motion of the pharynx with swallowing. Although these techniques for detecting swallowing are useful, they have not been widely adopted in carotid VWI. Plaque volume measurements in the carotids have been shown to not benefit from cardiac gating (Mani et al. 2005) and most carotid MRI protocols are not cardiac gated. Peripheral artery imaging in general also does not require cardiac gating for vessel wall measurements. In addition, motion artifacts are also less common with peripheral artery imaging than other arterial beds making peripheral artery imaging relatively straightforward.

9.2.5 Protocols for Vessel Wall Imaging

Plaque morphology, such as lumen and outer wall boundaries, must be visualized clearly for accurate plaque burden measurement. Plaque components such as IPH, LRNC, loose matrix, and calcification should also be assessed for identification of plaque at risk of rupture. Direct assessment of plaque rupture or fibrous cap thickness is an area of research interest in plaque imaging. Proper assessment of plaque thus requires high-resolution images with high SNR obtained without motion degradation. VWI protocols are optimized to address these requirements in a short scan time to avoid excessive subject discomfort and motion artifacts. Recent advances in VWI have addressed these challenges to provide high-quality vessel wall images in multiple vascular beds. Imaging techniques developed for the carotid arteries can be extended to other arterial beds with proper attention to cardiac and respiratory gating requirements. DIR, multislice DIR, and MSDE have been used for aortic imaging. DIR, T_2-IR, and SPACE have been shown useful for peripheral artery imaging. Due to lower blood velocities in the peripheral arteries, T_2-IR can provide better blood suppression than other flow velocity-dependent methods. Coronary VWI has been demonstrated with DIR and MSDE. Special modifications of DIR such as pencil beam reinversion and oblique reinversion slab can also be used for the coronaries. Gating requirements are very stringent for coronary VWI. Methods to reduce scan time such as radial trajectories and subject-specific trigger delays are often used for coronary MRI. Due to the small size of the coronary vessels, resolution requirements are also higher. This, coupled with their position deep within the thorax, has prompted research into better ways to improve coronary vessel wall SNR such as post-contrast imaging.

9.2.6 New Trends in Vessel Wall MRI Techniques

Although the two-dimensional (2D) acquisition schemes for atherosclerotic vessel wall MRI remain the mainstay of many VWI sequences, 3D imaging provides a number of advantages that may represent the future of vessel wall MRI. The utilization of a second phase encoding direction in 3D acquisition schemes provides dramatic SNR advantages as the imaging volume expands in the third dimension. This SNR gain can also be used to improve the spatial resolution in that direction, typically the slice selection direction on the 2D images, to achieve isotropic voxel size in all directions. In addition, 3D dataset and isotropic voxel size can also potentially provide advantages in reducing registration and segmentation errors for quantitative and longitudinal vessel wall measurements.

3D black-blood MR VWI, however, represents significant technical challenges for IS- and DIR-based techniques. Although feasibility of DIR-based 3D BB MRI has been demonstrated (Balu et al. 2008), such techniques are significantly limited by the blood replenishing requirements posed by IS and DIR. In 3D imaging, the blood in a large volume, compared to a thin slice in the 2D case, needs to be replaced to achieve a satisfactory BB effect. This requirement can be very challenging when the image volume is extended. To address this issue, 3D turbo spin echo with variable-flip-angle refocusing RF pulses (SPACE) technique was proposed as a candidate for effective blood suppression when imaging a large-coverage 3D volume (Zhang et al. 2009). The multiple IR pulses used by the SPACE saturate blood signal thus provide automatic blood signal suppression.

Although the SPACE technique provides some black-blood effect, its blood suppression efficiency is still limited due to the lack of a dedicated BB module. The utilization of MSDE-based BB techniques provides an ideal solution for 3D imaging due to its effective, large-volume, flow-direction-independent blood suppression. The blood efficiency has also been shown to be superior than simply relying on the SPACE sequence (Fan et al. 2010). MSDE has been combined with different acquisition schemes, that is, spin echo (Wang et al. 2007; Fan et al. 2010), gradient echo (Balu 2011), steady-state free procession (Koktzoglou and Li 2007), and so on, for large-volume 3D acquisitions. They have also been applied in a number of vascular beds such as carotid, peripheral, and coronary arteries. The elimination of flow replenishing requirements of MSDE technique also allows the imaging volume to be planned in parallel to (as opposed to only perpendicular to) the targeted artery so that more time-efficient acquisition can be achieved.

As shown in Figure 9.12, a 3D MSDE-prepared rapid gradient echo (MERGE) sequence was used to acquire high-resolution carotid artery images. With the SNR benefit from 3D acquisition and efficient black-blood effect from the MSDE sequence, high spatial resolution ($0.7 \times 0.7 \times 0.7$ mm^3) and large-coverage (250 mm foot–head direction) imaging volume was acquired in only 2 min. Owing to its time efficiency, 3D-MERGE is highly suitable for applications such as carotid disease screening.

A remaining challenge of the 3D vessel wall MRI is its vulnerability to potential motion artifacts. Unlike in 2D acquisition where the motion-corrupted data impacts only individual images, any corrupted data in 3D volumes contributes to the whole volume. Potential solutions for such issues include switching to motion robust k-space acquisition schemes like radial or spiral acquisition, and combining with special motion compensation schemes. 3D acquisition schemes have not only been applied to plaque burden measurement but also for detecting plaque components such as hemorrhages and/or calcifications. MP-RAGE, SPI, and UTE imaging techniques for IPH and calcium detection described earlier are essentially 3D imaging techniques and have been successfully applied in patient studies. With the advent of fast and robust 3D VWI with good blood suppression, large coverage high-resolution imaging of atherosclerotic lesions that can be applied for patient imaging directly in the clinic is gaining traction. While such techniques for carotid MRI are mature and are being applied in clinical studies, their translation to the coronaries remains challenging and is an active area of research.

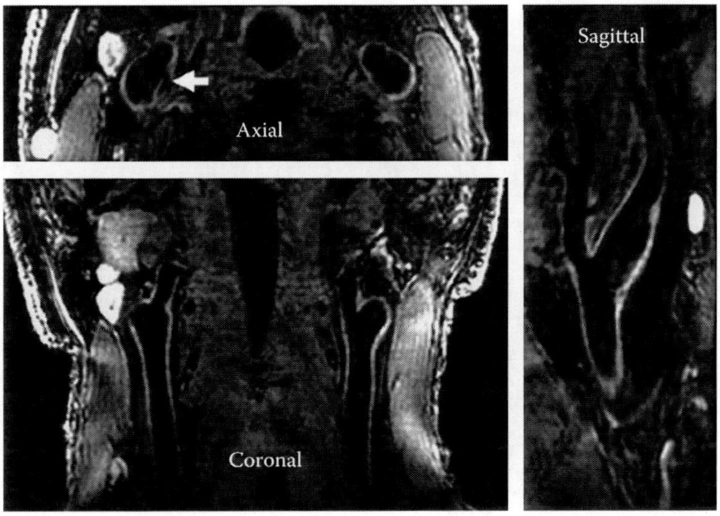

FIGURE 9.12 3D isotropic high-resolution ($0.7 \times 0.7 \times 0.7$ mm^3) carotid artery image acquired with 3D MERGE sequence. Arrow on axial reformat shows a small piece of calcification. Vessel wall boundaries are clearly visible on all reformats. (From Balu N et al. 2011. *Magn. Reson. Med.* 65:627–637.)

9.3 Image Processing in Vessel Wall MRI

The end goal of vessel wall MRI is to extract qualitative and quantitative descriptors of the atherosclerotic plaque that can be used for diagnostic risk assessment, monitoring of therapy, or scientific studies of plaque pathophysiology. The volume and complexity of available imaging information—characterized by multiple contrast weightings and multiple, high-resolution image cross sections—mandates the use of image processing tools for robust and efficient analysis. This section describes the major goals and methods for image processing of vessel wall MRI.

Typically, vessel wall MRI yields serial cross-sectional images perpendicular to the axis of the vessel. Other useful views include longitudinal cross sections and reformatted images, such as curviplanar reformats, and maximum intensity projections (MIPs). These basic image formats are illustrated in Figure 9.13. All are useful for aspects of vessel wall image processing.

A major advantage of MRI for VWI is the availability of multiple contrast weightings and techniques for both flow enhancement and suppression. As illustrated in Figure 9.9, this permits image contrasts with both bright and dark vessel lumens, and with varying intensities for internal components of the atherosclerotic plaque. Utilization of the multiple contrast data is both an advantage and a challenge for image processing of vessel wall MRI.

This section is organized as follows. First, methods for assessment of plaque burden are presented, where burden refers to the general size and shape of the vessel wall and lumen. Second, methods for assessing plaque composition are presented, in which the internal structure of the plaque itself is determined. Finally, the section closes with a description of emerging 3D techniques in image processing for vessel wall MRI.

9.3.1 Plaque Burden

The term "plaque burden" is used to denote the severity of atherosclerosis as measured by the size of the plaque. As such, it is used interchangeably to denote a variety of measurements, including thickness, cross-sectional area, volume, or ratios of these measurements. All of these measurements share

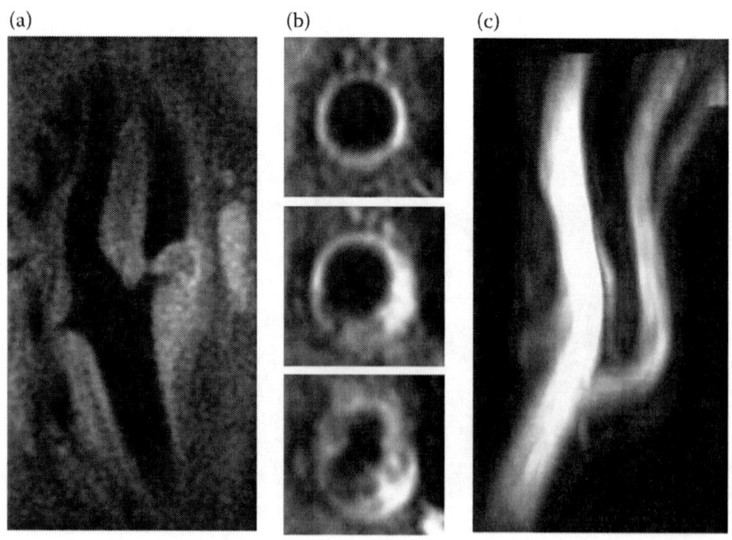

FIGURE 9.13 Useful image viewing options include (a) longitudinal cuts along the vessel axis, (b) axial cuts perpendicular to the vessel axis, and (c) reformatted images, such as maximum intensity projections (MIPs) of 3D image volumes.

TABLE 9.1 Common Plaque Burden Measurements

Measurement	Description
Max NWI	Maximal value of normalized wall index over all slices
Mean NWI	Mean value of normalized wall index over all slices
Max thickness	Maximal thickness over all slices
Max wall area	Maximal wall area over all slices
Wall volume	Total volume of wall over all slices
Max eccentricity	Maximal value of eccentricity over all slices

the common need to determine the inner (lumen) and/or outer boundaries of the vessel wall. When discussing plaque burden, it is important to be explicit regarding which measurement is being made. Common examples are summarized in Table 9.1.

9.3.1.1 Common Burden Measurements

Stenosis, which is the standard clinical assessment of disease severity, can also be viewed as a measure of plaque burden. Stenosis is defined as the difference between the normal and most narrowed vessel diameters divided by the normal diameter, and expressed as a percentage

$$\text{Stenosis} = 100\% \times (d_{\text{normal}} - d_{\text{min}})/d_{\text{normal}}$$

Stenosis of 0% indicates an apparently normal artery, whereas highly diseased arteries approach stenosis of 100%. Most commonly, the normal diameter is measured distal to the point of maximal narrowing, but can also be estimated from the point of maximal narrowing by estimating the apparent original diameter. In the carotid artery, these two approaches are referred to as the "NASCET" and "ACAS" criteria, respectively (ACAS et al. 1995, Barnett et al. 1998). To a lesser extent, "area stenosis" may also be used, in which the diameter measurements are replaced by cross-sectional area measurements.

A limitation of stenosis measurements is that they often fail to accurately depict the severity of disease, in part due to a phenomenon known as "Glagovian remodeling" in which the vessel wall expands with plaque growth (Glagov et al. 1987). As a result, large plaques have been reported even in vessels with 0% stenosis (Dong et al. 2010a). To address this shortcoming, the "normalized wall index" (NWI) was introduced, which is defined as the ratio of the vessel wall area to the total vessel (wall plus lumen) area

$$\text{NWI} = A_{\text{wall}}/(A_{\text{lumen}} + A_{\text{wall}})$$

NWI appears similar in definition to stenosis, but studies have shown little association between maximal NWI and stenosis. Values of NWI range from 0.4 or less in apparently normal arteries to approaching 1.0 for nearly occluded arteries. NWI has also been called "plaque index," "wall/outer wall ratio," and "percent atheroma."

9.3.1.1.1 Thickness

Vessel wall thickness has also emerged as an important indicator of plaque burden. Ultrasound studies have established that intima-media thickness (IMT) along with proximal segments of the carotid arteries reflects global atherosclerosis risk (Lorenz et al. 2007). MRI techniques for thickness measurement have been shown to agree closely with ultrasound measurements (Underhill et al. 2006). Furthermore, MRI provides omnidirectional thickness measurements throughout the carotid bulb and into the internal carotid artery. Thickness measurement does, however, present some challenges in advanced plaques due to the lack of an objective definition of thickness.

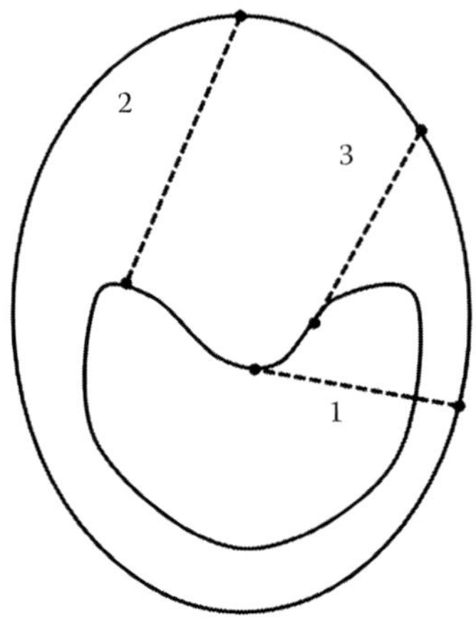

FIGURE 9.14 Challenges for thickness measurement encountered in vessel wall imaging with three approaches: (1) closest point on outer contour, (2) closest point on inner contour, and (3) radial rays from inner contour centroid.

In general, thickness can be defined as the length of a line connecting corresponding points on the inner and outer boundaries. Additionally, the set of lines of thickness circumscribing the vessel wall does not cross one another and each line of thickness should be as close to perpendicular to the vessel wall boundaries as possible. Simple thickness measurement techniques have been proposed based on casting rays from the centroid of the lumen or identifying the closest point on the second contour for each point on the first contour. These approaches work well for relatively normal arteries, but for more complicated shapes in highly diseased arteries, the methods break down (Figure 9.14).

More robust thickness measurement algorithms have been developed that provide appropriate results on any encountered morphology. One approach identifies points of mutual correspondence where each point is mutually closest to the other point on the opposite contour. For points where mutual correspondence is not achieved, corresponding points are identified via interpolation. A second approach is to identify a contour midway between the inner and outer contours. Then, lines of thickness are cast perpendicular to this midline. Finally, an approach using Delaunay triangulation has been proposed that ensures nearly perpendicular lines of thickness through the angle maximization property of Delaunay triangulation (Kerwin et al. 2007).

9.3.1.2 Eccentricity

Another useful but undefined concept for characterizing plaque burden is eccentricity, which captures the extent to which the plaque differentially accumulates along only one portion of the vessel circumference. One method to characterize eccentricity is to compute the ratio of the maximal and minimal lesion thickness. However, the minimal thickness is sensitive to error making the measurement unstable. For nondiseased vessels, this sensitivity can lead to erroneously high eccentricity indices. A better index, based on the centroids c_{wall} and c_{lumen} of the two contours, is given by

$$\text{Eccentricity} = \frac{\sqrt{\pi}}{A_{\text{lumen}} + A_{\text{wall}}} \| c_{\text{wall}} - c_{\text{lumen}} \|$$

where the multiplier out front normalizes the difference in centroids by the effective radius of the vessel wall.

9.3.1.3 Boundary Detection

The preceding discussion of measurements of plaque burden presumes that contours have been drawn defining the inner and outer boundaries of the vessel wall. Boundary detection is thus the central image processing concern for plaque burden analysis. In general, detection of the lumen boundary is comparatively easier than detection of the outer wall boundary.

Typically, the lumen exhibits a reasonably uniform signal with a sharp contrast with the vessel wall. This contrast can be either the result of flow enhancement in time-of-flight acquisitions or black blood in flow-suppressed images. Strategies for detecting the boundary of this region generally depend on region growing techniques and/or active contours.

One such active contour approach that has been used is a B-spline snake (Kerwin et al. 2007). In this approach, the boundary is represented as a contour with x and y components given by closed B-splines

$$c_x(u) = \sum_{k=1}^{K} \xi_k \beta(\lfloor u - k \rfloor_K)$$

$$c_y(u) = \sum_{k=1}^{K} \psi_k \beta(\lfloor u - k \rfloor_K)$$

where ξ_k and ψ_k are spline coefficients, K is the number of control points in the spline

$$\lfloor u \rfloor_K = \begin{cases} u + K & u < 2 - K \\ u - K & u > K - 2 \\ u & \text{otherwise} \end{cases}$$

and

$$\beta(u) = \begin{cases} \frac{3}{4}|u|^3 - \frac{2}{3}u^2 + 1 & 0 \le |u| < 1 \\ \frac{1}{4}(2 - |u|)^3 & 1 \le |u| < 2 \\ 0 & \text{otherwise} \end{cases}$$

The optimal boundary contour minimizes an energy functional based on the underlying image, such as

$$E = \frac{1}{l(C)} \int_0^K \left\| \frac{dC(u)}{du} \right\| (n(u) \cdot \nabla I(C(u)))du$$

where $C(u) = [c_x(u), c_y(u)]$, $l(C)$ is the total length of the contour, $n(u)$ is the unit normal to the contour at the point u, and ∇I is the image gradient. The contour that minimizes this energy can be found by gradient descent given an initial estimate of the contour.

For the outer boundary of the vessel, boundary detection can be substantially more challenging. Contrast of the vessel wall with surrounding tissues can be minimal. Furthermore, the vessel wall does not have a uniform intensity, which can range from hypointense calcified plaque to hyperintense. Finally, both within and adjacent to the plaque, additional boundaries can lead a boundary detection algorithm to converge to the wrong boundary. To overcome these challenges, shape-based constraints, such as an elliptical prior or active shape model, are advisable (Kerwin et al. 2007).

9.3.1.4 Measurement Accuracy Considerations

The fundamental concern in the accuracy of plaque burden measurements is the accuracy with which an edge can be located. In this regard, the edges are the inner and outer boundaries of the vessel wall. Surprisingly, accuracy of edge detection is not strongly dependent on image resolution. The standard deviation of the estimated location of a step edge can be shown to be approximated by

$$\sigma_d = \frac{p}{\text{CNR}}$$

where p is the fundamental pixel size (resolution) and CNR is the CNR of the edge. Notably, CNR scales with p, so any attempt to reduce σ_d by improving resolution is counteracted by a corresponding reduction of CNR. Furthermore, imaging at higher resolutions increases acquisition time, whereas that time could alternatively be used to acquire more signal averages and reduce σ_d by increasing CNR.

The resolution should, however, be set high enough to ensure the salient features of the vessel wall are captured. For plaque burden measurements, this requires that any Gibbs ringing from one edge does not significantly impinge upon the other. In general, a resolution of at most half the minimum thickness of the vessel wall will ensure any bias in plaque burden measurements is negligible.

MRI scanners also offer two features that should be considered in obtaining accurate measurements of plaque burden: magnitude detection and zero-filled interpolation. Magnitude detection is a standard step used on all scanners for MR image formation in order to eliminate random phase variations. For edges adjacent to a low-signal region, however, magnitude detection slightly perturbs the apparent location of the edge. Zero-filled interpolation is used to reconstruct images with pixel dimensions smaller than the fundamental resolution. Although zero-filled interpolation does not add information, its use does improve accuracy of edge detection based on subsequent linear interpolation and eliminates bias due to magnitude detection. The effects of zero-filled interpolation and magnitude detection on edge position errors are illustrated in Figure 9.15.

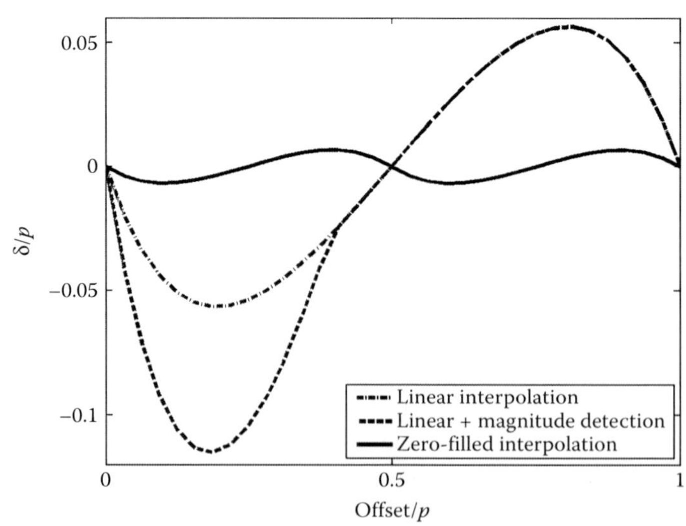

FIGURE 9.15 Measurement error due to interpolation and magnitude detection. Plot shows the error encountered relative to the fundamental image resolution p versus the location of the edge relative to the location of the pixels (i.e., zero and one are the locations of adjacent pixels and 0.5 is midway between).

9.3.2 Plaque Composition

Although plaque burden appears to provide additional clinical insight over and above stenosis, still further insight is provided by examining the components of the plaques themselves. Stable morphologies include greater amounts of fibrous tissue and possibly calcification, whereas components of the soft necrotic core appear to increase risk of clinical events as they increase in size. In particular, IPH within the necrotic core has recently been implicated as a leading marker of risk in carotid atherosclerosis. These components can be further subdivided as illustrated by the hierarchy in Figure 9.16. Notably, fibrous tissue may be subdivided into both dense fibrous tissue and loose matrix. Hemorrhage has been differentiated into type I with intact red blood cells and older type II with lysed red blood cells.

Numerous studies have shown that regions consisting of these different components can be identified by human interpretation of the relative image intensity levels on MRI. These studies have also shown the importance of using multiple contrast weightings. Given the need to integrate information from multiple contrast weightings, this is an especially attractive application for multicontrast segmentation algorithms.

9.3.2.1 Image Registration

Before information can be shared among contrast weightings, the images must be coregistered to account for movement of the patient between multiple contrast acquisitions. VWI poses some challenges for registration in that arteries of interest, such as the carotid arteries, are embedded in soft, deformable tissues in highly mobile areas of the body. Furthermore, adjacent structures can undergo independent motions, such as compression of the jugular veins. Because of these complications, registration techniques that rely on alignment of rigid landmarks or smoothness of the deformation field tend to fail for vessel wall image registration.

Fortunately, the hardening of the arteries in atherosclerosis lends the arteries themselves substantial rigidity. Therefore, in-plane translations are usually sufficient for image registration of the vessel wall itself. This has been successfully accomplished by optimizing the alignment of the previously identified lumen and wall boundaries with the edge features in the images (Kerwin et al. 2007). Simple, in-plane shifts are also conducive to human adjustment of the registration results, if needed.

9.3.2.2 Plaque Segmentation

The central function of segmentation is to estimate the probability of a given pixel being tissue type T_i given the set of observed intensities x in the multiple contrast-weighted images. Using Bayes' rule, this probability can be expressed as

$$\Pr(T_i|x) = \frac{p(x|T_i)\Pr(T_i)}{\Sigma p(x|T_j)\Pr(T_j)}$$

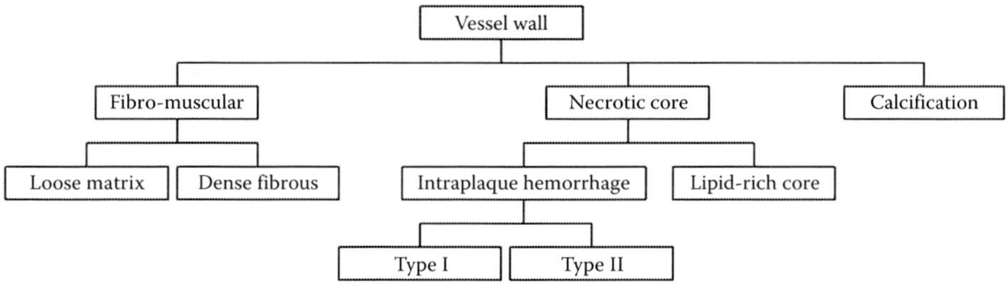

FIGURE 9.16 Hierarchy of common plaque components targeted for segmentation.

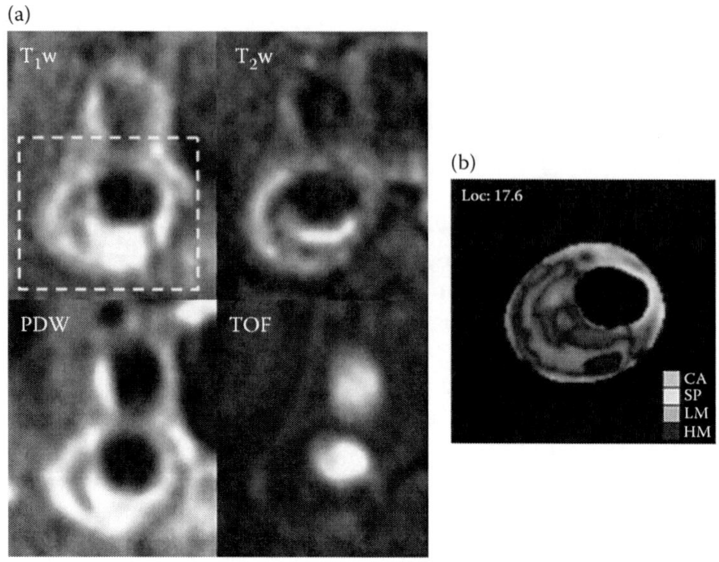

FIGURE 9.17 Segmentation based on multiple contrast-weighting MRI: (a) carotid artery immediately after bifurcation obtained with four indicated contrast weightings; (b) probabilities (high probabilities are dark) for each tissue type: fibrous tissue (Fib), calcification (CA), loose matrix (LM), hemorrhage (Hem), and lipid-rich core (LC). Final contours are shown on the T_1w image.

where the sum is carried out over all tissue types *j*. This probabilistic framework has been further expanded based on the observation that there are morphological cues regarding the likelihood of a given tissue being present. Specifically, this likelihood is dependent on the local thickness *t* of the vessel wall and the depth *d* of the pixel relative to the lumen. Under an assumption of conditional independence, these factors are entered into the morphology-enhanced probabilistic plaque segmentation (MEPPS; Liu et al. 2006) algorithm as

$$\Pr(T_i|t,d,x) = \frac{p(t,d|T_i)p(x|T_i)\Pr(T_i)}{\sum_{j=1}^{4} p(t,d|T_j)p(x|T_j)\Pr(T_j)}$$

For any pixel in the image, this model is able to determine the probability that it is any of the available tissue types (Figure 9.17). The probability distributions are trained based on a histological gold standard.

Once the probabilities for each pixel are available, the final step is to classify each pixel, which can be based on a maximum *a posteriori* assignment to the highest probability tissue class. Alternatively, regions of each tissue type can be captured by competing active contours in order to facilitate editing of the contour regions. In the MEPPS approach, this is done by maximizing the total probability contained within the set of contours, subject to the constraints that each pixel is classified as a single tissue and the contours are smooth.

9.3.3 Three-Dimensional Image Processing

The preceding descriptions of image processing adhere to the common approach of treating 2D cross-sectional images as mutually independent when, in reality, they correspond to closely spaced slices of the same anatomical structures. Thus, a subsequent consideration in VWI is how to account for and take advantage of the 3D information in the images.

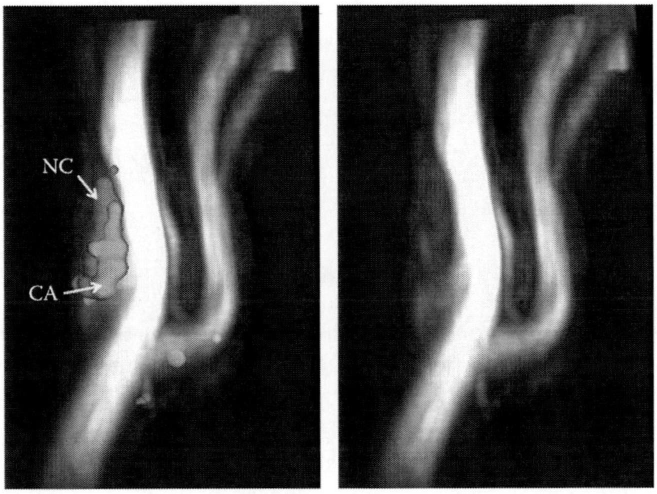

FIGURE 9.18 Three-dimensional rendering of plaque components in relationship to MIP view of angiogram.

9.3.3.1 Angular Measurement

One advantage of a 3D approach to VWI is that additional morphology metrics become available. One set of such metrics has been introduced for describing lumen morphology. By defining geometric positions using a series of maximal inscribed spheres, a scale-independent geometry is introduced. The center of each enumerated sphere is located on the surface of its preceding sphere. Then, metrics are defined based on specific sphere combinations on the set. Example metrics include bifurcation angle, planarity of the bifurcation, and tortuosity (Thomas et al. 2005).

9.3.3.2 Mesh Modeling

A common theme of these 3D approaches is a need to accurately reconstruct the 3D geometry of the vessel, often from sequential 2D contour data. For this, common approaches yield either tetrahedral volume meshes or surface patch models. Applications include mechanical analysis (see below) and rendering of plaque structures in relationship to traditional MRA visualizations (Figure 9.18).

9.3.3.3 3D Imaging

Finally, MRI methods are enabling high-resolution isotropic data to be obtained in 3D. This has the potential to allow image processing to account not only for in-plane information, but also through-plane continuity and smoothness. Image processing techniques to directly extract 3D surfaces corresponding to vessel wall features, as opposed to serial 2D contours, should be forthcoming.

9.4 Contrast Agent Application in Atherosclerosis

As we have established, MRI is a modality that is capable of providing structural as well as functional information on the human body. For VWI, MRI can provide high-resolution imaging of vessel lumen, outer wall boundary, and plaque components. Such information, however, tells only part of the story of the atherosclerotic plaque. Functional and active processes, such as angiogenesis and inflammation, also have significant effects in the initiation, progression, and rupture of plaque (Ross et al. 1999, Libby 2002). Fortunately, MRI has the ability to assess such processes through the use of contrast agents.

Currently, most MRI contrast agents utilize gadolinium, which can shorten T_1 and T_2 times. Accumulation of the gadolinium-based agent can be characterized as bright and dark regions in T_1- and T_2-weighted images, respectively. Chelates of gadolinium, such as Gd-DTPA, are currently available for clinical use. Except for MR angiography, techniques utilizing such agents in plaque imaging include late enhancement imaging (Wasserman et al. 2002, Yuan et al. 2002), which characterizes the agent diffused into the extracellular space of the tissue, and dynamic imaging, which observes the transfer of the agent from the blood stream to the tissue over time. The contrast agent delivery mainly takes place through leaky microvessels in the atherosclerotic plaque—a feature thought to be associated with inflammation. As a result, these techniques can quantify the angiogenesis and inflammation of the atherosclerotic plaque, as well as provide structural information with improved image contrast.

While established MRI contrast agents have proven valuable in plaque imaging, the development of new contrast agents with specific enhancements is a rapidly expanding research area. In plaque imaging, agents that bind to specifically targeted cells or molecules can provide functional information on atherosclerotic plaques (such as the plaque inflammation and angiogenesis), or better identification of key compositional features in atherosclerotic plaque (such as the thrombus), providing a way to better study and understand atherosclerotic disease.

This section describes the state of the art in contrast-enhanced MRI of atherosclerosis. It begins with description of late enhanced MRI and applications in plaque imaging. The majority of this section will cover quantitative techniques for dynamic contrast-enhanced imaging of plaque. Finally, a brief overview of other contrast agents used in plaque imaging, including blood pool and tissue-specific agents, will be presented.

9.4.1 Contrast-Enhanced MRI

Using clinically available contrast agents, such as Gd-DTPA, Lin et al. (1997) and Aoki et al. (1999) observed that atherosclerotic plaques showed significant enhancement in contrast-enhanced (CE) T_1-weighted MR images. The angiogenesis of the vessel wall and endothelial permeability are likely related to the mechanisms of the enhancement. Wasserman et al. (2002), Weiss et al. (2001), Yuan et al. (2002), and Cai et al. (2005) have all shown that CE-MRI can provide unique information regarding plaque composition, function, and vulnerability.

9.4.1.1 CE Imaging Techniques

Blood signal suppression is a common requirement for VWI MRI because black-blood imaging techniques provide excellent vessel wall contrast. As covered in Section 9.2, typically, spatial saturation bands and DIR techniques (Shen et al. 1993) are used to suppress the blood signal by dephasing the magnetization of inflowing blood.

However, as covered in Section 9.2.2.2, with gadolinium-based contrast agents, the DIR technique has a number of limitations, and Yarnykh and Yuan (2002) proposed the QIR technique to overcome them. QIR technique contains two double inversion recovery pulse set to introduce nearly perfect blood suppression over a wide range of T_1 values. Figure 9.6 shows the blood suppression comparison between QIR and DIR before and after contrast agent injection, clearly showing the advantages of QIR for CE-MRI of atherosclerotic plaque.

As covered in Section 9.2.2.3, Wang et al. (2007) recently proposed an improved motion-sensitized-driven equilibrium (iMSDE) sequence, in which the blood signal suppression is T_1 independent. This provides another potential option for blood suppression in CE-MRI.

9.4.1.2 Tissue-Specific Characteristic

By analyzing the enhancement patterns in T_1-weighted CE-MRI, several researchers have demonstrated a relationship between enhancement and plaque characteristics. Both Yuan et al. (2002) and Wasserman et al. (2002) found higher enhancement associated with fibrous tissue in CE-MRI, suggesting that the contrast-enhanced images have the potential to improve differentiation of necrotic core from fibrous tissue. Cai et al. (2005) further proved that *in vivo* high-resolution CE-MRI is capable of quantitatively

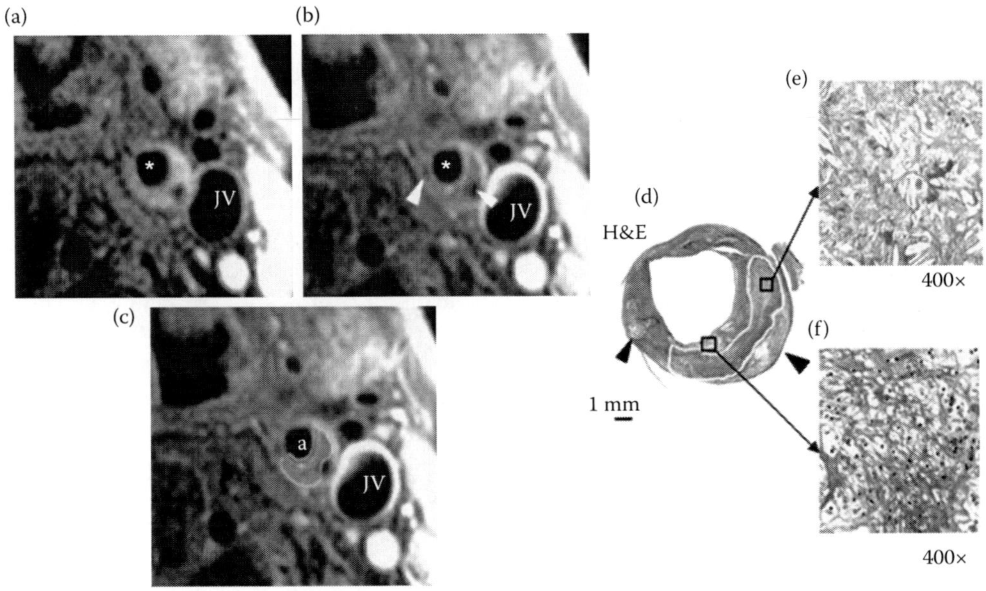

FIGURE 9.19 Corresponding precontrast and postcontrast MR and histological images, showing delineation of the fibrous cap (green contour) and lipid rich/necrotic core (yellow contour) in the left carotid artery. (a) Precontrast T_1-weighted image. (b) Postcontrast T_1-weighted image. (c) Illustration of the measurement method for length of fibrous cap (orange line a). The fibrous cap shows strong enhancement. Histology image (d) shows a matched section with green and yellow contours. High-power photomicrograph shows necrotic debris and cholesterol clefts taken from an area in the lipid rich/necrotic core (e) with no enhancement and loose matrix taken from an area in the fibrous cap (f) with corresponding strong enhancement in postcontrast MR images (b) and (c). Calcification is visible on all images (arrow head). * indicates lumen; JV, jugular vein; H&E, hemotoxylin–eosin staining. (From Cai J et al. 2005. *Circulation*. 22:3437–3444.)

measuring the dimensions of intact LRNC and evaluating the status of fibrous cap, both important parameters to evaluate plaque vulnerability (Underhill et al. 2010). Figure 9.19 shows an example of corresponding CE-MRI and histological images to present how CE-MRI are used in quantitatively measuring the lipid-rich/necrotic core and fibrous cap.

Many researchers have also demonstrated the values of CE-MRI in evaluating plaque inflammation. Inflammation is thought to play an important role in atherosclerotic plaque initiation, progression, and disruption, and is currently a major target in atherosclerosis research and treatment. Yuan et al. (2002) found that contrast-enhanced MRI can identify high neovascularization—thought to be a feature of inflammation and a contributor to plaque destabilization (Ross et al. 1999). Weiss et al. (2001) found that contrast-enhanced MRI characteristics were associated with elevated serum marker of inflammation, likewise suggesting the value CE-MRI for evaluating plaque inflammation. Also, in a study using a rabbit model, Hur et al. (2010) found that the macrophage area and microvessel density were associated with enhancement ratio in CE-MRI, further indicating the CE-MRI may be an efficient method to detect plaque inflammation.

9.4.2 Dynamic Contrast-Enhanced Imaging

Although CE-MRI has been shown to provide useful information for evaluation of the plaque inflammation, it is mainly used in characterizing the plaque composition. CE-MRI is not suitable for quantitative assessment of plaque inflammation because the enhancement of CE-MRI only provides limited information on the amount of remaining contrast agent within the tissue. Furthermore, the amount of enhancement varies depending on factors such as dose, scanning parameter, and timing. Dynamic contrast-enhanced (DCE) MRI, which acquires images over the course of contrast agent injection, can provide detailed information on the transfer of the contrast agent from plasma into tissue. Compared with CE-MRI, DCE-MRI has the advantage of showing not just "what" enhances but also "how" it enhances. For example, both neovasculature and loose matrix have strong enhancement patterns in CE-MRI. However, in DCE-MRI, the enhancement of neovasculature is faster than that of loose matrix because the contrast agent transfer in loose matrix relies on diffusion rather than direct perfusion from vascular system as in neovasculature. To quantitatively assess the enhancement pattern of DCE-MRI, kinetic modeling is used to generate quantitative parameters, presenting the permeability and perfusion characteristics of the plaque.

9.4.2.1 DCE Imaging Technique

DCE-MRI typically utilizes multiple T_1-weighted image acquisitions over a period of time. The imaging sequence of DCE-MRI requires a short acquisition time to capture the rapidly changing enhancement patterns. Quantitative DCE-MRI of atherosclerotic plaque also differs from CE-MRI in that it requires bright-blood imaging. Bright-blood imaging is useful for DCE because the blood signal can be used to extract the changes of contrast agent concentration in the blood, also referred as the arterial input function (AIF), a key function for kinetic modeling (see Section 9.4.2.3). To acquire an accurate AIF, the requirement of the DCE-MRI sequence then is to generate a T_1-dependent blood signal which should be proportional to contrast agent concentration.

The most commonly used imaging technique in bright-blood DCE-MRI of atherosclerotic plaque is the 2D spoiled gradient-recalled echo (SPGR) T_1-weighted sequence (Kerwin et al. 2003). Images usually need to be obtained at 15 time points or more, for a duration of about 4 min. Coincident with the second image in the sequence, the gadolinium-based contrast agent (0.1 mmol/kg for human carotid imaging) should be injected at a constant rate (2 mL/s for human carotid imaging) by a power injector. Spatial saturation bands should be placed at the proximal and distal ends of the imaging block to induce a T_1-dependent blood signal by suppressing the effect of untipped inflowing blood, resulting in dark blood on images prior to contrast bolus arrival. See Figure 9.20.

Although bright-blood DCE-MRI has some advantages, such as the availability of AIF, it creates difficulty in separating the signal in the vessel wall from the adjacent signal in the lumen. Consequently, bright-blood is only suitable for imaging moderate and advanced plaques, which are thick enough to negate the signal influence from a bright lumen. To overcome this challenge, Calcagno et al. (2008) proposed a black-blood DCE-MRI protocol for small lesion imaging. The requirements of black-blood DCE-MRI are similar to CE-MRI of plaque, except for a higher temporal resolution. The previously described DIR, QIR, and MSDE sequences (see Section 9.2) can fulfill the requirements for black-blood DCE-MRI by limiting acquisition coverage to reduce the scan time. To further minimize acquisition time, Yarnykh and Yuan (2006) proposed a small field of view quadruple inversion recovery (sfQIR) technique, which allows rapid dynamic imaging and black-blood preparation without wrap-around artifacts. Figure 9.21 shows an example of black-blood DCE-MRI of rabbit aorta obtained by using sfQIR.

9.4.2.2 DCE Image Processing

To generate quantitative parameters from DCE-MRI, kinetic modeling of enhancement pattern is necessary. To accurately carry out kinetic modeling, image processing of the dynamic images is essential. The main goal of the image processing is to correct image misalignment in the sequence caused by patient motion during the scan. This is very important for kinetic modeling as even small shifts can cause severe bias in the generated quantitative parameter due to the small size of the vessel wall. Additionally, image processing may correct for noise attenuation. Kerwin et al. (2002) proposed a Kalman filtering registration and smoothing (KFRS) algorithm to simultaneously automatically register and reduce the

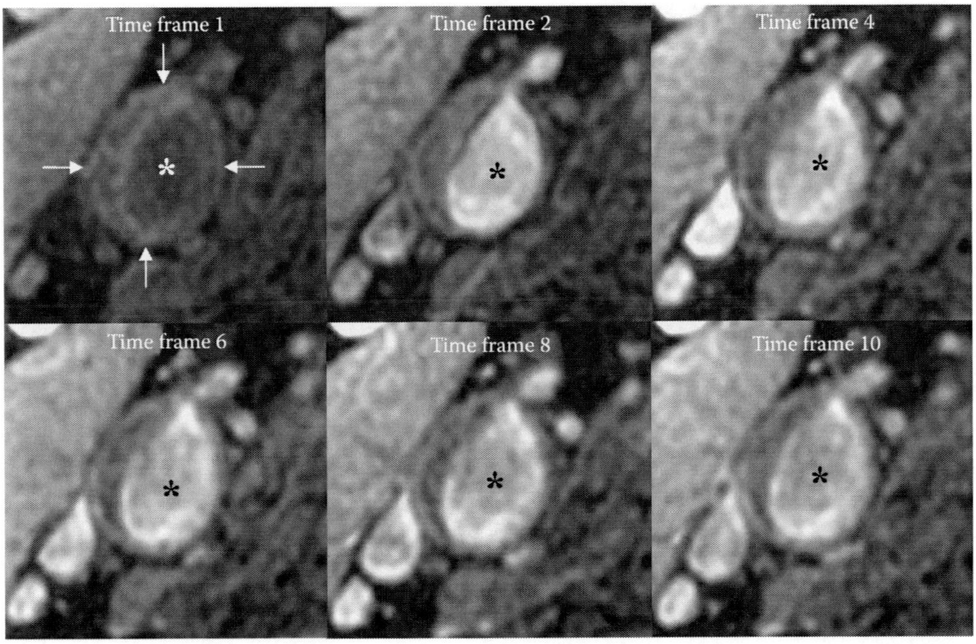

FIGURE 9.20 An example of bright-blood DCE-MRI of a human carotid artery near bifurcation, showing images at six different time points. The time frame 1 shows the precontrast image. * indicates the carotid lumen. The arrows show the outer wall boundary of carotid artery.

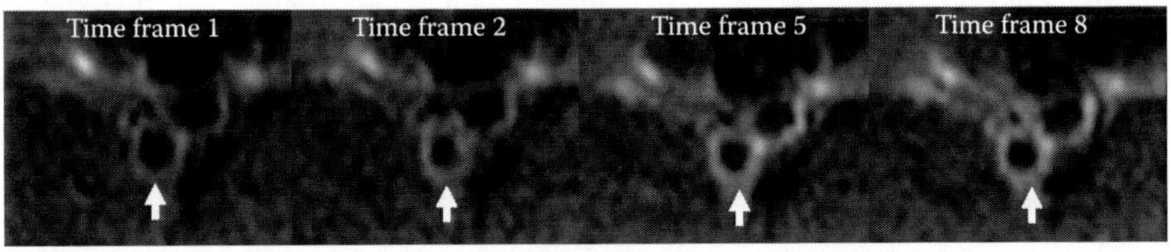

FIGURE 9.21 An example of black-blood DCE-MRI of an injured rabbit aorta, showing images at four time points in the dynamic sequence. Time frame 1 is the precontrast image. The arrows indicate the aorta.

noise for DCE-MRI of atherosclerotic vessels. Registration is performed by searching for displacement of the image at time frame j ($\mathbf{d}_j$) in a user-defined range of displacement that minimizes the cost function:

$$\sum_{\mathbf{x}} [I_j(\mathbf{x} + \mathbf{d}_j) - I_{j-1}(\mathbf{x})]^2$$

where I_j is the image at time frame j. This registration method attempts to find relative movement that minimizes the difference between adjacent time frames. It is also suitable for most DCE-MRI studies. The Kalman filtering method is based on a state model that estimates the true state of a linear dynamic system from a series of noisy observations over time. The Kalman filtering method can also denoise a series of DCE-MRI images by modeling the image with Gaussian noise.

Figure 9.22 shows an example of the processing result of KFRS algorithm. Detailed information of the KFRS algorithm can be found in Kerwin et al. (2002).

After registration and noise attenuation, the change in contrast concentration needs to be extracted from the processed serial images of DCE-MRI. First, change in contrast concentration can be converted

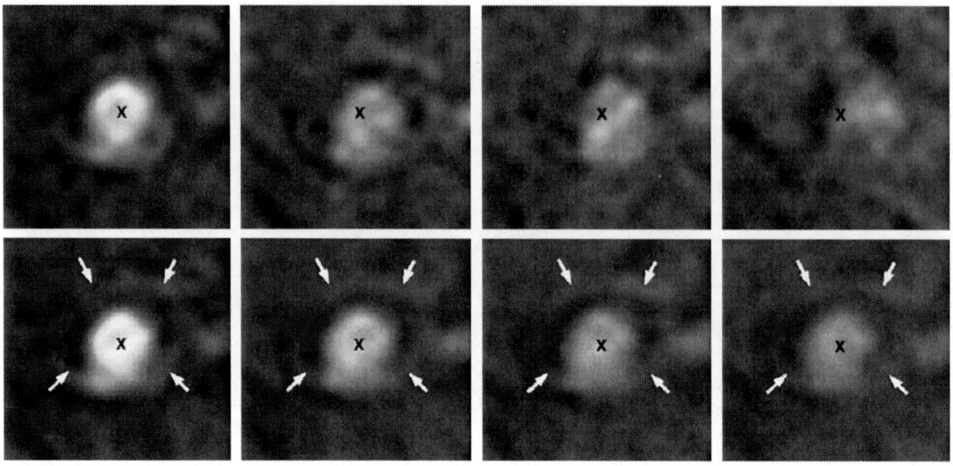

FIGURE 9.22 Four misaligned images in a DCE-MRI series with poor initial image quality before (top) and after (bottom) application of the KFRS algorithm. The filtered images show less motion relative to the fixed "X" and the presence of an enhancing outer wall (arrow). (Adapted from Kerwin W, Cai J, and Yuan C. 2002. *Magn Reson Med.* 47:1211–1217.)

from the MR intensity curves (Cron et al. 1999) based on an obtained precontrast T_1 map and the relationship between intensity and concentration known for the specific imaging sequence. This conversion conforms to the assumption of kinetic modeling and is widely used for relatively large tissues. However, this technique requires additional T_1 mapping, resulting in a longer acquisition time and more image processing and modeling challenges, especially for small tissues prone to motion. Alternately, an intensity curve can be used as a surrogate for the concentration curves for kinetic modeling because the linear relationship between the contrast agent concentration and the MR intensity can be assumed for low contrast concentrations in most DCE-MRI sequences. This technique simplifies both the acquisition and analysis, but can introduce bias in the kinetic modeling because its assumptions are not valid for high contrast concentrations.

9.4.2.3 Kinetic Modeling of DCE

The goal of kinetic modeling is to extract biologically relevant parameters from the contrast agent concentration curves. The contrast concentration change in the tissue of interests is dependent on the blood supply and permeability of the neovasculature. By properly modeling the contrast agent transfer in the targeted tissue after agent injection, the parameters can be calculated by fitting the mathematical model to reflect tissue permeability and blood supply.

The two-compartment model is the most popular model for kinetic modeling of DCE-MRI. As shown in Figure 9.23, this model describes the kinetic transfer of clinically available gadolinium extravascular extracellular agents from the plasma to extravascular extracellular space (EES), but not through the cellular membrane. The tissue in contrast kinetic can therefore be considered to be composed of two compartments: plasma and EES. There are several models proposed based on two-compartment model by multiple researcher groups.

A commonly used two-compartment model is the Kety/Tofts model (Tofts 1997). This model only considers the contrast agent in EES. Also, the bidirectional transfer of contrast agent between the plasma and EES is assumed. The equation of the Kety/Tofts model is

$$C_t(t) = K^{\text{trans}} \int_0^t C_p(\tau) e^{-\frac{K^{\text{trans}}}{v_e}(t-\tau)} d\tau \qquad (9.2)$$

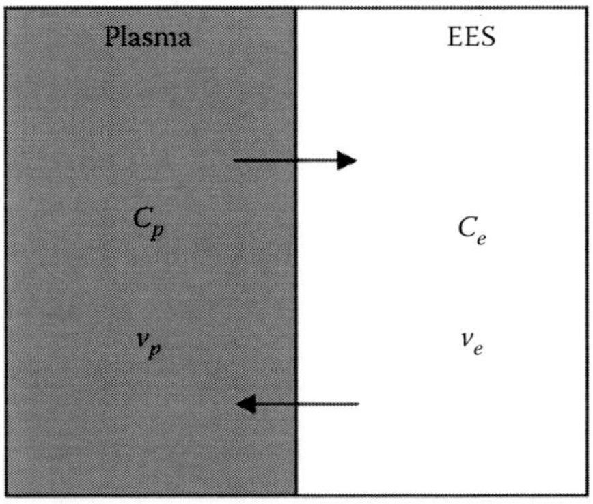

FIGURE 9.23 Illustration of the two-compartment model. Tissue concentration is composited by plasma and EES, v_p and v_e are the fractional volumes of plasma and EES, and C_p and C_e are the contrast agent concentrations in the plasma and EES. Contrast transfer from plasma to EES and can efflux from EES back to plasma.

where K^{trans} is the transfer constant, v_e is the fractional volumes of EES, and C_p and C_t are the contrast agent concentrations in the plasma and tissue. Based on this model, a modified Kety/Tofts model (Tofts, 1997) has been proposed that includes the contrast agent in plasma into consideration. The equation is

$$C_t(t) = v_p C_p(t) + K^{\text{trans}} \int_0^t C_p(\tau) e^{-\frac{K^{\text{trans}}}{v_e}(t-\tau)} d\tau \tag{9.3}$$

where v_p is the fractional volumes of plasma.

Another popular two-compartment model is the Patlak model (Patlak et al. 1983), which assumes that the efflux of the contrast agent from the EES back to plasma is negligible. The equation is

$$C_t(t) = v_p C_p(t) + K^{\text{trans}} \int_0^t C_p(\tau) d\tau \tag{9.4}$$

By mathematically analyzing the relationship between the modified Kety/Tofts model (Equation 9.3) and the Patlak model (Equation 9.4), an extended graphical model was proposed by partially modeling the contrast efflux effect from EES back to plasma (Chen et al. 2011):

$$C_t(t) = v_p C_p(t) + K^{\text{trans}} \int_0^t C_p(\tau) d\tau - \frac{K^{\text{trans}^2}}{v_e} \int_0^t \int_0^t C_p(\tau_2) d\tau_2 d\tau_1 \tag{9.5}$$

In these models, K^{trans}, v_e, and v_p are the parameters related to the physiology of the tissue. Usually, the transfer constant K^{trans} is considered important because it is associated with blood flow, capillary surface area, and permeability, making it a good indicator of blood supply and neovasculature permeability in the tissue.

As we can see from these models, given temporal measurements of C_p and C_t, physiology-related parameters can be estimated that provide the best fit between the observed measurements and the model. C_t can be easily extracted from the tissue of interest. However, extraction of C_p, which is also known as arterial input function (AIF), is typically difficult due to the high concentration of contrast agent in blood and the high temporal resolution requirement. In kinetic modeling of black-blood DCE-MRI, this is more challenging because the blood signal has been suppressed.

As a result, many studies fall back on calculating nonmodel-based parameters instead of the model-based parameters, such as the area under enhancement curve (AUC) (Patlak et al. 1983, Calcagno et al. 2008). These model-free parameters are easily calculated. However, their relationship to the physiological properties is neither intuitive nor obvious.

Many researchers have made efforts to develop kinetic modeling methods without the knowledge of the AIF. A population-derived AIF (Tofts et al. 1999) has been proposed to standardize AIF for an entire study population. However, to determine this population-derived AIF, extra experiments are needed, which increases the study cost. Also, the assumption that all individuals have the same AIF may introduce bias in the analysis. An alternative approach is to use the reference region model (Yankeelov et al. 2005). This technique can calculate the AIF by using the reverse function of the kinetic model from a reference region. For example, the reverse function of Patlak model is

$$C_p(t) = \frac{1}{v_{PRR}} C_{RR}(t) - \frac{K^{\text{trans}}{}_{RR}}{v_{PRR}^2} \int_0^t C_{RR}(\tau) e^{-\frac{K^{\text{trans}}{}_{RR}}{v_{PRR}}(t-\tau)} d\tau \tag{9.6}$$

With the concentration curve C_{RR} and known parameters K_{RR}^{trans}, v_{pRR} of the reference region, the AIF can be calculated. Then, the parameters of the tissue of interest (ROI) can be computed by fitting Equation 9.4. Furthermore, by substituting Equation 9.6 into Equation 9.4, we can get

$$C_t(t) = \alpha C_{RR}(t) - \alpha\beta \int_0^t C_{RR}(t) e^{-\beta(t-\tau)} d\tau + \gamma \int_0^t C_{RR}(\tau) d\tau - \gamma\beta \int_0^t \int_0^{\tau_1} C_{RR}(\tau) e^{-\beta(t-\tau)} d\tau d\tau_1 \tag{9.7}$$

where $\alpha = \frac{v_p}{v_{PRR}}, \beta = \frac{k_{RR}^{\text{trans}}}{v_{PRR}}, \gamma \frac{k^{\text{trans}}}{v_{PRR}}$. In this manner, K_{RR}^{trans} and v_{pRR} are not necessarily preknown. They can be estimated together with K^{trans} and v_p by fitting Equation 9.7 with measured C_{RR} and C_t. However, the complicated integer and the nonlinearity of the equation in these two methods increase the complexity of the estimate, resulting in higher noise sensitivity and poor stability of the fitting algorithm.

Alternatively, instead of directly using the reversion function, the AIF can be eliminated to generate an equation allowing linearized fitting algorithm (Wu 2008):

$$v_{PRR} C_t(t) + K_{RR}^{\text{trans}} \int_0^t C_t(\tau) d\tau = v_p C_{RR}(t) + K^{\text{trans}} \int_0^t C_{RR}(\tau) d\tau \tag{9.8}$$

The reference region model for the original and modified Kety/Tofts model can be found in Yankeelov et al. (2005) and Faranesh and Yankeelov (2008).

9.4.2.4 DCE-MRI Studies of Arterial Atherosclerosis

By utilizing the previously described MRI, image processing, and kinetic modeling techniques, many studies have proven that the DCE-MRI can provide unique information on revascularization and inflammation of atherosclerotic plaque.

Kerwin et al. conducted serial studies (Kerwin et al. 2003, 2006, 2008) to investigate quantitative evaluation of the plaque vascularity and inflammation using bright-blood DCE-MRI acquired by SPGR sequence and quantitative parameters derived from Patlak model (v_p and K^{trans}). In the first study (Kerwin et al. 2003), by analyzing 16 patients, v_p was found to be correlated ($r = 0.80$, $p < 0.001$) with neovessel area obtained in histological sections, indicating that v_p was an effective marker of plaque vascularities. In a subsequent study (Kerwin et al. 2006) involving 30 patients, measurements of K^{trans} and v_p were found to have a high correlation with macrophage (K^{trans}: $r = 0.75$, $p < 0.001$; v_p: $r = 0.54$,

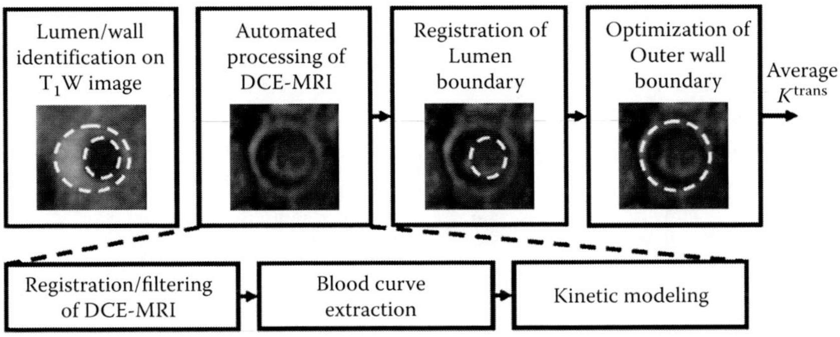

FIGURE 9.24 Illustration of the steps in creating a vasa vasorum image: boundaries drawn on a conventional contrast (T_1-weighted) define a region of interest for processing. Within this region, DCE-MRI images are run through the KFRS algorithm, then a blood curve (AIF) is automatically extracted and used for kinetic modeling. Finally, the boundaries are mapped to the resulting vasa vasorum image and K^{trans} is measured in the wall. (From Kerwin W et al. 2006. *Radiology.* 241:459–468.)

$p < 0.004$) and neovasculature (K^{trans}: $r = 0.71$, $p < 0.001$; v_p: $r = 0.68$, $p < 0.001$) content in 27 patients with analyzable MR images and histological results. Additionally, K^{trans} was significantly correlated with high-density lipoprotein levels ($r = -0.66$, $p < 0.001$) and was elevated in smokers compared with nonsmokers (mean, 0.134 min^{-1} vs. 0.074 min^{-1}; $p < 0.01$). Another study (Kerwin et al. 2008) has shown that K^{trans} measurements of the adventitial region of the carotid were significantly correlated with serum markers of inflammation. In 20 patients with moderate carotid lesions, adventitial K^{trans} was significantly correlated with the log of C-reactive protein (CRP) levels ($r = 0.57$; $p = 0.01$) and was elevated in active smokers compared to nonsmokers (0.141 vs. 0.111 min^{-1}; $p = 0.02$). This study also proposed a complete processing chain that produces colorized parametric images called "vasa vasorum images" (V-V image) (Figure 9.24). In V-V image, v_p are presented in red and K^{trans} is presented in green, a convenient and intuitive way to present the Patlak model-derived parameters. Based on this image acquisition and processing protocol, Chen et al. (2010) further showed that bright-blood DCE-MRI can exhibit differences in kinetic parameters within different plaque components, indicating the DCE-MRI can differentiate plaque components. More significantly, this study suggests that DCE-MRI has the potential to assess the inflammatory burden in specific regions.

By using a rabbit model, Calcagno et al. (2008) investigated the relationship between DCE-MRI-derived model-free parameter AUC and plaque vascularization. In this study, a DIR sequence was used for black-blood DCE-MRI acquisition. AUC was found to have a positive correlation between neovessels count in atherosclerotic plaques ($r = 0.89$, $p = 0.016$). In a following study, good reproducibility of the black-blood DCE-MRI-based AUC analysis was reported.

Table 9.2 is a summary of the significant findings from these studies. As we can see, these studies establish the basis for *in vivo* quantitative assessment of plaque neovasculature and inflammation using DCE-MRI. Inflammation is thought to play an important role in atherosclerotic plaque initiation, progression, and rupture, and there is a strong interest in treatments with anti-inflammatory effects. Recent studies have assessed the effectiveness of anti-inflammatory, antiatherosclerotic treatments using DCE-MRI. By using black-blood DCE-MRI, Lobatto et al. (2010) successfully revealed early changes in plaque permeability after liposomal glucocorticoid treatment with anti-inflammation effect in a rabbit model.

9.4.3 Other Contrast Agents

Small molecular extravascular extracellular gadolinium contrast agents have been the major agents used in clinical MRI research of arterial atherosclerosis. Currently, many researchers have put efforts

TABLE 9.2 Summary of Associations between DCE-MRI Parameters and Inflammation-Related Factors

	v_p Association with		K^{trans} Association with					AUC Association with
Study	Neovessel	Macrophage	Neovessel	Macrophage	Smoking	HDL	Log CRP	Neovessel
Kerwin 03	0.80[a]							
Kerwin 06	0.68[a]	0.54[a]	0.71[a]	0.75[a]	0.13 vs. 0.07[a]	−0.66[a]		
Kerwin 08			0.41[a]	0.49[a]	0.14 vs. 0.11[a]		0.57[a]	
Calcagno 08								0.89[a]

[a] Indicates statistical significance, $p < 0.05$.

into developing new contrast agents sensitive to specific tissues, cells, and even molecules to detect vulnerable atherosclerotic plaque in preclinical or clinical studies.

The challenge of developing new specific agents to target specific molecules or cells is that often these cells are present in very small numbers (Briley-Saebo et al. 2007). To be visible on MRI, contrast agents may be designed so that the agent has a high payload of the magnetic entity (Winter et al. 2003a), or high concentration when interacting with the targets (Sirol et al. 2009), or specify a target with a large number of cells or molecules in the diseased vessel wall (Winter et al. 2003a, Botnar et al. 2004a,b). There are two major nanoparticle platforms for new MR contrast agents development: first, the use of amphiphiles possessing both hydrophilic and lipophilic properties that can self-assemble into larger nanoparticles to deliver large payloads of magnetic entity, such as microemulsions (Briley-Saebo et al. 2007), micelles (Briley-Saebo et al. 2008, Howarth et al. 2009), liposomes (Briley-Saebo et al. 2007), and lipoproteins (Briley-Saebo et al. 2007); second, ultrasmall superparamagnetic iron oxide particles (USPIOs) (Johansson et al. 2001, Kooi et al. 2003, Trivedi et al. 2006, Howarth et al. 2009) with a size of 18–30 nm.

9.4.3.1 Gadolinium-Based Agents

Many new agents are still gadolinium based but possess unique characteristics that facilitate VWI, such as the ability to bind to blood albumin. Gadofosveset is an agent that binds to albumin with an 80–96% bound fraction in human plasma, resulting in a dramatic increase in both relaxivity and elimination half-life (Lauffer et al. 1998). The increased relaxivity can be characterized by signal enhancement in T_1-weighted images. By binding with the albumin, Gadofosveset should be restricted in the blood pool in normal vessels. However, in atherosclerosis, increased content of microvessel and permeability of the microvessel, enable albumin-bound Gadofosveset to enter the vessel wall. Therefore, Gadofosveset accumulation inside the vessel wall is a sign of atherosclerosis (Lobbes et al. 2009). Another albumin-binding agent under intensive investigation for atherosclerotic plaque MR imaging is the Gadoflurine M. Gadoflurine M is a lipophilic, water-soluble gadolinium chelate complex with a perfluorinated side chain (Meding et al. 2007). Besides binding to albumin, it can also bind to collagens, proteoglycans, and tenascin inside the plaque (Meding et al. 2007). As a result, Gadoflurine M can enter the atherosclerotic plaque by increased highly permeable neovasculature. Gadoflurine M can also accumulate inside the fibrous parts of the plaque containing collagens, proteoglycans, and tenascin. In preclinical studies, Barkhausen et al. (2003) found this accumulation in the aortic wall of Watanabe heritable hyperlipidemic (WHHL) rabbits at 48 h after injection. Gadoflurine M accumulation was also found to be related to degree of inflammation and extent of core neovascularization in atherosclerotic plaque (Sirol et al. 2009). Other gadolinium-based contrast agents have relative large molecules. They cannot be transferred easily inside normal vessel walls, but will enhance in diseased vessel walls with leaky microvessels. Carboxymethyl-dex-tran derivative CMD-A2-Gd-DOTA (P717) is a contrast agent with large molecules and has been investigated in a mouse model of atherosclerosis (Chaabane et al. 2004). This kind of macromolecular agent may be useful in kinetic modeling of DCE-MRI because the slow transfer rate of the large molecular agents reduces the need for rapid imaging (Pradel et al. 2003).

9.4.3.2 Iron-Based Agents

Another important type of the specific contrast agent are the iron-based agents, such as USPIOs. USPIOs are composed of an iron crystalline core and a biocompatible coating. Unlike gadolinium-based contrast agents which create bright signal in MR images, USPIOs usually are called negative contrast agents because they can significantly enhance T_2* relaxation, resulting in signal void in T_2*-weighted images (Hofmann-Amtenbrink et al. 2010). USPIOs can be taken up by macrophages in the whole body because of their long half-life in blood. In atherosclerotic plaque imaging, this nature of USPIOs has made them good contrast agents specific to inflammation, featured with macrophages infiltration inside the vessel wall. Kooi et al. (2003) found that USPIO accumulation in human aortic atherosclerotic vessel walls can be observed *in vivo* 24 h after administration, and ruptured and rupture-prone plaque showed more accumulation than stable lesions. Trivedi et al. (2006) showed that signal loss induced by USPIOs corresponded to macrophage uptake in the fibrous cap of human carotid artery. Howarth et al. (2009) investigated the role of *in vivo* inflammation detection in risk assessment of vulnerable plaque using USPIOs. They found symptomatic patients had more quadrants with signal drop than asymptomatic patients, indicating more inflammatory regions in symptomatic patients.

9.4.3.3 Molecular Imaging Agents

Although the previously mentioned gadolinium-based agents and USPIOs are macrophage or tissue specific, these agents are not targeted to specific molecules or receptors. Application of specifically targeted molecular contrast agents is a promising technique that allows molecular imaging of MRI to study the functional process of atherosclerosis.

The most popular goals in molecular imaging of plaque are plaque inflammation and angiogenesis. Inflammation is a complicated process in plaque which has not been fully defined, although it typically means inflammatory cells, such as macrophages, infiltrate the plaque. This infiltration process is accompanied by angiogenesis and endothelium dysfunction, which reinforces the infiltration of the inflammatory cells. To image inflammation in plaque, an immunomicelle targeted to the mouse macrophage scavenger receptor (MSR) A has been proposed for atherosclerotic plaque imaging of apoE – /– mouse aorta (Lipinski et al. 2006). The oxidized low-density lipoprotein, which can stimulate the process of inflammation and atherosclerosis, has been visualized with molecular MR imaging (Briley-Saebo et al. 2008) by using an agent that binds unique oxidation-specific epitopes. The matrix metalloproteinases (MMPs) produced in macrophages and released into plaque in macrophage apoptosis are considered to contribute to plaque rupture (Libby 2002) and to be viable targets in VWI. MMPs-targeted agent has been successfully applied for plaque MMPs MRI (Lancelot et al. 2008). Also, MR contrast agents targeting cell apoptosis can also be used in VWI (Strijkers et al. 2010). The dysfunctional endothelium of atherosclerotic lesions has also been characterized with MRI by contrast agent targeting VCAM-1, which is expressed by dysfunctional endothelium (Kelly et al. 2005). For angiogenesis MR imaging, Winter et al. (2003b) proposed an agent targeting $\alpha_v\beta_3$-integrin, which accompanies the process of angiogenesis. They found significantly greater enhancement in the aorta wall with expanded vasa vasorum in a rabbit model after injection of the new agent.

Another target of molecular plaque imaging is the thrombus. Thrombus may lead to a total occlusion of the vessel or lethal ischemia events. Therefore, early detection of thrombus in atherosclerosis has significant diagnostic value. USPIOs with a cyclic arginine–glycine–aspartic acid (RGD-USPIO) specific for the $\alpha_{IIb}\beta_3$-integrin, which is expressed on activated platelets, have been used to image the thrombus (Johansson et al. 2001). Small-molecule gadolinium-based contrast agents have also been proposed and successfully applied in the evaluation of acute and subacute coronary thrombus (Botnar et al. 2004a,b).

Winter et al. (2003a) also proposed a fibrin-targeted agent with larger molecule for thrombus imaging based on a different gadolinium formulation.

Another goal of molecular imaging is to improve the detection of plaque morphology. Recently, an elastin-specific MR contrast agent was proposed to assess the atherosclerotic plaque burden (Makowksi et al. 2011). Elastin is a key extracellular matrix protein in plaque. The authors showed the new agent can provide strong signal for accurate quantification of the plaque burden in a mouse model of atherosclerosis. This kind of contrast agent may be useful in the VWI of certain small but important arteries, which are difficult to image using traditional agents and MR imaging techniques, such as the coronary artery.

9.5 Imaging-Based Biomechanical Analysis of Atherosclerotic Plaques

The arterial wall must regenerate and remodel continuously to maintain its integrity and function while withstanding the hemodynamic stresses imposed by the pulsatile blood flow upon it. In fact, this remodeling process is the result of the response of the endothelial cells lining the luminal wall to those stresses. These biomechanical stimuli include blood flow-induced frictional stress or shear stress, pressure-induced cyclic stress or tensile stress, and a normal stress due to hydrostatic pressure differential across the wall. Unfortunately, the complex interrelationship between the endothelial response and the biomechanical stimuli might be altered, inducing endothelial dysfunction that will trigger pathophysiological processes leading to cardiovascular disease (Ku 1997, Taber 2001).

In particular, it is well accepted that the reaction of endothelial cells to the low and oscillatory shear stress resulting from the distinct complex patterns of blood flow in arterial bifurcations and bends (Figure 9.25) is responsible for the local susceptibility to atherosclerosis initiation in vessels such as the carotid or coronary arteries (Taber 2001, Ku 1997, Slager et al. 2005, Lee et al. 2008). Hypertension has also been identified as a risk factor in atherosclerosis. Increased pressure induces endothelial dysfunction increasing the mass transport from the lumen to the vessel wall, as well as endothelial cell proliferation and apoptosis (VanEpps and Vorp 2007). Several studies have indicated that the biomechanical stimuli also contribute to the lipid accumulation and inflammatory cell recruitment and adhesion occurring in atherogenesis (VanEpps and Vorp 2007).

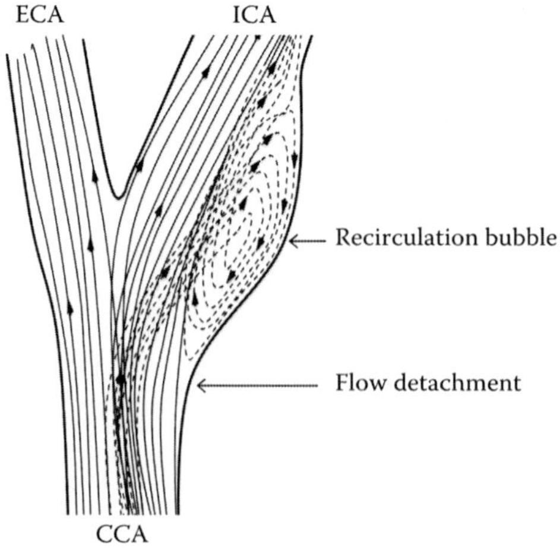

FIGURE 9.25 Depiction of some of the flow features developing with pulsatile flow through the common carotid artery (CCA) bifurcation into the external and internal carotid arteries (ECA and ICA, respectively). (Modified from Dong L et al. 2010a. *AJNR Am J Neuroradiol.* 31:311–316.)

(a) (b)

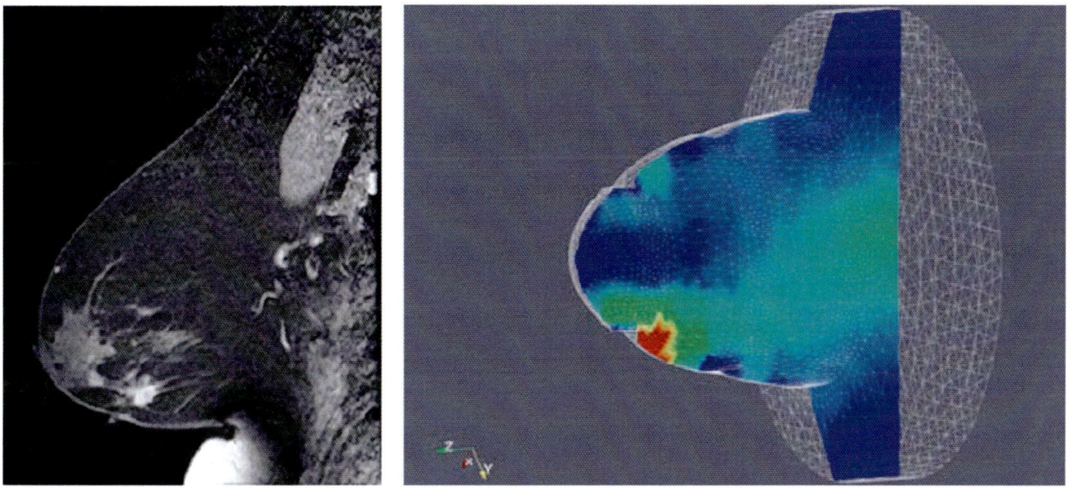

FIGURE 8.5 (a) Sagittal contrast-enhanced breast MRI images. 1.3 cm cancerous mass visible at 5 cm from nipple. (b) 3D EIT reconstruction of breast. (Adapted from Tizzard, A. et al. 2010. *J. Phys. Conf. Ser.* 224(1): 012034.)

(a) (b)

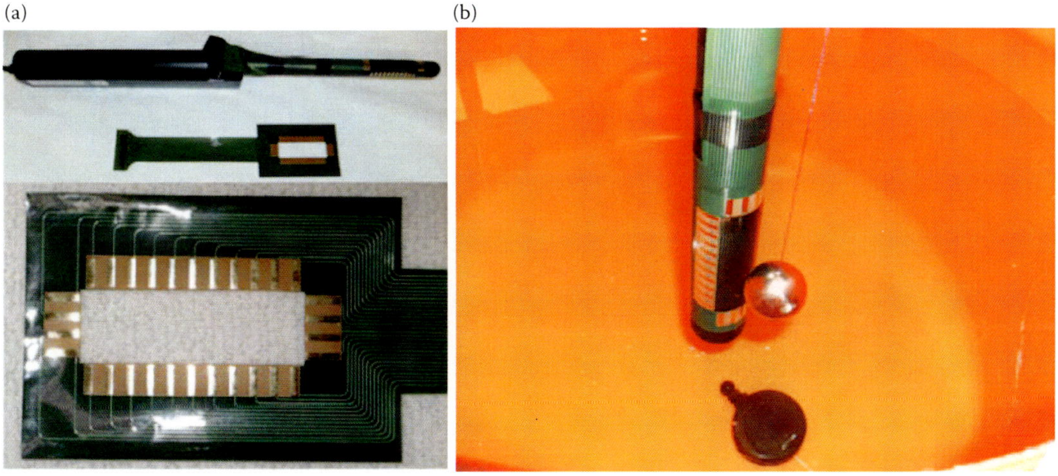

FIGURE 8.6 (a) Top: flex circuit of electrodes wrapped around the ultrasound probe. (a) Bottom: magnified view of flex circuit of electrodes laid flat. (b) Imaging configuration; an inclusion (steel ball) is placed immediately in front of the acoustic window. (Adapted from Wan, Y. et al. 2010. *J. Phys. Conf. Ser.* 224(1): 012067.)

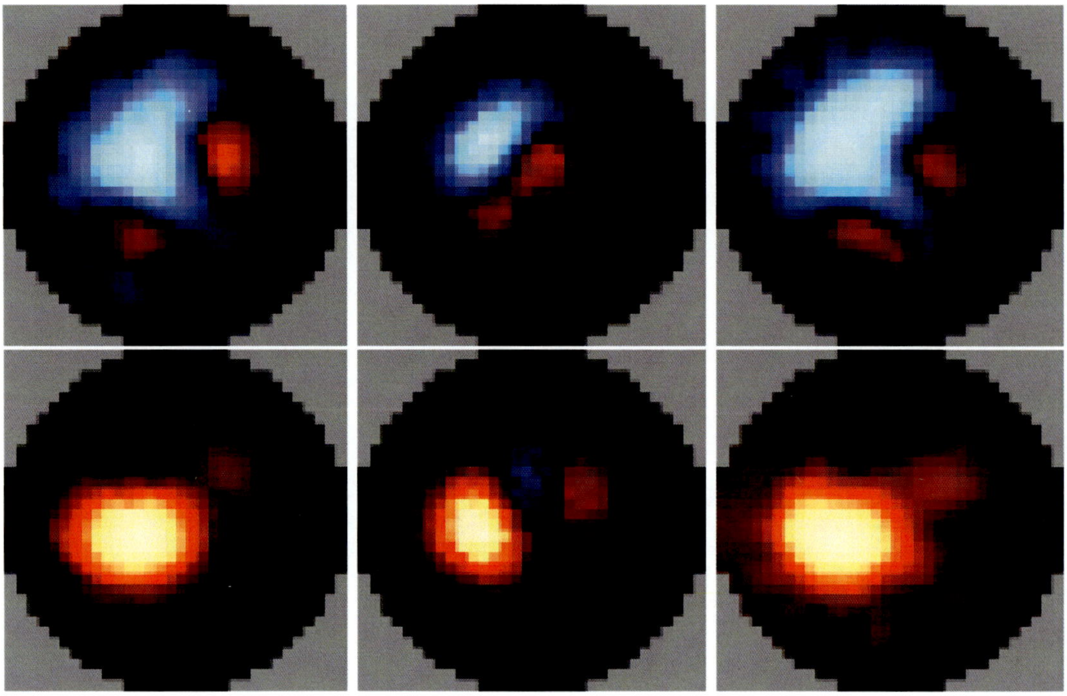

FIGURE 8.7 Images of localized (top row) pneumothorax and (bottom row) pleural effusion in an anesthetized, ventilated pig. Left: GREIT (R_{GN}), center: Gauss–Newton (R_{GN}), right: Sheffield backprojection (R_{SBP}). (Adapted from Adler, A. et al. 2009. *Physiol. Meas.* 30: S35–S55.)

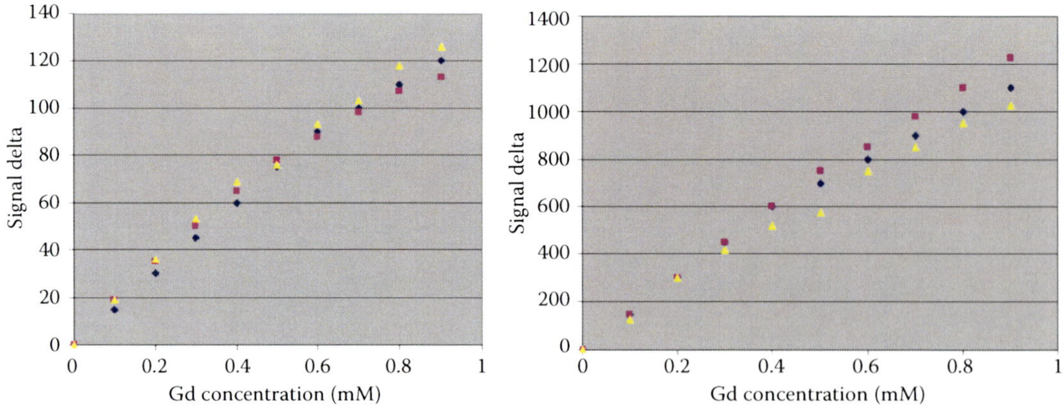

FIGURE 10.6 Plot of signal change versus gadolinium concentration for GE (l) and Siemens (r) systems. Baseline T_1 values were 90 ms (yellow), 300 ms (pink), and 700 ms (blue). Note that the Siemens system shows roll-off in signal response at significantly longer T_1 than the GE system.

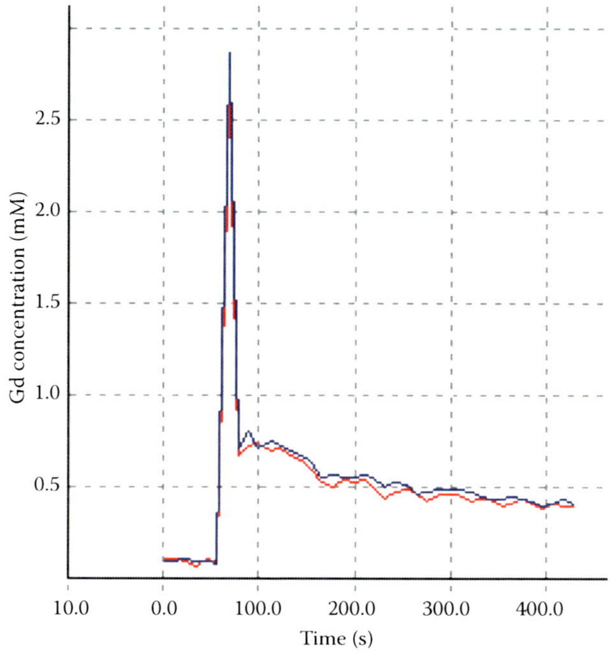

FIGURE 10.7 Typical AIF obtained using T_1-weighted imaging with a 0.1 mmol/kg contrast injection and 6 s temporal resolution.

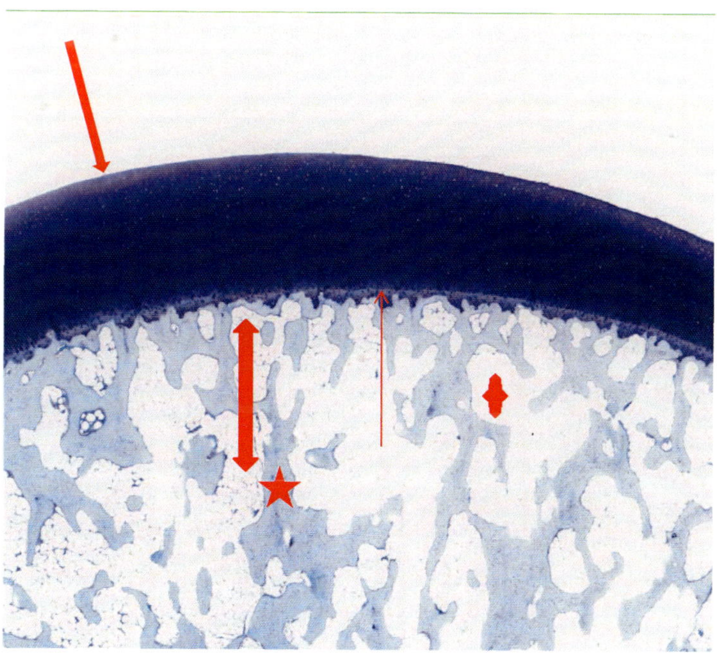

FIGURE 12.1 Histology section of cartilage showing dark blue noncalcified cartilage (thick arrow), light blue calcified cartilage underneath it (thin arrow), tide mark, a thin line between two cartilage types, subchondral bone plate underneath the calcified cartilage (double ended arrow), trabecular bone (star), and bone marrow space (cross) between them. The dark blue color of the cartilage indicates normal cartilage with high proteoglycan concentration.

FIGURE 12.21 A surface rendering of automatically segmented femur and tibia bones seen from behind. Note the widened and flattened tibia and femur in this OA knee.

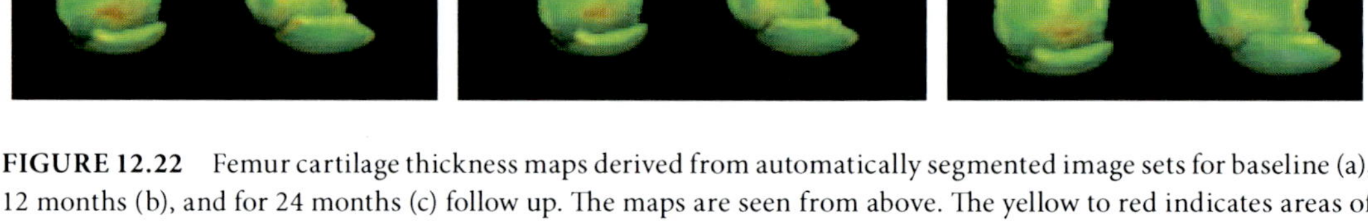

FIGURE 12.22 Femur cartilage thickness maps derived from automatically segmented image sets for baseline (a), 12 months (b), and for 24 months (c) follow up. The maps are seen from above. The yellow to red indicates areas of increasingly thinner cartilage.

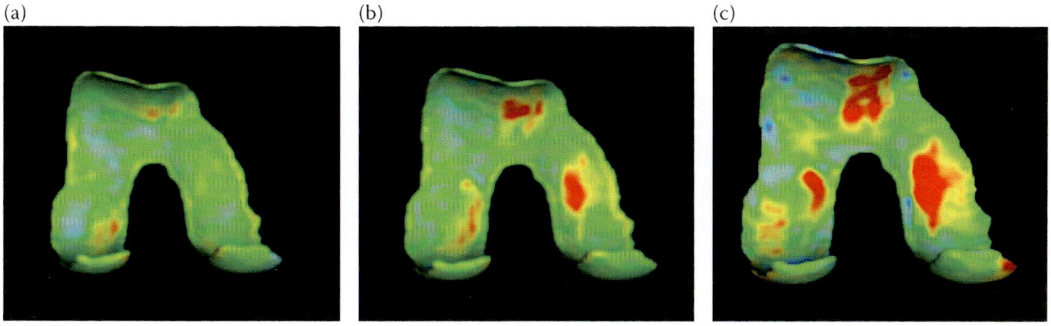

FIGURE 12.23 Results of mapping the cartilage thickness maps from Figure 12.6 to the reference atlas. The maps are seen from above. The baseline cartilage thickness (a) is compared to the 24 month follow-up visit (b) to compute the change in cartilage thickness over time (c). Note the locality of the changes; the defects are local and the local defects are getting deeper and bigger. However, at the same time, new local areas of defects as well as thickening appear.

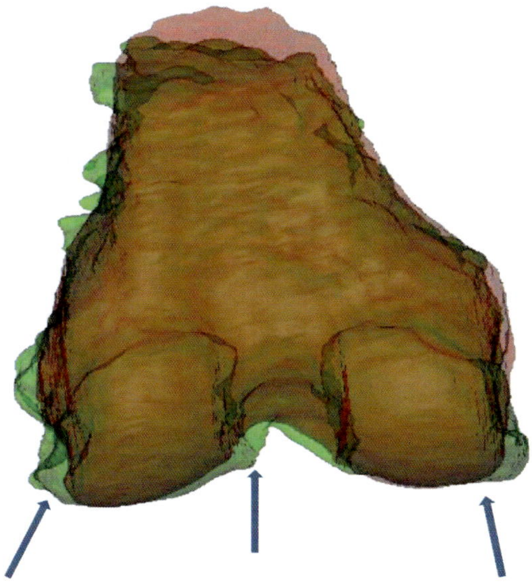

FIGURE 12.28 Surface-rendered results of mapping the segmented femur of the OA subject (green) to the segmented femur of the reference atlas (red). Note the flattened and widened articulating surfaces in the subject's femur (thick arrows).

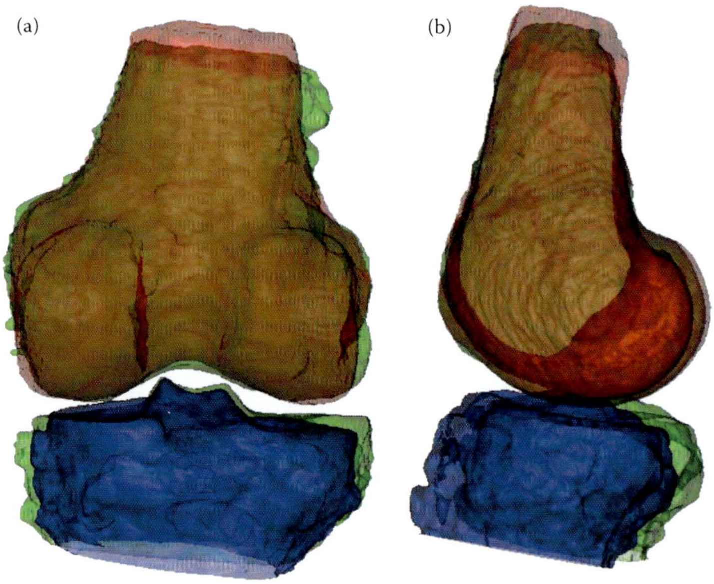

FIGURE 12.30 Posterior (a) and lateral (b) Views of the 3D relationships. The 3D relationship between the subject's tibia and atlas' tibia. The subject's segmented tibia and femur were coregistered to the atlas' tibia and femur. The images show subject's tibia as green and the location of atlas tibia as blue. The atlas femur is red and the subject's femur is green.

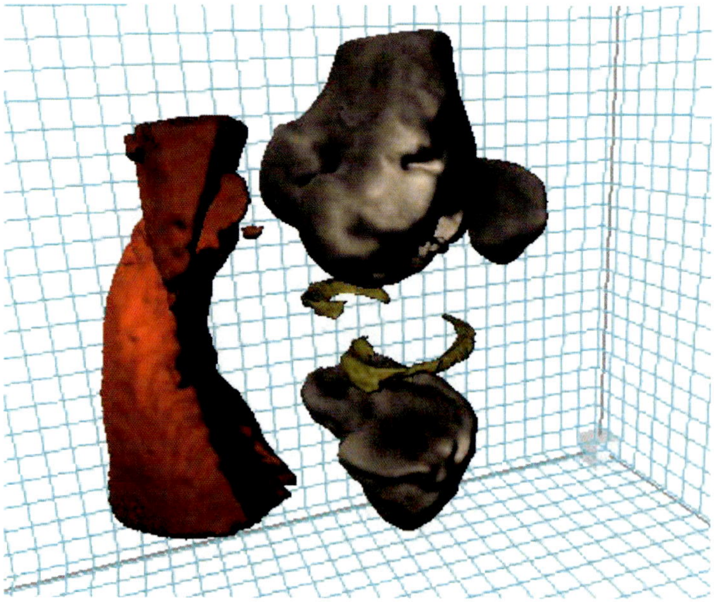

FIGURE 12.38 The normal knee anatomy can be automatically segmented from MRI data sets using advanced statistical analysis tools.

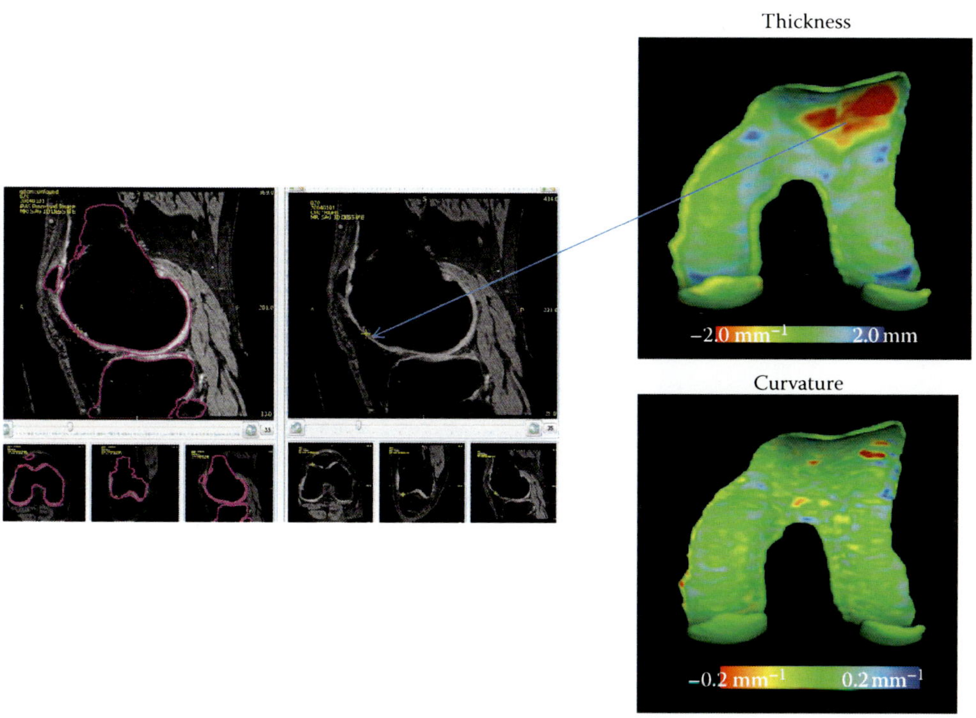

FIGURE 12.39 In the presence of OA, atlas-based approaches are a practical alternative to the segmentation of thousands of MRI images. Left, slice through the DESS images. (Courtesy of OAI.) Top right, 3D reconstruction of the cartilage thickness in the atlas space. Red values represent areas of cartilage loss. Bottom right, the curvature of the cartilage surface can be computed and measured. Drastic changes in cartilage topography can be measured by the curvature values and are seen as red and blue spots in the curvature map.

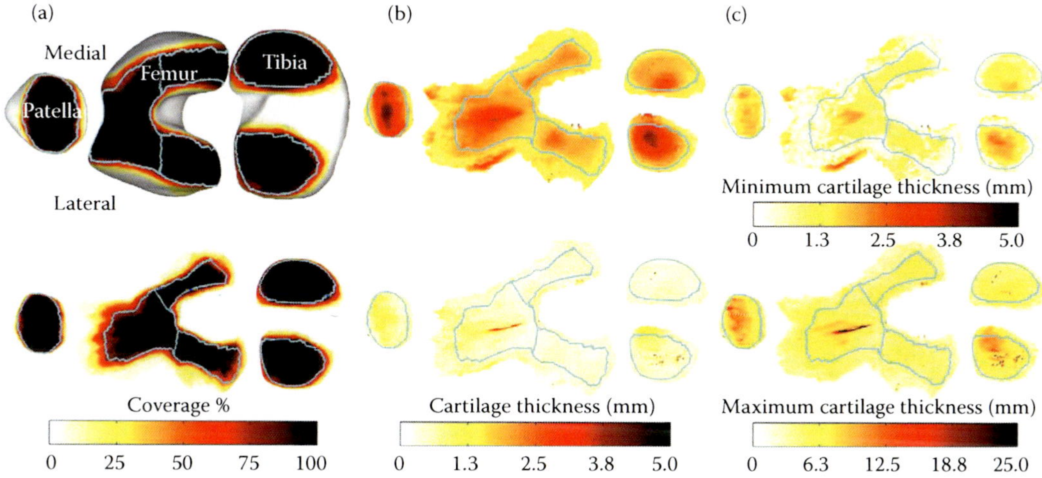

FIGURE 12.40 Statistical maps of changes in cartilage thickness. The thickness maps are a very useful tool for the visualization of the localized changes and heterogeneity in the patterns of cartilage loss in subjects with OA. (a) Top, 3D Rendering of knee bones with coverage analysis. (b) Mean (top) and standard deviation (bottom) of cartilage thickness. (c) Minimum and Maximum of the thickness. (This picture has copyrights by the IEEE TMI.)

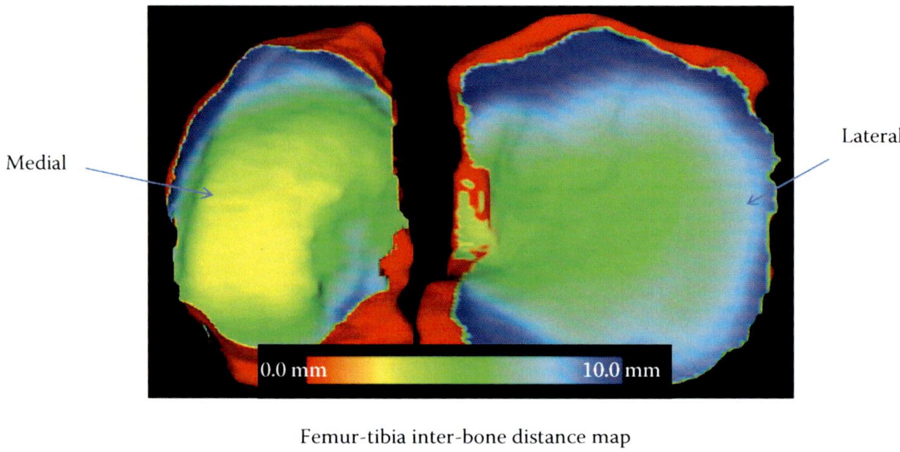

FIGURE 12.41 The MRI images can be used to compute the interjoint distance. Distance maps can automatically be computed and they are an indirect measurement of changes in the soft tissues in joint anatomy.

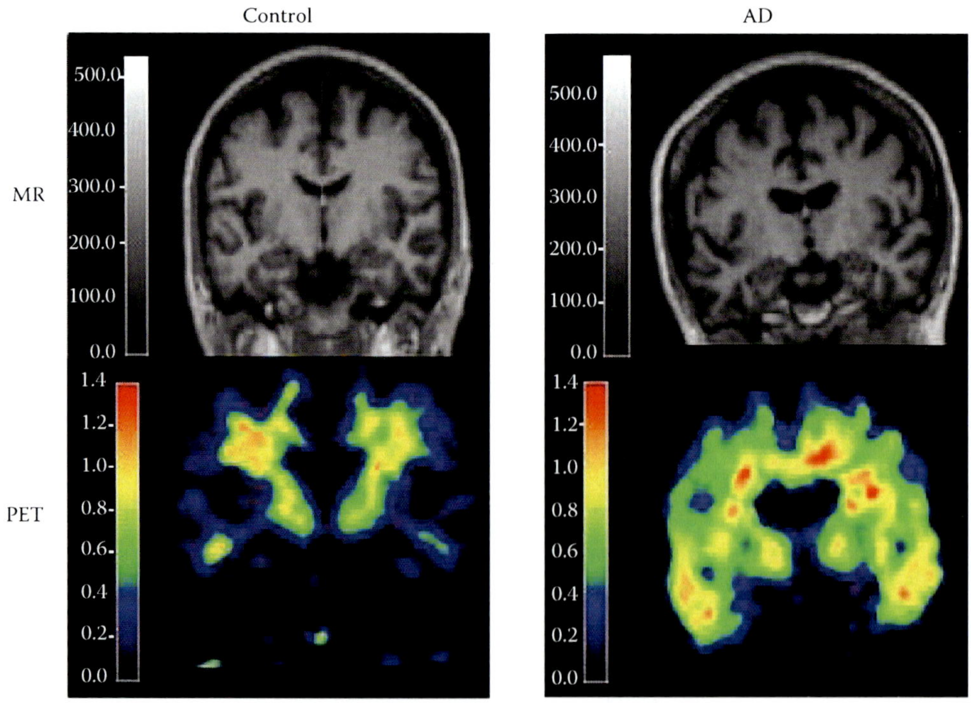

FIGURE 13.3 MRI (top row) and PET tracer-binding images obtained with the amyloid tracers [11C]-PIB (bottom) row for a healthy control (first column) and AD patient (second column). A clear increase in tracer binding is seen in the AD subject compared to the healthy control.

Even though the hemodynamic stresses involved in the initiation of the disease and the pathological processes that they trigger have been extensively described in the literature, their role in plaque growth and rupture has not yet been established. Endothelial cells react to shear stress modifying the balance between fibrous cap-reinforcing extracellular matrix synthesis by smooth muscle cells and extracellular matrix degradation by metalloproteinases secreted by infiltrating macrophages (Cheng et al. 2006, Clowes and Berceli 2000). This imbalance seems to be a major contributor to plaque rupture. Focal regions of high strain in the fibrous cap are linked to the location of plaque rupture (Li et al. 2006) indicating that the failure of the wall biological tissues to compensate the effect of the increased tensile (pressure-induced) stress (Arroyo and Lee 1999) could also be a major factor in plaque rupture.

As mentioned in previous sections, current criteria to identify plaques at risk of rupture mainly rely on morphological and compositional factors, and do not take into consideration the biomechanical factors. Thus, plaque vulnerability assessment could be considerably improved by incorporating hemodynamic stress information.

MRI has successfully been used to characterize the atherosclerotic plaques in terms of their morphological and compositional features. This unique quality of MRI is currently used to determine the patient-specific hemodynamic stress distributions that could be responsible for, or the result of patient-specific, atherosclerotic plaque features. This section will examine the combined use of MRI and mechanical engineering tools to obtain a comprehensive characterization of the atherosclerotic lesion in terms of both its morphological and compositional features, as well as of its hemodynamic factors. This comprehensive evaluation of the atherosclerotic vessels would allow advancing the knowledge of the role of biomechanics in plaque progression and rupture and would lead, in turn, to a more precise assessment of plaque vulnerability.

9.5.1 Image-Based Numerical Models to Characterize Hemodynamic Stresses in Atherosclerosis

The ability of MRI to define the vessel lumen has allowed reconstructing the geometry of atherosclerotic plaques to analyze the flow features present in the diseased vessel with the aid of computational fluid dynamics (CFD) models.

9.5.1.1 Computational Fluid Dynamics Models

The luminal boundary obtained from the segmentation of MR images is used to render a 3D surface that is discretized into a finite-element mesh constituting the fluid domain as shown in Figure 9.26. The size and distribution of the mesh elements is set by the compromise between the accuracy of the computed velocity field and the computational expense. Unstructured meshes with varying mesh density are most commonly used when dealing with atherosclerotic vessels for adequately resolving the complex velocity field that develops in those bent and/or bifurcating vessels.

9.5.1.1.1 Boundary Conditions and Assumptions Needed to Solve for Flow

Once the fluid domain is defined, special attention should be paid to specifying the physical properties and boundary conditions needed to solve the equations that govern blood flow. Due to the complexity of the problem, image-based CFD models have traditionally made simplifying assumptions regarding both blood properties and flow boundary conditions.

Blood is known to be non-Newtonian, that is, it exhibits a nonlinear relationship between shear stress and shear strain rate. However, the behavior of blood can be described as Newtonian (constant viscosity at all rates of shear) at the high shear strain rates present in large arteries, such as the ones where atherosclerosis is most often found (Ku 1997). Thus, in image-based CFD modeling of flow in atherosclerosis, blood is commonly assumed to be Newtonian. This assumption is even more reasonable after considering that the uncertainty of geometry reconstruction has a greater effect on the distribution of wall shear stresses (Lee and Steinman 2007).

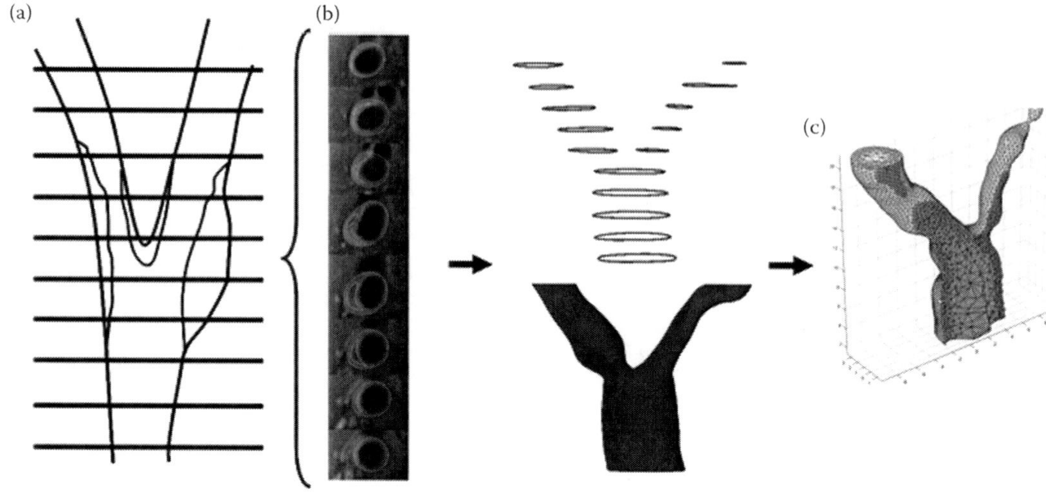

FIGURE 9.26 Summary of typical steps required for the reconstruction of a finite-element mesh from serial MR images. (a) Segmentation of the stack of cross-sectional MR images along the bifurcation; (b) three-dimensional reconstruction of the luminal volume from the extracted lumen contours; (c) three-dimensional computational mesh of the luminal volume or fluid domain.

The equations governing the flow of a Newtonian viscous fluid are known as the Navier–Stokes equations, which are the result of applying Newton's second law to fluid motion. Assuming that blood is an incompressible fluid, these equations are simplified to

$$\rho\frac{Du}{Dt} = \rho X - \frac{\partial p}{\partial x} + \mu\nabla^2 u$$
$$\rho\frac{Dv}{Dt} = \rho Y - \frac{\partial p}{\partial y} + \mu\nabla^2 v$$
$$\rho\frac{Dw}{Dt} = \rho Z - \frac{\partial p}{\partial z} + \mu\nabla^2 w$$

where ρ is the density of blood, normally set at 1045 kg/cm^3; u, v, and w are the velocity components along the x-, y-, and z-axis directions, respectively; X, Y, and Z are the body force per unit mass components along the x-, y-, and z-axis directions, respectively; p denotes pressure; and μ is the dynamic viscosity of blood, a typically chosen value is 0.0035 Pa · s.

Blood flow must also satisfy the continuity equation, derived from the conservation of mass. For an incompressible fluid, the equation of continuity becomes

$$\frac{\partial u}{\partial x} + \frac{\partial v}{\partial y} + \frac{\partial w}{\partial z} = 0$$

These equations are solved using finite-element methods to obtain the velocity profile. However, in order to solve these equations, boundary conditions must be specified. Blood flow is not constant, but varies with the cardiac cycle. The importance of the unsteady forces resulting from the flow pulsatility is characterized by the dimensionless Womersley number (α), which accounts for the relationship between unsteady inertial forces and viscous forces (Womersley 1955):

$$\alpha = R\sqrt{\frac{w}{v}}$$

where R is the arterial radius, ω is the angular frequency, and ν is the kinematic viscosity (μ/ρ). In the case of a healthy carotid bifurcation, α is approximately 4, indicating that the viscous forces dominate the flow (Ku 1997). Womersley analytical solution for fully developed flow is usually imposed as inlet boundary condition to account for the pulsatile nature of blood flow (Womersley 1955). This analytical solution for pulsatile flow in a straight or tapered tube expands the physiologic pressure or flow waveform as a Fourier series, and the harmonic components of the velocity are summed to yield the unsteady velocity profiles. The flow waveform used to obtain the Fourier coefficients is often determined from cine phase-contrast images or ultrasound centerline velocity measurements. However, this inflow velocity assumption should be accompanied by a sufficient entrance length to allow the flow to adapt to the specific vessel geometry, which might not always be straight, so that the results are not that dependent on the inlet condition, guarantying nonnegligible errors in the prediction of the distribution of hemodynamic stresses. At least three diameters of common carotid artery length prior to the bifurcation have been proposed when considering carotid atherosclerosis (Hoi et al. 2010). Again, a compromise is needed since the additional entrance length would require more 2D axial slices for the geometry reconstruction, and thus more scanning time. An alternate solution is to measure the time-varying axial velocity using cine phase-contrast MRI (PC-MRI) at the slice corresponding to the inlet surface of the computational mesh and interpolate the image grid onto the inlet CFD nodes (Figure 9.27). This solution is still not fully adopted since it requires high-quality cine PC-MR images at the expense of scan time. Approaches similar to the inlet boundary condition definition have been used for the outlet conditions. In the case of the carotid bifurcation, where there are two outlets, a combination of velocity profile and zero normal traction is normally prescribed as outlet boundary conditions.

Flow in a normal healthy carotid is considered to be laminar. In fact, the mean Reynolds number (dimensionless number that measures the ratio between inertia and viscous forces) in this artery is

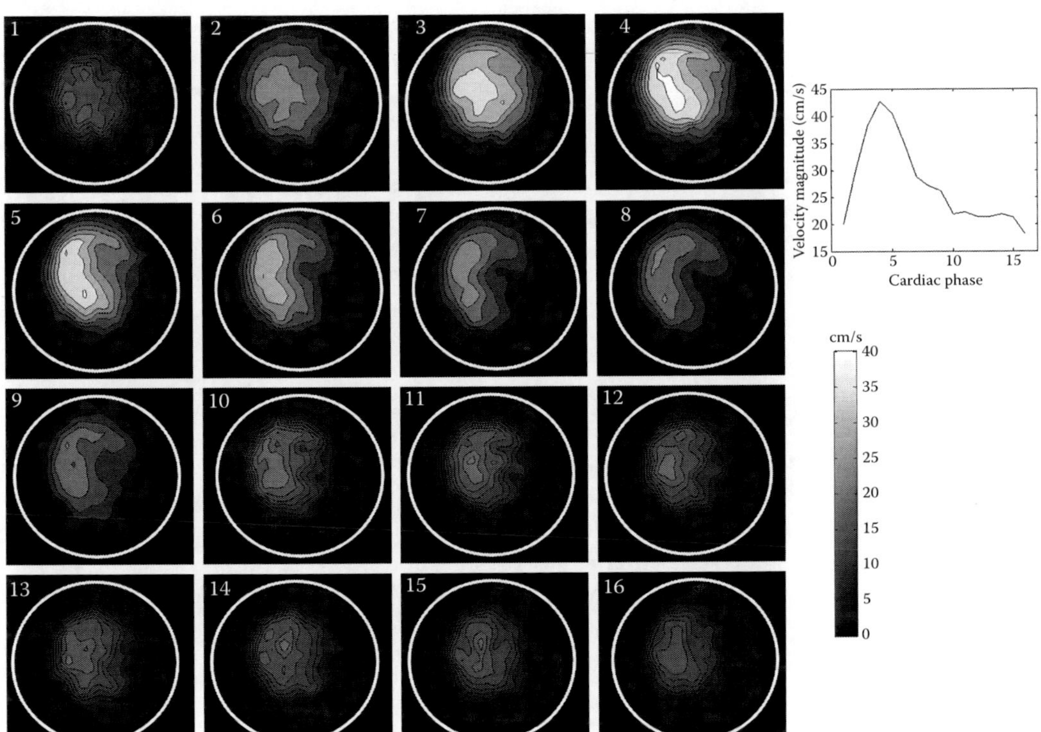

FIGURE 9.27 Flow field calculated from three-directional encoding cine PC-MR images of a cross-sectional plane of a human healthy common carotid artery (16 cardiac phases).

approximately 300, a value too low to be considered as turbulent. However, the stenosis can lead to an altered chaotic flow that could be described as in transition to turbulence or even as turbulent flow in most severe stenotic cases. Still, there is controversy in the literature on which model best describes the flow downstream of the stenosis (Banks and Bressloff 2007, Tan et al. 2008).

Another typical assumption made in image-based CFD simulations is the use of rigid-walled models; thus, no-slip boundary condition is applied to the rigid walls. Most numerical studies targeted to determine the distribution of wall shear stress make this assumption. Though it results in a modest overestimation of wall shear stress distribution compared to a compliant wall model, it is considered an appropriate approximation when analyzing atherosclerotic vessels due to their small wall motion. However, this approximation can be avoided by performing simulations of the fluid–structure interaction (FSI) between the flowing blood and vessel wall.

9.5.1.2 Fluid–Structure Interaction Models: Flow and Solid Mechanics Involved in Atherosclerosis

FSI models require solving the equations of motion for the flow and the wall simultaneously. The solution of these equations provides both the velocity profile as well as the distribution of tensional stresses, achieving a more comprehensive evaluation of the hemodynamic stresses present at the diseased vessel. The computational mesh includes both the fluid and solid domains; thus, it is built from segmenting the lumen, outer-wall, and wall components.

The 3D geometry reconstructed from MRI needs to be shrunk axially and circumferentially *a priori* to obtain the no-load starting geometry for the computational simulation that accounts for the fact that, under *in vivo* conditions, arteries are axially stretched and pressurized. This shrinkage is determined guarantying conservation of the mass of the vessel and that the loaded vessel geometry after the stretch underwent with pressurization matches the MRI-reconstructed geometry.

Although the arterial wall exhibits a viscoelastic behavior, FSI models usually assume the artery wall to be hyperelastic. Other assumptions commonly made are considering the wall to be isotropic, incompressible, and homogeneous. The effects of these assumptions on the resulting hemodynamic stresses have not been analyzed in the same detail as in the case of the approximations made in CFD modeling.

The nonlinear stress/strain dependency of the wall tissue is accounted for by describing the material properties of the vessel wall with a Mooney–Rivlin hyperelastic model using material parameters estimated from *in vitro* models of the wall and plaque tissues to define the strain-energy function. Several strain-energy functions have been proposed for the vessel wall (Fung 1994) and used in various FSI models of carotid atherosclerosis (Kock et al. 2008, Tang et al. 2009, Gao et al. 2009, Leach et al. 2010).

The arbitrary Lagrangian–Eulerian formulation is often used for modeling the fluid–solid coupling. Inlet and outlet are often fixed in the axial direction, but allowed to expand/contract in response to the pulsatile flow.

9.5.1.3 Hemodynamic Stresses: Wall Shear Stress and Tensile Stress

The hemodynamic stresses are extracted from the solution of the numerical model. In the case of only solving for the blood flow, the focus is on the distribution of wall shear stress (WSS). WSS is a vector ($\vec{\tau}_w$) proportional to the gradient of the velocity at the wall and blood viscosity. However, the challenge in analyzing the WSS data extracted from numerical simulations is to determine which WSS-derived parameters are most closely linked to the underlying pathology. The parameters most often reported are the peak systole and cycle-averaged WSS magnitude, and the oscillatory shear index (OSI):

$$\text{OSI} = \frac{1}{2}\left(1 - \frac{\left\| \int_0^T \vec{\tau}_w dt \right\|}{\int_0^T |\vec{\tau}_w| dt}\right)$$

where T is the period of the cardiac cycle (inverse of hear rate); this index describes the temporal change in direction of the shear stress vector with respect to its mean direction.

Tensile stress results can be extracted together with WSS when using an FSI technique. The first principal stress, representing the circumferential component of the stress tensor, is chosen to represent the wall tensile stress. As in the case of WSS, different quantities can be derived from the distribution of wall tensile stress. The relevance of each one with respect to the response of the endothelium and other wall cellular components is still to be determined. The tensile stress distribution and its temporal gradients resulting from the pulsatile nature of blood flow (normally defined in FSI models as the fibrous tissue covering the lipid region) are commonly considered to be closely related to plaque rupture, not only for the cellular response but also for the fatigue damage that could cause on the fibrous cap.

The challenge of image-based modeling is to assess how restricting assumptions on blood rheology, boundary conditions, vessel wall material properties, or blood–vessel interactions (among others) may affect the accuracy of the results. Recent advances in MRI technology are allowing MRI to be used as a tool to overcome the limitations of the numerical models by either improving the boundary conditions and assumptions on material properties by *in vivo* measurements, or estimating the hemodynamic stresses acquiring *in vivo* blood flow and vessel wall deformation measurements. The use of MRI in this direction is discussed in the following section.

9.5.2 *In Vivo* MR Quantification of Blood Flow (4D Flow MRI) and Wall Properties of Atherosclerotic Vessels

Cine phase contrast MRI (PC-MRI) provides flow rate wave forms measured at the inlet and outlet sections as boundary conditions for image-based CFD models. However, there is an increasing interest in advancing this and other MR sequences to provide *in vivo* information of blood flow and flow-derived parameters, as well as material properties and deformation data of the arterial wall.

9.5.2.1 WSS Estimation from 4D Flow MRI

Time-resolved gated cine PC-MRI with three-directional encoding, known as 4D flow MRI, is emerging as a tool to provide full hemodynamic information on 3D blood flow in large arteries.

The accuracy of the estimations of WSS is constrained by the limitations in time and spatial resolutions, partial volume artifacts and the numerical derivation of WSS from discrete velocity data. Indeed, the major challenge of 4D flow MRI is to obtain detailed measurement of the time-varying velocity near the wall in the branching and curved arteries where atherosclerosis develops, where the flow can be highly skewed; or in stenotic regions where the lumen is not circular. However, current advances in cine PCMRI indicate that the results can be sufficient to characterize pathological spatiotemporal distributions of WSS and OSI in atherosclerotic prone arteries, or at regions of atherosclerotic arteries other than the highly stenotic, where WSS estimations will suffer from large numerical errors due to the above-mentioned technical limitations.

Deriving WSS from the velocity data measured in atherosclerotic arteries needs to take into consideration the complexity of the flow through those arteries were atherosclerosis occurred. The secondary flow developing in branching and curved arteries, such as helical flow, can lead to shear stresses along the lumen circumference in addition to stresses along the main flow. Therefore, the 3D paraboloid model initially proposed to estimate WSS from cine PC-MRI velocity data (Oyre et al. 1998) does not seem accurate enough when studying the hemodynamics in atherosclerotic vessels. Markl and collaborators proposed to obtain an analytical expression for the velocity derivatives using a cubic B-spline interpolation that compensates for the limited spatial resolution of the PCMR data (Stalder et al. 2008, Markl et al. 2010).

9.5.2.2 Evaluation of Biomechanical Properties of Tissue Components in Atherosclerotic Plaques

The arterial wall is composed of three layers: the intima, composed of a monolayer of endothelial cells covering a layer of connective tissue, elastic fibril, and collagenous fibers; the media, consisting of a number of elastic sheets and fibrils and collagenous fibers and accounting for the elasticity of the wall; and the adventitia, formed mainly by ground substances, collagen fibers, nerves, and small blood vessels (vasa vasorum). The thickness, structure, and composition of these layers change with age and also with disease.

Release of nitric oxide due to endothelial dysfunction produces vasoconstriction that, in turn, increases stiffness. The arterial elasticity is further decreased as the media layer thickens with atherosclerosis progression. Risk factors associated to atherosclerosis such as hypertension and diabetes also accelerate arterial stiffness. This increase of stiffness in atherosclerotic arteries leads to a reduction of the pressure-driven cyclic stretching or strain.

Current imaging techniques are not able to monitor the changes in cellular function and structure that occur with aging and lead to alterations in the biomechanical properties of the wall (i.e., increase in stiffness). However, they can detect and characterize the strain (deformation resulting from the hemodynamic forces) and measure the biomechanical properties.

Displacement encoding with stimulated echoes (DENSE) pulse sequence can measure the 2D displacement vectors of the arterial wall and surrounding tissue. The circumferential strain distribution is computed from the tissue deformation matrix derived from linear regression of the displacement vectors (Lin et al. 2008).

Gated cine imaging MRI is used to measure the arterial wall distensibility by assessing the maximal change in luminal diameter during systole and diastole and calculating the distensibility coefficient, DC:

$$\mathrm{DC} = \frac{2 \times \Delta d / d_D}{\Delta P}$$

where Δd is the change in luminal diameter, d_D is the end diastolic diameter, and ΔP is the pulse pressure, calculated as the difference between maximal systolic and diastolic blood pressures.

Although the higher field strength of 3 T scanners have been shown to be a promising tool to measure strain and distensibility due to its substantially increase of the signal-to-noise, further increase of spatial and temporal resolution is needed to provide reliable measurements of distensibility in diseased carotid arteries (Harloff et al. 2009).

9.5.2.3 Concluding Remarks

MRI has shown its potential as a tool able to provide a comprehensive characterization of hemodynamic stresses and plaque features. In particular, MRI techniques are currently used in the evaluation of atherosclerotic plaque vulnerability by noninvasively characterizing the plaque in terms of its morphology and composition. Recently, hemodynamic stress information provided mainly by image-based computational models is being investigated to advance the knowledge of the hemodynamic stress role in plaque progression and rupture. Cine PC-MRI has emerged as a promising tool in the estimation of the hemodynamic stresses by either providing improved boundary conditions to numerical models, or on its own evaluation of WSS, wall strain, and vessel wall material properties. However, much research is still needed to improve current spatiotemporal resolution to reliably calculate WSS and wall strain and in highly stenotic or slow recirculation flow regions where few pixels are available for WSS estimation.

Furthermore, prospective studies are still warranted to investigate in depth the complex interplay of hemodynamic stresses and pathological processes underlying the morphological and compositional changes that occur as the disease progresses. Still, it is undoubtful that plaque assessment would

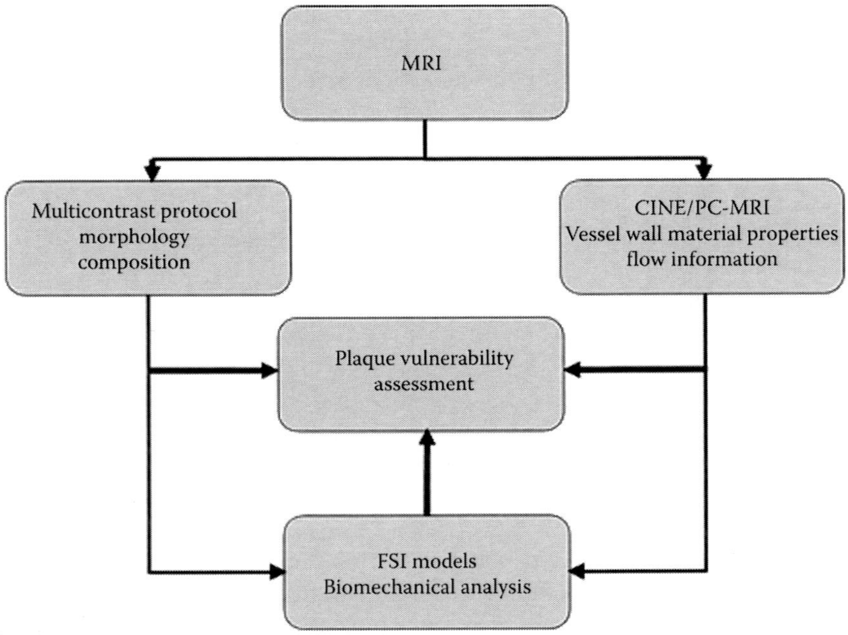

FIGURE 9.28 Integration of hemodynamic stress and morphological and compositional features in the assessment of atherosclerotic plaque vulnerability.

extremely benefit by considering the multifactorial nature of the disease integrating morphological and compositional features and hemodynamic factors (Figure 9.28).

9.6 Summary of the Chapter

VWI and VWI have gone through rapid technical development and validation in recent years. These technologies are used in many ways: for atherosclerotic disease risk assessment, in understanding the vascular pathophysiology, in monitoring the effects of atherosclerotic lesions under medical treatment, and in assessing the hemodynamical effects of plaque progression. These techniques are likely to be applied in clinical applications to present comprehensive evaluations of all aspects of atherosclerosis and improve cardiovascular risk assessment and guide treatment options. As discussed in this chapter, VWI also provides a unique opportunity in technological advancement in imaging hardware software, especially for the ultimate goal of coronary imaging.

Defining Terms

AIF: Abbreviation of arterial input function. In DCE-MRI, it refers to the contrast agent concentration curve over time inside plasma and is essential input function for kinetic modeling.

Angiogenesis: A physiological process involving the growth of new blood vessels from preexisting vessels. In atherosclerosis, it usually refers to the growth of new microvessel inside the vessel wall. It is also considered to accompany with plaque inflammation process.

AUC: Area under enhancement curve, a nonmodel-based parameter used in quantification of DCE-MRI.

Black-blood (BB) imaging: An image acquisition technique that suppresses the usually bright-blood signal for improved vessel wall visualization. It is frequently used in vessel wall imaging applications.

CE-MRI: Abbreviation of contrast-enhanced MRI, refers to an MR technique to acquire image after contrast agent injection, characterizing the agent diffused into the tissue.

CFD: Computational fluid dynamics (CFD) is a branch of fluid mechanics that uses numerical methods and algorithms to solve and analyze problems that involve fluid flows.

DCE-MRI: Abbreviation of dynamic contrast-enhanced MRI, refers to the MR imaging technique to consequently acquire series of images before and after the contrast agent injection, observing the contrast agent transfer process into the tissue over time.

Diamagnetic effect: An effect used to describe an object that causes a magnetic field in opposition to the external magnetic field.

Diffusion-weighted imaging: MRI technique to detect incoherent motion of water molecules by use of large diffusion gradients.

Double inversion recovery (DIR): A black-blood imaging technique that suppresses the blood signal by taking the advantage of both the inflow effect and the blood's unique T_1 relaxation time.

Finite-element method: A numerical technique for finding approximate solutions of partial differential equations.

FSI: Fluid–structure interactions (FSI) refer to the coupling of fluid and solid.

Image registration: A process of transforming different set of image into one coordinate system. In DCE-MRI of atherosclerosis, it refers to the process of correcting the relative movement between images obtained at different time due to patient motion.

Inflow effect: An effect that describes the blood flowing nature in vascular imaging: the blood segment at the start of an MR sequence is usually different from the blood segment at the time of image acquisition. Inflow effect a major cause of flow artifact in vessel wall imaging but it has also been used to achieve black-blood contrast.

Inflow suppression: A black-blood imaging technique that suppresses the blood signal by taking the advantage of the inflow effect.

Juxtaluminal: Close to the lumen of the vessel. IPH and calcification may both be found in juxtaluminal locations.

Kinetic modeling: A mathematical modeling technique for predicting the absorption, distribution, metabolism, and excretion of synthetic or natural chemical substances in humans and other animal species. In atherosclerotic plaque imaging, it is used to characterize the blood supply and permeability of the targeted tissue from DCE-MRI images.

K^{trans}: Transfer constant, a parameter in two-compartment model. In atherosclerotic plaque imaging, it is considered to characterize the blood supply and permeability.

Motion-sensitized-driven equilibrium (MSDE): A black-blood imaging technique that suppresses the blood signal by using the flow-dephasing effect in flowing blood.

Navier–Stokes equations: Named after Claude-Louis Navier and George Gabriel Stokes, describes the motion of fluid substances. These equations arise from applying Newton's second law to fluid motion of a Newtonian viscous fluid.

Neovasculature or microvessel: In atherosclerosis, it refers to the small vessels grown from vasa vasorum, providing blood supply to the plaque.

Strain: normalized measure of deformation representing the displacement of particle in the body relative to a reference length.

Strain rate: rate of change in strain.

Stress: Force exerted per unit area.

Susceptibility: The degree of magnetization of an object in response to the external magnetic field.

Susceptibility-weighted imaging (SWI): An MR imaging technique that creates unique imaging contrast by using the susceptibility differences between tissues.

Tensile stress: Force per unit area exerted normal to the body and results in compression or expansion.

Vasa vasorum: Blood vessels of small caliber that supply oxygen to the vessel wall. They may arise from surrounding vessels or from the lumen of the same vessel.

v_e: Fractional volumes of extravascular extracellular space (EES).

v_p: Fractional volumes of plasma, a parameter in two-compartment model. In atherosclerotic plaque imaging, it is considered to characterize the blood supply.

Wall shear stress: Force per unit area exerted in the direction tangential to the body. In fluid mechanics, it depends on the fluid viscosity and flow velocity profile.

References

Abd-Elmoniem KZ, Weiss RG, and Stuber M. 2010. Phase-sensitive black-blood coronary vessel wall imaging. *Magn Reson Med.* 4:1021–1030.

ACAS. 1995. Endarterectomy for asymptomatic carotid artery stenosis. Executive Committee for the Asymptomatic Carotid Atherosclerosis Study. *JAMA*. 273(18):1421–1428.

Achenbach S, Kramer CM, Zoghbi WA, and Dilsizian V. 2010. The year in coronary artery disease. *JACC Cardiovasc Imaging.* 10:1065–1077.

Aoki S, Aoki K, Ohsawa S, Nakajima H, Kumagai H, and Araki T. 1999. Dynamic MR imaging of the carotid wall. *J Magn Reson Imaging.* 9:420–427.

Arroyo LH and Lee RT. 1999. Mechanisms of plaque rupture: Mechanical and biologic interactions. *Cardiovasc Res.* 41:369–375.

Balu N, Chu B, Hatsukami TS, Yuan C, and Yarnykh VL. 2008. Comparison between 2D and 3D high-resolution black-blood techniques for carotid artery wall imaging in clinically significant atherosclerosis. *J Magn Reson Imaging.* 4:918–924.

Balu N, Yarnykh V, Chu B, Wang J, and Yuan C. 2009a. Carotid plaque assessment using fast 3D isotropic-resolution black-blood MRI. Poster presentation. *ISMRM*, April, pp. 18–24, Honolulu, Hawaii.

Balu N, Yarnykh VL, Scholnick J, Chu BC, Yuan C, and Hayes CE. 2009b. Improvements in carotid plaque imaging using a new eight-element phased array coil at 3T. *J Magn Reson Imaging.* 30(5):1209–1214.

Balu N, Yarnykh VL, Chu B, Wang J, Hatsukami T, and Yuan C. 2011. Carotid plaque assessment using fast 3D isotropic resolution black-blood MRI. *Magn. Reson. Med.* 65:627–637.

Banks J and Bressloff NW. 2007. Turbulence modeling in three-dimensional stenosed arterial bifurcations. *J Biomech Eng—Trans ASME.* 129:40–50.

Barkhausen J, Ebert W, Heyer C, Debatin Jf, and Weinmann HJ. 2003. Detection of atherosclerotic plaque with gadofluorine-enhanced magnetic resonance imaging. *Circulation.* 108:605–609.

Barnett HJ, Taylor DW, Eliasziw M et al. 1998. Benefit of carotid endarterectomy in patients with symptomatic moderate or severe stenosis. North American Symptomatic Carotid Endarterectomy Trial Collaborators. *N Engl J Med.* 339:1415–1425.

Botnar RM, Buecker A, Wiethoff AJ et al. 2004a. *In vivo* magnetic resonance imaging of coronary thrombosis using a fibrin-binding molecular magnetic resonance contrast agent. *Circulation.* 110:1463–1466.

Botnar RM, Perez AS, Witte S et al. 2004b. *In vivo* molecular imaging of acute and subacute thrombosis using a fibrin-binding magnetic resonance imaging contrast agent. *Circulation.* 109:2023–2029.

Boussel L, Herigault G, De La Vega A, Nonent M, Douek PC, and Serfaty JM. 2006. Swallowing, arterial pulsation, and breathing induce motion artifacts in carotid artery MRI. *J Magn Reson Imaging.* 3:413–415.

Briley-Saebo KC, Mulder WJ, Mani V et al. 2007. Magnetic resonance imaging of vulnerable atherosclerotic plaques: Current imaging strategies and molecular imaging probes. *J Magn Reson Imaging.* 26:460–479.

Briley-Saebo KC, Shaw PX, Mulder WJ et al. 2008. Targeted molecular probes for imaging atherosclerotic lesions with magnetic resonance using antibodies that recognize oxidation-specific epitopes. *Circulation.* 117:3206–3215.

Brown R, Nguyen TD, Spincemaille P et al. 2010. Effect of blood flow on double inversion recovery vessel wall MRI of the peripheral arteries: Quantitation with T_2 mapping and comparison with flow-insensitive T_2-prepared inversion recovery imaging. *Magn Reson Med.* 3:736–744.

Cai J, Hatsukami T, Ferguson M, Small R, Polissar N, and Yuan C. 2002. Classification of human carotid atherosclerotic lesions with *in vivo* multicontrast magnetic resonance imaging. *Circulation.* 106:1368–1373.

Cai J, Hatsukami TS, Ferguson MS et al. 2005. *In vivo* quantitative measurement of intact fibrous cap and lipid-rich necrotic core size in atherosclerotic carotid plaque—Comparison of high-resolution, contrast-enhanced magnetic resonance imaging and histology. *Circulation*. 22:3437–3444.

Calcagno C, Cornily J, Hyafil F et al. 2008. Detection of neovessels in atherosclerotic plaques of rabbits using dynamic contrast enhanced MRI and 18f-fdg pet. *Arterioscler Thromb Vasc Biol*. 28:1311–1317.

Chaabane L, Pellet N, Bourdillon MC et al. 2004. Contrast enhancement in atherosclerosis development in a mouse model: *In vivo* results at 2 tesla. *MAGMA*. 17:188–195.

Chen H, Cai J, Zhao X et al. 2010. Localized measurement of atherosclerotic plaque inflammatory burden with dynamic contrast-enhanced MRI. *Magn Reson Med*. 64:567–573.

Chen H, Li F, Zhao X, Yuan C, Rutt B, and Kerwin WS. 2011. Extended graphical model for analysis of dynamic contrast-enhanced MRI. *Magn Reson Med*. 66:868–878.

Cheng C, Tempel D, Van Haperen R et al. 2006. Atherosclerotic lesion size and vulnerability are determined by patterns of fluid shear stress. *Circulation*. 113:2744–2753.

Chu B, Ferguson MS, Chen H, Hippe DS, Kerwin WS, Canton G, Yuan C, and Hatsukami TS. 2009. Cardiac magnetic resonance features of the disruption-prone and the disrupted carotid plaque. *J Am Coll Cardiol Img*. 2009(2):883–896.

Clowes AW and Berceli SA. 2000. Mechanisms of vascular atrophy and fibrous cap disruption. *Ann. N. Y. Acad. Sci*. 902(1):153–162.

Cron GO, Santyr G, and Kelcz F. 1999. Accurate and rapid quantitative dynamic contrast-enhanced breast MR imaging using spoiled gradient-recalled echoes and bookend t(1) measurements. *Magn Reson Med*. 42:746–753.

Dong L, Underhill HR, Yu W et al. 2010a. Geometric and compositional appearance of atheroma in an angiographically normal carotid artery in patients with atherosclerosis. *AJNR Am J Neuroradiol*. 31:311–316.

Dong L, Wang JN, Yarnykh VL et al. 2010b. Efficient flow suppressed MRI improves interscan reproducibility of carotid atherosclerosis plaque burden measurements. *J Magn Reson Imaging*. 32:452–458.

Edelman RR, Mattle HP, Wallner B et al. 1990. Extracranial carotid arteries: Evaluation with "black blood" MR angiography. *Radiology*. 1:45–50.

Edelman RR, Chien D, and Kim D. 1991. Fast selective black blood MR imaging. *Radiology*. 3:655–660.

Fan Z, Zhang Z, Chung YC et al. 2010. Carotid arterial wall MRI at 3T using 3D variable-flip-angle turbo spin-echo (TSE) with flow-sensitive dephasing (FSD). *J Magn Reson Imaging*. 3:645–654.

Faranesh AZ and Yankeelov TE. 2008. Incorporating a vascular term into a reference region model for the analysis of DCE-MRI data: A simulation study. *Phys Med Biol*. 53:2617–2631.

Fung YC. 1994. *A First Course in Continuum Mechanics*. New Jersey: Prentice Hall.

Gao H, Long Q, Sadat U, Graves M, Gillard JH, and Li ZY. 2009. Stress analysis of carotid atheroma in a transient ischaemic attack patient using the MRI-based fluid-structure interaction method. *Br J Radiol*. 82:S46–S54.

Glagov S, Weisenberg E, Zarins CK, Stankunavicius R, and Kolettis GJ. 1987. Compensatory enlargement of human atherosclerotic coronary arteries. *N Engl J Med*. 22:1371–1375.

Harloff A, Zech T, Frydrychowicz A, Schumacher M, Schollhorn J, Hennig J, Weiller C, and Markl M. 2009. Carotid intima-media thickness and distensibility measured by MRI at 3 T versus high-resolution ultrasound. *Eur Radiol*. 19:1470–1479.

Herzka DA, Nezafat R, Chan R, Liu W, and Boernert P. 2008. high-resolution ultra-short TE imaging of ex vivo human carotid plaques correlates with CT. Proceedings 16th ISMRM Scientific Meeting and Exhibition, Toronto.

Hofmann-Amtenbrink M, Hofmann H, and Montet X. 2010. Superparamagnetic nanoparticles—A tool for early diagnostics. *Swiss Med Wkly*. 140:w13081.

Hoi Y, Wasserman BA, Lakatta EG, and Steinman DA. 2010. Carotid bifurcation hemodynamics in older adults: Effect of measured versus assumed flow waveform. *J Biomech Eng—Trans ASME*. 132 (7):071006.

Howarth S, Tang T, Trivedi R, Weerakkody R, U-King-Im J, Gaunt M, Boyle J, Li Z, Miller S, Graves M, and Gillard J. 2009. Utility of USPIO-enhanced MR imaging to identify inflammation and the fibrous cap: A comparison of symptomatic and asymptomatic individuals. *Eur J Radiol.* 70:555–560.

Hur J, Park J, Kim YJ, Lee HJ, Shim HS, Choe KO, and Choi BW. 2010. Use of contrast enhancement and high-resolution 3D black-blood MRI to identify inflammation in atherosclerosis. *JACC Cardiovasc Imaging.* 3:1127–1135.

Itskovich VV, Mani V, Mizsei G, Aguinaldo JG et al. 2004. Parallel and nonparallel simultaneous multislice black-blood double inversion recovery techniques for vessel wall imaging. *J Magn Reson Imaging.* 4:59–67.

Johansson LO, Bjørnerud A, Ahlström HK, Ladd DL, and Fujii DK. 2001. A targeted contrast agent for magnetic resonance imaging of thrombus: Implications of spatial resolution. *J Magn Reson Imaging.* 13:615–618.

Kelly KA, Allport JR, Tsourkas A, Shinde-Patil VR, Josephson L, and Weissleder R. 2005. Detection of vascular adhesion molecule-1 expression using a novel multimodal nanoparticle. *Circ Res.* 96:327–336.

Kerwin W, Cai J, and Yuan C. 2002. Noise and motion correction in dynamic contrast-enhanced MRI for analysis of atherosclerotic lesions. *Magn Reson Med.* 47:1211–1217.

Kerwin W, Hooker A, Spilker M et al. 2003. Quantitative magnetic resonance imaging analysis of neovasculature volume in carotid atherosclerotic plaque. *Circulation.* 107:851–856.

Kerwin W, O'Brien K, Ferguson M, Polissar N, Hatsukami T, and Yuan C. 2006. Inflammation in carotid atherosclerotic plaque: A dynamic contrast-enhanced MR imaging study. *Radiology.* 241:459–468.

Kerwin W, Xu D, Liu F et al. 2007. Magnetic resonance imaging of carotid atherosclerosis: Plaque analysis. *Top Magn Reson Imaging.* 18(5):371–378.

Kerwin W, Oikawa M, Yuan C, Jarvik G, and Hatsukami T. 2008. MR imaging of adventitial vasa vasorum in carotid atherosclerosis. *Magn Reson Med.* 59:507–514.

Kim SE, Jeong EK, Shi XF, Morrell G, Treiman GS, and Parker DL. 2009. Diffusion-weighted imaging of human carotid artery using 2D single-shot interleaved multislice inner volume diffusion-weighted echo planar imaging (2D ss-IMIV-DWEPI) at 3T: Diffusion measurement in atherosclerotic plaque. *J Magn Reson Imaging.* 5:1068–1077.

Kock SA, Nygaard JV, Eldrup N et al. 2008. Mechanical stresses in carotid plaques using MRI-based fluid-structure interaction models. *J Biomech.* 41:1651–1658.

Koktzoglou I and Li D. 2007. Diffusion-prepared segmented steady-state free precession: Application to 3D black-blood cardiovascular magnetic resonance of the thoracic aorta and carotid artery walls. *J Cardiovasc Magn Reson.* 1:33–42.

Kooi M, Cappendijk V, Cleutjens K et al. 2003. Accumulation of ultrasmall superparamagnetic particles of iron oxide in human atherosclerotic plaques can be detected by *in vivo* magnetic resonance imaging. *Circulation.* 107:2453–2458.

Ku DN. 1997. Blood flow in arteries. *Annu Rev Fluid Mech.* 29:399–434.

Lancelot E, Amirbekian V, Brigger I et al. 2008. Evaluation of matrix metalloproteinases in atherosclerosis using a novel noninvasive imaging approach. *Arterioscler Thromb Vasc Biol.* 28:425–432.

Lauffer RB, Parmelee DJ, Dunham SU et al. 1998. Ms-325: Albumin-targeted contrast agent for MR angiography. *Radiology.* 207:529–538.

Leach JR, Rayz VL, Soares B et al. 2010. Carotid atheroma rupture observed *in vivo* and FSI-predicted stress distribution based on pre-rupture imaging. *Ann Biomed Eng.* 38:2748–2765.

Lee SW, Antiga L, Spence JD, and Steinman DA. 2008. Geometry of the carotid bifurcation predicts its exposure to disturbed flow. *Stroke.* 39:2341–2347.

Lee SW and Steinman DA. 2007. On the relative importance of rheology for image-based CFD models of the carotid bifurcation. *J Biomech Eng—Trans ASME.* 129:273–278.

Li ZY, Howarth S, Trivedi RA et al. 2006. Stress analysis of carotid plaque rupture based on *in vivo* high resolution MRI. *J Biomech.* 39:2611–2622.

Libby P. 2002. Inflammation in atherosclerosis. *Nature.* 420:868–874.

Lin AP, Bennett E, Wisk LE, Gharib M, Fraser SE, and Wen H. 2008. Circumferential strain in the wall of the common carotid artery: Comparing displacement-encoded and cine MRI in volunteers. *Magn Reson Med.* 60:8–13.

Lin W, Abendschein DR, and Haacke EM. 1997. Contrast-enhanced magnetic resonance angiography of carotid arterial wall in pigs. *J Magn Reson Imaging.* 7:183–190.

Lipinski MJ, Amirbekian V, Frias JC et al. 2006. MRI to detect atherosclerosis with gadolinium-containing immunomicelles targeting the macrophage scavenger receptor. *Magn Reson Med.* 56:601–610.

Liu F, Xu D, Ferguson MS, Chu B et al. 2006. Automated *in vivo* segmentation of carotid plaque MRI with morphology-enhanced probability maps. *Magn Reson Med.* 55:659–668.

Lobatto ME, Fayad ZA, Silvera S et al. 2010. Multimodal clinical imaging to longitudinally assess a nanomedical anti-inflammatory treatment in experimental atherosclerosis. *Mol Pharm.* 7:2020–2029.

Lobbes MB, Miserus RJ, Heeneman S et al. 2009. Atherosclerosis: Contrast-enhanced mr imaging of vessel wall in rabbit model—Comparison of gadofosveset and gadopentetate dimeglumine. *Radiology.* 250:682–691.

Lorenz MW, Markus HS, Bots ML, Rosvall M, and Sitzer M. 2007. Prediction of clinical cardiovascular events with carotid intima-media thickness: A systematic review and meta-analysis. *Circulation.* 115(4):459–467.

Makowski MR, Wiethoff AJ, Blume U et al. 2011. Assessment of atherosclerotic plaque burden with an elastin-specific magnetic resonance contrast agent. *Nat Med.* 17:383–388.

Mani V, Itskovich VV, Aguiar SH et al. 2005. Comparison of gated and nongated fast multislice black-blood carotid imaging using rapid extended coverage and inflow/outflow saturation techniques. *J Magn Reson Imaging.* 5:628–633.

Markl M, Wegent F, Zech T et al. 2010. *In vivo* wall shear stress distribution in the carotid artery effect of bifurcation geometry, Internal carotid artery stenosis, and recanalization therapy. *Circulation—Cardiovasc Imaging.* 3:647–655.

Meding J, Urich M, Licha K et al. 2007. Magnetic resonance imaging of atherosclerosis by targeting extracellular matrix deposition with gadofluorine m. *Contrast Media Mol Imaging.* 2:120–129.

Moody AR, Murphy RE, Morgan PS et al. 2003. Characterization of complicated carotid plaque with magnetic resonance direct thrombus imaging in patients with cerebral ischemia. *Circulation.* 24:3047–3052.

Naghavi M, Libby P, Falk E et al. 2003. From vulnerable plaque to vulnerable patient: A call for new definitions and risk assessment strategies: Part I. *Circulation.* 14:1664–1672.

Ota H, Yarnykh VL, Ferguson MS et al. 2010. Carotid intraplaque hemorrhage imaging at 3.0-T MR imaging: Comparison of the diagnostic performance of three T_1-weighted sequences. *Radiology.* 2:551–563.

Oyre S, Ringgaard S, Kozerke S et al. 1998. Accurate noninvasive quantitation of blood flow, cross-sectional lumen vessel area and wall shear stress by three-dimensional paraboloid modeling of magnetic resonance imaging velocity data. *J Am College Cardiol.* 32:128–134.

Parker DL, Goodrich KC, Masiker M, Tsuruda JS, and Katzman GL. 2002. Improved efficiency in double-inversion fast spin-echo imaging. *Magn Reson Med.* 5:1017–1021.

Patlak C, Blasberg R, and Fenstermacher J. 1983. Graphical evaluation of blood-to-brain transfer constants from multiple-time uptake data. *J Cereb Blood Flow Metab.* 3:1–7.

Pradel C, Siauve N, Bruneteau G et al. 2003. Reduced capillary perfusion and permeability in human tumour xenografts treated with the vegf signalling inhibitor zd4190: An *in vivo* assessment using dynamic MR imaging and macromolecular contrast media. *Magn Reson Imaging.* 21:845–851.

Ross R. 1999. Atherosclerosis—An inflammatory disease. *N Engl J Med.* 2:115–126.

Shen Jf and Saunders JK. 1993. Double inversion recovery improves water suppression *in vivo*. *Magn Reson Med.* 29:540–542.

Sirol M, Moreno PR, Purushothaman KR et al. 2009. Increased neovascularization in advanced lipid-rich atherosclerotic lesions detected by gadofluorine-m-enhanced MRI: Implications for plaque vulnerability. *Circ Cardiovasc Imaging.* 2:391–396.

Slager CJ, Wentzel JK, Gijsen FJH et al. 2005. The role of shear stress in the generation of rupture-prone vulnerable plaques. *Nat Clin Pract Cardiovasc Med.* 2:401–407.

Stalder AF, Russe MF, Frydrychowicz A, Bock J, Hennig J, and Markl M. 2008. Quantitative 2D and 3D phase contrast MRI: Optimized analysis of blood flow and vessel wall parameters. *Magn Reson Med.* 60:1218–1231.

Strijkers GJ, Van Tilborg GA, Geelen T, Reutelingsperger CP, and Nicolay K. 2010. Current applications of nanotechnology for magnetic resonance imaging of apoptosis. *Methods Mol Biol.* 624:325–342.

Taber LA. 2001. Biomechanics of cardiovascular development. *Annu Rev Biomed Eng.* 3:1–25.

Takaya N, Cai J, and Ferguson MS. 2006. Intra- and interreader reproducibility of magnetic resonance imaging for quantifying the lipid-rich necrotic core is improved with gadolinium contrast enhancement. *J Magn Reson Imaging.* 1:203–210.

Tan FPP, Soloperto G, Bashford S et al. 2008. Analysis of flow disturbance in a stenosed carotid artery bifurcation using two-equation transitional and turbulence models. *J Biomech Eng—Trans ASME.* 130(6):061008-1-12.

Tang DL, Teng ZZ, Canton G et al. 2009. Sites of rupture in human atherosclerotic carotid plaques are associated with high structural stresses an *in vivo* MRI-based 3D fluid-structure interaction study. *Stroke.* 40:3258–3263.

Thomas JB, Antiga L, Che SL et al. 2005.Variation in the carotid bifurcation geometry of young versus older adults: Implications for geometric risk of atherosclerosis. *Stroke.* 36:2450–2456.

Tofts P. 1997. Modeling tracer kinetics in dynamic gd-dtpa MR imaging. *J Magn Reson Imaging.* 7:91–101.

Tofts P, Brix G, Buckley D et al. 1999. Estimating kinetic parameters from dynamic contrast-enhanced t(1)-weighted MRI of a diffusable tracer: Standardized quantities and symbols. *J Magn Reson Imaging.* 10:223–232.

Trivedi R, Mallawarachi C, U-King-Im J et al. 2006. Identifying inflamed carotid plaques using *in vivo* USPIO-enhanced MR Imaging to label plaque macrophages. *Arterioscler Thromb Vasc Biol.* 26:1601–1606.

Underhill HR, Hatsukami TS, Cai J et al. 2010. A noninvasive imaging approach to assess plaque severity: The carotid atherosclerosis score. *AJNR Am J Neuroradiol.* 31:1068–1075.

Underhill HR, Kerwin WS, Hatsukami TS, and Yuan C. 2006. Automated measurement of mean wall thickness in the common carotid artery by MRI: A comparison to intima-media thickness by B-mode ultrasound. *J Magn Reson Imaging.* 24:379–387.

Vanepps JS and Vorp DA. 2007. Mechanopathobiology of atherogenesis: A review. *J Surgical Res.* 142:202–217.

Wang J, Ferguson MS, Balu N, Yuan C, Hatsukami TS, and Börnert P. 2010a. Improved carotid intraplaque hemorrhage imaging using a slab-selective phase-sensitive inversion-recovery (SPI) sequence. *Magn Reson Med.* 64(5):1332–1340.

Wang J, Yarnykh VL, Hatsukami T, Chu B, Balu N, and Yuan C. 2007. Improved suppression of plaque-mimicking artifacts in black-blood carotid atherosclerosis imaging using a multislice motion-sensitized driven-equilibrium (MSDE) turbo spin-echo (TSE) sequence. *Magn Reson Med.* 58(5):973–981.

Wang J, Yarnykh VL, and Yuan C. 2010b. Enhanced image quality in black-blood MRI using the improved motion-sensitized driven-equilibrium (iMSDE) sequence. *J Magn Reson Imaging.* 31:1256–1263.

Wasserman BA, Smith WI, Trout HH, Cannon RO, Balaban RS, and Arai AE. 2002. Carotid artery atherosclerosis: *In vivo* morphologic characterization with gadolinium-enhanced double-oblique MR imaging initial results. *Radiology.* 223:566–573.

Weiss CR, Arai AE, Bui MN et al. 2001. Arterial wall MRI characteristics are associated with elevated serum markers of inflammation in humans. *J Magn Reson Imaging.* 14:698–704.

Wiggins G, Duan Q, Lattanzi R, Biber S, Stoeckel B, Mcgorty K, and Sodickson DK. 2009. 7 Tesla transmit-receive array for carotid imaging: Simulation and experiment. Proceedings of the 17th ISMRM Scientific Meeting and Exhibition, Honolulu.

Winter PM, Caruthers SD, Yu X et al. 2003a. Improved molecular imaging contrast agent for detection of human thrombus. *Magn Reson Med*. 50:411–416.

Winter PM, Morawski AM, Caruthers SD et al. 2003b. Molecular imaging of angiogenesis in early-stage atherosclerosis with alpha(v)beta3-integrin-targeted nanoparticles. *Circulation*. 108:2270–2274.

Womersley JR. 1955. Method for the calculation of velocity, rate of flow and viscous drag in arteries when the pressure gradient is known. *J Physiol—London*. 127:553–563.

Wu YG. 2008. Noninvasive quantification of local cerebral metabolic rate of glucose for clinical application using positron emission tomography and 18f-fluoro-2-deoxy-d-glucose. *J Cereb Blood Flow Metab*. 28:242–250.

Yang Q, Liu J, Barnes SR et al. 2009. Imaging the vessel wall in major peripheral arteries using susceptibility-weighted imaging. *J Magn Reson Imaging*. 2:357–365.

Yankeelov T, Luci J, Lepage M et al. 2005. Quantitative pharmacokinetic analysis of DCE-MRI data without an arterial input function: A reference region model. *Magn Reson Imaging*. 23:519–529.

Yarnykh VL, Terashima M, Hayes CE et al. 2006. Multicontrast black-blood MRI of carotid arteries: Comparison between 1.5 and 3 tesla magnetic field strengths. *J Magn Reson Imaging*. 5:691–698.

Yarnykh VL and Yuan C. 2002. T_1-insensitive flow suppression using quadruple inversion-recovery. *Magn Reson Med*. 5:899–905.

Yarnykh VL and Yuan C. 2003. Multislice double inversion-recovery black-blood imaging with simultaneous slice reinversion. *J Magn Reson Imaging*. 4:478–483.

Yarnykh VL and Yuan C. 2006. Simultaneous outer volume and blood suppression by quadruple inversion-recovery. *Magn Reson Med*. 55:1083–1092.

Yuan C, Kerwin WS, Ferguson MS et al. 2002. Contrast-enhanced high resolution MRI for atherosclerotic carotid artery tissue characterization. *J Magn Reson Imaging*. 15:62–67.

Zhang Z, Fan Z, Carroll TJ et al. 2009. Three-dimensional T_2-weighted MRI of the human femoral arterial vessel wall at 3.0 Tesla. *Invest Radiol*. 9:619–626.

Zhu DC, Ferguson MS, and Demarco JK. 2008. An optimized 3D inversion recovery prepared fast spoiled gradient recalled sequence for carotid plaque hemorrhage imaging at 3.0 T. *Magn Reson Imaging*.10:1360–1366.

Further Information

The following review papers provide more information about utilizing contrast agents in atherosclerotic plaque imaging:

Calcagno C, Mani V, Ramachandran S, and Fayad ZA. 2010. Dynamic contrast enhanced (DCE) magnetic resonance imaging (MRI) of atherosclerotic plaque angiogenesis. *Angiogenesis*. 13:87–99.

Cormode DP, Skajaa T, Fayad ZA, and Mulder WJ. 2009. Nanotechnology in medical imaging: Probe design and applications. *Arterioscler Thromb Vasc Biol*. 29:992–1000.

Wickline SA, Neubauer AM, Winter PM, Caruthers SD, and Lanza GM. 2007. Molecular imaging and therapy of atherosclerosis with targeted nanoparticles. *J Magn Reson Imaging*. 25:667–680.

Winter PM, Caruthers SD, Lanza GM, and Wickline SA. 2010. Quantitative cardiovascular magnetic resonance for molecular imaging. *J Cardiovasc Magn Reson*. 12:62.

For biomechanical analysis of atherosclerotic plaque, David Steinman and the members of the Biomedical Simulation Laboratory of the University of Toronto (http://www.mie.utoronto.ca/labs/bsl/) have provided over the years a thorough analysis of the effect of the various simplifying assumptions commonly made in image-based CFD simulations of the flow through atherosclerotic vessels. They have also characterized extensively the patterns of WSS and OSI in athero-prone regions of the carotid artery.

Serial MRI has been used to build FSI models to track the changes in the distribution of hemodynamic stresses as atherosclerosis evolves with time. The reader is referred to the work of Dalin Tang from the Department of Mathematical Sciences of Worcester Polytechnic Institute.

Michael Markl from the Department of Radiology, Medical Physics, of the University Hospital Freiburg and his team are one of the leading groups in advancing 4D flow MR technology.

The following review papers provide more information on the role of hemodynamic forces in atherosclerosis:

Makris GC, Nicolaides AN, Xu XY, and Geroulakos G. 2010. Introduction to the biomechanics of carotid plaque pathogenesis and rupture: Review of the clinical evidence. *Br J Radiol.* 83:729–735.

Slager CJ, Wentzel JJ, Gijsen FJH, Thury A, Van der Wal AC, Schaar JA, and Serruys PW. 2005. The role of shear stress in the destabilization of vulnerable plaques and related therapeutic implications. *Nat Clin Pract Cardiovasc Med.* 2:456–464.

Slager CJ, Wentzel JJ, Gijsen FJH, Schuurbiers JCH, Van der Wal AC, Van der Steen AFW, and Serruys PW. 2005. The role of shear stress in the generation of rupture-prone vulnerable plaques. *Nat Clin Pract Cardiovasc Med.* 2: 401–407.

Vanepps JS and Vorp DA. 2007. Mechanopathobiology of atherogenesis: A review. *J Surgical Res.* 142:202–217.

10

Dynamic Contrast-Enhanced Magnetic Resonance Imaging

Edward Ashton
VirtualScopics, Inc.

Vijay Shah
VirtualScopics, Inc.

10.1 Introduction

Dynamic contrast-enhanced magnetic resonance imaging (DCE-MRI) is a specialized technique for both imaging and quantitatively measuring changes in vascular function using standard magnetic resonance imaging systems as described in the previous chapter. DCE-MRI involves the periodic acquisition of T_1-weighted images before, during, and after injection of a gadolinium-labeled contrast agent such as gadopentetate dimeglumine. Contrast agents used in MRI affect the observed image signal by shortening the longitudinal (T_1) and transverse (T_2) relaxation time for protons that are in the vicinity of the contrast agent. The change over time in signal intensity in a voxel or region of interest in this time series can then be related to contrast agent concentration in that tissue. The observed enhancement curves in tissue and plasma can then be used to estimate various physiological parameters [1–4]. Figure 10.1 shows a typical time course for contrast agent uptake in sample tissue and in arterial plasma. Generally, contrast agents used with MRI are classified based on the effects they have on different compartments, including intravascular space (arterial-capillary and venous spaces), extracellular extravascular space (EES), the recticuloendothelial system, and the parenchymal cells of various organs [5]. In this chapter, we restrict our discussion to imaging and analysis techniques used for gadolinium-based contrast agents that distribute within the intravascular space and then rapidly diffuse to the EES of perfused tissues. A representative sampling of FDA-approved contrast agents available for clinical use in the United States is summarized in Table 10.1.

A number of models have been proposed for the derivation of useful parameters from DCE-MRI data. The most commonly used at this time was first proposed by Tofts and Kermonde [6] and is based on the Kety equations [7]. This model, like most others, assumes that the tumor vascular bed can be

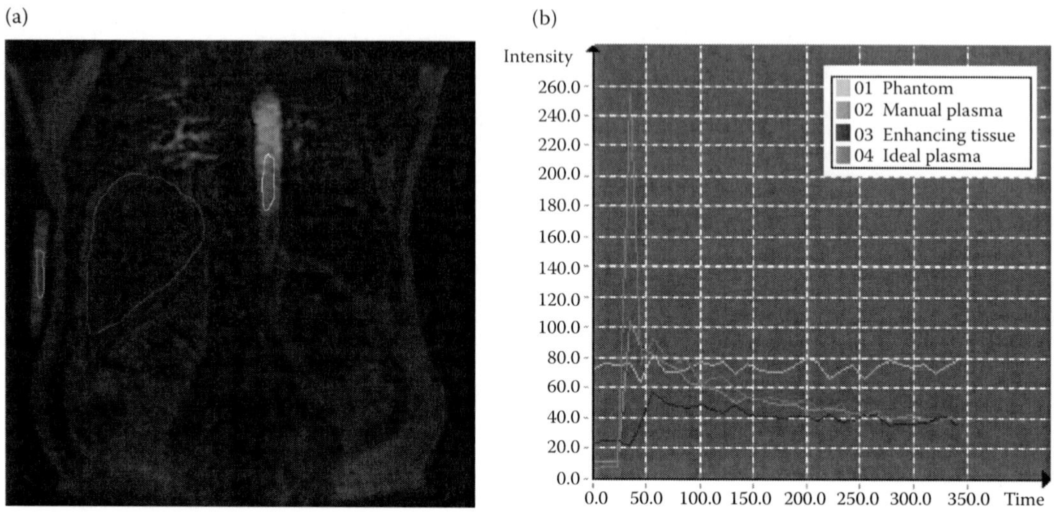

FIGURE 10.1 (a) MR image with contrast agent and (b) typical time course for enhancing tissue and plasma. Note that the y-axis shows signal intensity. This is generally converted to millimolar concentration of gadolinium prior to parameter estimation.

TABLE 10.1 FDA-Approved MRI Contrast Agents

Commercial Name	Active Ingredients	Vendor
Magnevist	Gadopentetate dimeglumine	Bayer HealthCare Pharmaceuticals, Wayne, NJ, USA
Eovist	Gadoxetate disodium	
Gadavist	Gadobutrol	
Optimark	Gadoversetamide	Covidien, Inc. Hazelwood, MO, USA
Omniscan	Gadodiamide	GE Healthcare, AS Oslo, Norway
Multihance	Gadobenate dimeglumine	Bracco Diagnostics, Milan, Italy
Prohance	Gadoteridol	

modeled as two compartments—intravascular plasma and EES. Gadolinium-based contrast agents do not cross cellular membranes, so intracellular space is ignored. A linear system model is assumed, with the input given by the contrast agent concentration in arterial plasma and the output given by the contrast agent concentration in tumor tissue. The transfer function $h(t)$ is parameterized by K^{Trans} and k_{ep}, where K^{Trans} is the volume transfer constant between arterial plasma and EES, k_{ep} is the volume transfer constant between EES and arterial plasma, and v_e is the fractional volume of EES. Both K^{Trans} and k_{ep} are composite parameters, which are affected by both blood flow and vascular permeability. Given these assumptions, it is a simple matter to fit an appropriately parameterized transfer function to any given set of plasma and tissue time–concentration curves. This model is described schematically in Figure 10.2.

The most common application of DCE-MRI currently is in the assessment of treatment effect for antiangiogenic and antivascular cancer therapies [8,9]. The technique is well suited to this task, as these therapies are generally designed to affect both blood flow and vascular permeability through a variety of biological mechanisms, including inhibition of vascular endothelial growth factor (VEGF) [8], inhibition of platelet-derived growth factor (PDGF) [10], and direct disruption of the cytoskeletons of tumor endothelial cells [9]. DCE-MRI and related techniques are able to directly assess the mechanistic effects of these therapies, allowing a rapid and precise estimate

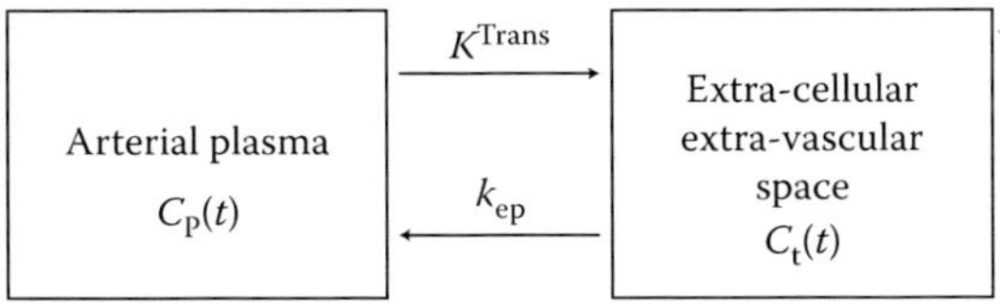

$$\frac{dC_t}{dt} = K^{Trans}[C_p(t) - C_t(t)/v_e]$$

FIGURE 10.2 Two-compartment model for the analysis of DCE-MRI data.

of pharmacodynamic (though not necessarily clinically significant) drug effect. Other clinical applications of DCE-MRI include the assessment of bone edema and synovial inflammation in rheumatoid arthritis patients [11,12], where observed changes are more likely related to changes in blood volume and flow rather than vascular permeability, and the assessment of neurological disorders such as multiple sclerosis, where observed changes are related primarily to disruptions in the blood–brain barrier which result in large changes in vascular permeability with limited effect on blood flow or volume.

The following sections will provide a detailed discussion of the various imaging techniques used to acquire DCE-MRI data, the analysis models that have been employed to derive biologically meaningful parameters from these images, and the clinical and experimental uses of this technique.

10.2 Imaging Techniques

10.2.1 Dynamic Contrast-Enhanced MRI

DCE-MRI refers specifically to techniques which make use of T_1-weighted images to measure changes in concentration of contrast agent in plasma and tissue over time. Time–concentration curves for contrast agent in arterial plasma (generally referred to as the arterial input function or AIF) and tissue are extracted from the image data. Modeling is then used to estimate biologically meaningful parameters such as blood flow, blood volume, and vascular permeability surface area product based on these curves. The development of an imaging protocol for DCE-MRI requires a complex trade-off among several competing priorities. In particular, as with many MRI protocols, DCE-MRI requires that the imager exchange both signal-to-noise ratio (SNR) and spatial coverage for acquisition speed.

Precisely how much speed will be required is heavily dependent on the analysis model that will be used to convert the observed time–concentration curves in arterial plasma and tissue into biologically meaningful parameters such as blood flow, blood volume, or vascular permeability–surface area product. These requirements will be discussed in detail in a subsequent section, but in general the temporal resolution required to extract meaningful information regarding tracer kinetics ranges from a minimum of 2 s/volume to a maximum of 15 s/volume. Protocol design is further constrained by the following requirements:

- Spatial coverage: The imaged volume for a DCE-MRI acquisition must be large enough to contain both the tissue of interest (target tumor, inflamed joint, brain lesion, etc.) and preferably one or more major arteries that can be used in the derivation of an AIF.
- 3-D acquisition: Because it is necessary to estimate changes in contrast concentration in arterial plasma, it is very desirable to minimize inflow and other artifacts. This is most easily done by prescribing a 3-D acquisition with the plane of imaging aligned as nearly as possible with the direction of flow in the target artery.
- Motion correction: Because images are acquired continuously over a period of several minutes, DCE-MRI protocols generally do not recommend breath-holding, opting instead for a so-called quiet breathing technique. As a result, there may be substantial movement of organs such as the

liver due to respiratory motion between acquisitions. This motion is most easily corrected if it is confined to the plane of imaging rather than between images. DCE-MRI of the chest and abdomen are therefore best acquired in either the coronal or sagittal planes.

- Conversion from signal delta to contrast concentration: We wish to extract the change over time in millimolar concentration of contrast agent in tissue from DCE-MRI images. However, this information cannot be directly measured, but must rather be inferred from observed changed in signal intensity in the images. This requirement limits both the selection of pulse sequence and the selection of acquisition coil in ways that will be discussed in detail in a subsequent section.

The most commonly used MR sequences are gradient-echo, turboFLASH, or echoplanar sequences. This choice is driven largely by the necessity for rapid acquisition time. For similar reasons, echo time (T_E) and repetition time (T_R) are typically kept as short as is practically possible, with flip angle (α) set to 20°–30°. Typical DCE-MRI sequence implementations for General Electric and Siemens 1.5 T systems are given in Table 10.2. Figure 10.3 provides sample MR images for both manufacturers.

If these sequence options are used in conjunction with a 3-D acquisition, a 256 × 192 in-plane acquisition matrix, and a slab thickness of 96 mm, they will yield a temporal resolution on the order of 6–9 s per volume. As will be seen in subsequent sections, this resolution will be adequate for some models but far too slow for others. If faster acquisition is required, it is possible to trade spatial coverage, spatial resolution, and SNR in this protocol for additional speed. Reducing the repetition time and echo time to the minimum allowable, for example, may reduce acquisition time substantially on more advanced systems. Reducing the in-plane acquisition matrix to 128 × 128 or even 64 × 64 and/or reducing the thickness of the acquired slab will further accelerate the acquisition.

TABLE 10.2 Differences between GE and Siemens Implementations of a Basic DCE-MRI Protocol

Parameter	GE	Siemens
Sequence type	SPGR	SPGR
Sequence variants	SS\SP\SK	SP\OSP
Sequence options	FAST\VB\EDR\MP\PFF	PFP
T_R/T_E/FA	4.62/1.05/30	5.00/1.57/30

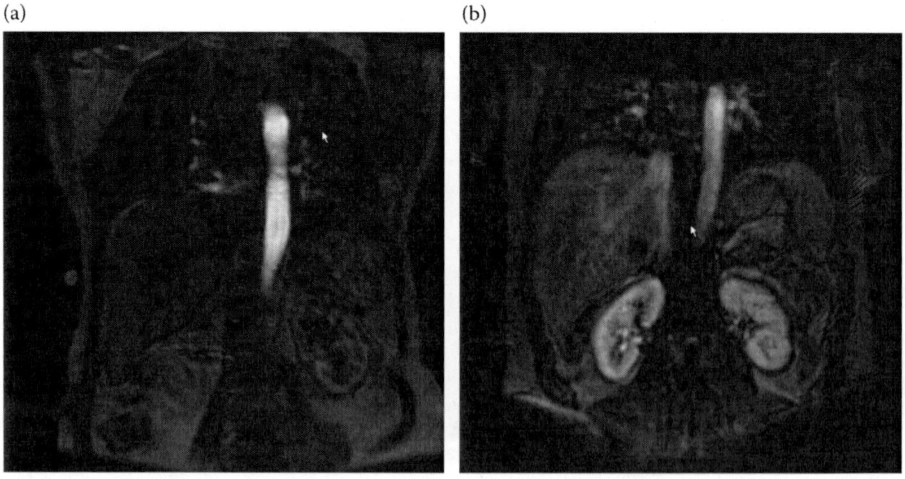

FIGURE 10.3 Image acquired using 3D-squences on 1.5 T scanners: (a) GE Signa (SP/SK with PFF option), slice thickness of 8 mm, $T_R = 5.9$ ms, $T_E = 1.2$ ms. (b) Seimens Espree scanner with slice thickness 5 mm, $T_R = 5$ ms, $T_E = 1.69$ ms. Both images have inplane resolution of 1.56 × 1.56 mm and FA = 30.

In addition to the methods outlined above, higher temporal resolution in T_1-weighted imaging can be achieved through switching to a small number (typically 1–3) of 2-D slices, switching to a so-called semi-keyhole acquisition [13,14], or a combination of the two [15]. Each of these approaches has significant drawbacks, however, and should be undertaken with caution.

Issues associated with 2-D acquisition are fairly obvious. Because the resulting acquired volume is very thin, it will often not contain the entirety of the target tissue. As an example, a target tumor that is well-suited to DCE-MRI analysis will typically be between 3 and 7 cm in diameter. A 3-slice acquisition with 8 mm slice thickness will only be able to capture the central portion of such a tumor, and repeating the precise positioning of the slices at later acquisitions may be very difficult. As a result, pretreatment and posttreatment images may in fact cover substantially different portions of the tumor. Additionally, as mentioned previously, 2-D acquisitions are frequently subject to severe artifacts within large arteries due to varying flow rates as individual slices are acquired at different points in the cardiac cycle. This makes it very difficult to accurately characterize the AIF using 2-D acquisitions. Comparison of images acquired with 2-D and 3-D acquisition techniques is shown in Figure 10.4.

The issues associated with semi-keyhole acquisition are less obvious and less well established, but are significant enough that the imaging community has generally been wary of this technique. Semi-keyhole acquisition techniques achieve improved acquisition speed through some combination of subsampling k-space in frequency and time. As a result, images acquired using this technique are subject to cross-talk between pixels and between volumes. The pixel values observed at a given time, therefore, may include signal contribution from prior times as well as the current one. As a result, contrast concentrations observed in arterial plasma during the period of first bolus passage, when concentration levels are changing very rapidly, are likely to be inaccurate. Because accurate characterization of the AIF during the first bolus passage is critical for most high-speed models, this acquisition technique is not an ideal one for these applications.

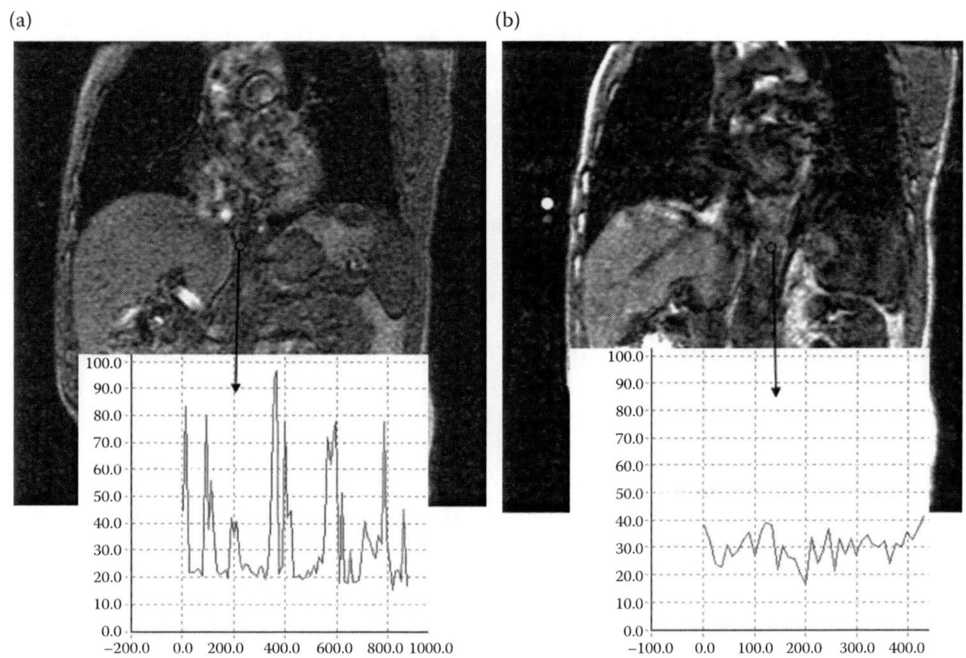

FIGURE 10.4 Comparison of images acquired using 2D (a) and 3D (b) techniques. The given plots show signal intensity changes due to flow artifacts in the descending aorta. Note that artifacts are much more severe in the 2D image.

As will be seen, many analysis models require temporal resolution of 2 s per volume. This is achievable with sufficient trade-offs along the lines described above. However, it must be understood that the loss of information on a per-volume basis in terms of coverage, spatial resolution, and SNR will be significant, and that these losses will translate directly into increased variability in the measured parameters which are the output of this process. High-speed models should therefore only be implemented with T_1-weighted acquisition protocols in cases where the more detailed parameter sets available with these models will provide information that is critical to the experiment being conducted. This issue will be addressed in greater detail in a later section.

10.2.2 Dynamic Susceptibility Contrast-Enhanced MRI

As we have seen, acquisition time is a severely limiting factor when working with T_1-weighted images. However, gadolinium-based contrast agents also affect both T_2- and T_2^*-weighted images. T_2^*-weighted images in particular are an attractive option due to their potential for high acquisition speed relative to T_1-weighted images. Acquisition times of 0.2 s/slice for T_2^*-weighted images are easily achievable using echo-planar imaging (EPI) techniques. As should be expected, however, there are a number of differences in the way contrast concentration changes affect T_2^*-weighted images that must be understood if this technique is to be used for perfusion assessment.

First and most obviously, the presence of intravascular contrast in T_2^*-weighted images causes a reduction in observed signal intensity (darkening) rather than increased signal intensity (brightening) as in T_1-weighted images [16]. From a modeling perspective, this is not particularly relevant as long as methods are available to convert observed changes in signal intensity into millimolar concentration of contrast in tissues. These methods will be discussed in a later section.

More significantly, the induced change in signal relative to background noise levels at a given contrast concentration is substantially lower in T_2^*-weighted images than it is in T_1-weighted images. As a result, the contrast agent dose administered for a DSC-MRI acquisition is higher than that for a DCE-MRI acquisition [17]. Typical values are 0.1 mmol/kg for T_1-weighted images versus 0.2–0.3 mmol/kg for T_2^*-weighted images. Until 2006, this limitation was seen primarily as a cost issue. However, in that year Marckmann et al. identified gadolinium-based contrast agents as a causative factor in nephrogenic systemic fibrosis (NSF), a potentially serious disease affecting patients with renal failure [18]. While the occurrence of NSF is very rare (the original paper identified only 13 cases) and has been seen only in patients with severe preexisting kidney disease, these findings do raise some concerns about administering high doses of gadolinium without significant medical justification.

The most intractable issue associated with DSC-MRI, however, is that echo-planar imaging sequences are very strongly affected by magnetic susceptibility artifacts [19,20]. They are therefore ill-suited for use in regions of the body containing significant amounts of air space, such as the chest or abdomen. They are also ill-suited for use with extravascular contrast. This is a severe drawback insofar as gadolinium-labeled contrast agents generally are able to cross from the capillary bed into EES even in normal tissues—and even more so in tumors, whose capillaries typically have poorly formed and highly permeable endothelia. Gadolinium-labeled contrast agents are generally not, however, able to cross an intact blood–brain barrier, with the result that they are essentially intravascular agents within the brain.

As a result of these limitations, despite the apparent attractiveness of this technique from a speed standpoint, DSC-MRI is used almost exclusively for assessing perfusion within the brain at this time.

10.3 Analysis Methods

10.3.1 Arterial Input Function Estimation

One of the primary challenges in estimating perfusion parameters is identifying an accurate arterial input function (AIF)—defined as the time–concentration curve for contrast agent in arterial plasma and given by $C_p(t)$ in Figure 10.2. One common approach is to avoid this problem by making use of a general concentration–time curve such as that utilized by Weinmann et al. [21]. However, using a theoretical AIF ignores differences in injection rate and cardiac output, which may vary from subject to subject and even with a single subject over time [22]. These differences can greatly reduce measurement reproducibility. Galbraith et al. have reported thresholds for statistically significant change over time in K^{Trans} in a single tumor using this method of −45% and +83% [23].

A second option is for an analyst to draw a manual ROI within an artery, and use the mean enhancement curve within that ROI as the subject-specific AIF (see Figures 10.1 and 10.5), as described by Vonken et al. [24]. This approach is complicated by the fact that the MR signal in arteries is frequently corrupted by flow artifacts, with the result that regions of interest at different points in the same artery or in other nearby vessels can provide grossly different enhancement curves. It should be noted that these artifacts are sequence dependent, and are worse for 2D methods than for central regions of 3D methods.

Several groups have described methods for the automated or semiautomated identification of a patient-specific arterial input function using either information from large arteries present in the scan [25–27] or a model-based derived AIF using information taken from normal tissues such as muscle [28]. Scan–rescan variability data on these various methods is not widely available in the literature. However, available evidence indicates that systems using automated, patient-specific AIF derivation generally show lower variability, as would be expected [22,23,26,29,30].

Sampling the AIF from large blood vessels presents some challenges, as a result of both artifacts present in vessels due to flow effects and the lack of availability of suitable large vessels in some commonly scanned anatomical regions such as the pelvis or extremities. In order to avoid sampling of the AIF from blood vessels, an alternative approach has been proposed using a reference region model (RRM) [31]. This approach was expanded by Yang et al. to use two different reference tissues. They showed that despite the use of two potential error sources, the double-reference tissue method provided a reliable estimate of the AIF [32]. Currently, efforts are also being made to incorporate a vascular term into a reference region analysis to study its effect on the accuracy of the model in estimating tissue kinetics using systematic computer simulations. Faranesh et al. showed that the vascular term may have a significant effect on parameter estimation, but was inconclusive about the effect of the sampling period [33]. Hsiao et al. studied the effect of the AIF onset-time shift and the injection duration under various

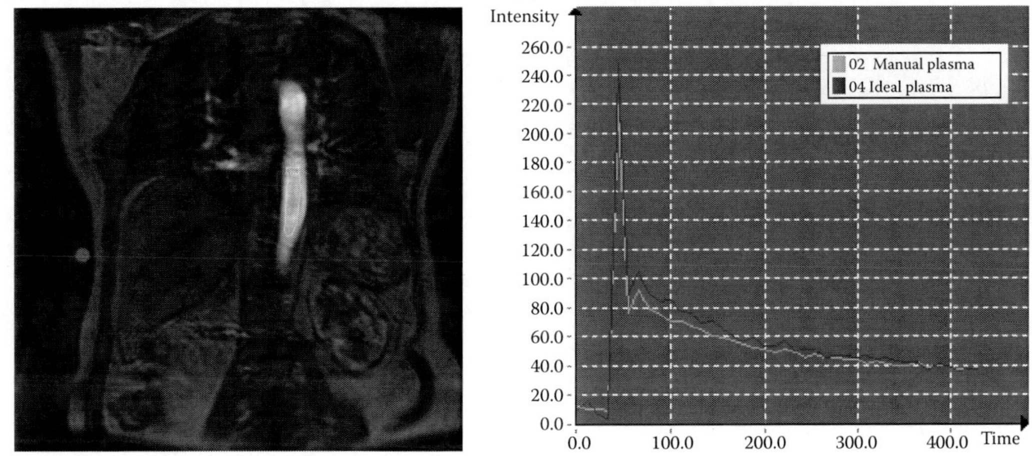

FIGURE 10.5 (a) Sample location for selection of the AIF in the descending aorta and (b) the resulting time-intensity plot.

sampling intervals, on physiological parameter estimation, using both the GKM and RRM. Their results suggested that with compromised temporal resolution, the RRM was relatively less sensitive to the AIF onset-time shift and the injection duration compared with GKM, but the coefficient of variance was higher because of lower SNR in the reference region [34].

The major limitations of the RRM approach are that it requires K^{Trans} and v_e (fractional volume of EES) for the reference region, and so inclusion of an additional parameter such as the vascular term v_p (fractional intravascular volume) means that it is a requirement to know this *a priori*. Additionally, the RRM method works well only when the reference region and tissue of interest share the same AIF. In practice, the measured concentration curves are noisy, and because of the dispersion effects, AIFs of the two tissues are not exactly the same and thus could produce errors in the AIF estimate [32].

10.3.2 Conversion from Signal Delta to Contrast Concentration

10.3.2.1 Conversion Using a Look-Up Table

As has been mentioned previously, most (though not all) DCE-MRI parameter calculation schemes are dependent first on the conversion of observed changes in signal intensity in individual image voxels or larger volumes of interest into millimolar concentrations of contrast agent. The simplest way to do this is to make use of a phantom-generated look-up table [35]. This method requires that a phantom containing a range of concentrations of gadolinium and a range of baseline T_1 values (generally obtained via different concentrations of copper sulfate or a similar compound) is scanned using the dynamic protocol on each scanner that will be used for the study. Data from these phantoms can then be used to construct a look-up table relating baseline T_1, signal delta, and gadolinium concentration (see Figure 10.6). This look-up table can then be used during the analysis process to directly convert signal delta to gadolinium concentration prior to parameter calculation.

This approach has the advantage of simplicity. Additionally, by avoiding the use of complex modeling, each step of which carries its own variability and therefore adds to the noise level in the calculated parameters, it holds the promise of improved measurement precision. This method does have some drawbacks, however. First, it requires a uniform set of suitable phantoms. Such phantoms are not commercially available, and can be difficult and expensive to design and build. Second, it assumes that the available phantom data are representative of the data being analyzed. This may not be valid unless a separate phantom

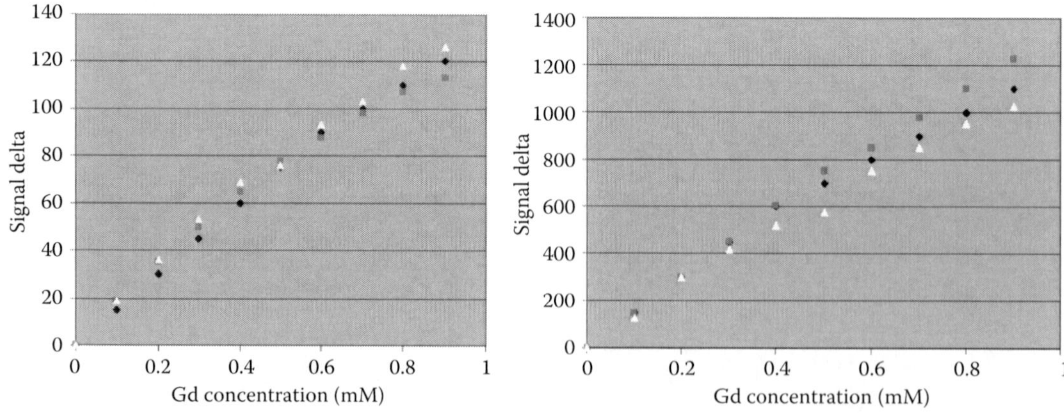

FIGURE 10.6 (**See color insert.**) Plot of signal change versus gadolinium concentration for GE (l) and Siemens (r) systems. Baseline T_1 values were 90 ms (yellow), 300 ms (pink), and 700 ms (blue). Note that the Siemens system shows roll-off in signal response at significantly longer T_1 than the GE system.

scan is acquired with each patient scan, since software or hardware upgrades may potentially change the relationship between signal and contrast concentration that the phantoms are designed to assess.

10.3.2.2 Conversion Using a Signal Formation Model

The more common approach to the problem of conversion from signal delta to gadolinium concentration is to make use of a signal formation model. The most frequently used model requires knowledge of the relaxivity of the contrast agent being used, as well as the baseline T_1 value at each voxel in the image, the baseline signal value at each voxel, and the postcontrast signal value at each voxel and at each postcontrast volume. Baseline T_1 must be calculated using a T_1 mapping sequence. There are a number of options for generating a T_1 map. However, many of them (e.g., acquiring sequences with multiple inversion recovery times) are very time consuming and so poorly suited for this application. A popular approach is to acquire two or more sequences with different flip angles, preferably covering a broad range above and below the Ernst angles for the acquired T_R and range of T_1 values expected in the human body. A typical protocol might call for five sequences with flip angles ranging from 5° to 30°. These images should be prescribed identically to the dynamic series so that the resulting T_1 map can be conveniently coregistered with the dynamic data, and should be acquired immediately prior to the dynamic series. The T_1 values at each voxel can then be calculated as follows:

1. Create a vector x containing the signal intensity at each flip angle divided by the tangent of the flip angle.
2. Create a vector y containing the signal intensity at each flip angle divided by the sine of the flip angle.
3. For the n acquired flip angles, create a set of points $(x_0,y_0)\cdots(x_{n-1},y_{n-1})$.
4. Fit a line with slope s to the set of points defined in Step 3.
5. $T_1 = -T_R/\log(s)$.

Given the resulting T_1 map, the gadolinium concentration at each image voxel at time t is given by

$$C(t) = \left(\frac{1}{T_1} - \frac{1}{T_{10}}\right)\Big/R_1$$

In this equation, T_{10} is taken from the T_1 map calculated above. R_1 is the relaxivity of the contrast agent, and T_1 is given by

$$\frac{1}{T_1} = \left(\frac{-1}{T_R}\right) * \log\left(\frac{1-(A+B)}{1-\cos(\alpha)*(A+B)}\right)$$

here T_R is the repetition time for the dynamic series, α is the flip angle for the dynamic series, and A and B are given by

$$A = (S(t) - S_0)/(M_0 * \sin(\alpha))$$
$$B = (1 - E_{10})/(1 - \cos(\alpha) * E_{10})$$

10.3.3 Subjective and Semiquantitative Analysis Methods

Because DCE-MRI images are inherently four-dimensional, they are not in their raw form particularly well suited to visual interpretation. As a result, a number of heuristically selected characteristics of the observed time–signal intensity curves have been used to color-code DCE-MRI images for visual inspection (see Section 10.3.4.6). Most common among these are maximum slope and peak enhancement. These parameters help to characterize the shape of the uptake curve, which may differ between, for example, malignant and benign breast lesions. False-coloring images according to these parameters, therefore, can provide a useful visual cue to an examining radiologist. These parameters are not well suited, however, to quantitative analysis.

A more common semiquantitative approach is to make use of some variation of the initial area under the tumor uptake curve (IAUC)—“initial” generally referring to the first 30, 60, 90, or 120 s postinjection. IAUC is not in itself biologically meaningful. It is dependent on the fractional volume of EES in addition to blood flow and endothelial permeability–surface area product. It is in general highly correlated with K^{Trans} (see Section 10.3.4.1), but is not dependent on any physiological model, and is therefore in some cases less sensitive to noise and data irregularities. IAUC has been used frequently as a marker of drug effect [3,36]. There is no generally agreed-upon definition for precisely what IAUC should refer to, however, so any use of this parameter should include as explicit a definition as possible. In particular, it is important to specify

- Duration: How many seconds postinjection will be analyzed? Shorter times will emphasize flow effects, while longer times place more emphasis on permeability effects.
- Signal type: Will the data be corrected to gadolinium concentration prior to analysis? Some investigators have designated this parameter taken from corrected data as IAUGC in order to distinguish it from the area under the signal intensity curve IAUC. However, this usage is not consistent.
- Normalization: One of the primary weaknesses of this parameter is that it is strongly affected by differences in injection volume or cardiac output. Some investigators have attempted to correct for this by dividing the IAUC or IAUGC for the tissue of interest by the IAUC or IAUCG for the arterial input function. The resulting parameter is generally referred to as IAUC_{BN}, where BN refers to blood normalization. This usage is also somewhat inconsistent.

These parameters are sometimes used in isolation. However, more commonly, they are used in conjunction with some subset of the more physiologically meaningful parameters described in Section 10.3.4.

10.3.4 Quantitative Analysis Methods

10.3.4.1 Generalized Kinetic Model

As was mentioned previously, the most commonly used model for analysis of DCE-MRI data at this time is the generalized kinetic model (GKM), frequently referred to as the standard Tofts model, first proposed by Tofts and Kermonde [6]. Nomenclature and usage for this model were standardized by Tofts et al. in 1999 [37]. This model is described schematically in Figure 10.2. There are several critical points which distinguish this model from those that will be described subsequently. First, relative to most other proposed models, the standard Tofts model is minimally parameterized. It assumes that the vascular bed can be modeled as a linear system in which the AIF (usually referred to as $C_p(t)$) can be related to the time–concentration curve in the tissue of interest (usually referred to as $C_t(t)$) via the transfer function $h(t)$, given by

$$h(t) = K^{\text{Trans}} \cdot e^{-(k_{\text{ep}} \cdot t)}$$

where K^{Trans} is the volume transfer constant between arterial plasma and EES, and k_{ep} is the volume transfer constant between EES and arterial plasma (both in units of 1/min). Given $C_p(t)$ and $C_t(t)$, which can be derived directly from the dynamic images, estimation of K^{Trans} and k_{ep} is an optimization problem, frequently carried out using the alternating estimation (AE) or another similar energy minimization scheme. Parameter estimation will be discussed in detail in a later section. The third parameter associated with this model, v_e, the fractional volume of EES, is given by K^{Trans}/k_{ep}. Because of this relatively straightforward formulation, parameter calculation using this model is computationally inexpensive relative to other models in this class.

A second distinguishing aspect of this model is that it does not attempt to separate changes in contrast uptake due to blood flow or blood volume from those due to vascular permeability. Both K^{Trans} and k_{ep} are composite parameters which are related to both flow and permeability. This is in some respects a weakness of this model, in that therapies which produce differential effects on flow and permeability may not produce a distinguishable signal using this model. Certain PDGF inhibitors, as an example, have been shown to selectively prune smaller, more tortuous and immature vessels within tumors. As a result, these therapies induce a regularization of the tumor vascular bed, which results in a net decrease in vascular permeability coupled with a net decrease in tumor interstitial pressure and corresponding increase in blood flow. These changes may manifest as either an increase or decrease in K^{Trans}, depending largely on whether the contrast uptake in the tumor pretherapy was more heavily dependent on flow or permeability. As a result, K^{Trans} calculated using this model is not a particularly useful measure of drug effect for this class of therapies.

Interestingly, however, this same property is also one of the primary strengths of this model. The information which allows a model to distinguish between flow effects and permeability effects is entirely contained in the first contrast bolus passage, when contrast agent concentration in the arterial plasma is both very high and rapidly changing (see Figure 10.7). Total injected contrast volume for a DCE-MRI examination is generally between 10 and 15 cc. This volume is followed with a saline flush of roughly 20 cc. Contrast is administered with a power injector at a rate of ~3 cc/s. As a result, the period between injection, when contrast concentration in plasma rises rapidly to a peak of 3 mM or more, and the end of first bolus passage, when contrast concentration reaches a "shoulder value" of approximately 0.5 mM and the rate of decline drops rapidly, is no more than 15 s. Distinguishing between flow effects and permeability effects requires an accurate characterization of both $C_p(t)$ and $C_t(t)$ during this period, and

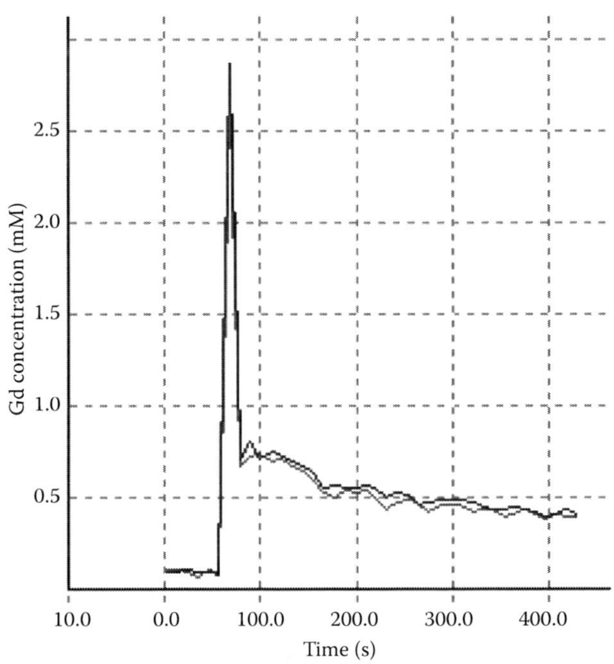

FIGURE 10.7 **(See color insert.)** Typical AIF obtained using T_1-weighted imaging with a 0.1 mmol/kg contrast injection and 6 s temporal resolution.

therefore requires that these curves must be sampled at their Nyquist rate or better [38]—in this case roughly 2 s per sample.

Because this model does not attempt to make this distinction, it can be implemented with sampling rates of 6 s per sample or slower. This allows the use of acquisition protocols (like that described previously) with relatively high SNR, spatial resolution, and spatial coverage. As a result, with other factors held constant, this model can be expected to provide higher precision than those requiring more rapid sampling. It can be argued (as detailed in the following section) that this increased precision is coupled with a systematic bias in estimation of K^{Trans}. However, in most applications, precision is in fact more important than absolute accuracy. If the intent of an experiment is to determine whether a particular antivascular therapy induces a 50% decline in K^{Trans}, for example, it makes no difference whether the absolute values measured for K^{Trans} are biased 20% high or 20% low, as long as an identical bias is applied to both pre- and posttherapy measurements.

In other applications, where the critical factor is not changed over time but rather the absolute value of K^{Trans} at a single time, however, this model may be a poor choice. An example of this is the use of K^{Trans} measurement to determine malignancy in breast lesions [39]. Such applications require one of the more biologically accurate models described below.

10.3.4.2 Extended Tofts Model

In the standard Tofts model described above, it is implicitly assumed that all observed signal changes are a result of contrast agent in the EES—contrast agent that remains within the vasculature is ignored. It should be clear that this assumption will lead to an overestimation of both K^{Trans} and k_{ep}, since the model will treat any contrast that remains intravascular during its passage through the capillary bed as if it had passed into the EES and back into the plasma compartment, thus contributing inappropriately to both K^{Trans} and k_{ep}. This observation has been confirmed experimentally [40].

The extended Tofts model [2] compensates for this error by separating $C_{\text{t}}(t)$ into two components—one intravascular and one extravascular, such that

$$C_{\text{t}}(t) = v_{\text{e}}C_{\text{e}}(t) + v_{\text{p}}C_{\text{p}}(t)$$

where v_{e} is the fractional extracellular extravascular volume, v_{p} is the fractional intravascular volume, $C_{\text{e}}(t)$ is the contrast concentration in EES, and $C_{\text{p}}(t)$ is the AIF. This yields a revised relationship between $C_{\text{p}}(t)$ and $C_{\text{t}}(t)$:

$$C_{\text{t}}(t) = K^{\text{Trans}}\int_0^t C_{\text{p}}(t')e^{-(k_{\text{ep}}(t-t'))}dt' + v_{\text{p}}C_{\text{p}}(t)$$

Given an observed $C_{\text{p}}(t)$ and $C_{\text{t}}(t)$, the vascular parameters K^{Trans}, k_{ep}, and v_{p} can be estimated using similar methods to those used for the standard Tofts model. This model is an improvement over the standard Tofts model insofar as it provides a more biologically accurate estimate of K^{Trans} and k_{ep}. It may also provide greater insight into the effects of therapies that induce differential effects on flow, blood volume, and vascular permeability.

The primary drawback of this model is that it does require temporal resolution of roughly 2 s per volume or better. When this model is applied to data which was acquired at a lower rate, the resulting aliasing manifests as parameter instability in K^{Trans} and v_{p}. In this case, given a particular $C_{\text{p}}(t)$ and $C_{\text{t}}(t)$, multiple equally valid solutions can often be obtained by increasing one parameter and decreasing the other, yielding nearly identical results when the resulting transfer function is convolved with $C_{\text{p}}(t)$. Because of the previously described difficulty in achieving the temporal resolution necessary for this model, it should be used only in applications where there is a reasonable expectation that the improved biological accuracy will compensate for the necessary decrease in SNR, spatial resolution, and spatial coverage.

10.3.4.3 Distributed Parameter Model

The distributed parameter model was developed by Ting-Yim Lee and others [41], originally for analysis of DCE-CT data of the brain. This model is also the most common one used to analyze DSC-MRI data of the brain. Because this model is heavily dependent on an accurate characterization of the first bolus passage, like the extended Tofts model, it requires temporal resolution of 2 s/volume or better. It is a two-compartment model, similar conceptually to the GKM described previously. Rather than assuming a parameterized transfer function as described in Section 10.3.4.1, however, the distributed parameter model attempts to estimate the transfer function directly through deconvolution of $C_p(t)$ and $C_t(t)$, such that

$$C_t(t) = \text{rBF}(C_p(t) * h(t))$$

where rBF is the regional blood flow. It is assumed that $h(t)$ will have a roughly biexponential form, with an early maximum followed by a rapid decline to a shoulder value, an increase to a second peak representing recirculation, and a much slower roll-off from there. From this function, we extract the maximum value M_1, the maximum value of the second peak M_2, and the area under the first peak A. We are then able to calculate the extraction fraction (E), representing the portion of contrast agent entering the EES on first pass, the mean transit time (MTT), regional blood flow (rBF), regional blood volume (rBV), and permeability–surface area product (PS) according to the following equations:

$$\begin{aligned} E &= M_2/M_1 \\ \text{MTT} &= A/M_1 \\ \text{rBF} &= M_1/(1.0 - E) \\ \text{rBV} &= A/(1.0 - E) \\ \text{PS} &= -\text{rBF} * \log(1.0 - E) \end{aligned}$$

Results obtained using this model have been shown to correlate reasonably well with K^{Trans}, and to correlate with drug exposure in some clinical trials [42]. This model is attractive in that it provides the most biologically meaningful measures of any described here. The major drawbacks associated with this model are that (1) it requires very high speed imaging and (2) it is reliant on deconvolution, which is both computationally expensive and very noise sensitive. For these reasons, this model has not been used extensively with T_1-weighted images in the body.

10.3.4.4 Shutter Speed Model

The standard GKM described previously yields pharmacokinetic parameters with the assumption that equilibrium transcytolemmal water exchange is effectively infinitely fast. Recent studies have shown that consideration of the transmembrane water exchange affects the accuracy of the tissue contrast agent concentration estimates [43–45]. These studies compared the accuracy of the standard GKM and shutter-speed model (SSM) to differentiate malignant and benign breast lesions, and found that K^{Trans} in malignant tumors is generally underestimated by the standard GKM model, while the estimation for benign tumors was similar. However, this model is still controversial, and needs to be evaluated in one or more large clinical trials to determine whether its application yields any significant benefits over older methods.

10.3.4.5 Computational Methods for Parameter Estimation

Quantitative estimates of tissue physiological parameters such as perfusion, capillary permeability, and the volume of extravascular–extracellular space (EES) as described above are obtained by curve-fitting and compartmental modeling, where the minimization of a merit function yields the "best fit" of model to data. The two most popular curve-fittings methods are Levenberg–Marquardt and Minpack-1. Ahearn et al. have investigated these methods with respect to the search start points, and found that both algorithms return unreliable results with a single starting point. This is a result of the fact that both of these methods are gradient-descent techniques, which are known to be vulnerable to local minima. Their experimental analysis showed that with multiple start points, both algorithms returned reliable parameters. Additionally, they also showed that the Minpack-1 method generally outperformed the Levenberg–Marquardt method [46].

In order to increase the speed of parameter estimation, several groups have proposed the use of linear-least square (LLSQ) estimation approaches by using the differential equation for kinetic behavior of contrast agents in tissue [4,47]. Computer simulation showed that in comparison with the nonlinear least-squares (NLSQ) method, the LLSQ approach is much quicker and more accurate than the NLSQ method at an SNR of <10 [4]. Horsfield et al. proposed more sophisticated schemes for data processing for a sparsely sampled AIF using three discrete approximation schemes that include the dirac delta function, piecewise constant, and piecewise linear to quantify the errors involved in estimating parameters. They concluded that more sophisticated schemes gave more accurate parameter estimates, with the piecewise constant representation of the AIF giving the best results when data are sparsely sampled [47]. However, the LLSQ approach has a severe limitation in that it does not provide any bounding condition to the estimated dataset. Hence, it is possible that the estimated parameters might not have any physiological meaning. Some examples are a negative plasma volume fraction for a three parameter fit, or an EES volume fraction that is >1. Such limitations should be accounted for when developing parametric maps from DCE-MRI data.

Recent studies have shown that incorporation of additional parameters such as bolus arrival delay and mean transit time can increase the accuracy of estimated parameters in some cases. Experimental results show that unless the AIF initial bolus passage sampling is accurate, more robust parameter estimates are achieved simply by a more slowly sampled biexponential fit. Small values of K^{Trans} can be accurately estimated, but this approach may underestimate blood volume [48].

10.3.4.6 Presentation of Analysis Results

As we have seen, calculation of a DCE-MRI parameter set requires time–concentration curves in both arterial plasma ($C_p(t)$ – the AIF) and in the tissue of interest ($C_t(t)$). Generally the AIF is defined on either a per-scan or population basis as described above. $C_t(t)$ can also be defined in one of two ways: by averaging the uptake curves for all pixels within a user-defined volume of interest (VOI), yielding a single $C_t(t)$ which can be used to calculate a single parameter set for that region, or on a voxel-by-voxel basis, yielding a unique parameter set for each voxel in the image. The VOI-based technique has the advantage of simplicity. Parameter estimation using energy minimization techniques is computationally expensive, and can be extremely so when the process must be carried out for every voxel in a large image. VOI-based estimation avoids this problem, but it does not help to visualize the heterogeneity in the target tissue. Tumors in particular are typically highly heterogeneous. Because they often have very high interstitial pressure, enhancement is generally much more rapid around the periphery of tumors and slower in the interior. As a result, parameter estimation using a single $C_t(t)$ averaged across a VOI may lead to erroneous results [49].

Several groups have proposed the use of voxel-by-voxel analysis to generate parametric images to overcome this shortcoming. With this approach, parametric kinetic enhancement maps are coregistered with high-resolution anatomic images to visualize the spatially heterogeneous distributions of kinetic parameters in tumors [50–52]. However, use of this approach provides significant challenges in terms of both data acquisitions and analysis. Because each voxel has a low SNR, signal quality becomes increasingly important. Furthermore, because $C_t(t)$ is estimated on a per-voxel basis, registration between volumes must be accurate to within a single pixel. This can be challenging in areas such as the chest and abdomen that have significant physiologic motion [53,54]. Development of specialized data acquisition techniques and postprocessing algorithms designed to reduce between-volume motion is an active area of research [55,56].

No uniform consensus is available on quantification methodology to understand the heterogeneity of parametric maps. The most common approach is to report out the mean and/or median value for a VOI for each estimated parameter. Recommendations have been made to use summary statistics, histogram analysis, and pixel scatter plots derived from the parameters. However, all these approaches still require clinical validation [43,57–60]. It is more challenging to analyze the heterogeneity changes in physiologic parameters that occur after therapeutic intervention. It is impractical to obtain voxel-by-voxel match between different visits of the same subject, so comparison is performed using histograms obtained from the parametric maps [26]. Recent efforts also include the use of texture analyses and analytic methods to generate functional response maps for use in monitoring the response to therapy [61,62].

Recently, nonphysiological approaches have been evaluated to analyze the high-dimensional data such as DCE-MRI datasets for visual exploration. Varini et al. have compared established data-reduction algorithms such as principal component analysis (PCA), self-organizing maps (SOM), and locally linearly embedded (LLE). They showed that LLE achieves higher separation between the benign and malignant data points compared to PCA, but that SOM provided the best separation. In another study, Twellmann et al. showed that image fusion can be performed using kernel-principal component analysis (KPCA) for gray or color image visualization for discrimination of malignant, benign, and normal tissues [63,64]. However, the accuracy of nonlinear algorithms such as KPCA and SOM depends on the initial set parameters, which may make its tuning difficult. This is a significant limiting factor for their use in clinical applications.

Another nonpharmacokinetic-based approach very commonly used involves categorizing kinetic curves based on different patterns such as nonenhancing, persistently enhancing (progressive), plateau type, and washout type [65–67]. With this approach, each voxel is color-coded based on uptake curve type for display. Such methods may be useful for visualization of DCE-MRI data to differentiate different types of regions in the tissue. However, their application to determine the effects of therapeutic intervention is still unknown.

10.4 Clinical Utility: Connection to Response in Clinical Trials

The most common practical use of DCE-MRI at this time is in clinical trials of cancer therapeutics which are designed to affect the tumor microvasculature. There are numerous cancer therapeutics either approved or currently in development that are either antiangiogenic or vascular disruptive agents [8,9,68]. The ability to accurately and precisely estimate parameters related to blood flow and vascular permeability is critical to the early evaluation of these compounds, as this gives the most direct window into their targeted biological effects (see Figure 10.8). DCE-MRI has been used frequently in Phase I trials of antiangiogenic and vascular disruptive agents [8,9,69] as a screen to determine whether the expected mechanistic changes in tumor microvasculature are being observed in study patients. In order for this technique to progress to later phase trials, however, two conditions must be met:

- It must be demonstrated that a DCE-MRI protocol can be implemented consistently and accurately across a large number of clinical sites in a single trial.
- It must be demonstrated that there is a significant connection between changes observed using DCE-MRI and other clinically significant endpoints such as changes in tumor size.

Adherence to a consistent protocol and a strict quality control process can largely address the first concern [70]. Numerous multisite clinical trials using DCE-MRI as a mechanistic biomarker have

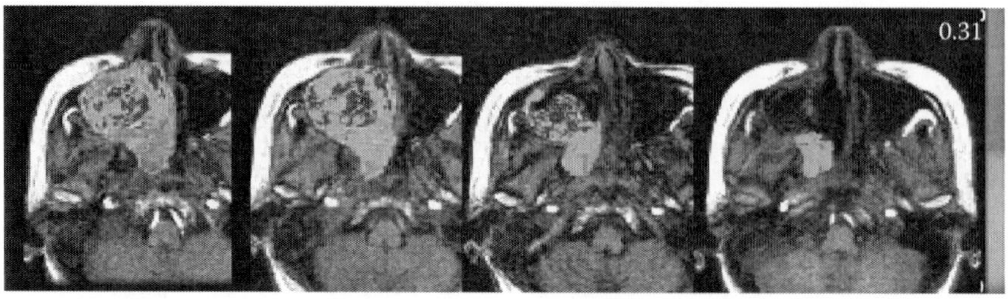

FIGURE 10.8 Overlay of the pharmacokinetic parameter K^{Trans} on a target tumor. Note the large reduction observed in K^{Trans} at Day 2, which precedes the morphological changes observed by Week 8.

been carried out successfully in the past several years using some variation of this approach [8,9,69,71]. However, in order to have serious utility in Phase II and beyond, changes detected using DCE-MRI must be predictive of clinical changes that might be used to obtain approval in Phase III. A number of studies have attempted to address this question, with somewhat mixed results. Flaherty et al. have shown early changes seen with DCE-MRI to be predictive of progression-free survival in renal cell carcinoma patients [72]. These results are contradicted by those of Hahn et al., who show no such relationship in a study of 44 renal cell carcinoma patients [73]. A meta-analysis of 13 Phase I and II clinical trials comprising a total of 156 subjects has shown that early DCE-MRI changes are highly predictive of later changes in tumor burden measured using CT [74]. These divergent results are most easily explained by differences in acquisition and analysis techniques, which can cause measurement variability (and thus predictive power) to vary widely among studies [26,75]. A useful overview of studies comparing DCE-MRI changes to clinical endpoints has been provided by O'Connor et al. [76].

10.5 Summary and Conclusions

DCE-MRI has demonstrated considerable utility over the past 20+ years in both the clinic and in the clinical trials arena. This technique provides information about blood flow, blood volume, and vascular perfusion *in vivo*, providing insight into the mechanistic effects of numerous therapeutic interventions as well as helping to identify viable tissue and to distinguish in some cases between benign and malignant lesions. While a great deal of work has gone into the standardization of nomenclature and general classification of acquisition and analysis techniques, there is not as yet any single widely accepted approach to either the appropriate imaging protocol or the optimal analysis method for DCE-MRI data. In part, this is a necessary consequence of the nature of MRI in general and DCE-MRI in particular, in that gains in one area typically must be balanced by sacrifices in another. So, improvements in acquisition speed allow the use of more biologically relevant analysis models, permitting the separation of flow and permeability effects. However, this comes at the price of a reduction in spatial coverage and/or SNR, decreasing the final reliability of measured parameters. Any DCE-MRI protocol, whether for use in the clinic or in a clinical trial, must therefore be designed with the specific application in mind, and with an understanding of the necessary trade-offs between precision and absolute accuracy outlined above.

References

1. Jackson A, Haroon H, Zhu X, Li K, Thacker N, Jayson G: Breath-hold perfusion and permeability mapping of hepatic malignancies using magnetic resonance imaging and a first-pass leakage profile model. *NMR Biomed* 2002; 15:164.
2. Tofts P: Modeling tracer kinetics in dynamic Gd-DTPA MR Imaging. *J Magn Reson Imag* 1997; 7:91.

3. Evelhoch J: Key factors in the acquisition of contrast kinetic data for oncology. *J Magn Reson Imag* 1999; 10:254.
4. Taylor J, Tofts P, Port R et al.: MR imaging of tumor microcirculation: Promise for the new millennium. *J Magn Reson Imag* 1999; 10:903.
5. Mitchell D, Cohen M: In: *MRI Principles II* (Ed.). Philadelphia, Pennsylvania, Saunders, 2003; p. 275.
6. Tofts P, Kermonde A: Measurement of the blood-brain barrier permeability and leakage space using dynamic MR imaging. 1. Fundamental Concepts. *Magn Reson Med* 1991; 17:357.
7. Kety S: Peripheral blood flow measurement. In: *Methods in Medical Research*, vol. 8. (Potter, VR ed.), Chicago, Year Book Medical Publishers 1960; p. 223.
8. Liu G, Rugo H, Wilding G et al.: Dynamic contrast-enhanced magnetic resonance imaging as a pharmacodynamic measure of response after acute dosing of AG-013736, an oral angiogenesis inhibitor, in patients with advanced solid tumors: Results from a phase I study. *J Clin Oncol* 2005; 23:5464.
9. Anderson H, Yap J, Miller M, Robbins A, Jones T, Price P: Assessment of pharmacodynamic vascular response in a phase I trial of combretastatin A4 phosphate. *J Clin Oncol* 2003; 21:2823.
10. Jayson G, Parker G, Mullamitha S et al.: Blockade of platelet-derived growth factor receptor-beta by CDP860, a humanized, PEGylated di-Fab', leads to fluid accumulation and is associated with increased tumor vascularized volume. *J Clin Oncol* 2005; 23:973.
11. Hodgson R, Grainger A, O'Connor P, Barnes T, Connolly S, Moots R: Dynamic contrast enhanced MRI of bone marrow oedema in rheumatoid arthritis. *Ann Rheum Dis* 2008; 67:270.
12. Malattia C, Damasio M, Basso C, Verri A, Magnaguangno F, Viola S, Gattorno M, Ravelli A, Toma P, Martini A: Dynamic contrast-enhanced magnetic resonance imaging in the assessment of disease activity in patients with juvenile idiopathic arthritis. *Rheumatology* 2010; 49:178.
13. Medic J, Tomazic S, Sersa I, Demsar F: Fast frequency selective keyhole MRI. In: *Proceedings of the 7th Annual Meeting of ISMRM*, Philadelphia, Pennsylvania, 1999 (abstract 2139).
14. McGrath D, Bradley D, Tessier J, Lacey T, Taylor C, Parker G: Comparison of model-based arterial input functions for dynamic contrast-enhanced MRI in tumor bearing rats. *Magn Reson Med* 2009; 61(5):1173.
15. Ashton E, Remick S, Tolcher A et al.: Assessment of pharmacodynamic effect in a phase I study of MN-029, and IV administered vascular disruptive agent, using dynamic contrast-enhanced MRI. In: *Proceedings of ECCO*, Barcelona, Spain, 2007 (abstract 1001).
16. Sorensen A, Tievsky A, Ostergaard L, Weisskoff R, Rosen B: Contrast agents in functional MR imaging. *J Magn Reson Imag* 1997; 7(1):47.
17. Bruening R: Effects of three different doses of a bolus injection of gadodiamide: Assessment of regional cerebral blood volume maps in a blinded reader study. *Am J Neuroradiol* 2000; 21(9):1603.
18. Marckmann P, Skov L, Rossen K et al.: Nephrogenic systemic fibrosis: Suspected causative role of gadodiamide used for contrast-enhanced magnetic resonance imaging. *J Am Soc Nephrol* 2006; 17:2359.
19. Rausch M, Scheffler K, Rudin M, Radu E: Analysis of input functions from different arterial branches with gamma variate functions and cluster analysis for quantitative blood volume measurements. *Magn Reson Imag* 2000; 18(10):1235–1243.
20. Collins D, Padhani A: Dynamic magnetic resonance imaging of tumor perfusion—Approaches and biomedical challenges. *IEEE Eng Med Bio Mag* 2004; 65–83.
21. Weinmann HJ, Laniado M, Mutzel W: Pharmacokinetics of Gd-DTPA/dimeglumine after intravenous injection into healthy volunteers. *Physiol Chem Phys Med NMR* 1984; 16:167.
22. Padhani A, Hayes C, Landau S, Leach M: Reproducibility of quantitative dynamic MRI of normal human tissues. *NMR Biomed* 2002; 15:143.
23. Galbraith S, Lodge M, Taylor N et al.: Reproducibility of dynamic contrast-enhanced MRI in human muscle and tumours: Comparison of quantitative and semi-quantitative analysis. *NMR Biomed* 2002; 15:132.
24. Vonken E, Osch M, Bakker C, Viergever M: Measurement of cerebral perfusion with dual-echo multi-slice quantitative dynamic susceptibility contrast MRI. *J Magn Reson Imag* 1999; 10:109.

25. Ashton E, McShane T, Evelhoch J: Inter-operator variability in perfusion assessment of tumors in MRI using automated AIF detection. *LNCS* 2005; 3749:451.
26. Rijpkema M, Kaanders J, Joosten F, Van Der Kogel A, Heerschap A: Method for quantitative mapping of dynamic MRI contrast agent enhancement in human tumors. *J Magn Reson Imag* 2001; 14:457.
27. Shah V, Turkbey B, Pang Y, Liu W, Choyke P, Bernardo M: Population-averaged arterial input function for dynamic-contrast enhanced MRI obtained with inflow suppression and B1 correction. In *In vivo Studies & Development/Novel Use of Imaging Probes*, World Molecular Imaging Congress, Kyoto Japan, September 8, 2010.
28. Walker-Samuel S, Leach M, Collins D: Reference tissue quantification of DCE-MRI data without a contrast agent calibration. *Phys Med Biol* 2007; 52:589.
29. Ashton E: Quantitative medical imaging. *J Imag Sci Technol* 2007; 51:117.
30. Ashton E, Raunig D, Ng C, Kelcz F, McShane T, Evelhoch J: Scan-rescan variability in perfusion assessment of tumors in MRI using both model and data-derived arterial input functions. *J Magn Reson Imag* 2008; 28:791.
31. Kovar D, Lewis M, Karczmar G: A new method for imaging perfusion and contrast extraction fraction: Input functions derived from reference tissues. *J Magn Reson Imag* 1998; 8(5):1126.
32. Yang C, Karczmar G, Medved M, Stadler W: Estimating the arterial input function using two reference tissues in dynamic contrast-enhanced MRI studies: Fundamental concepts and simulations. *Magn Reson Med* 2004; 52:1110.
33. Faranesh A, Yankeelov T: Incorporating a vascular term into a reference region model for the analysis of DCE-MRI data: A simulation study. *Phys Med Biol* 2008; 53:2617.
34. Hsiao I, Liao Y, Liu H: Study of onset time-shift and injection duration in DCE-MRI: A comparison of a reference region model with the general kinetic model. *NMR Biomed* 2010; 23(4):375.
35. Ashton E, Durkin E, Kwok E, Evelhoch J: Conversion from signal intensity to Gd concentration may be unnecessary for perfusion assessment of tumors using DCE-MRI. In: *Proceedings of the 15th Annual Meeting of ISMRM*, Berlin, 2007. (abstract 2813).
36. Leach M, Brindle K, Evelhoch J et al.: The assessment of antiangiogenic and antivascular therapies in early-stage clinical trials using magnetic resonance imaging: Issues and recommendations. *Br J Cancer* 2005; 92:1599.
37. Tofts P, Brix G, Buckley D et al.: Estimating kinetic parameters from dynamic contrast-enhanced T_1-weighted MRI of a diffusable tracer: Standardized quantities and symbols. *J Magn Reson Imag* 1999; 10:223.
38. Landau H: Sampling, data transmission, and the Nyquist rate. *Proc IEEE* 1967; 55:1701.
39. Huang W, Carney P, Tudorica L et al.: Approaching complete separation of benign and malignant breast lesions by DCE-MRI: Impact on healthcare costs. In: *Proceedings of the 18th Annual Meeting of ISMRM*, Stockholm, 2010. (abstract 987).
40. Buckley D: Uncertainty in the analysis of tracer kinetics using dynamic contrast-enhanced T1-weighted MRI. *Magn Reson Med* 2002; 47(3):601.
41. Koh T, Tan C, Cheong L, Lim C: Cerebral perfusion mapping using a robust and efficient method for deconvolution analysis of dynamic contrast-enhanced images. *NeuroImage* 2006; 32:643.
42. Thng C, Hartono S, Koh T et al.: Dynamic contrast enhanced MRI (DCE MRI) for Phase I anti-angiogenic tiral: Comparison of the transfer constant (Ktrans) to blood flow and permeability derived by a distributed parameter model. *J Clin Oncol* 2008; 26 (May 20 suppl; abstr 3514).
43. Li K, Wilmes L, Henry R et al.: Heterogeneity in the angiogenic response of a BT474 human breast cancer to a novel vascular endothelial growth factor-receptor tyrosine kinase inhibitor: Assessment by voxel analysis of dynamic contrast-enhanced MRI. *J Magn Reson Imag* 2005; 22:511.
44. Huang W, Lid X, Morris E, Tudorica L, Seshan V, Rooney W, Tagged I, Wang Y, Xu J, Springer, Jr. C: The magnetic resonance shutter speed discriminates vascular properties of malignant and benign breast tumors in vivo. *PNAS* 2008; 105(46):17,947.

45. Lowry M, Zelhof B, Liney G, Gibbs P, Pickles M, Turnbull L: Analysis of prostate DCE-MRI: Comparison of fast exchange limit and fast exchange regimen pharmacokinetic models in the discrimination of malignant from normal tissue. *Investigative Radiol* 2009; 44(9):577.
46. Ahearn T, Staff R, Redpath T, Semple S: The use of the Levenberg-Marquardt curve-fitting algorithm in pharmacokinetic modeling of DCE-MRI data. *Phys Med Biol* 2005; 50:85.
47. Horsfield M, Morgan B: Algorithms for calculation of kinetic parameters from T1-weighted dynamic contrast-enhanced magnetic resonance imaging. *J Magn Reson Imag* 2004; 20:723.
48. Cheng H: Investigation and optimization of parameter accuracy in dynamic contrast-enhanced MRI. *J Magn Reson Imag* 2008; 28:736.
49. Gribbestad I, Nilsen G, Fjosne H et al.: Comparative signal intensity measurements in dynamic gadolinium-enhanced MR mammography. *J Magn Reson Imag* 1994; 4:477.
50. Teifke A, Behr O, Schmidt M et al.: Dynamic MR imaging of breast lesions: correlation with microvessel distribution pattern and histological characteristics of prognosis. *Radiology* 2006; 239:351.
51. Gaustad J, Benjaminsen I, Graff B et al.: Intratumor heterogeneity in blood perfusion in orthotopic human melanoma xenografts assessed by dynamic contrast-enhanced magnetic resonance imaging. *J Magn Reson Imag* 2005; 21:792.
52. Jackson A, Kassner A, Annesley-Williams D et al.: Abnormalities in the recirculation phase of contrast agent bolus passage in cerebral gliomas: Comparison with relative blood volume and tumor grade. *Am J Neuroradiol* 2002; 23:7.
53. Jackson A, O'Connor JP, Parker GJ et al.: Imaging tumor vascular heterogeneity and angiogenesis using dynamic contrast-enhanced magnetic resonance imaging. *Clin Cancer Res* 2007; 13:3449.
54. Gribbestad I, Gjesdal K, Nilsen G et al.: An introduction to dynamic contrast-enhanced MRI in oncology. In: *Dynamic Contrast-Enhanced MRI in Oncology* (Jackson A, Buckley DL, Parker GJ, eds.), 1st edition. Berlin/Heidelberg (NY): Springer; 2005; 1:3.
55. Orton M, Miyazaki K, Koh D, Collins D, Hawkes D, Atkinson D, Leach M: Optimizing functional parameter accuracy for breath-hold DCE-MRI of liver tumors. *Phys Med Biol* 2009; 54:2197.
56. Melbourne A, Atkinson D, White M, Collins D, Leach M, Hawkes D: Registration of dynamic contrast-enhanced MRI using a progressive principal component registration (PPCR). *Phys Med Biol* 2007; 52:5147.
57. Parker G, Suckling J, Tanner S et al.: MRI: Parameteric analysis software for contrast-enhanced dynamic MR imaging in cancer. *Radiographics* 1998; 18:497.
58. Parker G, Buckley D: Tracer kinetic modeling for T1-weighted DCE-MR imaging. In: *Dynamic Contrast-Enhanced Magnetic Resonance Imaging in Oncology* (Jackson A, Parker GJ, Buckley DL, eds.), 1st Edition. Berlin/Heidelberg (NY): Springer; 2005; 1:81.
59. de Lussanet Q, Backes W, Griffioen A et al.: Dynamic contrast-enhanced magnetic resonance imaging of radiation therapy-induced microcirculation changes in rectal cancer. *Int J Radiat Oncol Biol Phys* 2005; 63:1309.
60. Rose C, Mills S, O'Connor J et al.: Quantifying heterogeneity in dynamic contrast-enhanced MRI parameter maps. In: *Lecture Notes in Computer Science* (Ayache N, Ourselin A, Maeder A, eds.), Verlag/Berlin/Heidelberg (NY): Springer; 2007; 4792:376.
61. Rose C, Mills S, O'Connor J: Quantifying spatial heterogeneity in dynamic contrast-enhanced MRI parameter maps. *Magn Reson Med* 2009; 62:488.
62. Moffat B, Chenevert T, Lawrence T et al.: Functional diffusion map: a noninvasive MRI biomarker for early stratification of clinical brain tumor response. *Proc Natl Acad Sci USA* 2005; 102:5524.
63. Twellmann T, Saalbach A, Gerstung O, Leach M, Nattkemper T: Image fusion for dynamic contrast enhanced magnetic resonance imaging. *BioMedical Eng Online* 2004; 3:35.
64. Twellmann T, Meyer-Baese A, Lange O, Foo S, Nattkemper T: Model–free visualization of suspicious lesions in breast MRI based on supervised and unsupervised learning. *Eng Appl Artif Intell* 2008; 21(2):129.

65. Khouli R, Macura K, Jacobs M, Khalil T, Kamel I, Dwyer A, Bluemke D: Dynamic contrast-enhanced MRI of the breast: Quantitative method for kinetic curve type assessment. *Am J Radiol* 2009; 193:295.
66. Leinsinger G, Schlossbauer T, Scherr M, Lange O, Reiser M, Wismuller A: Cluster analysis of signal-intensity time course in dynamic breast MRI: Does unsupervised vector quantization help to evaluate small mammographic lesions? *Eur Radiol* 2006; 16:1138.
67. Kubassova O, Boesen M, Boyle R, Cimmino M, Jensen K, Bliddal H, Radjenovic A: Fast and robust analysis of dynamic contrast enhanced MRI datasets. *Med Image Comput Comput Assist Interv* 2007; 10(2):261.
68. Herbst R, Onn A, Sandler A: Angiogenesis and lung cancer: Prognostic and therapeutic implications. *J Clin Oncol* 2005; 23:3243.
69. Mross K, Drevs J, Muller M, Medinger M, Marme D, Hennig J, Morgan B, Lebwohl D, Masson E, Ho Y: Phase I clinical and pharmacokinetic study of PTK/ZK, a multiple VEGF receptor inhibitor, in patients with liver metastases from solid tumors. *Eur J Cancer* 2005; 41:1291.
70. Ashton E: Quantitative MR in multi-center clinical trials. *J Magn Reson Imag* 2010; 31:279.
71. Collins J: Imaging and other biomarkers in early clinical studies: Ones step at a time or re-engineering drug development? *J Clin Oncol* 2005; 23:5417.
72. Flaherty K, Rosen M, Heitjan D, Gallagher M, Schwartz B, Schnall M, O'Dwyer P: Pilot study of DCE-MRI to predict progression-free survival with Sorafenib therapy in renal cell carcinoma. *Cancer Biol Therapy* 2008; 7:496.
73. Hahn O, Yang C, Medved M, Karczmar G, Kistner E, Karrison T, Manchen E, Mitchell M, Ratain M, Stadler W: Dynamic contrast-enhanced magnetic resonance imaging pharmacodynamic biomarker study of Sorafenib in metastatic renal carcinoma. *J Clin Oncol* 2008; 26:4572.
74. Ashton E: Early DCE-MRI findings predict tumor volume changes. In: *Proceedings of the 17th Annual Meeting of ISMRM*, Honolulu, 2009. (abstract 806).
75. Ashton E, Raunig D, Ng C, Kelcz F, McShane T, Evelhoch J: Scan-rescan variability in perfusion assessment of tumors in MRI using both model and data-derived arterial input functions. *J Magn Reson Imag* 2008; 28:791.
76. O'Connor J, Jackson A, Parker G, Jayson G: DCE-MRI biomarkers in the clinical evaluation of anti-angiogenic and vascular disrupting agents. *Br J Cancer* 2007; 96:189.

11

MRI of Myocardial Deformations: Imaging and Modeling

Hui Wang
University of Louisville

Amir A. Amini
University of Louisville

Globally, cardiovascular diseases (CVD) are the number one cause of death and are projected to remain so. An estimated 17 million people died from CVD in 2005, representing 30% of all global deaths. Among these deaths, 7.2 million were due to heart attacks and 5.7 million were due to stroke. If current trends continue, by 2030, an estimated 23.6 million people will die from CVD in the world [138]. In America, an estimated 80 million adults (more than one in three) have one or more types of CVD. In 2005, about 864,000 people died of CVD, accounting for 35.3% of all deaths [8].

Heart disease, such as myocardial ischemia, secondary to coronary artery disease, may be identified and localized through the analysis of the cardiac deformation. Early efforts for quantifying ventricular wall motion used surgical implantation and tracking of radiopaque markers with x-ray imaging in canine hearts [47]. Such techniques are invasive and affect the regional motion pattern of the ventricular wall during the marker tracking process and, clearly are not feasible clinically. Noninvasive imaging techniques are vital and have been widely applied to clinical use. Magnetic resonance imaging (MRI) is a noninvasive imaging technique with the capability to monitor and assess the progression of CVD so that effective procedures for the care and treatment of patients can be developed by physicians and researchers. MRI can provide three-dimensional (3D) analysis of global and regional cardiac function with great accuracy and reproducibility [23]. In the past few years, numerous efforts have been devoted to cardiac imaging techniques for deformation recovery. Many approaches have been proposed for tracking cardiac motion and for computing deformation parameters and mechanical properties of the heart from a variety of cardiac MR imaging techniques.

There are many techniques available nowadays for cardiac imaging which provide qualitative and quantitative measurements of cardiac function. MRI is a highly advanced and sophisticated imaging modality for cardiac motion assessment and quantitative analysis. In the realm of MRI, cardiac cine MRI is considered as the standard technique for measuring global function parameters, for example, ejection fraction, cavity volume, and myocardial mass [23]. It also can reveal important regional function parameters, such as wall thickness. MR tagging [17,145] is a well-known method to track local deformations. Local parameters such as twist, strain, and strain rate can be derived from tagged MRI. Other techniques, such as phase contrast MRI (PCMRI), displacement encoding with stimulated echoes imaging (DENSE), and strain encoding (SENC), can also reveal regional properties of the heart. Several review papers have previously been published in the field of deformation recovery from cardiac MR imaging [18,23,40,153]. This chapter describes cardiac MR image acquisition techniques as well as image analysis techniques for recovery of ventricular deformations from the acquired images.

The organization of this chapter is as follows. Section 11.1 gives a review of cardiac cine MRI pulse sequences and motion analysis techniques from cine MR images. Section 11.2 gives an overview of tagged MRI techniques and the associated image analysis methods. A novel acquisition technique, termed Orthogonal CSPAMM, is included in this section. Section 11.3 summarizes phase contrast MRI for cardiac deformation recovery. Sections 11.4 and 11.5 introduce two more recent techniques: DENSE and SENC. Strain is an important kinematic index, revealing fractional change in length of a continuously deforming body. Section 11.6 provides a review of prior work on strain analysis from cardiac MRI. Conclusions and future research trends are provided in Section 11.7.

11.1 Cine MRI

Cardiac cine MRI is widely used clinically due to its high soft-tissue contrast. The temporal resolution is also high enough to do motion tracking over the cardiac cycle. It is the gold standard for noninvasively quantifying global *in vivo* heart function and for measuring ejection fraction, cavity volume, and mass.

11.1.1 Cine MR Imaging

Cine gradient echo sequences, often known as bright blood sequences, are commonly used for evaluating cardiac function. Blood appears bright in these sequences due to the contrast properties of blood and its rapid flow. The technique can discriminate between blood and myocardium very well. Two imaging pulse sequences are most commonly used for acquiring cine MRI: spoiled gradient echo (GRE) and balanced steady-state free precession (b-SSFP). b-SSFP can yield better signal-to-noise ratio (SNR) and contrast-to-noise ratio (CNR) than spoiled GRE and is less sensitive to motion-induced signal loss. Therefore, cine MR images using b-SSFP appear more uniform within the blood and are darker in the myocardium. b-SSFP gradient echo sequences have largely replaced spoiled gradient echo sequences for this purpose. Different trade names for these b-SSFP sequences are TrueFISP (True Fast Imaging with Steady-State Precession; Siemens), FIESTA (Fast Imaging Employing Steady-State Acquisition; GE), and b-FFE (Balanced Fast-Field Echo; Phillips). These sequences are typically used in conjunction with segmented *k*-space acquisition. Segmented acquisition is the process of dividing the cardiac cycle into multiple segments (phases) to produce a series of images that can be displayed as a movie (cine). Each image in the cine is typically constructed from information gathered over several heart beats—within a breath hold of 10–20 s depending on the sequence. Figures 11.1 and 11.2 show a sequence of short-axis (SA) and long-axis (LA) cine images, respectively. Thirty frames were required using a b-FFE sequence on a Philips 3T Achieva scanner showing the contraction wave during the cardiac cycle. One-third of them have been displayed here. The first row shows systolic images and the second row shows diastolic images.

In cine MRI, there are multiple cardiac phases in one cardiac cycle, usually between 20 and 30, depending on the temporal resolution (30–50 ms per frame). Although it is possible to image the heart in real time with cine MRI, the resulting image quality is poor. Instead, most sequences use electrocardiogram

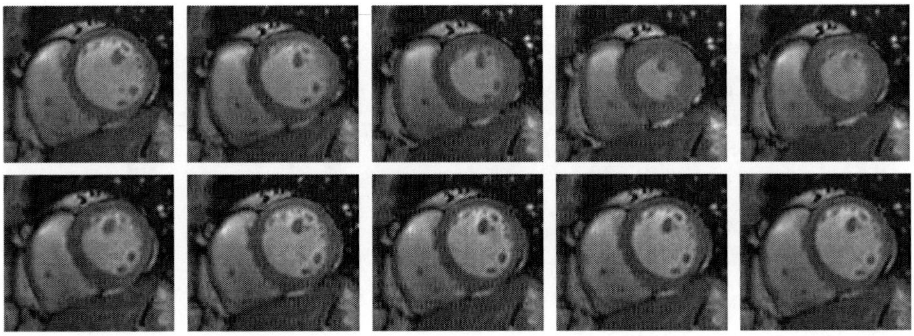

FIGURE 11.1 A sequence of short-axis cine images acquired with b-FFE on a Philips Achieva 3T scanner. Ten out of thirty phases are shown from the top left to bottom right. The first row shows systolic images and the second row shows the diastolic images. Imaging parameters were as follows: TE = 1.603 ms, slice thickness = 8 mm, spatial resolution = 1.25×1.25 mm^2, acquisition matrix = 160×187, flip angle = 45°. A phased-array thoracic coil with 16 elements was used.

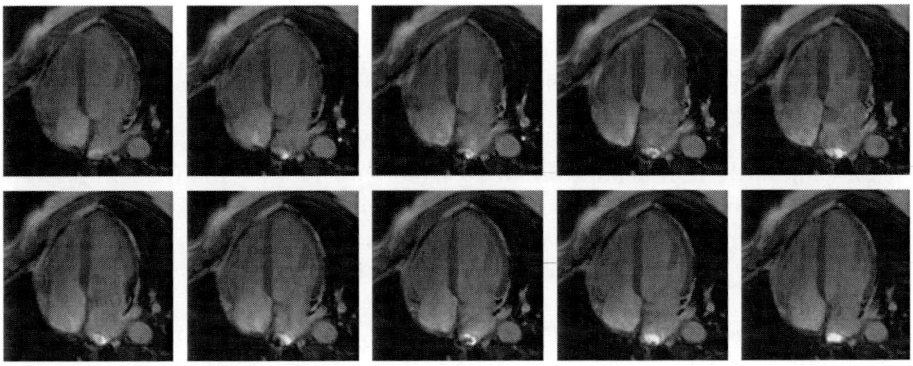

FIGURE 11.2 A sequence of long-axis cine images acquired with b-FFE on a Philips Achieva 3T scanner in the same volunteer whose short-axis images were shown in Figure 11.1. Ten out of thirty phases are shown from the top left to bottom right. The first row shows systolic images and the second row shows diastolic images. Imaging parameters were as follows: TE = 1.737 ms, slice thickness = 8 mm, spatial resolution = 1.21×1.21 mm^2, acquisition matrix = 176×211, flip angle = 45°. A phased-array thoracic coil with 16 elements was used.

(ECG) gating to acquire images at each stage of the cardiac cycle over several heart beats. This technique is the primary basis for functional assessment by cine MRI. During one cardiac cycle, ECG-gated cine MRI acquires one segment of image raw data for each cardiac phase, as shown in Figure 11.3. Multiple segments comprising the complete *k*-space raw data are acquired in multiple heartbeats. With the help of a phased-array radio frequency (RF) receiver coil and parallel image acquisition and reconstruction techniques, typical scan times of 6–7 heartbeats per slice are now routine. The in-plane spatial resolution of cine MRI is on the order of 1.5×2 mm^2. The slice thickness is between 5 and 10 mm [40]. Multiple 2D-imaged slices may be stacked to construct a 3D volume for each cardiac phase.

11.1.2 Deformation Recovery from Cine MR Images

Since to a large extent the myocardium appears homogeneous in cine MR images, cine MRI cannot directly provide information on myocardial motion patterns, for example, relative rotation of the endocardial layer relative to the epicardial layer (transmural shear), or of basal slices with respect to apical slices (torsion) [128]. The most commonly used features in cine MRI are endocardial and epicardial

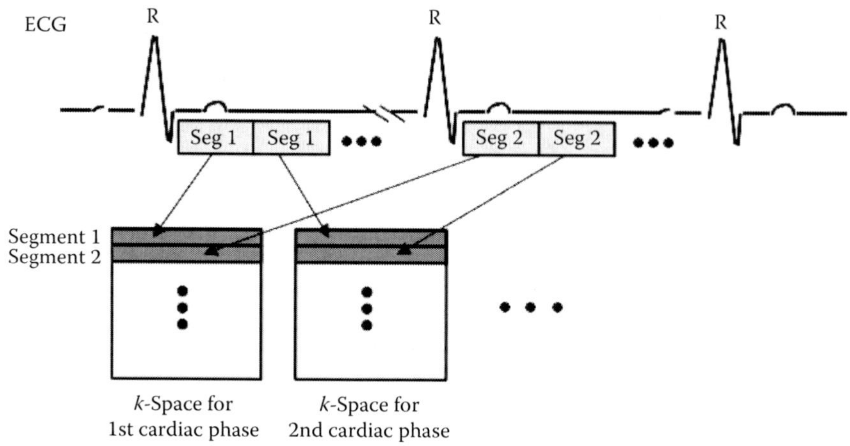

FIGURE 11.3 Schematic diagram of ECG-gated breath-hold cine MRI. There are multiple cardiac phases in one cardiac cycle. For each phase, a segment is acquired in one heart beat.

contours, or specific feature points on these contours. For motion analysis from cine MRI, a useful model that could be utilized is its incompressibility. One strategy for motion tracking may be to first track sparse feature points on the contours and subsequently perform dense motion field reconstruction. Reconstruction of the dense motion field from a sparse set of landmarks is an ill-posed problem and therefore additional constraints are needed to obtain a unique solution. The constraints previously proposed in the literature are either mathematical or mechanical, enforcing smoothness of the deformation field subject to additional requirements [20,21,140]. The other strategy is to use image registration [97,124,130,131], either for the whole image or for some feature points. Due to the substantial challenges in deformation recovery from cine MRI, including significant modeling needs, it is reasonable to assume that the accuracy of cine MRI deformation recovery techniques is not on par with the methods that will be described later in this chapter. In the rest of this section, a detailed review of cardiac motion tracking methods based on cine MRI is provided.

11.1.2.1 Incompressible Deformable Models

Aside from the effect of blood perfusion, which will alter myocardial volume, the myocardium is nearly incompressible. The common conclusion regarding myocardial volume change is that it is no more than 4% during the cardiac cycle [20]. Song et al. [121] were the first in applying divergence-free and incompressibility constraints to construct 3D velocity field inside the heart though utilizing 3D CT data. Based on the assumptions that myocardium is almost incompressible and that there is no transmural bending during myocardial movement, Bistoquet et al. [20] proposed an incompressible deformable model to recover the motion of the left ventricle from cine MRI data. The main contribution was that this method modeled the deformation using a new transformation which was the displacements of middle surface nodes between the reference frame and current frame. The framework was extended to biventricular model in Reference 21. Different from the model in Reference 20, the new model had the ability to represent the deformation of structures with arbitrary topologies. The nodes were approximately uniformly spaced over the left ventricle (LV) and right ventricle (RV) wall middle surfaces. A divergence-free matrix-valued radial basis function scheme was used to interpolate the displacements at the nodes.

11.1.2.2 Registration-Based Methods

Veress et al. [130,131] proposed a deformable registration technique, called hyperelastic warping, to determine left ventricular strains from mid-diastole to end-diastole using multiple slices SA cine MR images. A finite element (FE) model was deformed by the difference in image intensities as a body force and a hyperelastic strain energy based on continuum mechanics as the regularization force. The boundaries of the LV were manually segmented for both epicardium and endocardium. The myocardium was represented as transversely isotropic hyperelastic with fiber angles varying from −90° at the epicardial surface to 90° at the endocardial surface. Phatak et al. [98] validated hyperelastic warping for LV strain measurement during systole, by comparing circumferential and radial strains with results obtained from HARP [83]. Image-based forces driving deformations were combined with a computational model which included estimated myocardial material properties, fiber direction, and active fiber contraction. Different from diastole, where the forces acting on the myocardium are passive and are associated with diastolic filling, the forces in systole are active and are generated by myofiber contraction, making the model more complicated. A transversely isotropic hyperelastic constitutive model was utilized to represent the myocardium.

Shen and Sundar et al. [117,124] used a 4D image registration method to estimate cardiac motion. This framework considered registering the first frame to all other frames simultaneously. An attribute vector for every point in the image was used to contain information about intensity, boundary, and geometric moment invariants. Perperidis et al. [97] proposed two B-spline-based free-form registration methods for the spatiotemporal alignment of cardiac MR image sequences. It extended Ruckert's framework [108] to 4D B-spline tensor products and enabled comparison between corresponding anatomy at different temporal positions in the cardiac cycle. These two B-splines methods used the same transformation. The difference between them was the order of the optimization process. One performed a combined optimization with spatial and temporal components. The other optimized each component separately. Chenuoune et al. [30] proposed a method to assess relative circumferential shortening strain of endocardium by segmenting endocardial contours of a 2D + t SA cine MR data set using a level set method and then matched these contours using a morphing registration method. This method only tracked the endocardial contours for 2D SA images.

11.1.2.3 Feature-Based Tracking Methods

This category is usually involved in segmenting the ventricular wall first followed by tracking the movement of feature points, such as epicardial and endocardial boundaries [30], geometric measure as tracking tokens [66,67,120,140]. Quantifying the boundary motion of nonrigid body from feature points is often treated as a two-step process: first establishing correspondence between sampled points on the boundaries at two time points, then finding the transformation for every point on the boundaries according to the feature point correspondences.

Amini and Duncan [11] used principal surface curvatures to track local bending and stretching of the left ventricular endocardial wall. Surface patches around geometrically significant landmark points were modeled as thin plates. Subsequently, a smoothing procedure was employed to generate a dense motion field. McEachen et al. [74] proposed a method to track the LV endocardial contour. Presegmented endocardial contours at each frame were the input to this method. Correspondence of tokens with geometric properties on the contours were determined by shape-based tracking. An adaptive filtering scheme was used to produce a set of sinusoidal parameters that represent quantified motion trajectories. Three-dimensional motion trajectory of the endocardial surface can be recovered by locating and matching differential surface properties [120]. These trajectories were then used to deform a mesh that represented the left ventricular myocardial surfaces.

Lin et al. proposed a generalized robust point matching (G-RPM) framework [67] to track the motion of left ventricle and extended this framework using extended free-form deformation (EFFD) as regularization model [66]. Different from free-form deformation (FFD) which uses parallelepipedical lattice, the EFFD model employed an arbitrarily shaped lattice. Feature points were extracted according to the curvatures. Under G-RPM framework, Pan et al. [140] proposed a new regularization model based on the boundary element method (BEM) to estimate dense displacement fields and strains from 3D cine MRI sequences of the LV. This method was also applicable to 4D echocardiographic image sequences. The most attractive feature of BEM over FEM is that it only requires discretization of the surface rather than the volume. In order to automatically obtain the endocardial and epicardial contours and feature points, Zhu et al. [149,152] proposed an integrated Bayesian framework to segment LV and estimate its motion. The feature points were extracted from the narrowband of segmented contours and the motion estimation was accomplished by BEM-GRPM algorithm [140].

Shi et al. [119] used both image-derived information from myocardial surfaces and the mid-wall from phase contrast MRI. This method combined image-derived information and mechanical modeling of the myocardium, using FEM to solve for deformations. Papademetris et al. [90] validated strains derived from 3D cine MRI and 3D echocardiography images using shape-based tracking. The dense motion field was obtained from integrating a transversely linear elastic model and sparse displacements using a shape-tracking method. Papademetris et al. [91] utilized a symmetric nearest neighbor algorithm for initialization in the shape-tracking step, instead of nearest neighbor [120].

Remme et al. [105] proposed a modality-independent method to analyze LV motion from anatomic modalities, using fiducial marker fitting in conjunction with a parameter distribution model. A FEM mesh was constructed according to the geometry of the myocardium and deformed by optimizing the nodal parameters to the motion of a sparse number of manually tracked fiducial markers. In order to overcome the sparsity of the fiducial markers, *a priori* information from tagged MRI database on the probability distribution of possible LV motion was incorporated.

11.2 Tagged MRI

11.2.1 Acquisition

Myocardial tagging was first introduced by Zerhouni et al. [145] and Axel et al. [17] in 1988 and 1989, respectively. It uses spin tagging prepulses to produce noninvasive markers in the myocardial tissues. The main reason why tagged MRI can image motion is that when the local magnetization of a material point is altered, the material point maintains the altered magnetization when it moves within the limits of the T_1 relaxation time. The process of MR tagging uses a special pulse sequence to spatially modulate the longitudinal magnetization, prior to image acquisition using conventional imaging, as shown in Figure 11.4. The varying magnetization produces alternating light and dark pattern on the image. Spatial modulation of magnetization (SPAMM) [17] is the most commonly used technique to produce sinusoidal tag patterns. Optimal tagging and acquisition of MR images for cardiac motion analysis were investigated by Nguyen et al. [81]. Pai and Axel [87] gave a thorough review of tagged cardiac MR imaging methods, including advances in pulse sequence development, image acquisition, high temporal and spatial resolution imaging, high field strength imaging, and 3D whole heart tagging. A review covering the current clinical applications of myocardial tagging can be found in Reference 116.

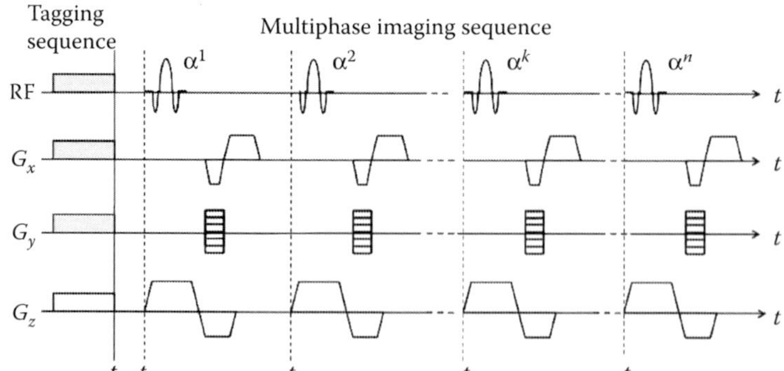

FIGURE 11.4 Timing diagram of a typical tagging experiment. A tagging sequence is followed by a standard multiphase imaging sequence. t_i corresponds to the start of the ith phase in the cardiac cycle.

11.2.1.1 Spatial Modulation of Magnetization

The SPAMM myocardial tagging technique is widely available commercially [17,145] and is the most commonly used technique for producing sinusoidal tag patterns. Figure 11.5 (left) shows a one-dimensional 1-1 SPAMM sequence with two 90° RF pulses, an interspersed tagging gradient in the readout direction, and a spoiler gradient. Before the first RF pulse, magnetization in rotating frame is initially all polarized along the main magnetic field (in the z direction) (Figure 11.5a). The first RF pulse flips the initial longitudinal magnetization into the transverse plane (Figure 11.5b). The gradient G_t produces a periodic spatial modulation of the phase of the transverse magnetization along the gradient direction (Figure 11.5c). The second RF pulse produces modulated longitudinal magnetization (Figure 11.5d). A tag grid can be produced by following the second RF pulse with a second gradient in the direction orthogonal to the first gradient and then with another RF pulse. Some tagged and untagged images from a canine are shown in Figure 11.6. In the subsequent imaging step, the spatially modulated longitudinal magnetization is made visible by the RF pulse which flips it to the transverse plane, followed with phase-encoding and read-out. The tag lines are not sharp, but have sinusoidal variation of intensity in the image (Figure 11.5d). Sharper tagging stripes can be obtained by using more binomially distributed RF pulses that are each separated by dephasing gradients [16]. The effect of the SPAMM pulse sequence is to produce a series of stripes in the acquired images. As shown in Figure 11.7, the tag lines fade at the end of the cardiac cycle.

11.2.1.2 Complementary Spatial Modulation of Magnetization

Complementary SPAMM (CSPAMM) was introduced by Fischer et al. to improve tagging contrast in later phases of the cardiac cycle [42]. CSPAMM is based on the subtraction of two images with complementary signed tagging modulation. Compared to SPAMM, CSPAMM has the ability to eliminate the DC interference of the off-center peaks, resulting in significant improvement of tagging contrast in later phases of the cardiac cycle [42]. The diagram for the CSPAMM pulse sequence is shown in Figure 11.8. The tagging pulses in (a) and (b) are out of phase by 180°. By subtracting them, CSPAMM reduces tagline intensity fading and consequently allows longer net tag persistence throughout the heart cycle.

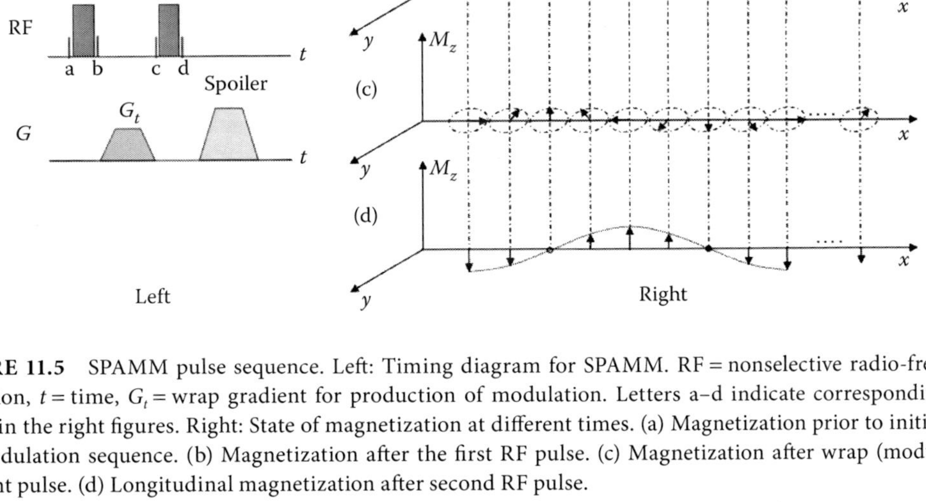

FIGURE 11.5 SPAMM pulse sequence. Left: Timing diagram for SPAMM. RF = nonselective radio-frequency excitation, t = time, G_t = wrap gradient for production of modulation. Letters a–d indicate corresponding time points in the right figures. Right: State of magnetization at different times. (a) Magnetization prior to initiation of the modulation sequence. (b) Magnetization after the first RF pulse. (c) Magnetization after wrap (modulating) gradient pulse. (d) Longitudinal magnetization after second RF pulse.

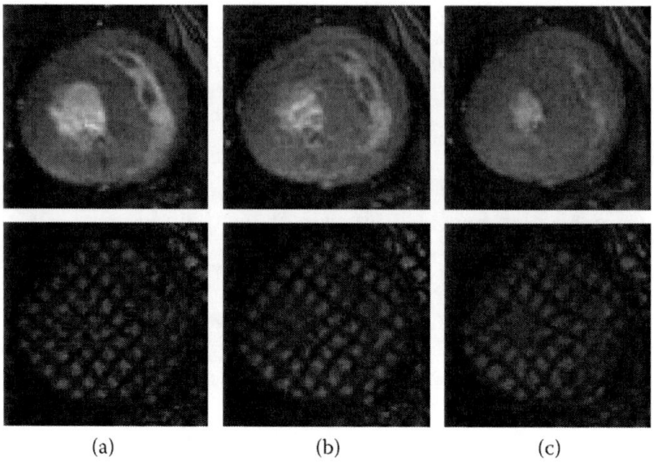

FIGURE 11.6 The top row shows a sequence of short-axis mid-ventricular untagged images acquired in a dog at three different time points in the cardiac cycle. (a) End-diastole, (b) mid-systole, and (c) end-systole. The bottom row shows tagged images at corresponding times and locations. Imaging parameters were as follows: TR/TE = 32/2.8 ms, slice thickness = 8 mm, spatial resolution = 1.17 × 1.17 mm², flip angle = 20°.

In CSPAMM, the longitudinal magnetization M_z is decomposed into two terms: one for tagging information Q_T, the other for the relaxation part Q_R. A timing diagram of a typical tagging experiment is shown in Figure 11.4. At time t_0 right after the SPAMM tagging sequence, the modulated longitudinal magnetization is

$$M_z(t_0) = M_{ss}TAG(x, y) \tag{11.1}$$

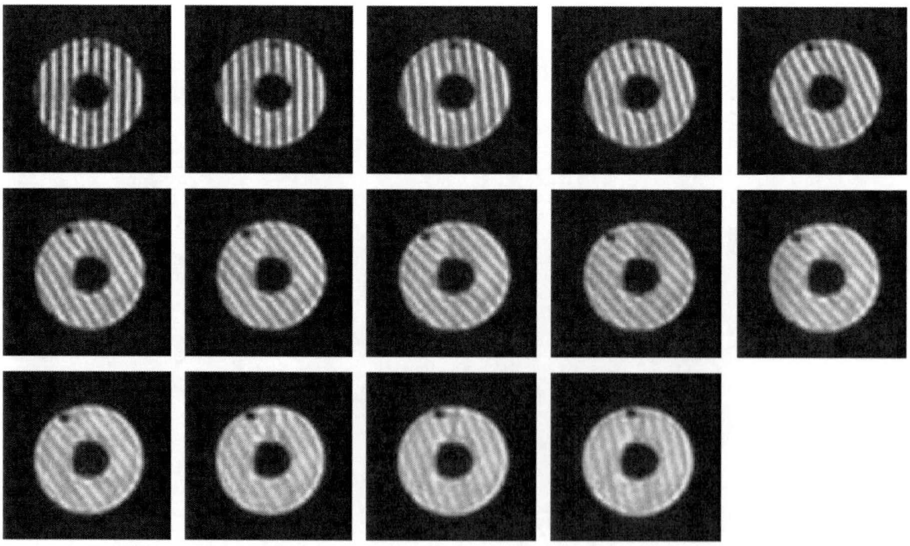

FIGURE 11.7 Fourteen cardiac phases for a rotating phantom imaged during an entire "cardiac cycle" (see Figure 11.15) using the SPAMM pulse sequence: starting at the top left and to the right and bottom. As may be seen, the tags fade as a function of time due to the T_1 effect. Imaging parameters were as follows: TR/TE = 7.9/4.0 ms, slice thickness = 8 mm, spatial resolution = 1.17 × 1.17 mm², acquisition matrix = 102 × 112, tag spacing = 6.22 mm, flip angle = 10°.

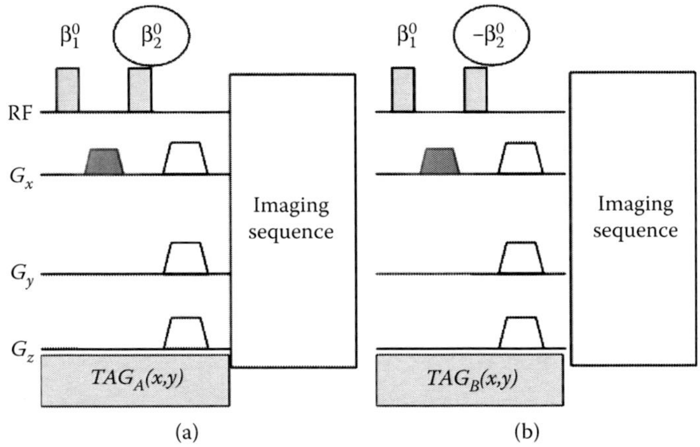

FIGURE 11.8 Timing diagram of a 1–1 CSPAMM sequence. (a) Measurement with positive tagging pattern $TAG_A(x,y)$. (b) Measurement with negative tagging pattern $TAG_B(x,y)$.

where M_{ss} is the steady-state magnetization before tagging and $TAG(x,y)$ represents the spatial modulation of magnetization introduced by tagging sequence. At time t_1

$$\begin{aligned} M_z(t_1) &= (M_z(t_0) - M_0)e^{-t_1/T_1} + M_0 \\ &= (M_{ss}TAG(x,y) - M_0)e^{-t_1/T_1} + M_0 \\ &= \underline{M_{ss}TAG(x,y)e^{-t_1/T_1}} + \underline{M_0(1 - e^{-t_1/T_1})} \\ &= Q_{T_1} + Q_{R_1} \end{aligned} \tag{11.2}$$

where M_0 is the equilibrium magnetization and T_1 is the longitudinal relaxation time. At time t_k

$$\begin{aligned} M_z(t_k) &= (M_z(t_{k-1}) - M_0)e^{-(t_k - t_{k-1})/T_1} + M_0 \\ &= [(Q_{T_{k-1}} + Q_{R_{k-1}})\cos\alpha_{k-1} - M_0]e^{-(t_k - t_{k-1})/T_1} + M_0 \\ &= (Q_{T_{k-1}}\cos\alpha_{k-1} + Q_{R_{k-1}}\cos\alpha_{k-1} - M_0)e^{-(t_k - t_{k-1})/T_1} + M_0 \\ &= \underline{Q_{T_{k-1}}\cos\alpha_{k-1}e^{-(t_k - t_{k-1})/T_1}} + \underline{(Q_{R_{k-1}}\cos\alpha_{k-1} - M_0)e^{-(t_k - t_{k-1})/T_1} + M_0} \\ &= Q_{T_k} + Q_{R_k} \end{aligned} \tag{11.3}$$

Therefore, the two components of the longitudinal magnetization just before the kth RF pulse are

$$\begin{aligned} Q_{T_k} &= Q_{T_{k-1}}\cos\alpha_{k-1}e^{-(t_k - t_{k-1})/T_1} \\ &= (Q_{T_{k-2}}\cos\alpha_{k-2}e^{-(t_{k-1} - t_{k-2})/T_1}\cos\alpha_{k-1}e^{-(t_k - t_{k-1})/T_1} \\ &= \ldots \\ &= M_{ss}TAG(x,y)e^{-t_k/T_1}\prod_{j=0}^{k-1}\cos\alpha_j \\ Q_{R_k} &= (Q_{R_{k-1}}\cos\alpha_{k-1} - M_0)e^{-(t_k - t_{k-1})/T_1} + M_0 \end{aligned} \tag{11.4}$$

where Q_{T_k} is the tagging component, while Q_{R_k} is the relaxed term. After the kth RF imaging pulse of flip angle α_k, the longitudinal magnetization is rotated to the xy plane which contributes to the kth image.

$$I_k = M_z(t_k)\sin\alpha_k e^{-TE/T_2^*} = (Q_{T_k} + Q_{R_k})\sin\alpha_k e^{-TE/T_2^*} \tag{11.5}$$

The basic idea of CSPAMM is to eliminate the relaxation term Q_{R_k} while only keeping the tagging information term Q_{T_k} by acquiring two images A_k and B_k using the same parameters except for their respective tagging patterns $TAG_A(x,y)$ and $TAG_B(x,y)$ (see Figure 11.8). The subtraction of the kth pair of images leads to

$$A_k - B_k = M_{ss}[TAG_A(x,y) - TAG_B(x,y)]e^{-t_k/T_1}\left(\prod_{j=0}^{k-1}\cos\alpha_j\right)\sin\alpha_k e^{-TE/T_2^*} \tag{11.6}$$

In order to obtain the maximum grid amplitude, the tagging pattern $TAG_A(x,y)$ and $TAG_B(x,y)$ should satisfy the following condition:

$$TAG_A(x,y) + TAG_B(x,y) = 0 \tag{11.7}$$

which will produce the kth tagged image

$$I_k = 2Q_{T_k}\sin\alpha_k e^{-TE/T_2^*} \tag{11.8}$$

Visualization of k-space for an image modulated by a cosine tagging function in the horizontal direction is shown in Figure 11.9 and for the vertical direction is shown in Figure 11.10. In both figures,

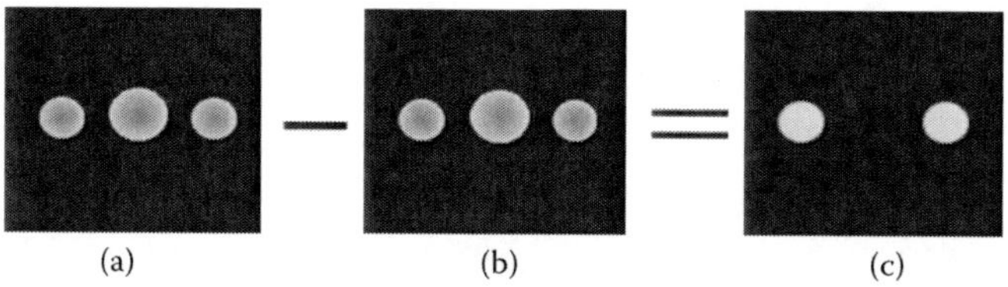

FIGURE 11.9 Visualization of k-space for an image modulated by a cosine in the horizontal direction. (a) k-Space for one SPAMM with positive tagging pattern. (b) k-Space for the other SPAMM with negative tagging pattern. (c) k-Space for CSPAMM.

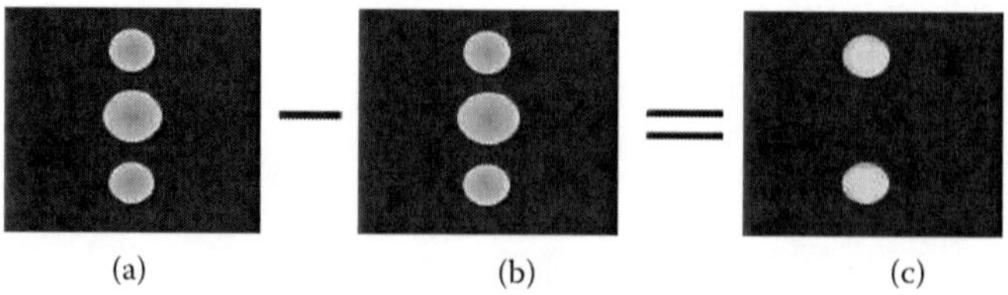

FIGURE 11.10 Visualization of k-space for an image modulated by a cosine in the vertical direction. (a) k-Space for one SPAMM with positive tagging pattern. (b) k-Space for the other SPAMM with negative tagging pattern. (c) k-Space for CSPAMM.

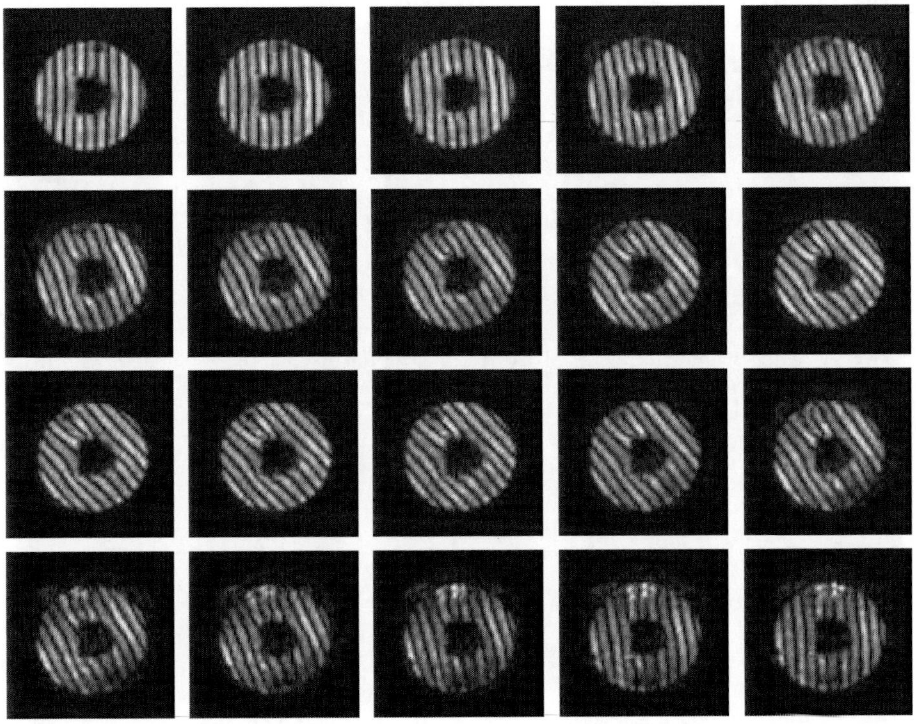

FIGURE 11.11 Twenty cardiac phases during an entire cardiac cycle imaged for a rotating phantom (see Figure 11.15) using CSPAMM pulse sequence. Starting at the top left and traversing to the right and bottom. Imaging parameters were as follows: TR/TE = 12/6.5 ms, slice thickness = 8 mm, spatial resolution = 1.17 × 1.17 mm^2, acquisition matrix = 102 × 112, tag spacing = 6.22 mm, flip angle = 10°.

(a) shows the k-space for the SPAMM with positive tagging pattern, (b) shows the k-space for the other SPAMM with negative tagging pattern, and (c) is the k-space for CSPAMM which is the subtraction of (b) from (a). Figure 11.11 shows a cine sequence of vertical CSPAMM tagging lines on a rotating phantom for the entire cardiac cycle.

11.2.1.3 Orthogonal CSPAMM

In this section, a new tagging pulse sequence, Orthogonal CSPAMM (OCSPAMM), is proposed which acquires tagged data in two orthogonal directions within the same time as the common SPAMM acquisition procedure for creating grid tags. However, relative to SPAMM, OCSPAMM has the advantage of eliminating the DC peak which contributes significantly to the loss of tag-myocardium contrast.

As seen in Figure 11.8, in CSPAMM, two SPAMM tagging sequences 180° out of phase are placed in the same direction, either in frequency encoding or in phase encoding direction, resulting in the need for four separate acquisitions. In the proposed OCSPAMM sequence, the second SPAMM tag orientation is rotated 90° relative to the first so that tag lines in two directions are combined (through subtraction) after only two acquisitions, therefore achieving removing of the central DC peak in k-space. The OCSPAMM sequence timing diagram is shown in Figure 11.12. As may be seen in this figure, the first tagging gradient is in G_x direction, while the second tagging gradient is in G_y direction. A TFE-EPI sequence is used to image the modulated magnetization, as shown in Figure 11.13.

Visualization of k-space for OCSPAMM sequence is shown in Figure 11.14. Similar to the CSPAMM sequence of Figure 11.9, two SPAMM images are subtracted, but unlike the original CSPAMM technique, the second tagged acquisition is orthogonal to the first one. This approach eliminates the DC

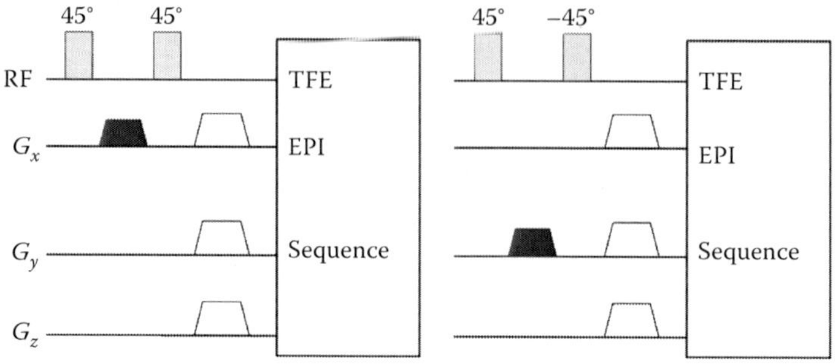

FIGURE 11.12 Timing diagram for the OCSPAMM sequence. The first pair of 45° RF pulses with an interspersed tagging gradient are used to define the tags in G_x direction. The second pair of 45° RF pulses with an interspersed tagging gradient orthogonal to the first tagging gradient are used to define the tags in G_y direction. A TFE-EPI sequence is used for imaging as shown in Figure 11.13.

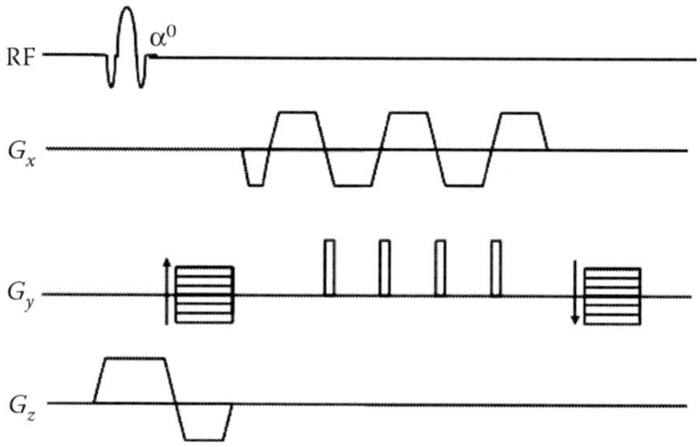

FIGURE 11.13 Timing diagram for the TFE-EPI sequence used for imaging the modulated magnetization with EPI factor 5. After each RF pulse, five k-space profiles are acquired with the help of the blip gradients in phase encoding direction.

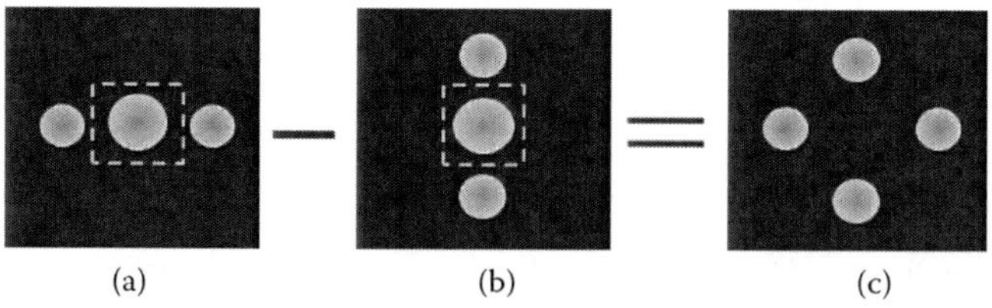

FIGURE 11.14 Visualization of k-space for the OCSPAMM sequence. (a) k-Space for tagged image with positive tagging pattern. (b) k-Space for tagged image with negative tagging pattern in orthogonal direction to (a). (c) k-Space for OCSPAMM.

component while leaving the off-center peaks in k-space intact and reduces the acquisition time by a factor of two when compared to the original CSPAMM technique.

All imaging experiments were conducted on a 3T Achieva MR scanner (Philips Healthcare, Best, NL) using a two-element receive coil. The tag line distance is 7 mm. The phantom was imaged using a turbo field echo-echo planar imaging (TFE-EPI) cine pulse sequence (Figure 11.13) with the following parameters: TR/TE = 9.1/4.7 ms, 10° flip angle, turbo factor 7, FOV 225 × 225 mm², in-plane voxel size 2 × 2 mm², reconstruction resolution 1.25 × 1.25 mm², slice thickness 8 mm, 14 heart phases, 112 × 85

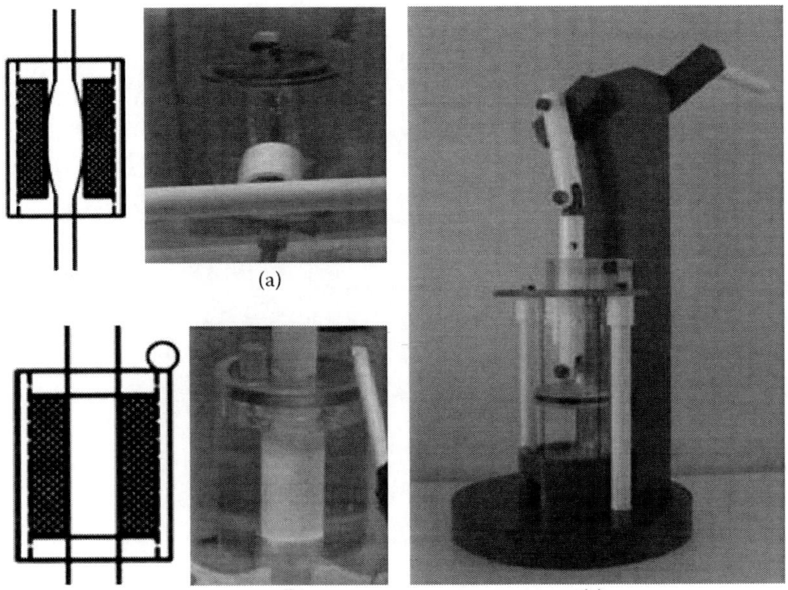

FIGURE 11.15 Phantom components. (a) Contraction motion phantom. (b) Rotation motion phantom. (c) Air pump.

acquired matrix, EPI factor 5. The cardiac cycle was approximately 1000 ms and T_1 of dielectric gel in the phantom was 728 ms (similar to myocardial tissue, *in vivo*).

A cardiac motion phantom, which independently models myocardial wall thickening and rotation in the human heart, was utilized to test the proposed OCSPAMM pulse sequence. The main elements of the LV motion phantom are the air pump, two phantoms within a common enclosure, trigger circuit, and rotation motion actuator. The two phantoms and the air pump are shown in Figure 11.15. Other than the triggering circuit, all materials used in construction of the phantom were nonferromagnetic and MR compatible (polycarbonate, wood, latex, sponge, dielectric gel) [41].

Two experiments were conducted on the cardiac phantom to validate the proposed OCSPAMM tagging sequence: one for rotation as shown in Figure 11.16 and the other for contraction as shown in Figure 11.17. The OCSPAMM tagged images showed good image quality and persistent tag contrast for the entire duration of the cardiac cycle. For comparison to SPAMM sequence, Figure 11.18 shows a cine sequence of SPAMM tagging on the same rotation phantom during an entire cardiac cycle. Notice that there are some artifacts in the upper and lower portion of the last frame in Figure 11.17. These were due to susceptibility from the air inside the center of the phantom and are unrelated to the OCSPAMM pulse sequence.

11.2.2 Analysis of Tagged MR Images

Although MR tagging provides visually interesting data reflecting myocardial motion, fast and accurate image analysis methods are required before tagged MRI data can be used for routine quantitative analysis. Since the development of tagged MRI, many methods have been developed to detect tag features and track the heart motion. This should be seen in the backdrop of more recent work on frequency-based analysis of tagged data (such as HARP and SinMod), or DENSE, and SENC techniques which will be presented later in this chapter, all of which present the possibility for more automated analysis. In the rest of this section, different image analysis methods using tagged MRI are discussed.

There are two different categories of cardiac motion analysis methods using tagged MRI. The first category is feature-based motion tracking methods, that is, those that measure the deformations by tracking

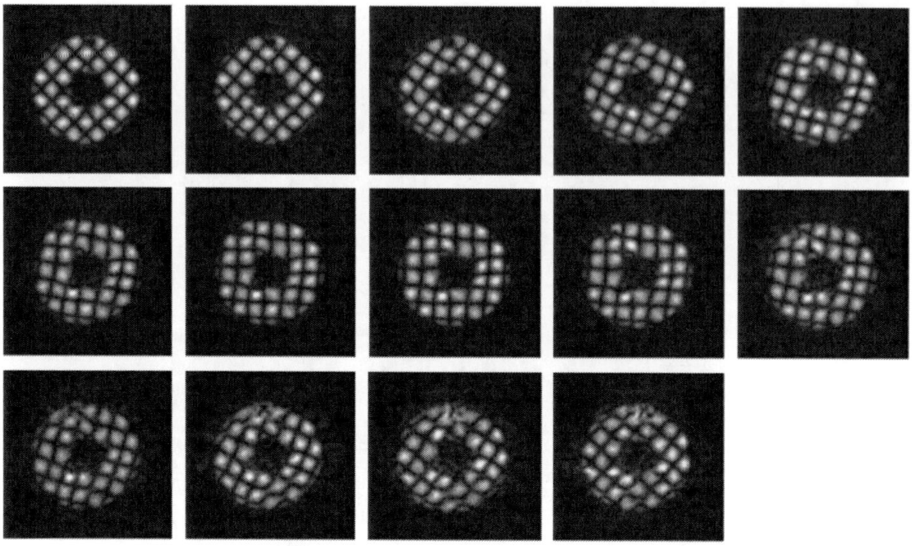

FIGURE 11.16 Fourteen cardiac phases imaged for a rotating phantom (see Figure 11.15) using OCSPAMM pulse sequence. Starting at the top left and traversing to the right and bottom. Imaging parameters were as follows: TR/TE = 9.1/4.7 ms, slice thickness = 8 mm, spatial resolution = 1.17 × 1.17 mm^2, acquisition matrix = 85 × 112, tag spacing = 10 mm, flip angle = 10°.

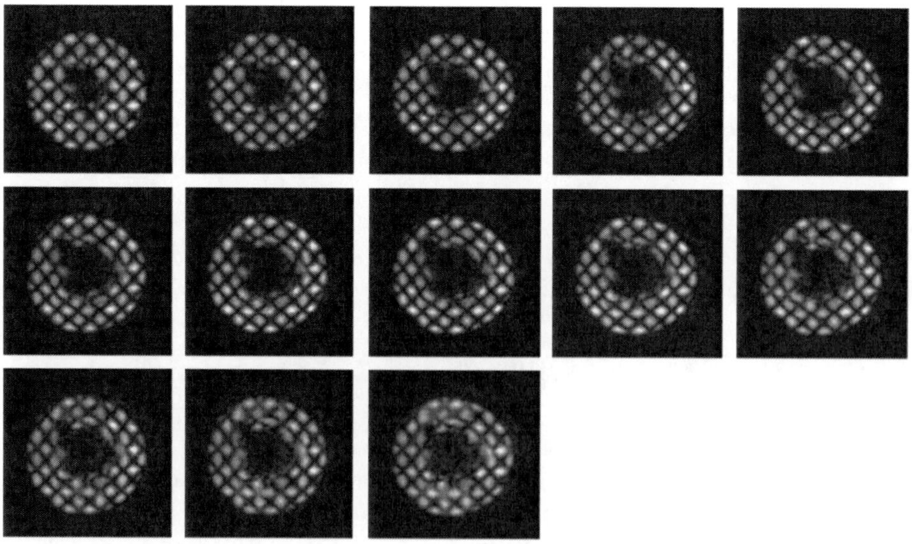

FIGURE 11.17 Thirteen cardiac phases imaged for a contraction phantom (see Figure 11.15) using OCSPAMM pulse sequence. Starting at the top left and traversing to the right and bottom. Imaging parameters were as follows: TR/TE = 9.1/4.7 ms, slice thickness = 8 mm, spatial resolution = 1.17 × 1.17 mm^2, acquisition matrix = 85 × 112, tag spacing = 10 mm, flip angle = 10°.

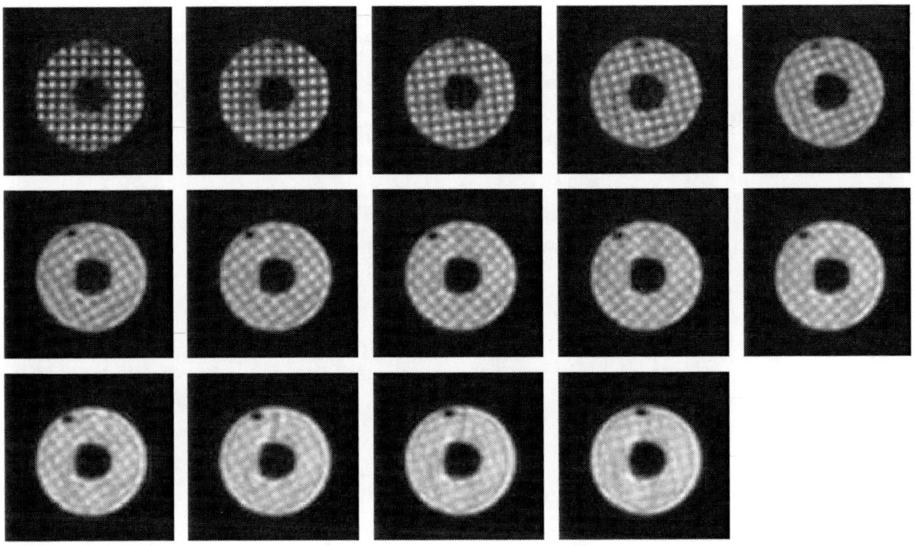

FIGURE 11.18 Fourteen cardiac phases during an entire cardiac cycle imaged for a rotating phantom (see Figure 11.15) using SPAMM pulse sequence. Starting at the top left and traversing to the right and bottom. TR/TE = 11/5.4 ms, slice thickness = 8 mm, spatial resolution = 1.25×1.25 mm^2, acquisition matrix = 144×160, tag spacing = 7 mm, flip angle = 10°.

the movement of tagging features. The most commonly used features are sparse tag lines, geometrically salient land markers, tag line intersections, as well as epicardial and endocardial contours. A number of papers have been published on how to detect tag lines and contours [26,37,48]. For tracking, two sets of techniques have emerged. One set of techniques is to track the features without constructing a dense motion field. The other set constructs a dense motion field from the sparse motion field obtained by tracking the tag features. The accuracy of feature-based image analysis methods using tagged MRI depends highly on the quality of the image and the spacing of tag lines. Tagged MR images with higher spatial resolution will provide more information and constraints on the models. Any advances and improvements in MR imaging, tagging techniques, tag detecting methods, and modeling methods will have impact on another. The second category aims to obtain the deformation field directly from tagged MR images, without extracting tag features. Three different types of methods fall in this category: optical flow methods, registration-based methods [25,107], and frequency-based methods, for example, HARP [83] and SinMod [14].

11.2.2.1 Tracking Myocardial Beads/Landmark-Based Methods

Myocardial beads, defined as the intersection points of orthogonal tagging planes, can be used as noninvasive markers [10]. Compared to physical implanted markers, tagged MRI offers a larger number of MR markers in the myocardium, with the clear advantage that they are placed noninvasively. Tracking beads is based on the fact that intersections encode a unique 3D position in the myocardium which move along with the deforming tissue during the cardiac cycle. Because in MRI imaging planes are fixed relative to the magnet's coordinate system throughout the cardiac cycle, as the myocardial tissue deforms, the intersections in the reference frame may move in and out of the imaging plane.

Kerwin and Prince [59] proposed a method for measuring 3D material points displacements by tracking the intersection points of a 3D grid of applied tag surfaces. Amini et al. [10] proposed a fast method to track myocardial beads by utilizing the B-spline surface to represent tag planes in 3D explicitly. The parametric representation of the tag surfaces leads to an easy way to compute the position of 3D myocardial beads by minimizing the summation of distances between any two reconstructed tag surfaces. Instead of extracting MR beads by reconstructing tag planes, Sampath et al. [114] proposed an automatic 3D cardiac beads tracking method. A true 3D trajectories (instead of apparent motion) of the tag beads were obtained by combining two 2D pathlines that were computed on both SA and LA image planes using 2D HARP tracking techniques. The slice-following CSPAMM technique was used to take through-plane motion into account.

11.2.2.2 Optical Flow Methods

Optical flow is a motion tracking technique first proposed in computer vision. The fundamental assumption is that image intensity remains constant along a motion trajectory. However, in tagged MR images, the intensity constancy condition is not satisfied because of the relaxation of the magnetization of the spins throughout the cardiac cycle, which causes the intensity and contrast to change from image to image. There are different ways to deal with the variable intensity in tagged MRI. Prince and McVeigh [100] used a variable brightness optical flow (VBOF) method to overcome the intensity variation in tagged MRI by introducing a term which accounts for the variable brightness of the stripes using the MR parameters T_1, T_2, and initial magnetization M_0. Dougherty et al. [38] utilized a Laplacian filter to compensate for intensity and contrast loss in myocardial tags. Xu et al. proposed a 3D optical flow method to extract tissue displacements from 3D tagged images [139]. For harmonic phase images (which will be introduced later) the phase of each material point undergoing motion is preserved—the idea is based on phase-based optical flow proposed by Fleet and Jepson [44]. Harmonic images are produced by filtering the spectral peaks in Fourier domain and extracting the spatial phase information from the inverse Fourier transform of the filtered images. Florack et al. [45] applied a multiscale optical flow framework using HARP images. A spatiotemporal Gaussian filter was applied as a preprocessing step. It can simultaneously capture 0th- and 1st-order structure of the motion field. LV rotational analysis was performed using this method [129].

11.2.2.3 Deformable Models

In the deformable modeling approach, a model deforms to fit the data using energy minimization or based on classical physics-based equation of motion [75]. How to deform models and apply constraints to the deformable model vary with application. The force to drive the model deformation can originate from tag intersections, tag lines, or tag surfaces. Because of the sparsity of feature information, vector field interpolation is needed to be incorporated to obtain a dense motion field for every point within the myocardial region. Interpolation methods used to date include mathematical regularization [120], FEM [118,136], BEM [140], deformable model with parameter functions [92], B-splines [10,127], incompressibility assumptions [20], and continuum-mechanics-based energy minimization [91].

Clarysse et al. [31] proposed a cosine series-based model to get a mathematical expression of the reconstructed displacement field for 2D + t series of tagged MR images. The coefficients of cosine model were obtained by minimizing the Euclidean distances between projection of deformed tag lines and undeformed ones.

Park et al. proposed a surface deformable model with parameter functions for mid-wall analysis [94] and volumetric deformable model with parameter functions [92]. Continuously varying functions of parameters were used, instead of fixed parameters. Reference 92 was one of the first papers that used 3D FEMs (prismatic FEMs) and converted the FEM nodal deformations into parameter functions. It used tag intersection points in the image plane (three orthogonal sets) and boundary points to estimate the volumetric deformation of the LV. The tag line intersections were automatically computed and were used to deform the model's tagging lines. In order to study the cardiac motion in its complete cycle, Park et al. were the first to use cascaded SPAMM data acquisition within a deformable model framework to avoid the problem of SPAMM tags not lasting throughout the entire cardiac cycle [93]. Haber et al. [49] modified this framework to estimate the motion and deformation of right ventricular free wall and septum. It used volumetric FEMs and tag line intersections in the plane, as well as points within the tag intersections. The 3D motion was reconstructed by constructing a volumetric FEM model from the tag lines and the estimated boundaries of the RV and the LV. A physics-based FEM approach with Poisson's ratio 0.05 and Young's modulus 0.02 was used to estimate the strains throughout the LV and the RV.

However, in this work, the focus was on RV disease. In addition, an *in vivo* validation method was first proposed based on the agreement between the model's tagging lines and the imaging-based tagging lines. Hu et al. [51] presented a statistical model to estimate *in vivo* material properties and strain and stress distributions in both ventricles, using the displacements calculated from tagged MRI data based on Haber's work [49]. It investigated both kinematics and material properties.

Amini et al. [9] proposed an efficient thin-plate spline warp method that warped an area in the plane such that two embedded snake grids from two tagged frames were brought into registration, interpolating a dense displacement vector field. A major problem for techniques relying on 2D short-axis data [9,135] is that in general they do not take into account the through-plane motion. Extended to 3D and 4D (3D B-spline solid plus 1D B-spline interpolation over time), B-spline cubic deformable models [52,103] were introduced for analysis of cardiac motion. Tustison [127] improved the 4D B-spline model framework in Reference 52 by using an internal energy function and enabling inserting additional control points. Ozturk and McVeigh [86] also used the 4D B-spline model to analyze cardiac motion for both ventricles. In Reference 126, Tustison extended the 4D B-spline models proposed in References 127 and 52 to NonUniform Rational B-spline (NURBS) models. The proposed NURBS models were implemented in both polar and cylindrical coordinates and were restricted to the ventricular wall. Due to the spatial resolution of MRI tags, only sparse displacement information in both short and long axis images can be used. In order to make use of more information, Chen [28] extended Tustison's work [126] by introducing dense virtual tag lines obtained from phase-based displacement estimates. In References 132 and 133, a multilevel B-spline (MBS) model was used to integrate displacement information obtained from tagged images in the spatial domain with displacement information obtained from spectral peaks in the frequency domain in order to improve the accuracy of motion tracking. MBS uses both real and virtual tag intersections to derive motion fields and has several advantages: First, B-spline inherent characteristics guarantee the smoothness of the recovered motion fields. Second, it results in fast, accurate fitting. Third, incorporation of phase information (see the section on frequency-based methods) helps overcome the sparsity of real tag intersections.

Instead of B-spline models parameterizing the Cartesian space, Deng and Denney [35,36] proposed a 3D B-spline cylindrical parameterized model. This was further improved on by Tustison and Amini [126] who showed that Cartesian-based cylindrical and prolate-spheroidal B-spline parameterization improves on accuracy of circumferential strains. Young [141] proposed a FEM model to reconstruct 3D myocardial motion from tagged MRI. As with other methods using deformable models and tag lines, for example, References 49 and 92, the model deformation was driven by the distance between model stripe points and image stripe points in the direction orthogonal to the tags. Wang et al. [136] proposed a method to determine the passive diastolic mechanics. The 3D deformation field was constructed from tagged MRI data and material properties were estimated by matching the observed deformations with the FE model.

11.2.2.4 Registration-Based Methods

Cardiac motion tracking can also be considered as a 4D intramodality registration problem [69]. In general, image registration is a process to find the optimal transformation that can transform one image to the other, maximizing a similarity metric between them. The advantages of registration methods are: (1) tag detection and extraction steps are not required and (2) they are automatic without the need for user supervision. The disadvantages of the method involve getting stuck in local minima and potential misalignment due to image noise and artifacts. Additionally, the computational time for the registration-based approaches in general is long.

Numerous methods have been proposed for image registration, for example, nonrigid registration using FFD. In Rueckert's approach [108], global motion was modeled by an affine transformation, while local motion was described by an FFD based on multilevel B-splines. Registration was achieved by optimizing a cost function measuring the similarity between two images as well as the smoothness of the deformation needed to align the images. Chandrashekara et al. [25] made use of the nonrigid registration algorithm first proposed in Reference 108 and applied it to the analysis of myocardial deformations. The algorithm in Reference 108 can only be applied to either SA or LA images, which means that through-plane motion could not be accounted for. In order to obtain complete 3D motion of the myocardium, the cost function was modified to be the sum of the normalized mutual information (NMI) between the registered SA and LA images. The main advantage of this method is that tag localization and deformation field reconstruction were performed simultaneously. Another advantage is that no assumptions about the nature of the tag pattern are made so that the proposed method could potentially be applicable to untagged cine MR images or ultrasound images. However, the fact that no tag position information is incorporated could be considered a limitation. Rougon et al. [107] extended the similarity measure based on NMI to generalized information measures in order to quantitatively assess ventricular wall function from tagged MRI. Different from Chandrashekara's method, Ali-Silvey class of generalized information measures was used, which is a superset of mutual information (MI) and NMI. For more information about medical image registration, refer to the survey paper [68,69].

11.2.2.5 Frequency-Based Methods

11.2.2.5.1 Harmonic Phase Imaging

Harmonic phase imaging (HARP) is a phase-based technique for rapid analysis of tagged cardiac MR images [83]. Rather than working on tagged images directly, HARP uses phase information in the frequency domain. It is based on the fact that the Fourier transform of a SPAMM tag image has a collection of distinct spectral peaks, each of which contains motion information in a certain direction. After bandpass filtering, one spectral peak can be extracted and an inverse Fourier transform applied, yielding the harmonic image. A harmonic image is a complex image that has magnitude and phase images, as given by Equation 11.9. This process is illustrated in Figure 11.19 for simulated data from a 3D solid model (Figure 11.20) and in Figure 11.21 for real dog data. The HARP method uses $\Phi_k(y,t)$, the phase of the harmonic image, which is called harmonic phase image or simply HARP image. The harmonic phase image is linearly related to a directional component of the true motion.

$$I_k(y,t) = D_k(y,t)e^{j\Phi_k(y,t)} \tag{11.9}$$

The principle of HARP tracking is that the harmonic phase of a point remains constant as it moves during the cardiac cycle. For a given material point (x,y) at image slice s at time t, which takes on

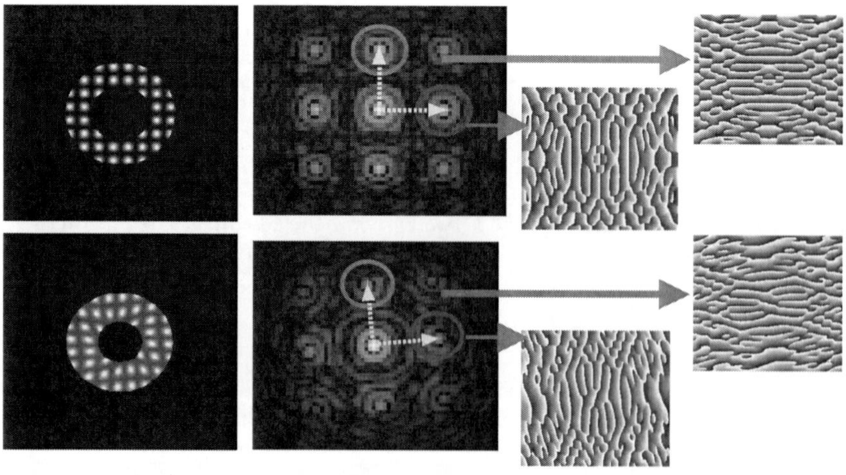

FIGURE 11.19 Tagged images, their Fourier peaks, and their corresponding phase images.

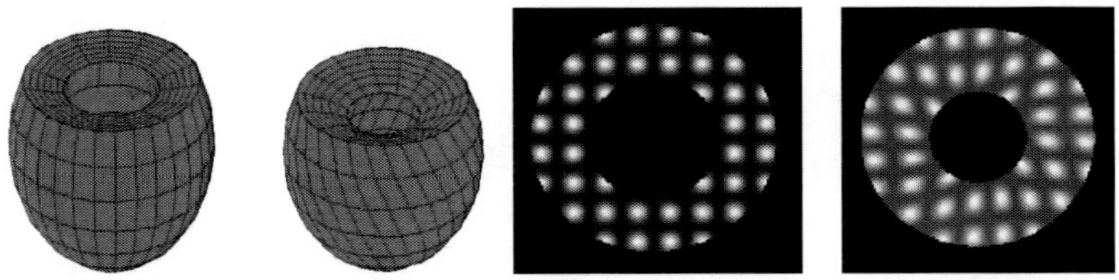

(a) (b) (c) (d)

FIGURE 11.20 A realistic simulation. A 3D solid model of the heart from Arts' 13-parameter kinematic model [13] at (a) end-diastole and (b) end-systole. Corresponding mid-ventricular short-axis simulated images at (c) end-diastole and (d) end-systole.

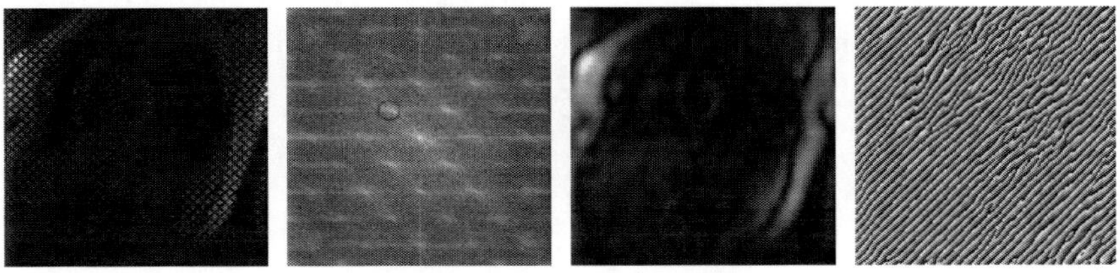

(a) (b) (c) (d)

FIGURE 11.21 (a) A tagged MR image. (b) The magnitude of its Fourier transform. By extracting the spectral peak inside the circle in (b), a complex image is produced with a magnitude (c) and a phase (d).

position (x',y') in the same slice at time $t + 1$, the phase time-invariance condition is expressed as follows:

$$\begin{cases} P^h(x',y',s,t+1) = P^h(x,y,s,t), \\ P^v(x',y',s,t+1) = P^v(x,y,s,t) \end{cases} \tag{11.10}$$

where x, y, s, and t stand for x and y coordinates of a point in image plane, the short-axis image slice location, and the time frame, respectively.

Although HARP is fast in postprocessing tagged MR images, the time needed to acquire a cine sequence of tagged images is on the order of 8–20 heartbeats during a breath-hold. The reason for long acquisition time is that traditional tagged MR imaging approach is to use the segmented k-space method with breath-hold [15]. The segmented k-space technique acquires the entire k-space over several cardiac cycle, where each cycle yields a part of the k-space. Based on the fact that only two spectral peaks are needed to compute 2D deformation, Sampath et al. proposed a pulse sequence to only acquire the region of k-space corresponding to the application of the selected spectral peaks, thereby reducing the acquisition time [112]. The region of k-space acquired for both a full k-space acquisition and a partial k-space acquisition are shown in Figures 11.22 and 11.23, respectively. The timing diagram for partial k-space acquisition is shown in Figure 11.24. Gradient B compensates for the dephasing of the gradients in the EPI read-out and the tagging gradient. The purpose of gradient B is to center the acquisition matrix on the frequency of one of the spectral peaks, and the sign of gradient B determines which peak is acquired. The pulse sequence used by Sampath was based on SPAMM and suffered from low spatial resolution. We have developed a partial k-space acquisition based on CSPAMM with the ability to remove the DC component in order to improve the spatial resolution compared to SPAMM (see Figure 11.8) [134]. Using the partial acquisition based on CSPAMM, a larger acquisition matrix was possible when compared to Reference 112 (60×56 compared to 32×32), thus allowing for higher spatial resolution (5×5 mm^2 compared to 9×9 mm^2) and larger deformations to be tracked within the larger acquisition window in k-space. Eight odd number frames out of 15 cardiac phases are displayed in Figure 11.25 for the rotation

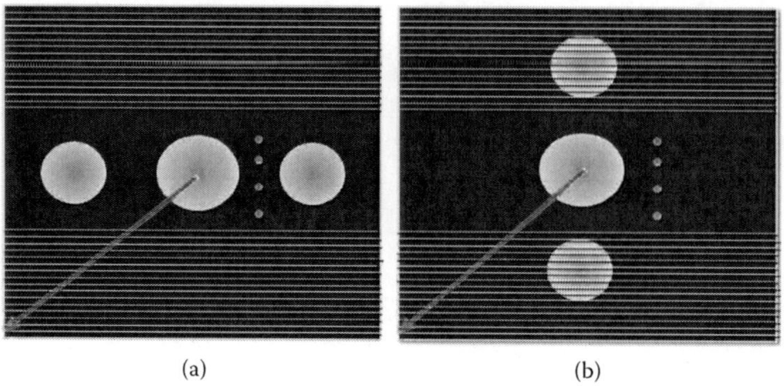

FIGURE 11.22 Full k-space acquisition for data acquisition in MR tagging. The arrow points to the k-space trajectory. (a) Tagging in horizontal direction and (b) tagging in vertical direction.

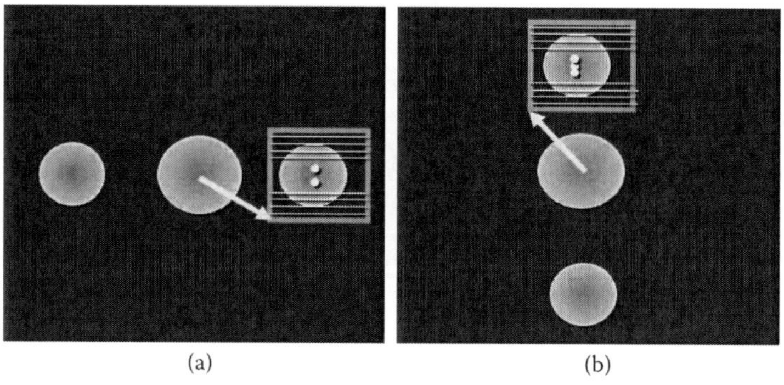

FIGURE 11.23 Partial k-space acquisition of off-center spectral peak. The arrow points to the k-space trajectory. (a) Tagging in horizontal direction and (b) tagging in vertical direction.

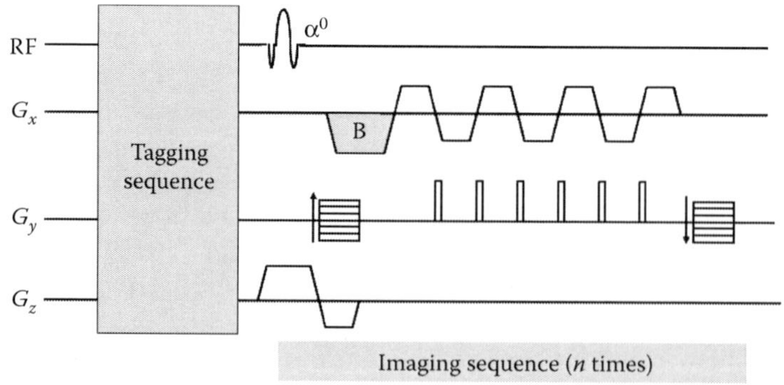

FIGURE 11.24 Timing diagram for partial k-space acquisition utilized in acquisition of images in Figure 11.25 [134]. A modified TFE-EPI sequence is used as the imaging sequence. Gradient B compensates for the dephasing of the gradients in the EPI read-out and the tagging gradient.

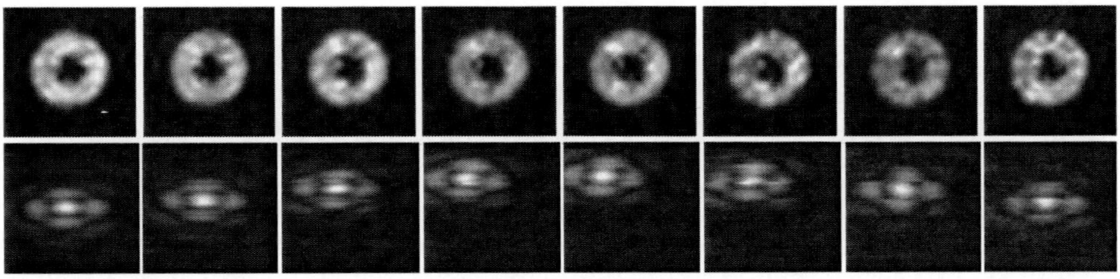

FIGURE 11.25 Eight temporal frames acquired with partial k-space acquisition using CSPAMM in the rotating phantom of Figure 11.15b [134]. Top row: magnitude images. Bottom row: corresponding k-space magnitude data. The imaging parameters were as follows: TR/TE = 7.2/3.8 ms, flip angle = 10°, turbo factor = 8, FOV = 300 × 300 mm^2, acquired voxel size = 5 × 5 mm^2, recon resolution = 5 × 5 mm^2, slice thickness = 8 mm, heart phases = 15, acquisition matrix = 60 × 56, EPI factor = 7, total scan duration = 2 s.

experiment. The top row displays the magnitude images which are blurred versions of the phantom anatomy. Since only the off-center peak is acquired, there are no tag lines visible on these images. The bottom row displays the corresponding k-space data. We can see clearly that the off-center peak rotates from the center to the upper-left corner and then back to the center, in conformance with the rotation of the moving phantom. The phases of the acquired data contain motion information and therefore can be utilized as part of frequency-based motion tracking techniques. Additionally, acquisition time decreases dramatically when compared to full k-space tagged MRI acquisition with CSPAMM.

Using the fast-HARP pulse sequence, real-time myocardial strains were computed in 2D by Abdelmoneim et al. [2]. By using chirp inverse Fourier transform, myocardial ROI was cropped from the raw k-space data so that image reconstruction and cardiac strain map computation could be accomplished in 11 ms. It was shown that fast-HARP can produce the same results as original HARP with a four times speedup.

By manipulating harmonic peaks, more desirable tag patterns from conventional SPAMM, CSPAMM, and fast-HARP tagged images can be synthesized [84]. In order to correct errors in harmonic phase images, Ryf et al. [111] employed both negative and positive peaks from CSPAMM to produce more accurate harmonic phase images. The main advantage of HARP is that it is fast and automatic with no need for extracting tag lines in a preprocessing step.

One limitation of HARP and fast-HARP is that they can only estimate the in-plane motion, typically using 2D SA tagged MR images. Several solutions for including the out-plane component of motion have been proposed. Pan et al. [88] extended the traditional HARP method for 3D cardiac motion tracking. A stack of SA images with tagging grid as well as LA images with one horizontal set of tags were used. Similar to 2D HARP, phase invariant condition was used to track the motion. A material mesh model was used for interpolation from sparse phase information. In this study, only a single layer mesh inside the myocardium was created and tracked according to 3D phase invariant condition. The use of CSPAMM led to improved myocardial strain maps with HARP analysis which was achieved through the suppression of the center peak [64]. Using a true 3D tagging technique followed by a conventional fast 3D gradient echo imaging sequence, Ryf et al. [110] extended HARP from 2D to 3D. The phase invariant property in three directions was utilized after extracting spectral peaks in 3D Fourier space. True 3D motion can be captured, but the scan time was quite long and demanded a sophisticated breath-holding technique. Rutz et al. proposed a fast method for acquiring 3D CSPAMM data using localized tagging preparation and a hybrid multishot, segmented echo planar imaging sequence [109]. Abd-Elmoniem et al. [1] developed a new method, zHARP pulse sequence, which can track both in-plane and through-plane motion from a single image plane. It is similar to slice-following CSPAMM [43] and combines tagged MRI with through-plane displacement phase encoding without affecting the image acquisition time. A z-encoding gradient is added to the refocusing lobe of the slice-selection gradient and produces a phase term depending on the displacement in through-plane direction in the final data. From two or more images in one orientation, 3D strain can be calculated [3]. zHARP can compute 3D strain tensor over the whole LV with no spatial interpolation. Another method to obtain through-plane strain as well as in-plane strain is to combine fast-HARP with SENC which will be discussed later [113]. Using the HARP-SENC pulse sequence, a single slice of SA image can be acquired in six heartbeats and multiple slices can be acquired in a sequential fashion.

11.2.2.5.2 Local Sine Wave Modeling

SinMod is also a frequency-based method to analyze the heart displacement and deformation from tagged MRI sequences using phase information [14]. The main difference between SinMod and HARP is that SinMod detects both local spatial phase shift and local spatial frequency from band-pass filtered images, while HARP uses the phase invariant condition and tracks local spatial phase. The speed of SinMod method is as fast as HARP but SinMod method has advantages in accuracy, noise reduction, and lack of artifacts [14]. In SinMod, the intensity distribution around each pixel (p, q) is modeled as a cosine wave front:

$$
\begin{aligned}
I_1(p,q) &= A_1 \cos\left(\omega_p\left(p + \frac{u}{2}\right) + \varphi\right) + n_1(p,q) \\
I_2(p,q) &= A_2 \cos\left(\omega_p\left(p - \frac{u}{2}\right) + \varphi\right) + n_2(p,q)
\end{aligned}
\tag{11.11}
$$

where ω_p and φ are the spatial frequency and phase of the wave, respectively. A_1 and A_2 are wave magnitudes for the first image I_1 and the second image I_2, while n_1 and n_2 are additive noise. u is the displacement between these two images at position (p, q) along the p direction.

The flow chart of SinMod algorithm is shown in Figure 11.26. The principle behind SinMod tracking is that both phase and frequency for each pixel are determined directly from the frequency analysis and the displacement is calculated from the quotient of phase difference and local frequency. After obtaining the Fourier Transform of the input images $I_1(p, q)$ and $I_2(p,q)$ (temporal frames at time t and time $t + 1$), the same band-pass filter is applied to both (similar to HARP) isolate corresponding spectral peaks and produce a pair of complex images in the Fourier domain (since the Fourier Transform will be complex to begin with). Let us refer to the two complex images in Fourier domain following band-pass filtering $I_{bf1}(\omega_p, \omega_q)$ and $I_{bf2}(\omega_p, \omega_q)$. Applying a low-frequency band-pass filter and a high-frequency one to both I_{bf1} and I_{bf2} followed by an inverse Fourier transform leads to four complex images $I_{bfLf1}(p,q)$, $I_{bfHf1}(p,q)$, $I_{bfHf2}(p,q)$, and $I_{bfHf2}(p,q)$. The reasoning behind application of an LPF and a HPF to I_{bf1} and I_{bf2} is to determine the local spatial frequency by power spectra. Then the displacement is the local quotient of

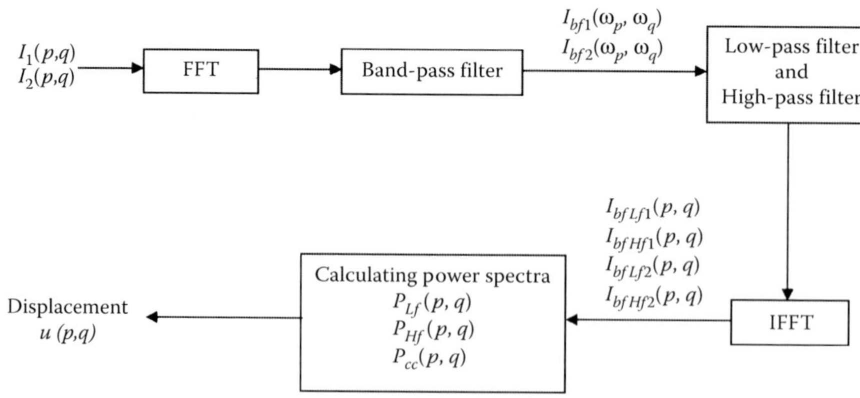

FIGURE 11.26 Flow chart for the SinMod method.

phase difference and local frequency at that position. The power spectra and cross power spectrum are given by

$$\begin{aligned} P_{Lf}(p,q) &= | I_{bfLf1} |^2 + | I_{bfLf2} |^2 \\ P_{Hf}(p,q) &= | I_{bfHf1} |^2 + | I_{bfHf2} |^2 \\ P_{cc}(p,q) &= I_{bfLf1}\overline{I}_{bfLf2} + I_{bfHf1}\overline{I}_{bfHf2} \end{aligned} \tag{11.12}$$

where $\overline{I}$ is the complex conjugation of I.

The local frequency ω_p and local displacement u can then be estimated from

$$\omega_p(p,q) = \omega_c\sqrt{\frac{P_{Hf}}{P_{Lf}}} \qquad u(p,q) = \frac{\arg(P_{cc})}{\omega_p} \tag{11.13}$$

where ω_c is the band-pass center-frequency.

11.2.2.5.3 Gabor Filter Banks

Gabor filter is a band-pass filter, with the form of a Gaussian multiplied by a complex sinusoid in the spatial domain, or equivalently, a shifted Gaussian in the spatial frequency domain. By choosing appropriate parameters, the magnitude response of the Gabor filter can be used to remove tags in the myocardium, and the phase response can be used to track tags. Once again, utilizing phase-based optical flow, the phase response of Gabor filter bank for a material point is assumed to be constant during the deformation [27].

A Gabor filter bank is a set of 2D Gabor filters with tunable parameters that represent the variable spacing and orientation of tag lines. By finding optimal parameters that maximize the Gabor filter response, Montillo et al. [77] extracted deformation information in a 2D simulated model. Qian et al. [101] extended the 2D Gabor filter bank method to 3D in order to extract and track deformed tag surfaces. Chen et al. [27] used the phase information from Gabor filters to track tag lines. To increase the accuracy of tag tracking, a combination of the response of Gabor filters, gradient information of original images, an intensity probabilistic model, and a spatiotemporal smoothness constraint was used in the deformable model. Chen et al. [29] proposed a three-step process for 3D cardiac motion tracking: extract tag intersections based on local phase analysis using Gabor filter bank, robust point matching (RPM) method to track the intersections movement, and meshless deformable modeling to generate a dense displacement field.

Instead of calculating the strain values in terms of gradient of the displacement by tracking the tag pattern, Qian et al. [102] developed a nontracking-based strain estimation method for tagged MRI. It is based on the extraction of tag's deformation gradient using Gabor filter, instead of tracking displacement prior to strain calculation. Subsequently, 2D strains can be obtained by using the strain formula in terms of the deformation gradient tensor.

11.3 Phase Contrast MRI

Another approach for cardiac motion tracking is through phase contrast MR imaging (PCMRI) of myocardium. PCMRI is based on the idea that the motion of tissue on blood produces a change in the MR signal phase that is proportion to the velocity of the tissue [96]. Therefore, PCMRI is also called velocity-encoded MRI.

11.3.1 Phase Contrast MR Imaging

Phase contrast methods are based on the fact that moving spins accumulate a different phase offset than static spins in the presence of a magnetic field gradient. The phase shift is proportional to the product

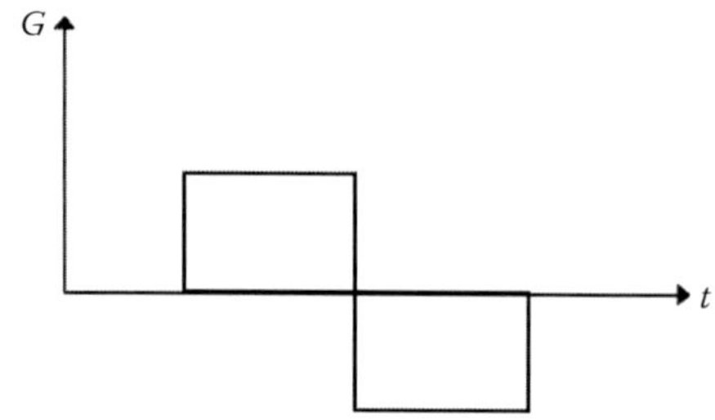

FIGURE 11.27 A bipolar gradient lobe.

of tissue velocity and the first moment of the magnetic field gradient in the direction of motion (see Equation 11.16). PCMRI sequence uses a bipolar gradient as shown in Figure 11.27, which is first positive at a particular amplitude and duration and then negative for an equal amplitude and duration to encode the velocity. For static spins, the accumulated phase under the bipolar gradient is zero. However, for spins that move at a constant velocity, there is an accumulated phase shift. The accumulated phase at time TE is

$$\Phi = \gamma \int_0^{TE} G(t) r(t) dt \tag{11.14}$$

where $G(t)$ is the gradient, $r(t)$ is the position in the direction of $G(t)$, and γ is the gyromagnetic ratio. The Taylor expansion of position vector $r(t)$ on initial position r_0 is

$$r(t) = r_0 + vt + a\frac{t^2}{2}\cdots \tag{11.15}$$

Under the condition that there are no phase shifts from initial position r_0 (which is the case for bipolar gradient) and the phase shifts due to higher-order terms are small, Equation 11.14 becomes

$$\Phi = v(\gamma M_1) \tag{11.16}$$

where M_1 is the first moment of the gradient waveform evaluated at TE.

$$M_1 = \int_0^{TE} G(t) t dt \tag{11.17}$$

This forms the foundation for PCMRI. In deriving this expression, we have assumed that the velocity is constant during the time (0,TE).

Unfortunately, the phase can be affected by many other factors, including magnetic field inhomogeneity, tissue magnetic susceptibility, pulse sequence tuning, and eddy currents due to gradient switching. Therefore, to accurately extract the motion of interest, phase shift due to these sources have to be eliminated. In order to do that, the procedure in PCMRI consists of a reference scan without velocity encoding, followed by two or three further acquisitions with velocity encoding along different directions [95,96]. The reference scan is used to correct undesired phase offsets of the second acquisition with velocity encoding. Since the phase of each pixel is modulated by velocity, one can obtain functional data with maximum resolution, the same spatial and temporal resolution as the underlying anatomical images. Velocity information in PCMRI can be used directly to calculate strain rate. The motion of the myocardium (i.e., displacements) may be found by integration with respect to time. Gradient system performance is critical in accurately extracting velocity information with phase contrast methods.

11.3.2 Analysis Techniques for Phase Contrast MR Images

Meyer et al. [76] used a deformable mesh guided by a Kalman filter to obtain myocardial deformation. Zhu et al. [147] developed a Fourier tracking algorithm which modeled trajectories as composed of Fourier harmonics and integrated the velocity data in the frequency domain. Deformation gradient tensor and strain can be calculated using a least-squares fit of the trajectory data to a local deformation model [148]. The accuracy of the forward-backward [33] and Fourier tracking methods was analyzed using a computer-controlled deformable phantom in Reference 39. The Fourier tracking algorithm in Reference 147 was extended to 3D volumetric data analysis [151]. Compared to method in Reference 147, a dynamic spatiotemporal FEM model was proposed to calculate deformation by taking into account all velocity information from the entire analyzed region [150].

Masood et al. [73] presented a B-spline-based deformable model, named virtual tagging framework, to track cardiac motion and deformation from PCMRI. The intersections of virtual tag lines served as control points which were determined by minimizing the difference between calculated velocity field from B-spline-based deformable model and the velocity field measured from PCMRI. The mass conservation constraint was incorporated into the proposed framework to ensure that the derived strain complied with physical constraints. Selskog et al. [115] proposed a method to calculate 4D strain rate from 3D PCMRI data by weighting of a forward and a backward difference and a 3D strain tensor visualization method using ellipsoids where the three axes of the ellipsoid denoted the directions of the strain rate and the length of the axes denoted the magnitude. Bergvall et al. [19] fitted a spatiotemporal spline model to track cardiac motion and deformation from PCMRI data. It extended the spatiotemporal model first proposed in Reference 150 by using TPS and Navier splines as spatial elements.

Reese et al. [104] developed a single-shot 3D PCMRI data acquisition sequence by acquiring two contiguous 2D images in *k*-space in a single echo train of an EPI readout. Markl et al. [72] developed a fast 2D PCMRI protocol based on a black blood *k*-space segmented gradient echo imaging. The in-plane velocity information can be obtained within a single breath-hold measurement by choosing optimal number of *k*-space lines and view-sharing technique. Jung et al. [58] proposed a high-temporal-resolution phase contrast MRI acquisition method using a prospective respiratory gating technique with two navigators in each cardiac cycle. The usage of dual navigator gating and view sharing [71] increased the temporal resolution to 13.8 ms. This method was applied to investigate details in LV motion patterns [56,57]. A schematic visualization model and correlation analysis on myocardial velocities in different segments and slices were done to a large number of 58 normal subjects in three different age groups and one patient with dilated cardiomyopathy [46].

11.4 Displacement Encoding with Stimulated Echoes

DENSE is a quantitative imaging technique that encodes tissue displacement in the phase of the acquired signal. In order to overcome the low resolution limit of tagged MRI and bypass the integration required in PCMRI in order to obtain displacements, Aletras et al. [5] were the first to apply stimulated echo techniques for quantification of myocardial strain, named DENSE, which combines many of the advantages of both tagging and phase velocity mapping. DENSE is a high-resolution myocardial displacement mapping method. Similar to PCMRI, it uses a bipolar gradient to encode motion of spins into the phase of the MRI signal. Figure 11.28 shows a timing diagram of DENSE in readout direction. Using stimulated echo, the phase is modulated and demodulated by the gradients applied between the first two RF pulses G_1 and the one after the third RF pulse G_2. The displacements occurring during the mix time *TM* (time between the second and third RF pulses) is recorded in the phase of stimulated echo image. Similar to PCMRI, in order to eliminate the phase contributions from other sources, the sequence is repeated once more with a different gradient (shown in red dots in Figure 11.28) and subtracted from the first acquisition. In other words, the phase of DENSE images is proportional to the displacement, therefore strain can be obtained by computing the derivatives of the phase. This technique encodes motion over

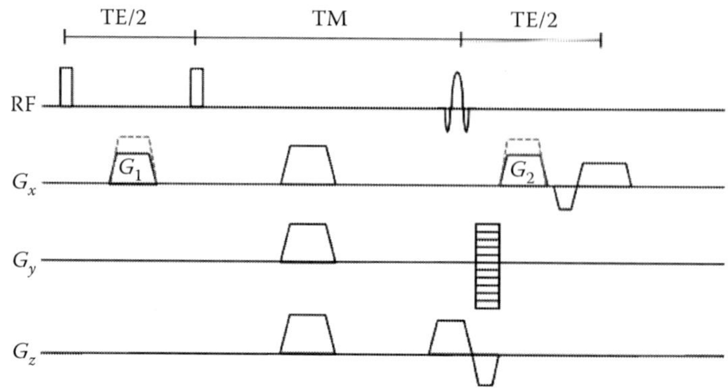

FIGURE 11.28 Timing diagram for DENSE in readout direction. Phase wrapping happens through the application of G_1, G_2 unwraps the phase. (From A. H. Aletras et al. *Journal of Magnetic Resonance*, 137(1):247–252, March 1999.)

long time intervals. The encoded displacements are large; therefore, improved phase contrast is obtained with moderate gradient strength. For more details of this technique, refer to Reference 5.

DENSE has also been used in detecting small focal regions of abnormal contraction in patients, in characterizing mouse ischemia models, and in tissue tracking of human hearts via 2D breath-hold imaging. In Reference 4, DENSE was combined with a segmented echo-planar imaging readout (fast-DENSE) and applied to normal volunteers. In order to improve SNR and reduce breath-hold time, mixed echo train acquisition DENSE (meta-DENSE) was proposed in [6]. Aletras et al. [7] accelerated the image acquisition process by incorporating SENSE into DENSE. This method can reduce the breath-hold time by a factor of two.

Initial DENSE sequence in Reference 5 imaged the heart at a single cardiac phase; Kim et al. [61] developed a cine DENSE imaging sequence to sample 2D DENSE images at multiple phases of the cardiac cycle. A SPAMM sequence was used to encode the magnetization at end-diastole, then the displacement was presented as the phase shift between position encoding and readout. Cine DENSE data provide a way to track every material point in myocardium through time in the cardiac cycle. In order to improve the SNR for cine DENSE [61], a signal-averaged DENSE (sav-DENSE) was proposed [60], which extracted a pair of DENSE images with uncorrelated noise from CSPAMM image, and combined them during image reconstruction. In order to decrease the phase shifts from sources other than the encoded displacement, a general n-dimensional balanced multipoint encoding strategy was used in DENSE [146]. Spatiotemporal phase unwrapping methods were used to process cine DENSE images in Reference 122. In Reference 123, slice following was incorporated into cine DENSE to track 3D cardiac motion. An adaptive phase-unwrapping and spatial filtering techniques were developed to process cine DENSE data and compared with conventional phase-unwrapping and spatial filtering techniques [137]. The novelty of this technique is that it incorporates the location of the myocardial wall into the quality map.

Since DENSE is based on stimulated echo, there is an inherent 50% signal loss. Although DENSE and HARP may seem to be quite different at first glance, these two techniques turn out to be different implementations of phase displacement encoding. A thorough comparison of these two techniques was given in Reference 63.

11.5 Strain Encoding MRI

Strain encoding, or SENC, is a relatively new MRI technique that can measure regional contraction, or relaxation, of the heart's myocardium. SENC pulse sequence was first developed by Osman et al. [85] to image longitudinal strain from short-axis images. SENC is built on the concept of MR tagging, but it uses tag planes parallel to the image plane. Strain is calculated from two images with different frequency modulation in the slice-selection direction. Longitudinal strain can be computed by measuring the local tag frequency components from SA images, while circumferential strain can be measured from the LA images. The unique advantage of SENC is that it directly encodes the regional strain of the heart into the acquired image without measuring the displacement or velocity first. Using SENC imaging, the existence of stiff masses can be detected in the acquired MR images without postprocessing [82]. Pan et al. [89] proposed a fast-SENC pulse sequence which can reduce the strain image acquisition time to one heartbeat by combining localized SENC, interleaved tuning, and spiral imaging. Youssef et al. [144] used SENC to estimate the circumferential strain of the RV free wall.

Ibrahim et al. [53] combined SENC with slice-following to correct for the through-plane motion in real time. SENC imaging with and without slice-following can result in significantly different strain values as expected with tagged MRI [53]. Ibrahim et al. [54] proposed a composite-SENC (C-SENC) pulse sequence which adjusted the tunings such that an additional image was acquired at zero frequency. Three consecutive SENC images can be obtained which contain both strain information and myocardial viability information. C-SENC produces three images: no-tuning (NT), low-tuning (LT), and high tuning (HT). Adding LT and HT images will result in an anatomical (ANAT) image of the myocardium. According to different intensity distribution in this NT-ANAT-HT 3D space, different tissue types can be classified using an unsupervised, multistage fuzzy clustering technique [55].

SENC is a 1D out-of-plane strain encoding method. In order to obtain 3D strain, Sampath et al. [113] integrated SENC with HARP, which can produce 2D in-plane strain. In this way, 3D strains can be obtained from a single slice of SA image. Hess et al. [50] combined DENSE and SENC to obtain 3D strain map in a single slice of the myocardium.

SENC has some limitations versus MR tagging and DENSE. First, it is not capable of tracking the motion of the heart. Because it only encodes the strain (contractility), the exact motion of the tissue cannot be followed. Therefore, it cannot be directly used to calculate quantities such as rotation and twisting. Second, SENC provides strain measurements only in the through-plane direction, and not the in-plane directions. The advantages of SENC include its simplicity, as images of strain can be directly produced in real time. It does not require image postprocessing. Also, SENC has higher in-plane resolution of strain than MR tagging, which can be very useful in assessing regional function of the RV free wall [65].

11.6 Strain Pattern Analysis

An important application of MRI motion-imaging methods such as tagging is in providing a better definition of the normal patterns of motion. Similarly, the effects of heart diseases such as hypertrophy, infarction, failure, and dyssynchrony can be well characterized and quantified with tagging.

Regional myocardial strains have direct or indirect relationship with cardiac diseases. Most of the motion tracking methods aim to extract strains in the heart, for the reason that strain encapsulates the basic mechanical function of the myocardium and has clinical potential. There are different types of strains. Normal strain is defined as the ratio of the length of the deformed line element to its original length. Due to the geometry of the ventricle, normal strains are usually calculated in radial, circumferential, and longitudinal directions. Shear strain is defined as the angular change between any two originally mutually orthogonal line elements that occurs as a result of a continuous deformation. These can be defined in terms of the off-diagonal terms of the Lagrangian strain tensor [12].

In order to analyze regional strain patterns in the myocardium, the LV and RV are typically divided into segments and each of the directional strains (e.g., radial, circumferential, and/or longitudinal) is averaged over the segment. Ideally, in 3D, the 16 segment and 17 segment models as recommended by the American Society of Echocardiography should be used [24]. In the 16 segment model, the LV is divided into six basal, six mid-ventricular, and four apical regions. In the 17 segment model, the apex comprises the 17th segment [35,78,79,88,126,127,139]. In 3D, the RV can be divided into three layers in the long-axis direction and each layer further divided into three segments [62,126]. However other finer

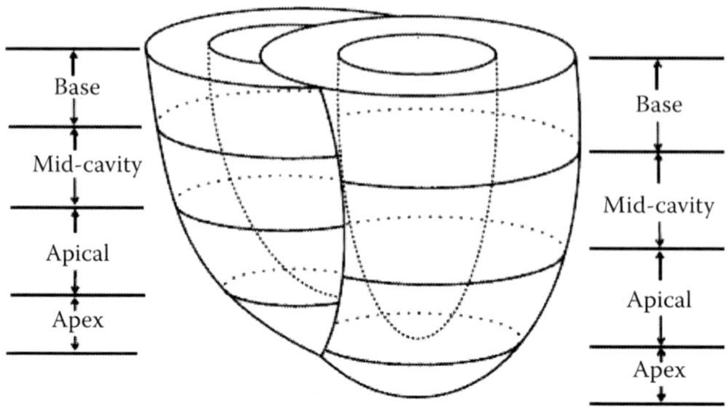

FIGURE 11.29 Geometrical representation of the regional divisions for strain analysis for both the left and right ventricles in the LA view. (From N. J. Tustison. *Biventricular Myocardial Strains with Anatomical NURBS Models from Tagged MRI.* PhD thesis, Washington University in St. Louis, August 2004.)

or sparser partitions have also been utilized. For 2D analysis, typically, mid-ventricular strains for four to eight regions are reported [102,112–114].

In Reference 125, the division along the long-axis consisted of four layers: basal, mid-cavity, apical, and apex as shown in Figure 11.29. Based on the location of the LV/RV junctions, the basal and mid-cavity portions of the left ventricle were each further divided into six regions in the short-axis view: antero-septal (AS), anterior (A), lateral (L), posterior (P), inferior (I), and inferoseptal (IS) as shown in Figure 11.30. Similar to the left ventricle, the right ventricle was divided into basal, mid-cavity, and apical layers and each layer was further divided into anterior, mid, and inferior regions. Tustison et al. [125] calculated both the Eulerian and Lagrangian strain values for LV data sets from three species: canine data, human data, and porcine data. The strain values from the RV for two canine data sets were also included. Results showed that across normal species, radial strains remain positive for most of the regions indicative of systolic thickening of the left ventricle while the circumferential and the longitudinal strains are negative. Circumferential shortening during left ventricular contraction results in the negative strain values in the circumferential direction while compression in the longitudinal direction results in negative longitudinal strains.

Strain analysis can also be done based on extracted model parameter functions [92] where quantitative differences between normal and abnormal hearts with hypertrophic cardiomyopathy (htcm) were extracted. Park et al. [92] showed that the contraction and the twisting deformation are more significant

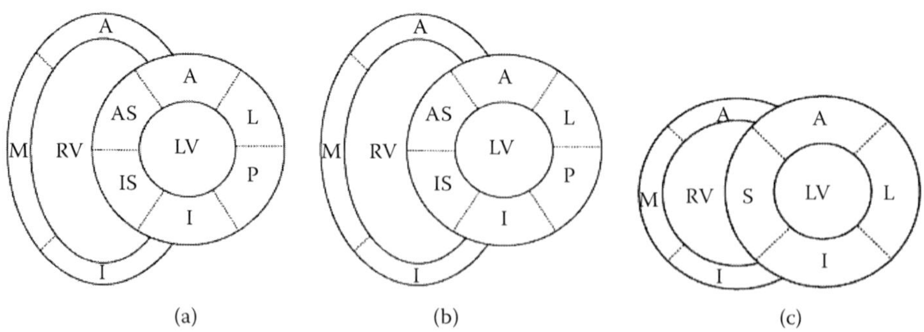

FIGURE 11.30 Geometrical representation of the regional divisions for strain analysis for both the left and right ventricles in the SA view. Illustrated are (a) the basal region, (b) the mid-cavity region, and (c) the apical region. (From N. J. Tustison. *Biventricular Myocardial Strains with Anatomical NURBS Models from Tagged MRI.* PhD thesis, Washington University in St. Louis, August 2004.)

on the endocardial wall compared to the epicardial wall. Due to the complexity of RV geometry, Haber et al. [49] utilized coordinate-system-independent principal strains and directions to describe regional deformations, which were derived by calculating the eigenvalues and eigenvectors of the Lagrangian strain tensor. The strains in normal hearts and the hearts with right ventricular hypertrophy (RVH) diseases were compared. Maximum and minimum principal strains were used to quantify normal sheep left ventricular functions [139].

Normal strains have been widely used, in radial, circumferential, and longitudinal directions, though the circumferential strain values have been reported most [29,83,88,112,126,127]. In Reference 114, circumferential strains were computed, not for every material points, but based on the percent change in the distance between every set of two adjacent markers with respect to the reference time frame. In [113], radial, circumferential, and longitudinal strains (in-plane and out-of-plane strains) were obtained simultaneously within a mid-ventricular SA slice. In Reference 35, circumferential, radial, and longitudinal strains for a normal volunteer and a patient with an antero-septal myocardial infarction were computed. In Reference 14, circumferential strains of the LV in five slices from apex to base and the mid-wall circumferential strain for a patient with LBBB and infarction in the lateral LV-free wall were investigated.

Moore et al. [78,79] reported 3D displacement and strain evolution of the normal LV in humans determined from tagged MRI. Such data can serve as a database of strain in normal hearts. Clarysse et al. [32] proposed a framework for analysis of spatiotemporal deformation parameters of the myocardium throughout the systolic phase from tagged MR data. After strain calculation, a functional data analysis method was applied to healthy subjects, which provided a normal contraction reference model. Pathological cases were then compared with the reference model. Statistical analysis of regional wall motion performance parameters in normals relative to disease processes was done in Reference 142. After the deformation was reconstructed based on FEM models fitted to tagged MR images [143], the deformation of LV can be decomposed into separate deformation modes such as longitudinal shortening, wall thickening, and twisting [106]. By doing this, healthy and diseased cardiac deformation patterns can be distinguished.

Cupps et al. [34] estimated LV wall stress distribution at rest in patients with chronic aortic insufficiency and normal ejection fraction within a FEM formulation. Differences in regional LV systolic stress in normal subjects and patients with aortic insufficiency were given. Moustakidis et al. [80] estimated 3D strain distribution at rest and with inotropic stimulation in patients with ischemic cardiomyopathy using tagged MR imaging and FE analysis. Similar studies were performed in a group of normal volunteers for comparison. Further study showed that circumferential strains at rest and with low-dose dobutamine could discriminate viable and nonviable myocardium in patients with ischemic cardiomyopathy [22]. Maniar et al. [70] compared circumferential strain in patients before and after 3 months after coronary artery bypass grafting using tagged MRI. Strain analysis from dobutamine-stressed tagged MRI was shown to be feasible to quantitatively predict myocardial recoverability after coronary artery bypass grafting [99]. Chen et al. [29] assessed the differences in myocardial strain distributions between hearts with left ventricular hypertrophy and normal hearts. They calculated statistics of 17 different variables related to the circumferential strain distribution in the LVH patients and the normal control.

11.7 Conclusions and Research Trends

In this chapter, we reviewed different imaging and image analysis methods to evaluate cardiac function from MRI. Cine MRI is critical in imaging the anatomy of the heart and calculating global indices, but with limited ability to track local motion. Traditional cardiac motion imaging pulse sequences, that is, tagged MRI, PCMRI, contain more information about the regional myocardial motion than cine MRI. It is felt that tagged MRI will continue to play an important role in cardiac motion analysis. Among different tagging methods, SPAMM is the most commonly used sequence. Given the current advances in gradient technology, image reconstruction techniques, and data analysis algorithms, MR tagging has become the reference modality for calculating strain in the heart [116]. Data in tagged MRI provides information with respect to muscle deformation, but suffers from low spatial resolution. Spatial resolution of phase contrast MR images is 1–3 mm, while the distance between tags in tagged MRI is around 4–8 mm [57]. In recent years, several new pulse sequences have been proposed to image cardiac motion, such as HARP, DENSE, and SENC, which have characteristics of high spatial resolution, and direct encoding of displacement or strain. HARP is largely limited to 2D in-plane strain computation, while SENC is for 1D through-plane strain calculation. A comparison between DENSE and HARP imaging may be found in [63]. Both SENC and HARP use tagging. The difference is the tagging in HARP is orthogonal to the image plane, while the tagging in SENC is parallel to the image plane. Thanks to the fast tagging and imaging method for cardiac motion tracking, deformation throughout the entire cardiac cycle can be tracked, instead of only systole or diastole.

The motion of heart is 3D and nonrigid. There are several strategies to capture its 3D motion. One way is to use a stack of SA images. Another one is to use multiview planar images, both in SA and in LA directions. For tagged MRI, the tag planes only provide information about motion in the direction perpendicular to the initial tag planes, three mutually perpendicular tag planes are necessary. For the fact that MR imaging plane is fixed, pixels in 2D images are not always the same material points in myocardium. Slice-following technique can overcome this limitation. Slice-following images a thicker slice which encompasses a thinner tagged slice to ensure imaging the same tagged slice throughout the cardiac cycle, despite tissue displacement in the through-plane direction. Slice-following (SF) technique is also used in other sequences, for example, SF-SENC [53] and SF-DENSE [123]. Another way is to use 3D imaging and analysis approaches to recover true motion [110].

Cardiac analysis has focused mostly on the myocardium of the left ventricle, since LV, as the primary heart chamber, controls systemic perfusion and is responsible for pumping blood throughout the entire body. Compared to LV, RV has thinner wall and more complex geometry. Besides, RV can have a large through-plane motion near the base of the ventricle, about 15 mm, which is more than two short-axis image planes [49]. All these factors increase the difficulty in motion analysis. But in recent years, RV function has also been investigated [21,126].

Better understanding of pathological mechanism of the heart can be achieved due to increasing performance of medical imaging techniques. Strain is considered to have more potential for clinical diagnosis. Researchers have already investigated alterations of strain patterns in diseased populations. Most analysis methods end up obtaining strains in different directions. Most commonly reported are radial, circumferential, and longitudinal strains. After strains are obtained, strain analysis approaches are emerging.

The trends in cardiac MR imaging and image analysis methods are as follows:

- Phase- and frequency-based methods, for example, SinMod [14], HARP [83]
- Integration of frequency-based and spatial analysis for improved measurement of myocardial strains [28,132,133]
- Direct displacement or strain imaging, for example, DENSE (Section 11.4) and SENC (Section 11.5)
- Real-time imaging (one heartbeat), for example, fast-HARP [112] and fast-SENC [89]
- Approaches to analysis of both LV and RV [21,51,126]
- 3D tagging and imaging [110]
- High resolution PCMRI [46,56]
- Clinically useful strain analysis methods [22,29,32,99,106,126,127,142]

The literature on cardiac motion analysis based on MRI has grown rapidly in the past decade. Cardiac motion analysis is continuing to be a hot and challenging research area. Research efforts on cardiac motion analysis are devoted to quantitative methods for automatic, fast, and robust image analysis. The trend in image analysis is toward phase-based methods. With the development of more advanced MR imaging techniques and progress in hardware, research in this area will continue to grow.

Acknowledgments

We are grateful to the following individuals for their help in various aspects of this work: Melanie Kotys and Stefan Fischer for useful discussions, and Mehmet Ersoy and Randy Setser for providing the cardiac phantom used in Section 11.2.1.3.

References

1. K. Z. Abd-Elmoniem, N. F. Osman, J. L. Prince, and M. Stuber. Three-dimensional magnetic resonance myocardial motion tracking from a single image plane. *Magnetic Resonance in Medicine*, 58(1):92–102, July 2007.
2. K. Z. Abd-Elmoniem, S. Sampath, N. F. Osman, and J. L. Prince. Real-time monitoring of cardiac regional function using fastHARP MRI and region-of-interest reconstruction. *IEEE Transactions on Biomedical Engineering*, 54(9):1650–1656, September 2007.
3. K. Z. Abd-Elmoniem, M. Stuber, and J. L. Prince. Direct three-dimensional myocardial strain tensor quantification and tracking using zHARP. *Medical Image Analysis*, 12(6):778–786, December 2008.
4. A. H. Aletras, R. S. Balaban, and W. Han. High-resolution strain analysis of the human heart with fast-DENSE. *Journal of Magnetic Resonance*, 140(1):41–57, September 1999.
5. A. H. Aletras, S. Ding, R. S. Balaban, and W. Han. DENSE: Displacement encoding with stimulated echoes in cardiac functional MRI. *Journal of Magnetic Resonance*, 137(1):247–252, March 1999.
6. A. H. Aletras and W. Han. Mixed echo train acquisition displacement encoding with stimulated echoes: An optimized DENSE method for *in vivo* functional imaging of the human heart. *Magnetic Resonance in Medicine*, 46(3):523–534, September 2001.
7. A. H. Aletras, W. P. Ingkanisorn, C. Mancini, and A. E. Arai. DENSE with SENSE. *Journal of Magnetic Resonance*, 176(1):99–106, September 2005.
8. American Heart Association. Heart disease and stroke statistics—2009 update (at-a-glance version). http://www.americanheart.org/presenter.jhtml?identifier=3037327.
9. A. A. Amini, Y. Chen, R. W. Curwen, V. Mani, and J. Sun. Coupled B-snake grids and constrained thin-plate splines for analysis of 2-D tissue deformations from tagged MRI. *IEEE Transactions on Medical Imaging*, 17(3):344–356, June 1998.
10. A. A. Amini, Y. Chen, M. Elayyadi, and P. Radeva. Tag surface reconstruction and tracking of myocardial beads from SPAMM-MRI with parametric B-spline surfaces. *IEEE Transactions on Medical Imaging*, 20(2):94–103, February 2001.
11. A. A. Amini and J. S. Duncan. Bending and stretching models for LV wall motion analysis from curves and surfaces. *Image and Vision Computing*, 10(6):418–430, July–August 1992.
12. A. A. Amini and J. L. Prince. *Measurement of Cardiac Deformations from MRI: Physical and Mathematical Models*. Kluwer Academic Publishers, Dordrecht, The Netherlands, 2001.
13. T. Arts, W. C. Hunter, A. Douglas, A. M. Muijtjens, and R. S. Reneman. Description of the deformation of the left ventricle by a kinematic model. *Journal of Biomechanics*, 25(10):1119–1127, 1992.
14. T. Arts, F. W. Prinzen, T. Delhaas, J. Milles, A. Rossi, and P. Clarysse. Mapping displacement and deformation of the heart with local sine wave modeling. *IEEE Transactions on Medical Imaging*, 29(5):1114–1123, May 2010.
15. D. J. Atkinson and R. R. Edelman. Cineangiography of the heart in a single breath hold with a segmented turboFLASH sequence. *Radiology*, 178(2):357–360, February 1991.
16. L. Axel and L. Dougherty. Heart wall motion: Improved method of spatial modulation of magnetization for MR imaging. *Radiology*, 172(2):349–350, 1989.
17. L. Axel and L. Dougherty. MR imaging of motion with spatial modulation of magnetization. *Radiology*, 171(3):841–845, June 1989.
18. L. Axel, A. Montillo, and D. Kim. Tagged magnetic resonance imaging of the heart: A survey. *Medical Image Analysis*, 9(4):376–393, August 2005.

19. E. Bergvall, E. Hedstrom, K. M. Bloch, H. Arheden, and G. Sparr. Spline-based cardiac motion tracking using velocity-encoded magnetic resonance imaging. *IEEE Transactions on Medical Imaging*, 27(8):1045–1053, August 2008.
20. A. Bistoquet, J. Oshinshi, and O. Skrinjar. Left ventricular deformation recovery from cine MRI using an incompressible model. *IEEE Transactions on Medical Imaging*, 26(9):1136–1153, September 2007.
21. A. Bistoquet, J. Oshinshi, and O. Skrinjar. Myocardial deformation recovery from cine MRI using a nearly incompressible biventricular model. *Medical Image Analysis*, 12(1):69–85, February 2008.
22. D. Bree, J. R. Wollmuth, B. P. Cupps, M. D. Krock, A. Howells, J. Rogers, N. Moazami, and M. K Pasque. Low-dose dobutamine tissue-tagged MRI with 3D strain analysis allows assessment of myocardial viability in patients with ischemic cardiomyopathy. *Circulation*, 114(1 Suppl):I33–I36, July 2006.
23. E. Castillo, A. C. Lima, and D. A. Bluemke. Regional myocardial function: Advances in MR imaging and analysis. *Radiographics*, 23:S127–S140, October 2003.
24. M. D. Cerqueira, N. J. Weissman, V. Dilsizian, A. K. Jacobs, S. Kaul, W. K. Laskey, D. J. Pennell, J. A. Rumberger, T. Ryan, and M. S. Verani. Standardized myocardial segmentation and nomenclature for tomographic imaging of the heart: A statement for healthcare professionals from the cardiac imaging committee of the council on clinical cardiology of the American Heart Association. *Circulation*, 105:539–542, 2002.
25. R. Chandrashekara, R. H. Mohiaddin, and D. Rueckert. Analysis of 3-D myocardial motion in tagged MR images using nonrigid image registration. *IEEE Transactions on Medical Imaging*, 23(10):1245–1250, October 2004.
26. Y. Chen and A. A. Amini. A MAP framework for tag line detection in SPAMM data using Markov random fields on the B-spline solid. *IEEE Transactions on Medical Imaging*, 21(9):1110–1122, September 2002.
27. T. Chen and L. Axel. Using Gabor filter banks and temporal-spatial constraints to compute 3D myocardium strain. In *Proceedings of IEEE 2006 International Conference of the Engineering in Medicine and Biology Society*, Volume 1, pp. 4755–4758, August 2006.
28. J. Chen, N. J. Tustison, and A. A. Amini. Accurate recovery of 4-D left ventricular deformations using volumetric B-splines incorporating phase based displacement estimates. In *Proceedings of SPIE Medical Imaging*, Volume 6143, February 2006.
29. T. Chen, X. Wang, S. Chung, D. Metaxas, and L. Axel. Automated 3D motion tracking using Gabor filter bank, robust point matching, and deformable models. *IEEE Transactions on Medical Imaging*, 29(1):1–11, January 2010.
30. Y. Chenoune, E. Delechelle, E. Petit, T. Goissen, J. Garot, and A. Rahmouni. Segmentation of cardiac cine-MR images and myocardial deformation assessment using level set methods. *Computerized Medical Imaging and Graphics*, 29(8):607–616, December 2005.
31. P. Clarysse, C. Basset, L. Khouas, P. Croisille, D. Friboulet, C. Odet, and I. E. Magnin. Two-dimensional spatial and temporal displacement and deformation field fitting from cardiac magnetic resonance tagging. *Medical Image Analysis*, 4(3):253–268, September 2000.
32. P. Clarysse, M. Han, P. Croisille, and I. E. Magnin. Exploratory analysis of the spatio-temporal deformation of the myocardium during systole from tagged MRI. *IEEE Transactions on Biomedical Engineering*, 49(11):1328–1339, November 2002.
33. R. T. Constable, K. M. Rath, A. J. Sinusas, and J. C. Gore. Development and evaluation of tracking algorithms for cardiac wall motion analysis using phase velocity MR imaging. *Magnetic Resonance in Medicine*, 32(1):33–42, June 1994.
34. B. P. Cupps, P. Moustakidis, B. J. Pomerantz, G. Vedala, R. P. Scheri, N. T. Kou-choukos, V. G. Davila-Roman, and M. K. Pasque. Severe aortic insufficiency and normal systolic function: Determining regional left ventricular wall stress by finite-element analysis. *Annals of Thoracic Surgery*, 76(3):668–675, September 2003.
35. X. Deng and T. S. Denney, Jr. Three-dimensional myocardial strain reconstruction from tagged MRI using a cylindrical B-spline model. *IEEE Transactions on Medical Imaging*, 23(7):861–867, July 2004.

36. X. Deng and T. S. Denney, Jr. Combined tag tracking and strain reconstruction from tagged cardiac MR images without user-defined myocardial contours. *Journal of Magnetic Resonance Imaging*, 21(1):12–22, January 2005.
37. T. S. Denney, Jr. Estimation and detection of myocardial tags in MR image without user-defined myocardial contours. *IEEE Transactions on Medical Imaging*, 18(4):330–344, April 1999.
38. L. Dougherty, J. C. Asmuth, A. S. Blom, L. Axel, and R. Kumar. Validation of an optical flow method for tag displacement estimation. *IEEE Transactions on Medical Imaging*, 18(4):359–363, April 1999.
39. M. Drangova, B. Bowman, Y. Zhu, and N. J. Pelc. *In vitro* verification of myocardial motion tracking from phase-contrast velocity data. *Magnetic Resonance Imaging*, 16(8):863–870, October 1998.
40. F. H. Epstein. MRI of left ventricular function. *Journal of Nuclear Cardiology*, 14(5):729–744, September/October 2007.
41. M. Ersoy, M. Kotys, X. Zhou, and R. M. Setser. A left ventricular motion phantom for cardiac MRI. In *BMES 2010 Annual Meeting*, October 2010.
42. S. E. Fischer, G. C. McKinnon, S. E. Maier, and P. Boesiger. Improved myocardial tagging contrast. *Magnetic Resonance in Medicine*, 30(2):191–200, August 1993.
43. S. E. Fischer, G. C. McKinnon, M. B. Scheidegger, W. Prins, D. Maier, and P. Boesiger. True myocardial motion tracking. *Magnetic Resonance in Medicine*, 31(4):401–413, April 1994.
44. D. Fleet and A. Jepson. Computation of component image velocity from local phase information. *International Journal of Computer Vision*, 5:77–104, 1990.
45. L. Florack, H. V. Assen, and A. Suinesiaputra. Dense multiscale motion extraction from cardiac cine MR tagging using HARP technology. In *ICCV Workshop on MMBIA*, pp. 372–375, October 2007.
46. D. Föll, B. Jung, F. Stachle, E. Schilli, C. Bode, J. Hennig, and M. Markl. Visualization of multidirectional regional left ventricular dynamics by high-temporal-resolution tissue phase mapping. *Journal of Magnetic Resonance Imaging*, 29(5):1043–1052, May 2009.
47. J. B. Garrison, W. L. Ebert, R. E. Jenkins, S. M. Yionoulis, H. Malcom, G. A. Heyler, A. A. Shoukas, W. L. Maughan, and K. Sagawa. Measurement of three-dimensional positions and motions of large numbers of spherical radiopaque markers from biplane cineradiograms. *Computers and Biomedical Research*, 15(1):76–96, February 1982.
48. M. A. Guttman, J. L. Prince, and E. R. McVeigh. Tag and contour detection in tagged MR images of the left ventricle. *IEEE Transactions on Medical Imaging*, 13(1):74–88, 1994.
49. I. Haber, D. Metaxas, and L. Axel. Three-dimensional motion reconstruction and analysis of the right ventricle using tagged MRI. *Medical Image Analysis*, 4(4):335–355, December 2000.
50. A. T. Hess, X. Zhong, B. S. Spottiswoode, F. H. Epstein, and E. M. Meintjes. Myocardial 3D strain calculation by combining cine displacement encoding with stimulated echoes (DENSE) and cine strain encoding (SENC) imaging. *Magnetic Resonance in Medicine*, 62(1):77–84, July 2009.
51. Z. Hu, D. Metaxas, and L. Axel. *In vivo* strain and stress estimation of the heart left and right ventricles from MRI images. *Medical Image Analysis*, 7(4):435–444, December 2003.
52. J. Huang, D. Abendschein, V. G. Davila-Román, and A. A. Amini. Spatio-temporal tracking of myocardial deformations with a 4-D B-spline model from tagged MRI. *IEEE Transactions on Medical Imaging*, 18(10):957–972, October 1999.
53. E. H. Ibrahim, M. Stuber, A. S. Fahmy, K. Z. Abd-Elmoniem, T. Sasano, M. R. Abraham, and N. F. Osman. Real-time MR imaging of myocardial regional function using strain-encoding (SENC) with tissue through-plane motion tracking. *Journal of Magnetic Resonance Imaging*, 26(6):1461–1470, December 2007.
54. E. H. Ibrahim, M. Stuber, D. L. Kraitchman, R. G. Weiss, and N. F. Osman. Combined functional and viability cardiac MR imaging in a single breathhold. *Magnetic Resonance in Medicine*, 58(4):843–849, October 2007.
55. E. H. Ibrahim, R. G. Weiss, M. Stuber, A. E. Spooner, and N. F. Osman. Identification of different heart tissues from MRI C-SENC images using an unsupervised multi-stage fuzzy clustering technique. *Journal of Magnetic Resonance Imaging*, 28(2):519–526, August 2008.

56. B. Jung, D. Fooll, P. Boottler, S. Petersen, J. Hennig, and M. Markl. Detailed analysis of myocardial motion in volunteers and patients using high-temporal-resolution MR tissue phase mapping. *Journal of Magnetic Resonance Imaging*, 24(5):1033–1039, November 2006.
57. B. Jung, M. Markl, D. Föll, and J. Hennig. Investigating myocardial motion by MRI using tissue phase mapping. *European Journal of Cardio-Thoracic Surgery*, 29S(1):S150–S157, April 2006.
58. B. Jung, M. Zaitsev, J. Hennig, and M. Markl. Navigator gated high temporal resolution tissue phase mapping of myocardial motion. *Magnetic Resonance in Medicine*, 55(4):937–942, April 2006.
59. W. S. Kerwin and J. L. Prince. Cardiac material markers from tagged MR images. *Medical Image Analysis*, 2(4):339–353, December 1998.
60. D. Kim, F. H. Epstein, W. D. Gilson, and L. Axel. Increasing the signal-to-noise ratio in DENSE MRI by combining displacement-encoded echoes. *Magnetic Resonance in Medicine*, 52(1):188–192, July 2004.
61. D. Kim, W. D. Gilson, C. M. Kramer, and F. H. Epstein. Myocardial tissue tracking with two-dimensional cine displacement encoded MR imaging: Development and initial evaluation. *Radiology*, 230(3):862–871, March 2004.
62. S. S. Klein, T. P. Graham, Jr., and C. H. Lorenz. Noninvasive delineation of normal right ventricular contractile motion with magnetic resonance imaging myocardial tagging. *Annals of Biomedical Engineering*, 26(5):756–763, September–October 1998.
63. J. P. Kuijer, M. B. Hofman, J. J. Zwanenburg, J. T. Marcus, A. C. van Rossum, and R. M Heethaar. DENSE and HARP: Two views on the same technique of phase-based strain imaging. *Journal of Magnetic Resonance Imaging*, 24(6):1432–1438, December 2006.
64. J. P. Kuijer, E. Jansen, J. T. Marcus, A. C. van Rossum, and R. M. Heethaar. Improved harmonic phase myocardial strain maps. *Magnetic Resonance in Medicine*, 46(5):993–999, November 2001.
65. Medical Imaging & Image Processing Lab. Strain encoded MRI (SENC). http://miip.nileu.edu.eg/senc.
66. N. Lin and J. S. Duncan. Generalized robust point matching using an extended free-form deformation model: Application to cardiac images. In *IEEE International Symposium on Biomedical Imaging: Nano to Macro*, pp. 320–323, April 2004.
67. N. Lin, X. Papademetris, A. Sinusas, and J. S. Duncan. Analysis of left ventricular motion using a general robust point matching algorithm. In *Medical Image Computing and Computer-Assisted Intervention*, pp. 556–563, November 2003.
68. J. B. A. Maintz and M. A. Viergever. A survey of medical image registration. *Medical Image Analysis*, 2(1):1–36, March 1998.
69. T. Mäkelä, P. Clarysse, O. Sipilä, N. Pauna, Q. C. Pham, T. Katila, and I. E. Magnin. A review of cardiac image registration methods. *IEEE Transactions on Medical Imaging*, 21(9):1011–1021, September 2002.
70. H. S. Maniar, B. P. Cupps, D. D. Potter, P. Moustakidis, C. J. Camillo, C. M Chu, M. K. Pasque, and T. M. Sundt, 3rd. Ventricular function after coronary artery bypass grafting: Evaluation by magnetic resonance imaging and myocardial strain analysis. *Journal of Thoracic and Cardiovascular Surgery*, 128(1):76–82, July 2004.
71. M. Markl and J. Hennig. Phase contrast MRI with improved temporal resolution by view sharing: *k*-space related velocity mapping properties. *Magnetic Resonance Imaging*, 19(5):669–676, June 2001.
72. M. Markl, B. Schneider, and J. Hennig. Fast phase contrast cardiac magnetic resonance imaging: Improved assessment and analysis of left ventricular wall motion. *Journal of Magnetic Resonance Imaging*, 15(6):642–653, June 2002.
73. S. Masood, J. Gao, and G. Yang. Virtual tagging: Numerical considerations and phantom validation. *IEEE Transactions on Medical Imaging*, 21(9):1123–1131, September 2002.
74. J. C. McEachen, A. Nehorai, and J. S. Duncan. Multiframe temporal estimation of cardiac nonrigid motion. *IEEE Transactions on Image Processing*, 9(4):651–665, April 2000.
75. T. McInerney and D. Terzopoulos. Deformable models in medical image analysis: A survey. *Medical Image Analysis*, 1(2):91–108, June 1996.

76. F. G. Meyer, R. T. Constable, A. J. Sinusas, and J. S. Duncan. Tracking myocardial deformation using phase contrast MR velocity fields: A stochastic approach. *IEEE Transactions on Medical Imaging*, 15(4):453–465, 1996.
77. A. Montillo, D. N. Metaxas, and L. Axel. Extracting tissue deformation using Gabor filter banks. In *Proceedings of SPIE Medical Imaging 2004: Physiology, Function, and Structure from Medical Images*, Volume 5369, February 2004.
78. C. C. Moore, C. H. Lugo-Olivieri, E. R. McVeigh, and E. A. Zerhouni. Three-dimensional systolic strain patterns in the normal human left ventricle: Characterization with tagged MR imaging. *Radiology*, 214(2):453–466, February 2000.
79. C. C. Moore, E. R. McVeigh, and E. A. Zerhouni. Quantitative tagged magnetic resonance imaging of the normal human left ventricle. *Top Magnetic Resonance Imaging*, 11(6):359–371, 2000.
80. P. Moustakidis, B. P. Cupps, B. J. Pomerantz, R. P. Scheri, H. S. Maniar, A. M. Kates, R. J. Gropler, M. K. Pasque, and T. M. Sundt, 3rd. Noninvasive, quantitative assessment of left ventricular function in ischemic cardiomyopathy. *Journal of Surgical Research*, 116(2):187–196, February 2004.
81. T. D. Nguyen, S. J. Reeves, and T. S. Denney, Jr. On the optimality of magnetic resonance tag patterns for heart wall motion estimation. *IEEE Transactions on Image Processing*, 12(5):524–532, 2003.
82. N. F. Osman. Detecting stiff masses using strain-encoded (SENC) imaging. *Magnetic Resonance in Medicine*, 49(3):605–608, March 2003.
83. N. F. Osman, W. S. Kerwin, E. R. McVeigh, and J. L. Prince. Cardiac motion tracking using CINE harmonic phase (HARP) magnetic resonance imaging. *Magnetic Resonance in Medicine*, 42(6):1048–1060, December 1999.
84. N. F. Osman and J. L. Prince. Regenerating MR tagged images using harmonic phase (HARP) methods. *IEEE Transactions on Biomedical Engineering*, 51(8):1428–1433, August 2004.
85. N. F. Osman, S. Sampath, E. Atalar, and J. L. Prince. Imaging longitudinal cardiac strain on short-axis images using strain-encoded MRI. *Magnetic Resonance in Medicine*, 46(2):324–334, August 2001.
86. C. Ozturk and E. R. McVeigh. Four-dimensional B-spline based motion analysis of tagged MR images: Introduction and *in vivo* validation. *Physics in Medicine and Biology*, 45(6):1683–1702, June 2000.
87. V. M. Pai and L. Axel. Advances in MRI tagging techniques for determining regional myocardial strain. *Current Cardiology Reports*, 8(1):53–58, 2006.
88. L. Pan, J. L. Prince, J. A. C. Lima, and N. F. Osman. Fast tracking of cardiac motion using 3D-HARP. *IEEE Transactions on Biomedical Engineering*, 52(8):1425–1435, August 2005.
89. L. Pan, M. Stuber, D. L. Kraitchman, D. L. Fritzges, W. D. Gilson, and N. F. Osman. Real-time imaging of regional myocardial function using fast-SENC. *Magnetic Resonance in Medicine*, 55(2):386–395, February 2006.
90. X. Papademetris, A. J. Sinusas, D. P. Dione, R. T. Constable, and J. S. Duncan. Estimating 3D strain from 4D cine-MRI and echocardiography: *In-vivo* validation. In *Medical Image Computing and Computer-Assisted Intervention*, pp. 196–205, October 2000.
91. X. Papademetris, A. J. Sinusas, P. Dione, R. T. Constable, and J. S. Duncan. Estimation of 3-D left ventricular deformation from medical images using biomechanical models. *IEEE Transactions on Medical Imaging*, 21(7):786–800, July 2002.
92. J. Park, D. Metaxas, and L. Axel. Analysis of left ventricular wall motion based on volumetric deformable models and MRI-SPAMM. *Medical Image Analysis*, 1(1):53–71, March 1996.
93. J. Park, D. Metaxas, L. Axel, Q. Yuan, and A. S. Blom. Cascaded MRI-SPAMM for LV motion analysis during a whole cardiac cycle. *International Journal of Medical Informatics*, 55(2):117–126, August 1999.
94. J. Park, D. Metaxas, A. A. Young, and L. Axel. Deformable models with parameter functions for cardiac motion analysis from tagged MRI data. *IEEE Transactions on Medical Imaging*, 15(3):278–289, June 1996.
95. N. J. Pelc, R. J. Herfkens, A. Shimakawa, and D. R. Enzmann. Phase contrast cine magnetic resonance imaging. *Magnetic Resonance Quarterly*, 7(4):229–254, October 1991.

96. N. J. Pelc, F. G. Sommer, K. C. Li, T. J. Brosnan, R. J. Herfkens, and D. R. Enzmann. Quantitative magnetic resonance flow imaging. *Magnetic Resonance Quarterly*, 10(3):125–147, September 1994.
97. D. Perperidis, R. H. Mohiaddin, and D. Rueckert. Spatio-temporal free-form registration of cardiac MR image sequences. *Medical Image Analysis*, 9(5):441–456, October 2005.
98. N. S. Phatak, S. A. Mass, A. I. Veress, N. A. Pack, E. V. R. Di Bella, and J. A. Weiss. Strain measurement in the left ventricle during systole with deformable image registration. *Medical Image Analysis*, 13(2):354–361, April 2009.
99. D. D. Potter, P. A. Araoz, K. P. McGee, W. S. Harmsen, J. N. Mandrekar, and T. M. Sundt, 3rd. Low-dose dobutamine cardiac magnetic resonance imaging with myocardial strain analysis predicts myocardial recoverability after coronary artery bypass grafting. *Journal of Thoracic and Cardiovascular Surgery*, 135(6):1342–1347, June 2008.
100. J. L. Prince and E. R. McVeigh. Motion estimation from tagged MR image sequences. *IEEE Transactions on Medical Imaging*, 11(2):238–249, 1992.
101. Z. Qian, D. N. Metaxas, and L. Axel. Extraction and tracking of MRI tagging sheets using a 3D Gabor filter bank. In *Proceedings of IEEE 2006 International Conference of the Engineering in Medicine and Biology Society*, Volume 1, pp. 711–714, August 2006.
102. Z. Qian, D. N. Metaxas, and L. Axel. Non-tracking-based 2D strain estimation in tagged MRI. In *IEEE International Symposium on Biomedical Imaging: Nano to Macro*, Volume 1, pp. 711–714, May 2008.
103. P. Radeva, A. A. Amini, and J. Huang. Deformable B-solids and implicit snakes for 3D localization and tracking of SPAMM MRI data. *Computer Vision and Image Understanding*, 66(2):163–178, May 1997.
104. T. G. Reese, D. A. Feinberg, J. Dou, and V. J. Wedeen. Phase contrast MRI of myocardial 3D strain by encoding contiguous slices in a single shot. *Magnetic Resonance in Medicine*, 47(4):665–676, April 2002.
105. E. W. Remme, K. F. Augenstein, A. A. Young, and P. J. Hunter. Parameter distribution models for estimation of population based left ventricular deformation using sparse fiducial markers. *IEEE Transactions on Medical Imaging*, 24(3):381–388, March 2005.
106. E. W. Remme, A. A. Young, K. F. Augenstein, B. Cowan, and P. J. Hunter. Extraction and quantification of left ventricular deformation modes. *IEEE Transactions on Biomedical Engineering*, 51(11):1923–1931, November 2004.
107. N. Rougon, C. Petitjean, F. Preteux, P. Cluzel, and P. Grenier. A non-rigid registration approach for quantifying myocardial contraction in tagged MRI using generalized information measures. *Medical Image Analysis*, 9(4):353–375, August 2005.
108. D. Rueckert, L. I. Sonoda, C. Hayes, D. L. G. Hill, M. O. Leach, and D. J. Hawkes. Nonrigid registration using free-form deformations: Application to breast MR images. *IEEE Transactions on Medical Imaging*, 18(8):712–721, August 1999.
109. A. K. Rutz, S. Ryf, S. Plein, P. Boesiger, and S. Kozerke. Accelerated whole-heart 3D CSPAMM for myocardial motion quantification. *Magnetic Resonance in Medicine*, 59(4):755–763, April 2008.
110. S. Ryf, M. A. Spiegel, M. Gerber, and P. Boesiger. Myocardial tagging with 3D-CSPAMM. *Journal of Magnetic Resonance Imaging*, 16(3):320–325, September 2002.
111. S. Ryf, J. Tsao, J. Schwitter, A. Stuessi, and P. Boesiger. Peak-combination HARP: A method to correct for phase error in HARP. *Journal of Magnetic Resonance Imaging*, 20(5):874–880, November 2004.
112. S. Sampath, J. A. Derbyshire, E. Atalar, N. F. Osman, and J. L. Prince. Real-time imaging of two-dimensional cardiac strain using a harmonic phase magnetic resonance imaging (HARP-MRI) pulse sequence. *Magnetic Resonance in Medicine*, 50(1):154–163, July 2003.
113. S. Sampath, N. F. Osman, and J. L. Prince. A combined harmonic phase and strain-encoded pulse sequence for measuring three-dimensional strain. *Magnetic Resonance Imaging*, 27(1):55–61, January 2009.
114. S. Sampath and J. L. Prince. Automatic 3D tracking of cardiac material markers using slice-following and harmonic-phase MRI. *Magnetic Resonance Imaging*, 25(2):197–208, February 2007.
115. P. Selskog, E. Heiberg, T. Ebbers, L. Wigstrom, and M. Karlsson. Kinematics of the heart: Strain-rate imaging from time-resolved three-dimensional phase contrast MRI. *IEEE Transactions on Medical Imaging*, 21(9):1105–1109, September 2002.

116. M. L. Shehata, S. Cheng, N. F. Osman, D. A. Bluemke, and J. A. Lima. Myocardial tissue tagging with cardiovascular magnetic resonance. *Journal of Cardiovascular Magnetic Resonance*, 11(55):1–12, December 2009.
117. D. Shen, H. Sundar, Z. Xue, Y. Fan, and H. Litt. Consistent estimation of cardiac motions by 4D image registration. In *Medical Image Computing and Computer-Assisted Intervention (MICCAI)*, Volume 3750, pp. 902–910, October 2005.
118. P. Shi and H. Liu. Stochastic finite element framework for simultaneous estimation of cardiac kinematic functions and material parameters. *Medical Image Analysis*, 7(4):445–464, December 2003.
119. P. Shi, A. J. Sinusas, R. T. Constable, and J. S. Duncan. Volumetric deformation analysis using mechanics-based data fusion: Applications in cardiac motion recovery. *International Journal of Computer Vision*, 35(1):87–107, November 1999.
120. P. Shi, A. J. Sinusas, R. T. Constable, E. Ritman, and J. S. Duncan. Point-tracked quantitative analysis of left ventricular surface motion from 3-D image sequences. *IEEE Transactions on Medical Imaging*, 19(1):36–50, January 2000.
121. S. M. Song and R. M. Leahy. Computation of 3D velocity fields from 3D cine CT images of a human heart. *IEEE Transactions on Medical Imaging*, 10(3):295–306, 1992.
122. B. S. Spottiswoode, X. Zhong, A. T. Hess, C. M. Kramer, E. M. Meintjes, B. M. Mayosi, and F. H. Epstein. Tracking myocardial motion from cine DENSE images using spatiotemporal phase unwrapping and temporal fitting. *IEEE Transactions on Medical Imaging*, 26(1):15–30, January 2007.
123. B. S. Spottiswoode, X. Zhong, C. H. Lorenz, B. M. Mayosi, E. M. Meintjes, and F. H. Epstein. 3D myocardial tissue tracking with slice followed cine DENSE MRI. *Journal of Magnetic Resonance Imaging*, 27(5):1019–1027, May 2008.
124. H. Sundar, H. Litt, and D. Shen. Estimating myocardial motion by 4D image warping. *Pattern Recognition*, 42(11):2514–2526, November 2009.
125. N. J. Tustison. *Biventricular Myocardial Strains with Anatomical NURBS Models from Tagged MRI.* PhD thesis, Washington University in St. Louis, August 2004.
126. N. J. Tustison and A. A. Amini. Biventricular myocardial strains via nonrigid registration of anatomical NURBS models. *IEEE Transactions on Medical Imaging*, 25(1):94–112, January 2006.
127. N. J. Tustison, V. G. Davila-Román, and A. A. Amini. Myocardial kinematics from tagged MRI based on a 4-D B-spline model. *IEEE Transactions on Biomedical Engineering*, 50(8):1038–1040, August 2003.
128. S. W. J. Ubbink, P. H. M. Bovendeerd, T. Delhaas, T. Arts, and F. N. van de Vosse. Towards model-based analysis of cardiac MR tagging data: Relation between left ventricular shear strain and myofiber orientation. *Medical Image Analysis*, 10(4):632–641, August 2006.
129. H. C. van Assen, L. M. J. Florack, J. J. M. Westenberg, and B. M. ter Haar Romeny. Tuple image multiscale optical flow for detailed cardiac motion extraction: Application to left ventricle rotation analysis. In *Proceedings of the MICCAI Workshop on Analysis of Functional Medical Images*, pp. 73–80, September 2008.
130. A. I. Veress, G. T. Gullberg, and J. A. Weiss. Measurement of strain in the left ventricle during diastole with cine-MRI and deformable image registration. *Journal of Biomechanical Engineering*, 127(7):1195–1207, December 2005.
131. A. I. Veress, J. A. Weiss, R. D. Rabbitt, J. N. Lee, and G. T. Gullberg. Measurement of 3D of left ventricular strains during diastole using image warping and untagged MRI images. *Computers in Cardiology*, 28:165–168, 2001.
132. H. Wang and A. A. Amini. Accurate 2-D cardiac motion tracking using scattered data fitting incorporating phase information from MRI. In *Proceedings of SPIE Medical Imaging 2010: Biomedical Applications in Molecular, Structural, and Functional Imaging*, Volume 7626, February 2010.
133. H. Wang and A. A. Amini. Cardiac motion tracking approach with multilevel B-splines and SinMod from tagged MRI. In *Proceedings of SPIE Medical Imaging 2011: Biomedical Applications in Molecular, Structural, and Functional Imaging*, February 2011.

134. H. Wang and A. A. Amini. A partial *k*-space acquisition pulse sequence for tagged MRI using CSPAMM. Technical report, Medical Imaging Lab, Electrical and Computer Engineering Department, University of Louisville, 2011.
135. Y. Wang, Y. Chen, and A. A. Amini. Fast LV motion estimation using subspace approximation techniques. *IEEE Transactions on Medical Imaging*, 20(6):499–513, June 2001.
136. V. Y. Wang, H. I. Lam, D. B. Ennis, B. R. Cowan, A. A. Young, and M. P. Nash. Modelling passive diastolic mechanics with quantitative MRI of cardiac structure and function. *Medical Image Analysis*, 13(5):773–784, October 2009.
137. H. Wen, K. A. Marsolo, E. E. Bennett, K. S. Kutten, R. P. Lewis, D. B. Lipps, N. D. Epstein, J. F. Plehn, and P. Croisille. Adaptive postprocessing techniques for myocardial tissue tracking with displacement-encoded MR imaging. *Radiology*, 246(1):229–240, January 2008.
138. World Health Organization. Cardiovascular diseases. http://www.who.int/entity/nmh/publications/fact_sheet_cardiovascular_en.pdf.
139. C. Xu, J. J Pilla, G. Isaac, J. H Gorman, A. S. Blom, R. C. Gorman, Z. Ling, and L. Dougherty. Deformation analysis of 3D tagged cardiac images using an optical flow method. *Journal of Cardiovascular Magnetic Resonance*, 12(19), March 2010.
140. P. Yan, A. Sinusas, and J. S. Duncan. Boundary element method-based regularization for recovering of LV deformation. *Medical Image Analysis*, 11(6):540–554, December 2007.
141. A. A. Young. Model tags: Direct three-dimensional tracking of heart wall motion from tagged magnetic resonance images. *Medical Image Analysis*, 3(4):361–372, December 1999.
142. A. A. Young. Assessment of cardiac performance with magnetic resonance imaging. *Current Cardiology Reviews*, 2(4):271–282, November 2006.
143. A. A. Young, D. L. Kraitchman, L. Dougherty, and L. Axel. Tracking and finite-element analysis of stripe deformation in magnetic resonance tagging. *IEEE Transactions on Medical Imaging*, 14(5):413–421, 1995.
144. A. Youssef, E. H. Ibrahim, G. Korosoglou, M. R. Abraham, R. G. Weiss, and N. F. Osman. Strain-encoding cardiovascular magnetic resonance for assessment of right-ventricular regional function. *Journal of Cardiovascular Magnetic Resonance*, 10(33):1–12, July 2008.
145. E. A. Zerhouni, D. M. Parish, W. J. Rogers, A. Yang, and E. P. Shapiro. Human heart: Tagging with MR imaging–a method for noninvasive assessment of myocardial motion. *Radiology*, 169(1):59–63, October 1988.
146. X. Zhong, P. A. Helm, and F. H. Epstein. Balanced multipoint displacement encoding for DENSE MRI. *Magnetic Resonance in Medicine*, 61(6):981–988, April 2009.
147. Y. Zhu, M. Drangova, and N. J. Pelc. Fourier tracking of myocardial motion using cine-PC data. *Magnetic Resonance in Medicine*, 35(4):471–480, April 1996.
148. Y. Zhu, M. Drangova, and N. J. Pelc. Estimation of deformation gradient and strain from cine-PC velocity data. *IEEE Transactions on Medical Imaging*, 16(6):840–851, December 1997.
149. Y. Zhu, X. Papademetris, A. J. Sinusas, and J. S. Duncan. Integrated segmentation and motion analysis of cardiac MR images using a subject-specific dynamical model. In *IEEE Computer Society Conference on Computer Vision and Pattern Recognition Workshops*, pp. 1–8, June 2008.
150. Y. Zhu and N. J. Pelc. A spatiotemporal model of cyclic kinematics and its application to analyzing nonrigid motion with MR velocity images. *IEEE Transactions on Medical Imaging*, 18(7):557–569, July 1999.
151. Y. Zhu and N. J. Pelc. Three-dimensional motion tracking with volumetric phase contrast MR velocity imaging. *Journal of Magnetic Resonance Imaging*, 9(1):111–118, January 1999.
152. Y. Zhu, P. Yan, X. Papademetris, A. J. Sinusas, and J. S. Duncan. Integrated segmentation and deformation analysis of 4-D cardiac MR images. In *IEEE International Symposium on Biomedical Imaging: Nano to Macro*, pp. 1437–1440, May 2008.
153. H. Wang and A. Amini. Cardiac motion and deformation recovery from MRI: A review. *IEEE Transactions on Medical Imaging*, 31(2):487–503, Feburary 2012.

12

MRI for OA Diagnosis and Drug Development

Saara Totterman
Qmetrics Technologies, LLC

José G. Tamez-Peña
Tec de Monterrey

12.1 Introduction to Osteoarthritis

Saara Totterman

12.1.1 What Is Osteoarthritis?

Osteoarthritis (OA), a degenerative disease of the joint, is loosely divided into different groups based on the joints involved. The form which includes multiple joints, hands, knees, hips, and spine, is considered to be hereditary; the form which includes only peripheral joints or a single joint such as knee or hip is not considered to be hereditary [1–4]. OA can be primary, purely degenerative, or secondary, typically to trauma, rheumatoid arthritis, or to infection [3]. In this chapter, we will concentrate on degenerative joint disease. Since many of the histopathological changes and other disease features in OA are similar in different joints; for the sake of simplicity, we will concentrate on the knee joint.

12.1.2 Etiology of Knee Osteoarthritis

One of the causes of OA is joint injury; however, in cases without clear injury, the etiology of OA is not well understood. Therefore, there are several schools of thoughts regarding the underlying cause

[1–4]. Since cartilage fractures heal through fibrocartilage formation and the cartilage is not able to produce new cartilage to fill the fracture site, joints with either chondral or osteochondral fracture have a very high risk of developing OA. Anterior cruciate ligament tears have high risk regardless of treatment to lead to the development of OA. Whether the underlying cause is changed in the knee biomechanics due to the tear itself or the acute cartilage injury caused by the impact of the femur on tibia cartilage is not clear [5].

There are different thoughts about the etiology. One is heavily leaning toward the changes in biomechanics either due to the microinjuries in the supporting structures or due to the microinjuries in the bone or cartilage itself. In these cases, the cartilage and/or bone fails to heal the repeated microinjury leading to the development of OA [3].

12.1.3 Anatomy of Knee

The anatomic structures of the knee include three bones, their articulating cartilage, two menisci, and four main ligaments [6]. The knee joint itself has three different components, the medial tibio-femoral joint, lateral tibio-femoral joint, and patello-femoral joint. The articulating components for the medial tibio-femoral joint are the medial tibial plateau and the medial femoral condyle; in the lateral tibio-femoral joint, they include the lateral tibial plateau and the lateral femoral condyle. In the patello-femoral joint, they include the patella and the femoral trochlea [6]. The articulating bone surfaces are covered by smooth hyaline cartilage. The medial and lateral tibial plateaus have their own separate hyaline articular cartilage plates. The lateral and medial femoral condyle cartilage plates are connected via the trochlear cartilage, which contributes to the patello-femoral joint [6].

The two menisci, the medial meniscus, and the lateral meniscus in their corresponding knee joint compartments are wedged between the articulating cartilage surfaces. The main ligaments include the anterior cruciate ligament (ACL) extending from the medio-posterior part of the lateral femoral condyle to the anterolateral part of the medial tibial spine and the posterior cruciate ligament (PCL) extending from the anterolateral part of the medial femoral condyle to the posterior part of the tibia and crossing in the middle of the knee. In addition, the joint is supported by the medial collateral ligament extending from the medial surface of the medial femoral condyle to the medial surface of the tibia and the lateral collateral ligament extending from the lateral surface of the lateral femoral condyle to the fibular head [6].

12.1.4 Histology and Magnetic Resonance Image Appearance of Knee Structures

In magnetic resonance image (MRI), signal behavior of the anatomic structures in MR images depends on the microscopic anatomy, structural architecture, and composition. The following will discuss the histology of the knee structures and then their MRI appearance. The discussion includes bones, articular cartilage, menisci, and the main knee joint ligaments.

12.1.4.1 Normal Bone Histology

The tibia and femur are long bones while the patella is a sesamoid bone. Histologically, the articulating surfaces are composed of the hyaline cartilage and its underlying bone. The subchondral bone, the bone adjacent to the cartilage, is comprised of the subchondral bone plate followed deeper by the trabecular bone, where the trabecular network is filled by hematogenous bone marrow and bone marrow fat (Figure 12.1) [7]. The cartilage is tightly connected, anchored to the subchondral bone plate by a zigzagging osteochondral junction. The deepest layer of cartilage, the thin layer adjacent to the bone plate, is calcified cartilage; the much thicker superficial layer is noncalcified cartilage (Figure 12.1). These two are separated from each other by a very thin tidemark [8]. Superficial cartilage is covered by a very thin lamina splendum.

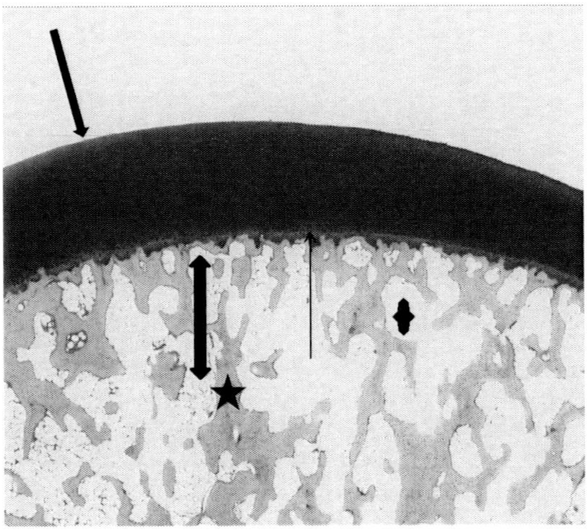

FIGURE 12.1 **(See color insert.)** Histology section of cartilage showing dark blue noncalcified cartilage (thick arrow), light blue calcified cartilage underneath it (thin arrow), tide mark, a thin line between two cartilage types, subchondral bone plate underneath the calcified cartilage (double ended arrow), trabecular bone (star), and bone marrow space (cross) between them. The dark blue color of the cartilage indicates normal cartilage with high proteoglycan concentration.

12.1.4.2 MRI Appearance of Normal Bones

Regardless of used pulse sequence, bones in MR images are outlined by a very low-signal-intensity structure. In peripheral bone, this is cortical bone. Underneath the cartilage, this is subchondral bone plate. The calcified cartilage due to its high calcium concentration is seen as part of the bone plate. In the trabecular bone, the trabeculae appear as a low-signal-intensity network separated by high-signal-intensity fatty marrow and/or medium-signal-intensity hematogenous marrow (Figure 12.2). In most MR imaging sequences, only the thickest trabecular structures can be appreciated.

12.1.4.3 Normal Articular Cartilage Histology

The knee hyaline cartilage is a well-organized structure, which is composed of a matrix with proteoglycans (PG) which have a glycosaminoglycan (GAG) component, well-organized collagen fibers, and well-organized chondrocytes all covered by lamina splendum. In normal cartilage, the collagen fibers are attached to the subchondral bone plate from where they radiate upwards toward the articular surface. The collagen orientation varies from perpendicular to the subchondral bone closest to the bone (the deep layer) to parallel to the articular surface closest to the articular surface (superficial layer).

12.1.4.4 MRI Features of Normal Hyaline Cartilage

The normal cartilage has three signal intensity zones: the thin high-signal-intensity zone, the intermediate-signal-intensity middle zone, and the low-signal-intensity deep zone with some correlation to the histology (Figure 12.3) [9–12]. The well-organized collagen bundles are a major component affecting the signal behavior of the cartilage in MR images. The tightly knit well-organized collagen fibers in the normal knee are reflected in low T2 values, especially in the lower cartilage layers [13]. The variance of the normal cartilage T2 values in normal population is small as is the intrasubject variance.

12.1.4.5 Behavior of Cartilage in dGEMRIC Images

Toluidine blue staining is used to histologically evaluate cartilage GAG (Figure 12.1). Those studies have shown that GAG concentration is high in normal cartilage [14]. In MRI, the cartilage GAG

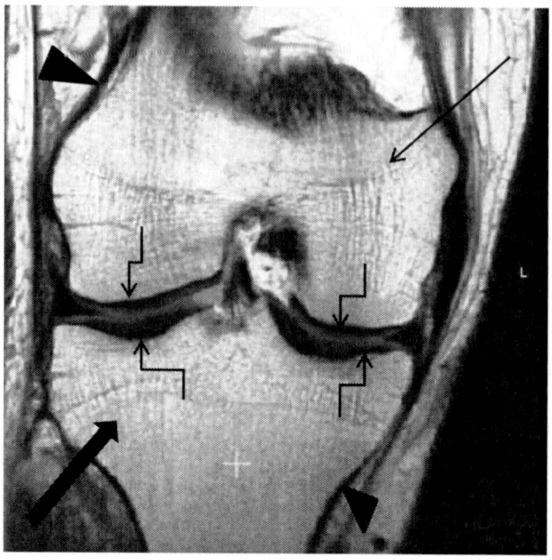

FIGURE 12.2 Proton density spin echo image of knee shows femur bone (thin arrow) and tibia bone (thick arrow) and the cartilage covering their ends (zigzag arrows). Note the low signal intensity of the deeper and high signal intensity of the superficial layers of the cartilage. The thin lines inside the bones are part of the trabecular network. Note the high signal intensity inside the bone coming from bone marrow fat and the low signal intensity of the enveloping cortical bone (triangles).

concentration is evaluated by measuring the delayed gadolinium-enhanced T1 relaxation time after an intravenous injection of an anionic paramagnetic contrast agent: gadolinium diethylene triamine penta-acid Gd(DPTA)2, Magnevist. This MRI study of GAG concentration is called delayed gadolinium-enhanced MR imaging of cartilage (dGEMRIC). Because the anionic agent, Magnevist, is repelled by the negatively charged GAG, it will concentrate in GAG-depleted areas. Furthermore, gadolinium concentration lowers T1. Therefore, T1 relaxation time is directly related to the cartilage GAG concentration in dGEMRIC studies [12]. Those studies have shown that, in normal knees, GAG concentration in normal population is age and physical activity dependent [15].

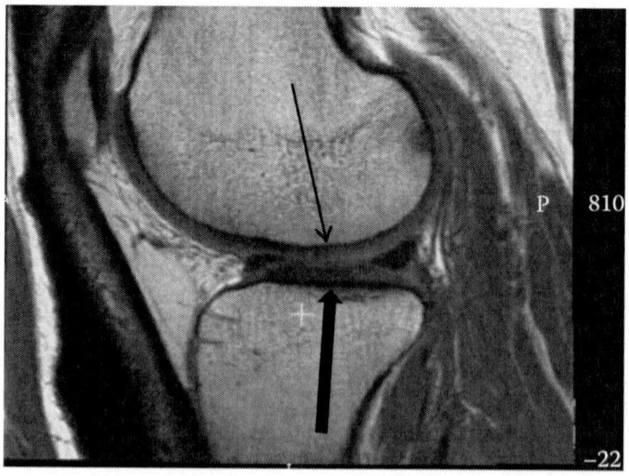

FIGURE 12.3 Sagittal proton density spin echo image shows femur (thin arrow) and tibia (thick arrow) cartilage plates. Note the low signal intensity in the deeper layers of tibia cartilage and the high signal intensity in the superficial layer.

12.1.4.6 Normal Menisci Histology

The lateral and medial menisci are fibrocartilagenous wedge-shaped crescents between the periphery of the articulating tibial and femoral cartilage surfaces. Their ends have ligamentous attachments to the tibia outside the articular cartilage. The anterior ends attach at the anterior edge of anterior tibia anterior to the tibial spines and the posterior ends to the posterior intercondylar fossa of tibia, posterior to the tibial spines. The menisci have ligamentous attachments to each other, to the corresponding collateral ligaments, to other small knee ligaments, and to the femur and tibia. Both menisci are divided based on the anatomical location into anterior horn, body, and posterior horn [6,16].

12.1.4.7 MRI Appearance of Normal Menisci

Most of the MR pulse sequences show menisci as very low-signal-intensity structures, which have high-signal-intensity lines traveling close to the periphery; these represent their vascular structures (Figure 12.4). Their ligamentous attachments to the tibia have more ligamentous appearance, higher SI (signal intensity), and parallel low-signal-intensity lines. Based on their appearance, they can be easily identified in MR images. The connecting meniscus to meniscus ligaments are well seen in MRI images. The higher-resolution images also show their smaller ligamentous attachments to the tibia and femur and to other knee ligaments [17].

12.1.4.8 Histology and MRI Appearance of Main Knee Ligaments

The main ligaments of the knee joint ACL, PCL, medial collateral ligament (MCL), and lateral collateral ligament (LCL) all have very different structures (Figure 12.5). ACL has two different components or bundles, the anteromedial (AM) and the posterolateral (PL) bundles [18,19]. The anterior bundle is more ligamentous and more vascular, while the posterior bundle is stronger and is composed of thicker, more tightly knit collagen bundles. In MR images, the AM bundle appears as a flatter higher-signal-intensity structure with parallel lower-signal-intensity lines, and the PL bundle appears with rounder well-defined low-SI structure [20]. The relative position of the two bundles is flexion dependent being parallel in full extension and being more and more crossed with increasing flexion. MCL and LCL are important for knee functions but are not discussed in detail in this chapter.

12.1.5 Functions of Knee Joint Structures

All anatomic structures of the knee have the following functions: load bearing, load distribution, stabilization, motion guiding, supporting, and geometry forming [21–23]. Although many of them have

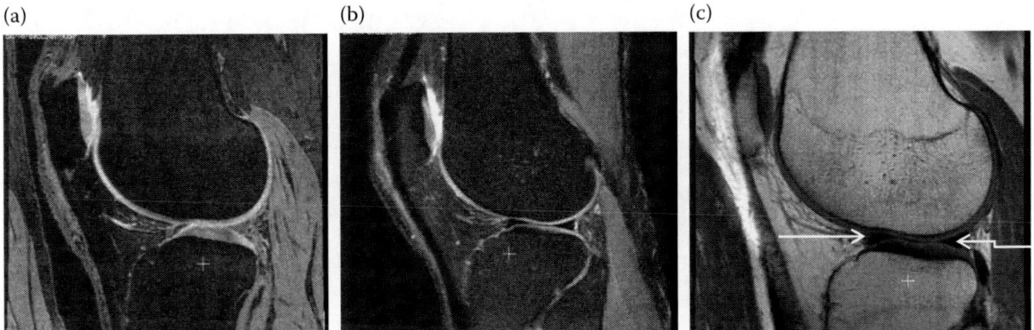

FIGURE 12.4 Anterior (thin straight arrow in c) and posterior (thin zigzag arrow in c) horn of lateral meniscus in (a) 3D water-exited DESS, (b) in fat-suppressed proton density spin echo, and (c) in proton density spin echo. Note the low signal intensity of bones in water-exited and fat-suppressed images. Also note the signal behavior of cartilage in different pulse sequences images.

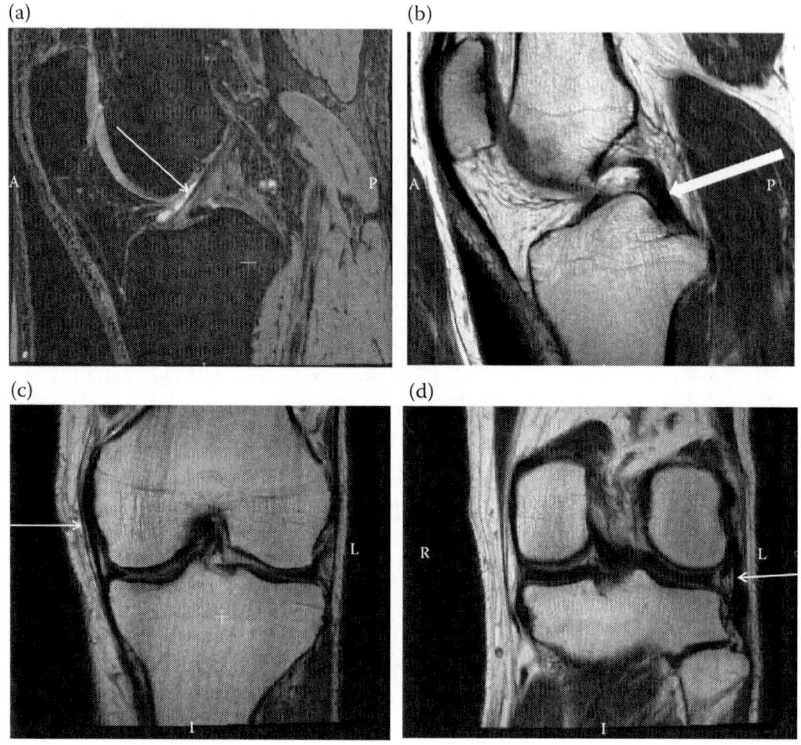

FIGURE 12.5 (a) ACL (thin arrow) in sagittal water-exited DESS and (b) PCL (thick arrow) in proton density spin echo. (c) MCL (thin arrow) and (d) LCL (thin arrow) in coronal proton density spin echo.

multiple functions, one is their primary function. The bones are the main supporting structures and form the structural platform of the joint, on which all other structures are built. Their shape forms the joint geometry, which provides the basis for the function of the joint and defines the path of the movements [21,23]. The articulating cartilage plates have their own geometry and they are guiding components of the movement mechanism (Figure 12.6). They are softer than bone and contribute to the load

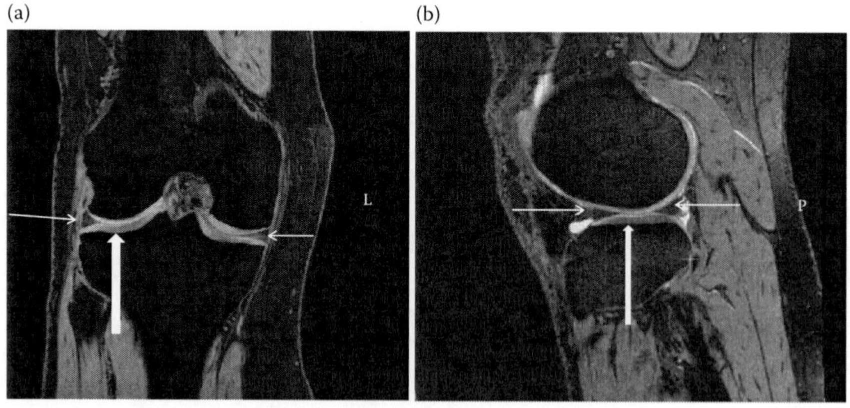

FIGURE 12.6 Images demonstrate the smoothness and geometry of the articulating cartilage surfaces (thick arrow in a and b). Also note that the curvature of the underlying bone is different from the curvature of the superficial surface of the corresponding cartilage. They also show how the geometry of the menisci (thin arrows in a and b) accommodates the shape of the cartilage plates.

bearing and load transmission and distribution to the tibial bone [22,23]. For smooth gliding motions, the articulating cartilage plates are lubricated by joint fluid.

The main function of all soft tissue structures of the joint is to stabilize the joint. In a normal knee, the smooth wedge shape of the menisci matches perfectly with the shape of the underlying tibia cartilage and the overlying femoral cartilage [24]. Their smooth surfaces and concave shapes contribute to the geometry of the articulating surfaces (Figure 12.6). As cushions between tibial and femoral cartilage, their main functions are load bearing, load transmission, and load distribution [25,26]. They are important components of the knee stabilization and their motion during knee flexion/extension guides and allows the rotation of the tibia [27–29]. The main ligaments, ACL, PCL, MCL, and LCL, are the primary knee stabilizers; the cruciate ligaments stabilize and limit the anteroposterior motion of the knee [30]. However, one important function of ACL is to guide and limit the rotation of the tibia during the knee flexion/extension. Even collateral ligaments also serve as anterior/posterior stabilizers; their main function is to provide varus valgus stability.

12.1.6 Knee OA Histopathology and MRI Findings

The knee OA can involve any or all of the three compartments of the knee joint. In the past, most of the studies have concentrated in the medial tibio-femoral joint; even the incidence in both lateral tibio-femoral OA and patello-femoral OA is relatively frequent.

Even though cartilage degeneration leading to cartilage loss is an important component of the knee OA, the use of MR imaging has shown that OA is a whole-joint disease. As such, it involves all soft tissue structures of the affected joint, not just cartilage and bone. Therefore, the next sections will describe the pathological OA changes in cartilage and bone as well as in menisci and ligaments. The picture of a knee with end-stage OA is very well known and understood. It is a knee with severe cartilage loss where two opposing bones are in contact with each other, forming the articulating surfaces. The articulating bones themselves have rough borders and, the articulating surfaces are wider and flattened, and the surface may also have attritions. The articulating bones have large peripheral and varying degrees of central osteophytes. The menisci are severely torn or totally macerated (Figure 12.7). The tibia and femur are malaligned and are either in varus or in valgus angulation (Figure 12.8). The following will try to describe the partially understood path from normal knee to knee with severe OA.

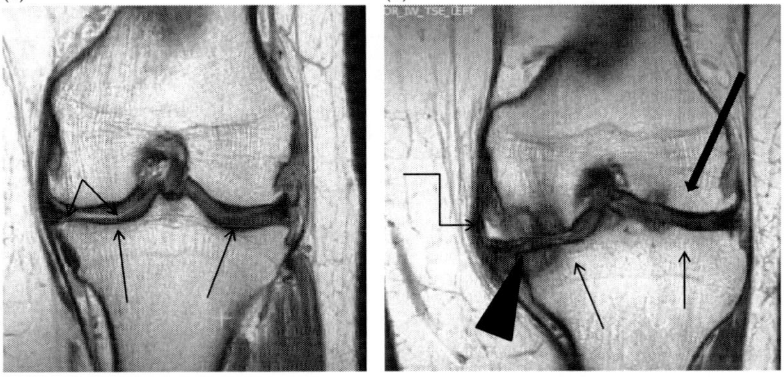

FIGURE 12.7 (a) Moderate and (b) an advanced OA MR images. Note the wide and flat ends of tibia in both cases (thin arrows), the flattened femur (thick arrow in b) and large osteophytes (zigzag arrow in b). There is advanced tibial cartilage thinning in a (dual arrows) and total cartilage loss and macerated meniscus in medial compartment in b (triangle in b).

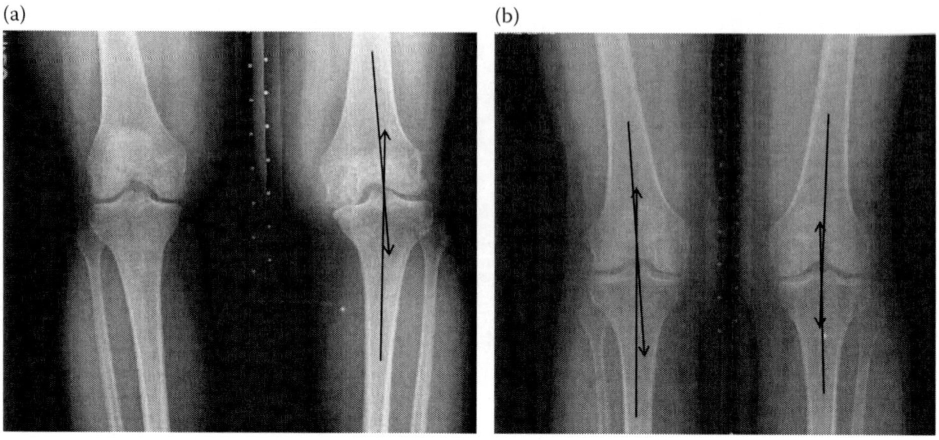

FIGURE 12.8 Plain film shows (a) varus alignment in left knee and (b) valgus alignment in both knees.

12.1.7 Bones

12.1.7.1 Histological OA Changes in Bones

12.1.7.1.1 Changes in Bone–Cartilage Interface

In the very early phase of the disease, far before cartilage defects appear, both the bone–cartilage interface and the osteochondral junction go through several changes [31].

12.1.7.1.2 Enchondral Ossification Leading to Cartilage Thinning and Subchondral Bone Thickening

Enchondral ossification, new bone formation at the bone–cartilage interface, causes many of OA-related changes. Their relationship to other cartilage changes and even the order of those changes are not well understood. Whether they are triggered by structural or compositional changes in other soft tissue structures, which alter the biomechanics or biochemical environments of the joint, or whether these changes are in the cartilage, is presently not well understood. In either case, in early OA, neurovascular bundles from the bone marrow located in the adjacent trabecular bone grow into the subchondral bone and calcified cartilage and advances further through the tidemark into noncalcified cartilage [32]. This phenomenon leads to enchondral ossification, resulting in multiple tidemarks as well as tidemark movement toward the cartilage surface. This leads to subchondral bone thickening from above and cartilage thinning from below (Figure 12.9) [33,34]. This phenomenon is probably at least partly responsible for the widening of the articulating surfaces of the bones (Figure 12.7), a common OA phenomenon seen also in plain films (Figure 12.9) [35–38].

12.1.7.1.3 Central Osteophytes

Enchondral ossification also leads to the formation of the so-called central osteophytes, where osteophyte seems to be growing from the underlying bone and slowly protruding through the cartilage into the joint space (Figure 12.10). Whether these bone–cartilage interface changes precede cartilage degeneration, occur simultaneously, or are a response to the biomechanical changes caused by cartilage deterioration is not clear.

12.1.7.1.4 Bone Shape Changes

The advancement of the tidemarks related to the enchondral ossification is not a global phenomenon. Both in the femur and in the tibia, it is more common at the periphery of the articulating surfaces but can also be seen in the middle of the cartilage, where it seems to favor certain anatomic locations. Central osteophytes are the most advanced form of that phenomenon (Figure 12.10). The local advancement of tidemark changes both the global and the local shape of the articulating bone surfaces.

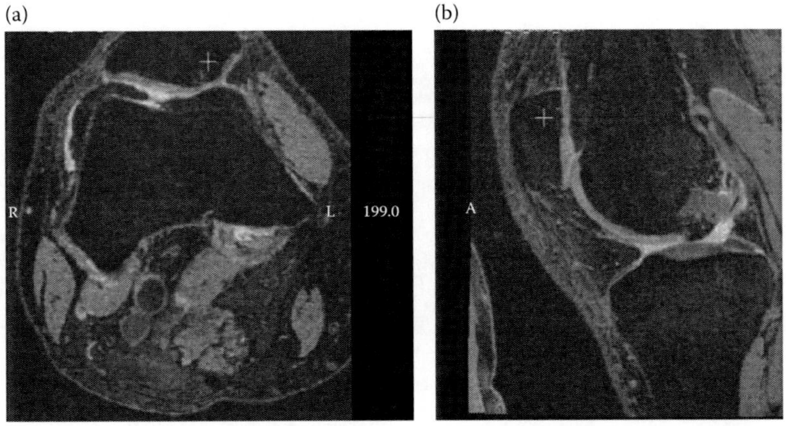

FIGURE 12.9 Axial (a) and Sagittal (b) slices showing the enchondral bone formation in patellar cartilage (cross-hair), which has led to irregular bone–cartilage interface where the bone advances into the cartilage. Also note the subsequent shape changes in the articulating surface of the patella.

The change of the shape of femoral condyles was noticed by Kellgren in 1957 [35]; however, it was ignored and forgotten for decades. During the last decade, MRI showed the widened tibial plateau in OA subjects and the widening of the plateau in increasing severity of the disease (Figure 12.7) [36,37]. More recently, besides widening articulating tibial plateaus, femoral condyles and trochlea were shown to be flattened in more advanced MRI study analysis (Figure 12.7) [38]. The articulating bone surfaces can also experience attrition in very advanced stages of OA.

12.1.7.1.5 Bone Marrow Lesions

Bone marrow edema lesions (BMEL), which were first identified as high-signal-intensity regions in T2-weighted fat-suppressed spin echo images [39], are located in the trabecular bone adjacent to the articular cartilage (Figure 12.11). Histologically, they are partly necrotic bone with vascularized fibrotic components; however, they do not have edema [40,41]. Their etiology is not yet clear.

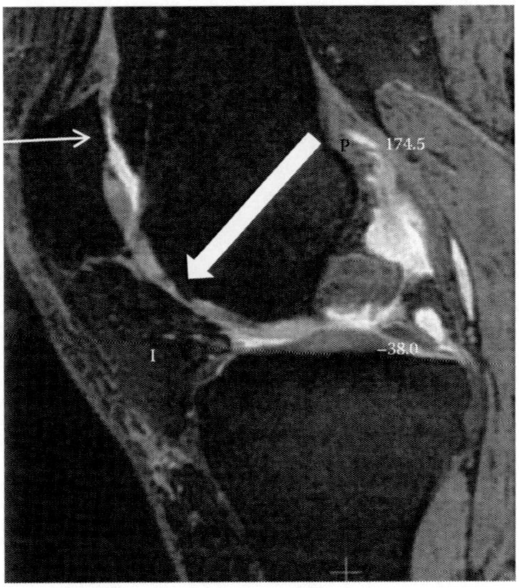

FIGURE 12.10 Central osteophyte in trochlea (thick arrow) and new bone formation in patella (thin arrow) have replaced the overlying cartilage and caused shape changes in corresponding articulating bone surfaces.

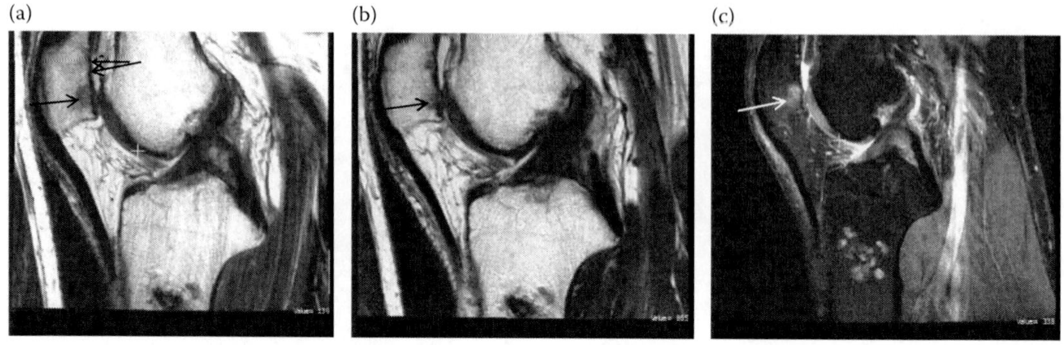

FIGURE 12.11 Bone marrow edema lesion (thin arrow) in proton density (a), in T2 (b), and in fat-suppressed proton density-weighted (c) image. In the fat-suppressed image, the lesion is well visualized while in the T2-weighted image, it blends with the surrounding tissue. Also note the advanced cartilage loss (two thin arrows) and shape changes in patella.

12.1.7.1.6 Trabecular Bone Architecture

The architecture of subchondral trabecular bone adjacent to the subchondral bone plate, to which it is directly connected, changes in OA. In OA, the subchondral bone plate itself may thicken or get thinner. Furthermore, the trabecular network is partially interrupted and the trabeculae themselves are thinner than in the contralateral normal knee [42–43].

12.1.8 Cartilage

12.1.8.1 Histological OA Cartilage Changes

Experimental animal studies and MR imaging in animals and in humans have indicated that local compositional cartilage changes precede the structural cartilage changes, cartilage thinning, and defects [44–48]. Histological cartilage studies in early OA have shown cartilage cell death and hypertrophy, edema, disappearance of lamina splendum, cartilage surface fibrillation, and decrease in cartilage GAG (Figure 12.12) [7]. The cell death and hypertrophy continue to the deeper layers and surface defects start appearing. This continues until there is full thickness defect extending to the underlying bone [7]. However, longitudinal MR studies have also shown that new areas will appear simultaneously in other regions of the same compartment or in other compartments; initially, as cartilage hypertrophy evolving into development of defects [46]. As a result, the same cartilage can have multiple hypertrophied areas and multiple defects with varying degrees of severity [46].

Collagen fibers, which are an integral part of the cartilage structure and play a significant role in the biomechanical properties of the cartilage, degenerate in OA. The degeneration changes the integrity of the collagen bundles and consequently the integrity of the cartilage [47]. This in turn changes the biomechanical properties of the cartilage [26]. The degeneration also leads to an increase in the water content of the cartilage [45].

12.1.9 MRI Findings in OA

12.1.9.1 Bones

The enchondral ossification, new bone formation at the bone–cartilage interface as mentioned before, causes many of the OA-related changes. The advancing tidemark can be seen in MRI as heterogeneity of signal intensity in the bone–cartilage interface, and later this leads to irregularity of the otherwise smooth bone–cartilage interface (Figure 12.10) [48]. The most advanced local case is a central osteophyte, which penetrates through the cartilage. The widening of the articulating ends of the bones has

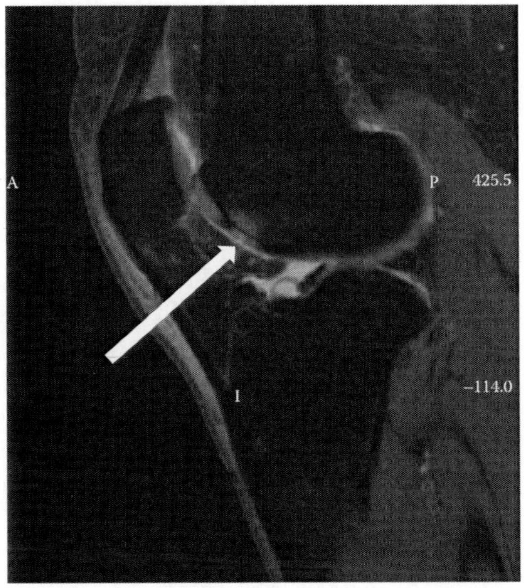

FIGURE 12.12 Fat-suppressed proton density spin echo image shows high-signal-intensity T2 cartilage lesion in trochlea cartilage (thick arrow). Note the underlying bone marrow edema lesion.

been shown by MRI (Figure 12.7) [36,37]. Besides widening of articulating tibial plateaus, femoral condyles and trochlea flatten in advanced OA [38]. The articulating bone surfaces can also experience attrition in very advanced stages of OA.

The MRI changes related to the changes in subchondral bone trabecular architecture have so far been studied only in research environment. However, there is promising research showing that the thinning of the trabeculae and disruption of the trabecular network is also identified in high-resolution MR images [43].

12.1.9.1.1 Bone Marrow Lesions

BMEL, which were first identified as high-signal-intensity regions in T2-weighted fat-suppressed spin echo images [49], are located in the trabecular bone adjacent to the articular cartilage (Figure 12.11). They are observed in all phases of OA. Small BMLs can be seen in any phase of the diseases and they have a tendency to come and go. Larger ones have a tendency to stay but their volume can fluctuate [39]. They are not specific to OA and can result from a multitude of other factors, including trauma, inflammation, infection, tumor, or osteonecrosis [39]. They can be transient and are also commonly seen in rheumatoid and psoriatic arthritis. Bone marrow lesions in OA are histologically partly necrotic bone with vascularized fibrotic components; however they do not have edema. Even though large bone marrow lesions predict disease progression and have been associated with pain, their role in early OA is not clear [40,41]. Nor do we know their temporal relationship to early cartilage changes or what leads to their disappearance.

12.1.9.2 Articular Cartilage

12.1.9.2.1 Compositional Changes

Cartilage changes in MR images can be divided into compositional and structural. The changes in GAG content, edema, and collagen degeneration are part of the compositional changes of the cartilage. The compositional changes change the MRI properties of the tissue, its signal behavior, T2 behavior and T1 values, and the dGEMRIC index. The increase in water content due to either edema or degeneration of collagen bundles changes cartilage signal intensity in MR images. This is well seen in proton density and in T2-weighted images and most probably is the source of the so-called T2 lesions (Figure 12.13), the high-signal-intensity areas in the deep layer of the cartilage with high T2 values [50,51]. In the same way, the

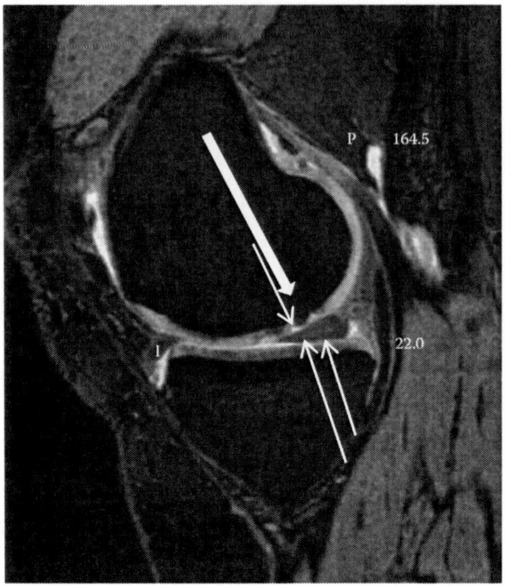

FIGURE 12.13 3D water-exited DESS image shows small very focal cartilage defects adjacent to each other. Note the good contrast between cartilage (thick arrow) and the adjacent high signal fluid (thin arrow) and meniscus (two thin arrows).

fibrillation of the cartilage surface will increase the water content and the signal intensity of the affected part of cartilage. The degeneration of collagen fibers per se affects the T2 of the cartilage; the decrease of well-organized tightly knit collagen fibers increases the cartilage T2 and increases its signal intensity [45,46]. The signal behavior of the cartilage is also affected by the enchondral ossification where the signal of the deeper layers due to the vascular invasion increases and later decreases due to the calcification of the noncalcified cartilage during the process of new bone formation as shown by advancing tidemark [48].

12.1.9.2.2 Structural Changes

The structural changes at the superficial surface of cartilage include surface fibrillation, fissures, and cartilage defects and thinning [7]. At the early phase of the disease, the superficial surface changes are detected in high-resolution MR images. In those images, the fibrillation caused distinct borders of the cartilage to disappear [44,45]. Since cartilage defects are filled by joint fluid, which has its own MRI SI, different from that of cartilage, the contrast between the cartilage and joint fluid allows their detection in MR images (Figure 12.13) [44]. In the same way, the small fissures are filled by fluid and can be seen in the early phase zigzagging inside the cartilage and in finally reaching the bone [44]. In later studies, those have led to cartilage defects and dislodged free pieces of cartilage in the joint space.

Cartilage hypertrophy or swelling and cartilage loss caused by thinning of the cartilage, either from below as a result of enchondral ossification or from above as a result of focal or larger cartilage defects, lead to changes in local cartilage thickness. Both of these phenomena also change the shape of cartilage plates and the smoothness of the corresponding surface.

12.1.9.3 Menisci

The lateral and medial menisci are crescent-shaped fibrocartilagenous structures between the periphery of the articulating tibial and femoral cartilage surfaces [6]. The menisci cushions between the tibial and the femoral cartilage bear part of the load. Their smooth surfaces and concave shapes contribute to the geometry of the articulating surfaces [24,25]. Their ends are attached to the tibia in front of and posterior to the tibial spines and have ligamentous attachments to the femur, controlling the path of

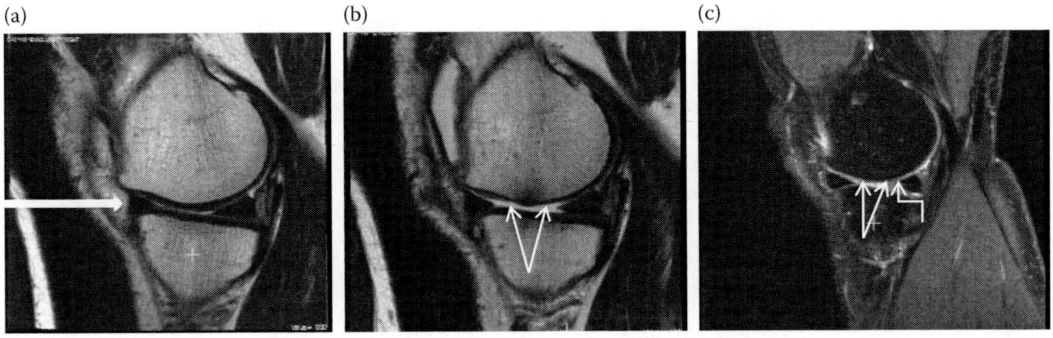

FIGURE 12.14 Low-signal-intensity anterior and posterior horn of medial meniscus (thick arrow) in proton density (a), in T2-weighted (b), and in fat-suppressed proton density spin echo (c) images. Note a tear seen in posterior horn in (c) (zigzag arrow). Note the high-signal-intensity of joint fluid in T2W and in fat-suppressed image (dual arrow).

movement during knee flexion. The end-stage OA is very much an end stage also for that compartment's meniscus; it is either severely torn, or macerated or entirely absent (Figures 12.14 and 12.15) [52]. Any structural change in meniscus will affect all of its functions [27,28,53]. Meniscus tears are common in advanced OA in the affected compartment, but they are less common in the early phase of the disease. They can affect any or all parts of the meniscus [54]. Tears in the meniscal body can extend to the anterior and/or posterior horn and vice versa. Meniscus tears change the hoop forces in the meniscus altering meniscus' load-absorbing ability and in thus changing the loading conditions in the knee [27,28]. On the other hand, the tears also affect the joint geometry and the ability of the meniscus to contribute to knee joint flexion. The meniscectomy, which also changes the geometry of the articulating surfaces and the load distribution over the joint, is known to lead to the development of knee OA [27,28,53]. Meniscus subluxation, where part of the meniscus is extruded outside the articulating tibia cartilage surface, changes the geometry of the articulating surface and the distribution of the load (Figure 12.15) [53]. Advanced meniscus tears have predicted the progression of the disease, defined as cartilage loss [54,55]. Meniscus subluxation has been associated with OA and has also correlated to and

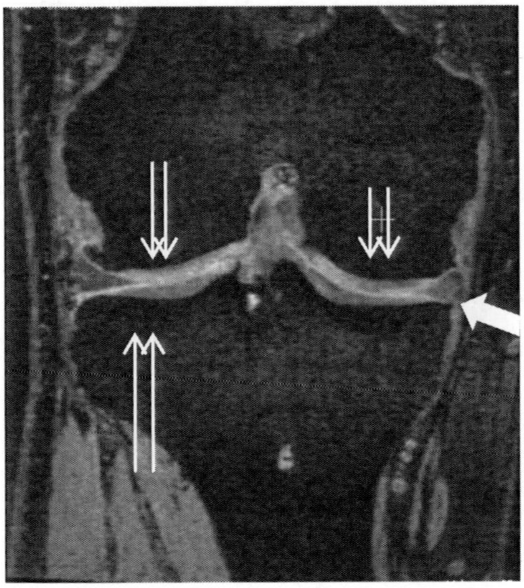

FIGURE 12.15 Coronal water-suppressed 3D DESS image shows subluxation of the body of medial meniscus (thick arrow). Also note the widened and flattened articulating bone surfaces of femur (two short arrows) and tibia (two long arrows).

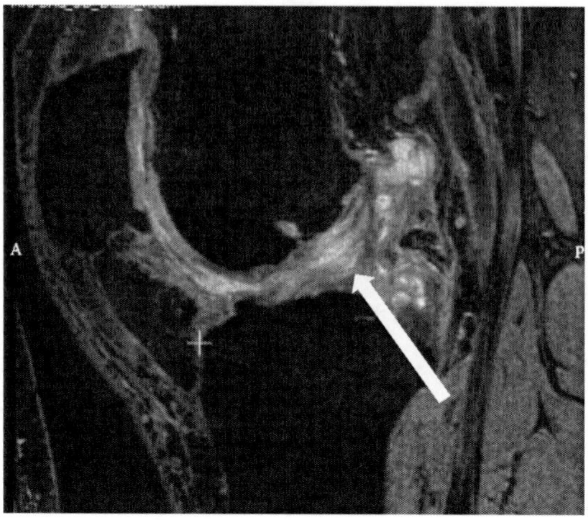

FIGURE 12.16 Sagittal image of knee with torn ACL (thick arrow).

predicted disease progression [56]. However, since meniscus tears are not as common in earlier OA, it is not clear how they are associated with the development of OA.

12.1.9.4 Knee Alignment

Changes in knee alignment, defined as the angle between the long axis of the distal tibia to the long axis of the distal femur at the level of knee joint, have been associated with the development of OA [57]. Several studies have shown that many knees with medial compartment OA have a greater degree of varus than normal knees and many knees with lateral compartment OA have a greater degree of valgus. These findings have led to an interpretation that varus or valgus malalignment is either associated with or is a risk factor for OA [57]. The change in knee alignment is considered to affect the biomechanics of the knee; the increase in varus or valgus is considered to change the load balance between the medial and lateral compartment. However, whether the malalignment is a result of OA or a result of changes in knee supporting structures leading to OA is not clear.

12.1.9.5 Ligaments

Acute ACL tears lead to a very high incidence of later OA regardless of the type of treatment.

On the other hand, an ACL tear can be a complication/consequence of the OA. Enlarging osteophytes at the edges of the intercondylar femoral notch can physically wear and consequently cause a rupture of weakened ACL. A torn ACL appears in MR as an ill-defined, greatly enlarged structure with a higher-SI band at the location of the ACL (Figure 12.16) [58].

In the acute post-ACL tear phase, the MR images have revealed bone marrow edema lesions and local compression fractures both in the lateral femur and in the tibia [59]. Even if all these injuries by themselves can lead to secondary OA, the changes in knee biomechanics are considered to be an equally important factor in the development of post-ACL tear OA [59,60].

12.2 MR Imaging of Osteoarthritis

Saara Totterman

MRI is currently the best imaging modality for musculoskeletal structures. It visualizes bones, including cortex and trabecular bone, bone marrow, cartilage, ligamentous structures, menisci, fat, muscles, tendons, vascular structures, and larger nerves in great detail.

Clinically, MR imaging is a diagnostic tool for different knee diseases and conditions. Both in late and acute post-traumatic phases, it is used for meniscal tears, ligament tears and injuries, cartilage and cartilage bone fractures, and ruptures and tears of tendon and muscles. Perhaps the most common indication for MR imaging outside of trauma is diagnosing the source of knee pain. Even if MR imaging is not used to make the OA diagnosis, it is now widely used by the medical device industry to develop patient-specific treatment options and tools to guide surgical procedures.

Since MR imaging shows the soft tissue structures of the knee in great detail, it is extensively used in cross-sectional and longitudinal OA progression studies to better understand the disease itself [61]. The hope is that this will lead to the discovery of new measurable OA MR imaging findings which are sensitive to disease progression. Their use in OA drug discovery should then facilitate more efficient and less costly development of new OA therapies.

The following will discuss the imaging sequences used for imaging bone and soft tissue structures in knee OA subjects, their MR image findings, and how to image them for quantitative measurements. The following discussion will concentrate on cartilage, bones, menisci, and ligaments.

12.2.1 MR Imaging of Cartilage in OA

12.2.1.1 MR Imaging Sequences for Cartilage

The most common cartilage imaging MR pulse sequences are: T1-weighted (T1W), proton density (PD), T2-weighted (T2W), spin echo (SE), multiecho SE sequences, and their fat-saturated or water-suppressed counterparts, 3D spoiled gradient recall echo (SPGR), 3D gradient recall echo (GRE), and fat-saturated or water-excited counterparts with newer faster modifications, all with their manufacturer-dependent variances, and 3D dual echo steady-state gradient recall echo sequence (DESS) [62,63]. Since the detection of changes, including small cartilage defects and local SI changes, greatly depends on the image resolution, many new pulse sequences have emerged with better signal-to-noise ratios and high isotropic resolution. Examples are the 3D fast spin echo (FSE) and the 3D FSE-Cube sequences [64,65]. To better evaluate the cartilage–bone interface considering the very short T2 time of calcified cartilage, new ultra-short-echo time imaging (UTE) sequences have been applied in cartilage imaging with and without fat saturation.

12.2.1.2 Imaging Planes

The MR images can be acquired in any of the three orthogonal planes or any in oblique modifications of them (Figure 12.17). The most qMRI studies evaluating tibio-femoral joint cartilage plates have acquired images in sagittal and/or in coronal planes. The 3D WE DESS imaging in the sagittal plane used for OAI data is more sensitive for cartilage loss than the 3D fat-saturated T1W SPGR sequence acquired in the coronal plane. Cartilage in the patello-femoral joint can also be seen in sagittal plane images [66]. However, axial plane imaging is better for knee cartilage due to its curved anatomy. The newer isotropic high-resolution sequences may make future imaging plane selection more or less irrelevant [64,65].

12.2.1.3 Cartilage in Different Pulse Sequences

MR appearance of cartilage depends on the pulse sequence used and the resolution. In highly T1-weighted (T1W) images like 3D spoiled gradient recall echo (SPGR) images, it has an almost homogenously high signal intensity (Figure 12.17a); in proton density (PD) images, for example, proton density SE images and proton density-weighted (PDW) 3D gradient recall echo (GRE) images, its SI changes gradually from lower signal intensity in deep layers to high signal intensity in the superficial layer [66,67]. In T2-weighted (T2W) images like T2W spin echo images and T*-weighted (T2*W) GRE images, the cartilage has lower SI (Figure 12.18). Higher-resolution images like 3D GRE and 3D DESS reveal more intricate inner structures of the normal cartilage, including a columnar appearance of normal cartilage whose origin is not well understood (Figure 12.19).

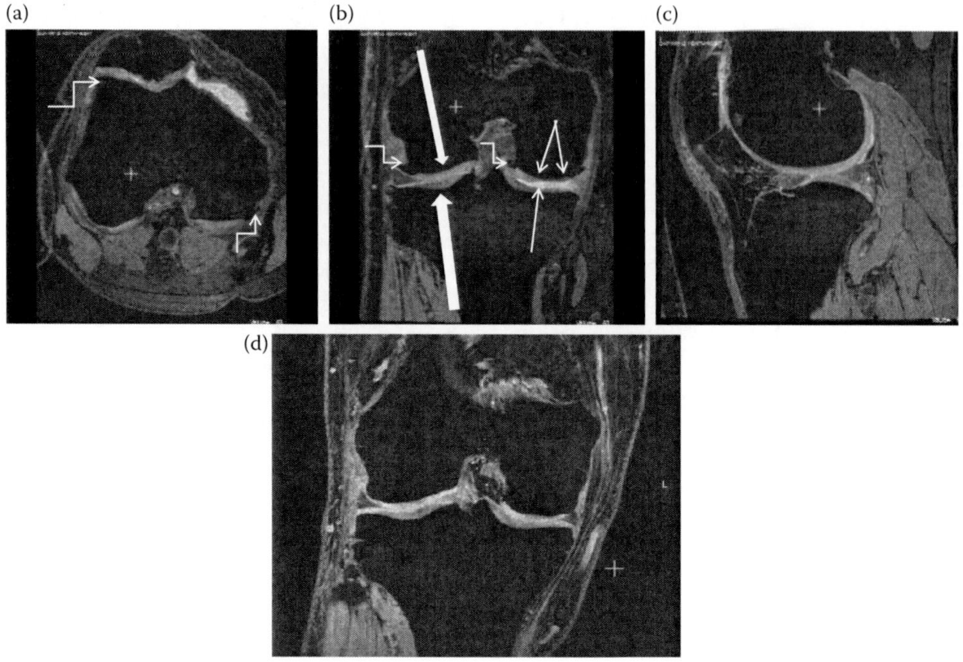

FIGURE 12.17 MR images of the osteoarthritic knee in three orthogonal planes, axial (a), coronal (b), and sagittal (c). The crosshair shows the same anatomic location in each plane. Note the widening and flattening of the tibial and femoral articulating surfaces (thick arrows), thinning and defect in lateral tibia and femur cartilage (thin arrows), and osteophytes (zigzag arrows). (d) 3D spoiled gradient recall echo (SPGR) images.

What imaging sequences and what planes are used depend on the purpose of the imaging. The purpose of clinical imaging is making a diagnosis and generally staging the disease. Progression studies have multiple purposes, including to detect how the diseased joint is different from a normal joint; in other words which MR findings reflect the existence of the disease, which show the progression of the disease, which predict disease progression, and which change most over time or are most sensitive to the progression. The clinical trials have predefined MR image findings, or imaging biomarkers, for which

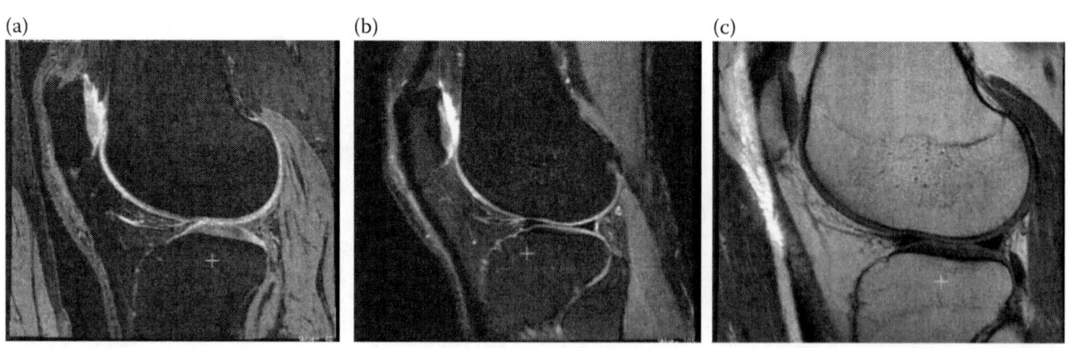

FIGURE 12.18 Cartilage signal behavior in water-exited 3D DESS (a), in fat-suppressed proton density spin echo (b), and in proton density spin echo (c) images. Note the good contrast between low-signal-intensity bone and high-signal-intensity cartilage in the water-exited 3D DESS image and the contrast between the low-signal-intensity deep layer and the higher-signal-intensity superficial layer in the spin echo images. The signal intensity difference is accentuated by fat suppression.

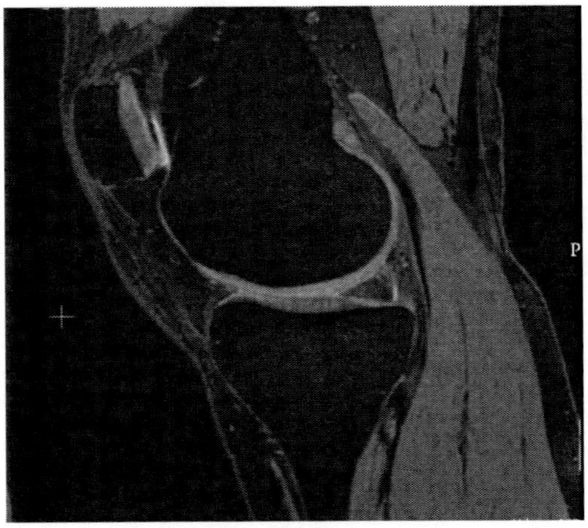

FIGURE 12.19 This high-resolution water-exited 3D DESS image demonstrates the columnar appearance of normal tibial cartilage.

the knee is imaged over the trial period. The image findings are then either scored or quantitatively measured to determine therapeutic efficacy.

12.2.2 Imaging of Structural Cartilage Changes

12.2.2.1 Quantitative Magnetic Resonance Imaging

Quantitative magnetic resonance imaging (qMRI) is widely used for cartilage, and less for bones, menisci, or muscles. The purpose of quantitative MRI is to provide numerical measurements for different OA-related structural changes. For that purpose, the cartilage, bones, menisci, and or muscles are segmented either by manual tracing or by semiautomated or automated segmentation methods (Figure 12.20) [67–70]. The segmentation provides a basis for both simple thickness and volume or more advanced curvature or other shape parameters' calculations. For visualization purposes, the segmented structures need to be surface rendered to create their 3D models (Figure 12.21) [71–74]. For complex measurements, the segmentations are treated as physical objects and all their physical parameters like

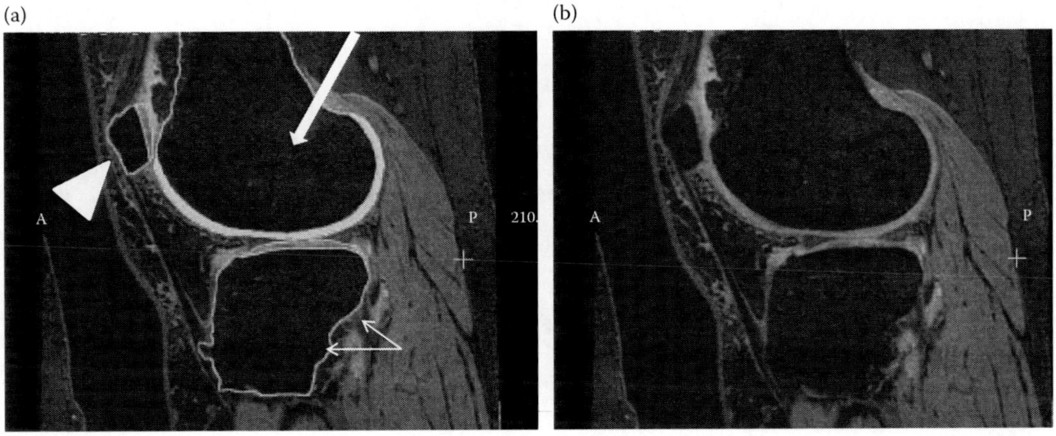

FIGURE 12.20 Segmented femur (thick arrow), tibia (thin arrows), and patella (triangle) bones and their cartilage in (a). These were automatically segmented from water-exited 3D DESS image series. The corresponding slice location is shown in (b).

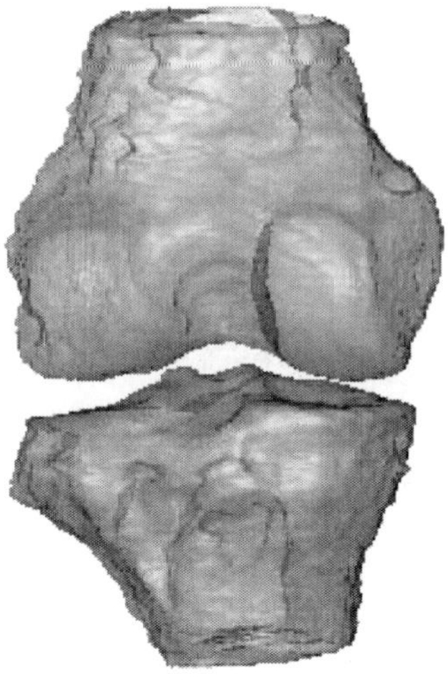

FIGURE 12.21 **(See color insert.)** A surface rendering of automatically segmented femur and tibia bones seen from behind. Note the widened and flattened tibia and femur in this OA knee.

thickness, curvature, and so on can be calculated (Figure 12.22) [38,75]. Information, for example, thickness, can be also visualized and displayed as a thickness map that visualizes the thickness measured at each point on the cartilage surface (Figure 12.23) [72–74]. Parameter maps such as thickness for different time points can be compared to create difference/change maps. Since the thickness maps calculate thickness for each surface point on the structure's surface, change maps calculate and visualize the change in thickness over time, either increase or decrease, for every surface point [72–74]. Newer, more advanced methods calculate numeric values representing the local or whole volume of cartilage lost due to defects and/or thinning or that gained in hypertrophic regions [73,74].

12.2.2.2 Cartilage Thickness and Volume Changes

Structural cartilage changes are caused by loss or gain [76]. Cartilage loss in OA is caused by degenerative defects and/or thinning from above or by advancing calcifying cartilage with an advancing tidemark from below [77–80]. Both of these lead to a decrease in cartilage thickness and volume. Cartilage hypertrophy and swelling, noted in the early phase of the disease, increase both thickness and volume [80].

In the OA knee, qMRI using manual, semiautomated, or automated segmentation is mostly used to study cartilage volume and thickness and corresponding changes in longitudinal and cross-sectional

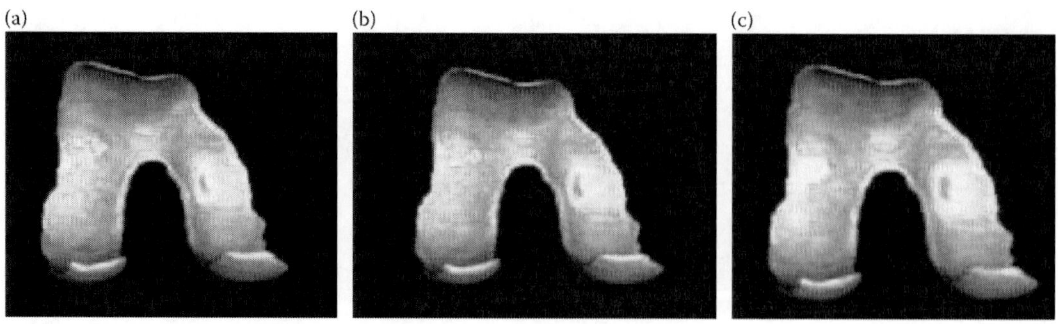

FIGURE 12.22 **(See color insert.)** Femur cartilage thickness maps derived from automatically segmented image sets for baseline (a), 12 months (b), and for 24 months (c) follow up. The maps are seen from above. The yellow to red indicates areas of increasingly thinner cartilage.

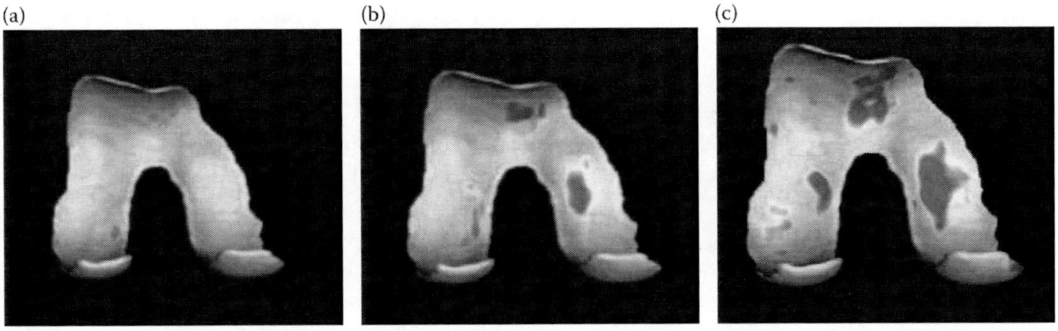

FIGURE 12.23 **(See color insert.)** Results of mapping the cartilage thickness maps from Figure 12.6 to the reference atlas. The maps are seen from above. The baseline cartilage thickness (a) is compared to the 24 month follow-up visit (b) to compute the change in cartilage thickness over time (c). Note the locality of the changes; the defects are local and the local defects are getting deeper and bigger. However, at the same time, new local areas of defects as well as thickening appear.

OA studies. However, since cartilage swelling and hypertrophy as well as cartilage defects are often colocated in the same cartilage regions, their combined effects on such gross measurements can obscure longitudinal change. The thickness maps capture the local bidirectionality of the changes with greater sensitivity (Figures 12.22 and 12.23) [71–74]. Therefore, particularly in early OA studies, thickness change maps are gaining more interest. Their use has improved the medical communities' understanding of the events related to the progression of the disease. Existing studies have shown that bidirectional changes continue through the different stages of the disease; they do not progress in the same location on a continuous scale, and instead new areas in different regions get involved and progress more rapidly [80]. This heterogeneity in the location and rate of progression makes it very difficult to follow disease progression with average thickness and volume measurements [80]. Therefore, more advanced algorithms have been developed to quantify changes related only to cartilage lost or gained from the thickness difference maps [72–74]. These will be discussed in detail in the following section.

Cartilage qMRI studies use mainly fat-saturated 3D SPGR, water-excited (WE) 3D DESS, and their fast counterparts (Figure 12.24) [62–66]. One group used both fat-saturated 3D SPGR and 3D GRE sequences, which they are coregistered and fused for segmentation [70]. The fat-saturated 3D SPGR provides good contrast between the saturated low-SI bone marrow fat and the high-SI cartilage [14].

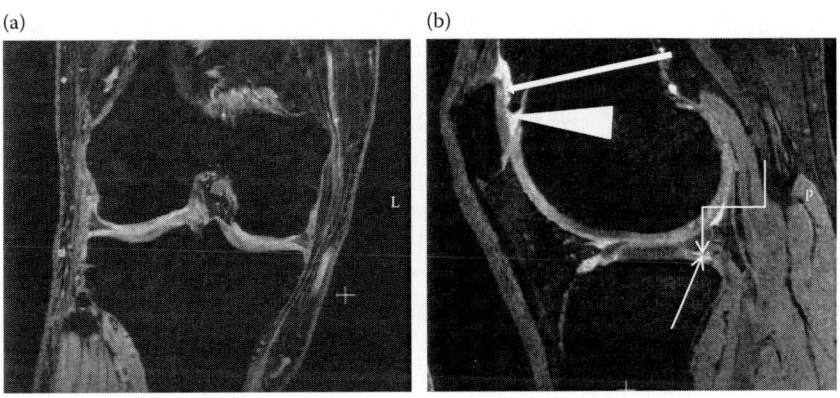

FIGURE 12.24 The two most common qMR images, a fat-saturated 3D SPGR (a) in coronal plane, and a water-exited 3D DESS in sagittal plane (b). Note the good contrast between bones and the cartilage in both images. Also note the good contrast in DESS series between the high signal intensity fluid (zigzag arrow) and a small cartilage defect in lateral tibia cartilage (thin arrow) and between fluid (triangle) and patella cartilage irregularities (thick arrow in b).

However, the contrast between the fluid and cartilage and between degenerated cartilage and degenerated menisci is not good [62–66]. When appropriate imaging parameters are used, the WE 3D DESS series provides good contrast between the medium-SI cartilage and the high-SI fluid (Figure 12.24) [62–66]. Both pulse sequences have problems separating the degenerating cartilage from the inflamed synovial tissue and separating the cartilage from the degenerated menisci.

The most important factor in early OA cartilage imaging besides tissue contrast is image resolution. In-plane resolution is crucial for detection of small cartilage defects and fissures, while slice thickness is important for accurately reflecting the thickness and MR properties of the curved cartilage [62–66]. The higher performance of the newer 3D pulse sequences and the newer receiver coil technology has considerably improved the odds of detecting earlier cartilage, bone–cartilage interface, and even bone changes [64,65].

12.2.3 Imaging of Compositional Cartilage Changes

12.2.3.1 T2 and Signal Intensity Changes

As mentioned before, osteoarthritic changes lead to cartilage T2 relaxation time (T2) and SI changes, and changes in GAG content of the cartilage. T2 is evaluated using both dual and multiecho spin echo sequences with 4–11 echoes [81,82]. Although the fast spin echo (FSE) dual echo sequences, due to stimulated echo, were not considered to provide reliable measurements, their fat-suppressed counterpart has performed satisfactorily well [62]. T1rho has also been used to evaluate compositional changes; however, since it requires special pulse sequences, it has not yet been widely applied in research [83]. The newer approach is to measure T2*, where the use of UTE T2 mapping has provided promising results to evaluate T2 changes in deep layers of cartilage [84]. The change is considered to reflect degeneration of the collagen bundles.

The signal intensity measurements have mainly used 3D fat-suppressed SPGR and 3D WE DESS sequences [85]. In T1W images, SI changes have roughly correlated to the K-L scores; in 3D WE DESS images, the correlation has been higher [85]. Due to the heterogeneity of OA pathology and that of its locations, both T2 and SI changes vary from local through multiple local to global changes [85]. The conventional 2D T2 measurements from one to two slices provided new information on OA. However, to fully understand the local and global changes during disease progression, studies require T2 information across the entire cartilage plates. Therefore, newer methods use a 3D approach by segmenting the cartilage either manually from T2 image sets or by coregistering the segmented 3D structural images to the T2 image sets and extracting the T2 values (Figure 12.25) as discussed in the following section. In OA populations, the 3D T2 measurements have proven to be reproducible. Even though they correlate to K-L scores, they are not in wide clinical use for diagnosis or for staging of the disease. However, there is a concerted effort by using the data collected in the NIH Osteoarthritis Initiative (OAI) to validate its utility as an early-phase diagnostic tool and its predictive value for disease progression.

Heterogeneity of cartilage SI is reflected in the variance of cartilage plate SI values. Cartilage SI changes are also reflected in the contrast between subchondral bone and the deep layers of cartilage. Normal cartilage has high contrast values; however, the vascularization of the subchondral bone plate, advancement of the tidemark, and increase in cartilage water content in osteoarthritic cartilage decrease the contrast. Since the nature of such changes in cartilage SI behavior is very sequence dependent, there is much less information available about local or global cartilage SI behavior in OA.

12.2.3.2 Cartilage dGEMRIC

Cartilage GAG content is measured using delayed gadolinium-enhanced MR imaging (dGEMRIC) [14]. Even though dGEMRIC images showed that cartilage T1 decreases in advancing OA, several confounding factors make defining the significance of the measurement and its change problematic. For example, body mass index (BMI), a known predictor and risk factor for OA, is directly related to the dGEMRIC index. Further, hypertrophic cartilage changes in degenerating cartilage may increase GAG content, also increasing the index. For that reason, though dGEMRIC is useful in cartilage research, it has less value in longitudinal progression studies or clinical trials.

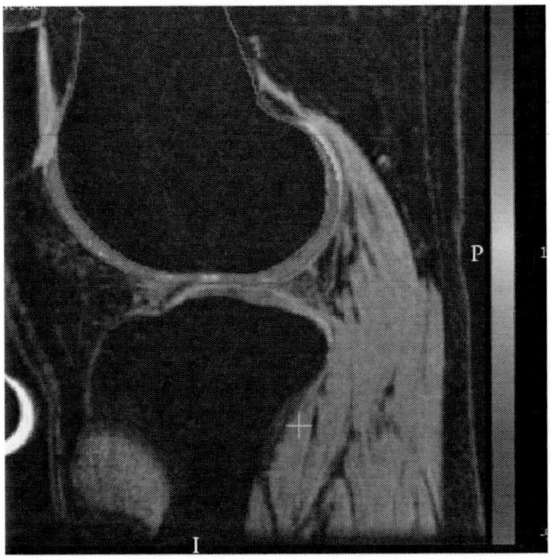

FIGURE 12.25 T2 map of segmented femoral cartilage.

12.2.4 MR Imaging of Bones in OA

12.2.4.1 Bones in Different Pulse Sequences

As mentioned before, bones are composed of two structurally different components, the very tightly woven cortical bone and the loosely connected trabecular network of inner trabecular, cancellous bone. Cortical bone envelops the inner trabecular bone and is directly connected to it. Due to the lack of water protons, cortical bone has very low signal intensity in all pulse sequences. Trabecular bone is seen as a low SI network visualized especially in more T2*W 3D GRE image series against the higher-SI red marrow and the high-signal-intensity fatty marrow, which fill the spaces in the trabecular network [79].

The commonly used qMRI sequences are suitable for segmenting the bones using both manual and automated approaches. The osteochondral junctions in tibia, femur, and patella are best seen in T1W fat saturated and WE 3D SPGR series, but can be easily seen in the other common qMR imaging series, WE 3D DESS series (Figure 12.26) [62–66]. In both of these series, the high-signal-intensity cartilage is well contrasted against the low-signal-intensity saturated/suppressed bone marrow fat. In

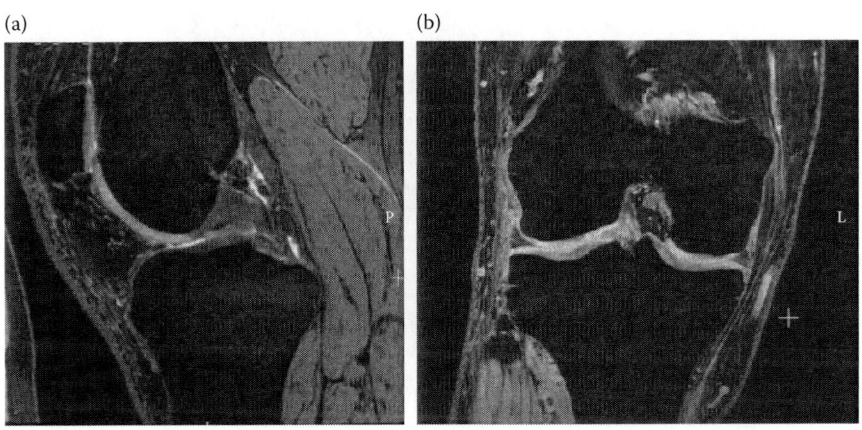

FIGURE 12.26 Water-exited 3D DESS (a) and fat-suppressed 3D SPGR (b) images. Both show good contrast between cartilage and bone. Cartilage to cartilage separation is not clear in SPGR sequence in (b).

these images, the cortical bone in the shaft, which is bone outside the articulating surfaces, is not as easily distinguished from the surrounding low-SI fat. Therefore, some groups have chosen to use both a fat-suppressed 3D SPGR sequence and a 3D GRE sequence, then coregister and fuse them and use the combined data for bone segmentation [70,73]. Automated atlas-based segmentation algorithms perform well in qMRI sequences [74].

12.2.4.2 Imaging of Structural Bone Changes

The structural bone changes induced by OA include flattening of the subchondral bone plate, widening of the cartilage–bone interface and the articulating ends of the bones, and emergence of peripheral and central osteophytes (Figure 12.27). All of these change the shape of the articulating end of the bones. Flattening of the subchondral bone plate changes the curvature and the spatial location of the bone–cartilage interface; widening of the articulating end of the bones increases the surface area of the subchondral bone plate and the curvature; the central osteophytes make the bone–cartilage interface irregular and bumpy; and formation of peripheral osteophytes widens the bone ends [71,75,86,87].

12.2.4.2.1 Curvature Changes

As mentioned above, flattening and widening of the subchondral bone plate in OA changes the curvature of the articulating bone surfaces [75,87]. Measuring the curvature or the bone–cartilage interfaces of the tibia, femur, or patella per se do not require segmenting the bones (Figure 12.20). The curvature can be evaluated from the segmented cartilage image data sets which are created during the routine articular cartilage qMRI processes. However, it does require surface rendering of the segmented cartilage and evaluating the bone-facing surface (Figure 12.28). Since the cartilage-only segmentation approaches do not include the cartilage at the osteophytes, this portion of bone–cartilage interface is excluded from surface area and curvature calculations. Many of the segmentation approaches divide the femur cartilage into medial and lateral weight-bearing regions, and medial and lateral posterior condyle and trochlea regions, and this enables the curvature change calculations to be performed on a per-region basis [75,89]. The commonly used measurement is average curvature; however, the standard deviation of curvature measurements reflects the variability of the local curvature measurements, indicating irregularity, the presence of central osteophytes, or other local changes in curvature [75].

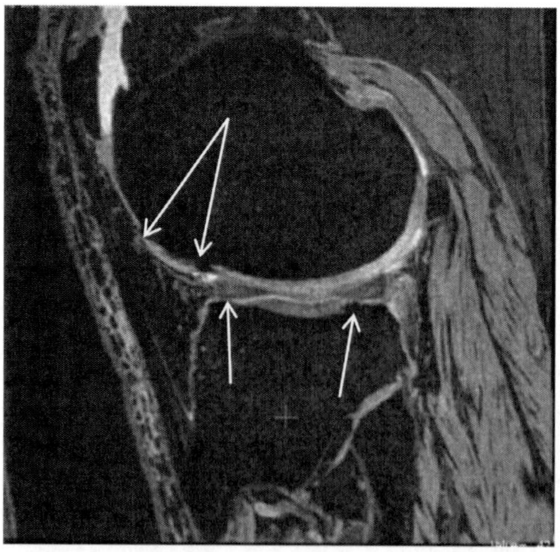

FIGURE 12.27 A sagittal 3D DESS image of advanced knee OA. Also note how the advancing bone in the form of central osteophytes and the lateral tibia and femoral trochlea (thin arrows) is changing the shape of the articulating surface of the lateral tibia and that of the femoral trochlea.

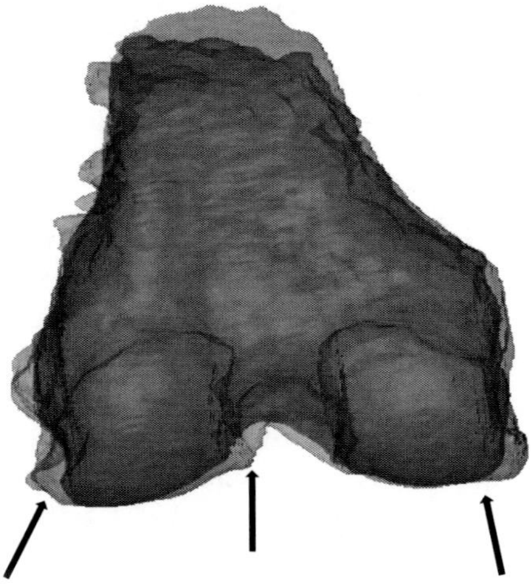

FIGURE 12.28 (See color insert.) Surface-rendered results of mapping the segmented femur of the OA subject (green) to the segmented femur of the reference atlas (red). Note the flattened and widened articulating surfaces in the subject's femur (thick arrows).

12.2.4.2.2 Increase in Bone–Cartilage Interface Surface Area

The processed qMRI cartilage data can also be used to compute the bone–cartilage interface surface area. Similar to curvature calculations, the bone-facing surface needs to be identified in the image set. The increase in bone–cartilage surface area is primarily caused by widening of the articulating ends of tibia and femur. The widening of the entire tibial end, including the proximal end of the shaft and formation of osteophytes, is easily identified in MR images. This phenomenon can be evaluated from processed cartilage qMRI image sets, in cases where the bone segmentation is part of the cartilage qMRI process. The surface-rendered segmented bone image data can also be used to create shape change maps to visualize the nature, location, and the degree of the widening either from time point to time point or in correlation to the normal population reference map. These maps can also be used to investigate the changes. These methods are explained in the following section.

12.2.4.3 Osteophytes

Peripheral osteophytes cannot be evaluated from segmented cartilage image data. Their evaluation requires segmenting the bones themselves. That could be done manually, semimanually, or using automated methods. The automated segmentation methods are considerably faster and more reproducible than the others. Most of the automated cartilage segmentation methods include bone segmentation as part of the process. Many of the methods also divide the bones into anatomic regions, which enables following local changes in the resulting numerical data. Using the same method employed for following cartilage and bone surface features, the segmented bone image sets can also be used to create osteophyte maps and osteophyte change maps. Those have shown a measureable increase in osteophytes volume over a 1-year period of time.

The bone and bone–cartilage segmentations from research subjects' and OA patients' MR images are commonly used by the medical device industry. For years, the industry used research subjects' image segmentations in total and partial knee prosthesis product development. Now, as the medical device industry is moving toward patient-specific device development, it uses patient-based bone and cartilage segmentations from MR images for surgical planning and for developing patient-specific guides for surgery or for developing patient-specific prostheses.

12.2.4.4 Imaging of Compositional Bone Changes

12.2.4.4.1 Bone Marrow Edema Lesions

Bone marrow edema lesions and changes in trabecular architecture are considered compositional bone changes. Bone marrow edema lesions are seen as high-signal-intensity areas in T2-weighted (T2W), fat-suppressed (FS), and water-excited (WE) spin echo (SE) images (Figure 12.29) [88]. However, due to their watery content, they are also seen in both PD and intermediate-weighted FS and WE spin echo images and PD and IW inversion recovery image series. However, the FS and WE gradient echo (GRE) images, either SPGR or GRASS or the WE 3D DESS series, do not visualize them well. Partially, this is caused by the magnetic susceptibility effects of the trabecular bone [88].

Quantitative BME lesion measurements require their segmentation [90]. Since the BME lesions can be located in any or all knee compartments or regions and since they have high variability, reliable clinically useful information of BML changes in volume, T2, and/or location requires also segmentation of all knee bones.

In theory, that could be accomplished just by manual segmentation of BME lesions and bones from T2 image sets. However, automated segmentation using the T2 image sets or coregistering automatically segmented structural image sets with T2 images and then segmenting the lesions and calculating their values is much faster and more reproducible [90]. In addition, the segmented data can be used to create local BME lesions' change maps. Furthermore, it can be used to correlate their spatial relationship to local cartilage defects and local compositional cartilage changes [90].

12.2.4.5 Changes in Trabecular Bone Architecture

The architectural changes in trabecular bone seen in OA have mostly been evaluated using fractal analysis and connectivity in CT and fractal analysis in plain films. However, since the high-resolution higher-field MR images show the trabecular bone structures in great detail, there is a growing interest and effort to extract the same parameters, bone volume fraction, trabecular bone thickness, number, and connectivity also from MR images and to develop more sensitive methods to detect the change [91].

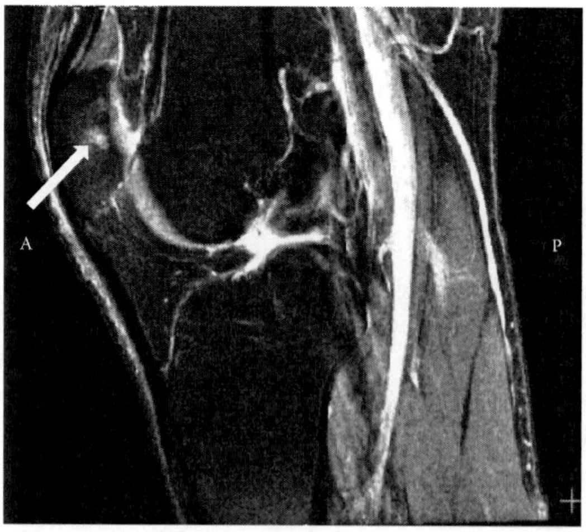

FIGURE 12.29 A bone marrow edema (thick arrow) in the patella seen in a fat-suppressed proton density spin echo image.

12.2.5 MR Imaging of Menisci in OA

12.2.5.1 Menisci in Different Pulse Sequences

The MR imaging appearance of menisci reflects their inner architecture of dense, mainly unidirectional tightly bound collagen bundles. This leads to their very low signal intensity in most of the pulse sequences, traversed by peripheral higher-signal-intensity vascular structures. Meniscus degeneration, tears, and maceration disturb or disrupt that architecture. Degeneration leads to local high-SI areas located in the middle of the meniscus; tears are seen as irregular lines traversing from the upper or lower surface into the meniscus or being contained inside the meniscus [92]. Commonly used pulse sequences for diagnosing meniscus tears are T1W, PDW spin echo (SE), sequences, T1W and PDW fast spin echo (SE) sequences and PD fat-suppressed (FS), or water-excited (WE) FSE, sequences. Cross-sectional and longitudinal OA studies use PD FSE or FS PD FSE sequences to score the meniscus changes [92]. The 3D WE SPGR qMRI cartilage sequence, which is heavily T1W, is not good for meniscus imaging. In those images, the higher-signal-intensity menisci blend with the surrounding structures. On the other hand, 3D WE DESS sequence demonstrates menisci well and, due to the high resolution, shows tears better than the SE sequences.

12.2.5.2 Imaging Composition and Volume of Meniscus

Since meniscus tears and degeneration eventually decrease the size of the menisci, some groups have developed methods to segment them for volume calculations [93,94]. Other groups use dGEMRIC, T2, or T1rho to detect the early degenerative meniscus changes or correlate the changes to the OA stage [95]. Meniscus subluxation has been shown to be a risk factor and/or predictor for OA progression. Most groups score the changes, even though some groups have developed methods to quantify subluxation. Much still needs to be done before those methods will be in common use.

12.2.6 Knee Alignment

Knee alignment measurements have long relied on plain films, where varus/valgus angulation between the long axis of the tibia and femur has been measured [96,97]. Flattening and widening of tibia and femur as well as changes in cartilage thickness changes the geometry of the articulating surfaces [96–98]. Furthermore, meniscus degeneration, tears, and maceration cause considerable changes to the location and amount of the local physical loads [99]. All of these factors have the potential to cause changes in the 3D spatial relationship between the bones by rotating, translating, and/or angulating them in different degrees. Therefore, it is highly likely that OA, which affects the geometry of all joint structures, will change the alignment of the bones three-dimensionally not only in plain film plane. Furthermore, both plain films and MR images show that in OA changes, the spatial relationship between the tibia and femur and between patella and femur has changed. Visually, the femur in OA knees has translated medially. Furthermore, for two decades, radiologists have used the posterior translation of the femur over the tibia as secondary for ACL tear diagnosis. As mentioned before, the segmentation of structures generates anatomically accurate 3D objects, whose physical parameters and their relationships to each other can then be evaluated [89]. Different groups have already used that approach to compute the distance between the closest point of femur and tibia and femur and patella [89]. Furthermore, they have computed and visualized the location for tibia and femur cartilage contact areas and calculated the size for the contact area [101]. Newer algorithms are in development to calculate the translational and rotational differences between normal and ACL-deficient knees and evaluate whether and to what degree the 3D spatial relationship in OA knees is different from that in normal knees (Figure 12.30) [101]. One reason, which has greatly increased the interest in 3D spatial relationship between the articulating bones in the knee joint, is the fact that anterior cruciate tears in the majority of cases lead to OA regardless of whether they are treated conservatively or surgically and regardless of the surgical method. That realization has led to a series of studies to develop methods to

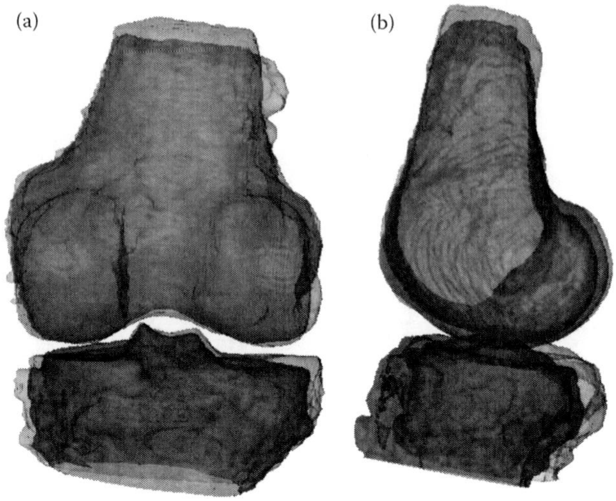

FIGURE 12.30 **(See color insert.)** Posterior (a) and lateral (b) views of the 3D relationships. The 3D relationship between the subject's tibia and atlas' tibia. The subject's segmented tibia and femur were coregistered to the atlas' tibia and femur. The images show subject's tibia as green and the location of atlas tibia as blue. The atlas femur is red and the subject's femur is green.

investigate how the tibio-femoral contact points and their movement during knee flexion are different in normal and ACL-deficient knees and how the ACL tear has changed their 3D relationship [100–103].

Some semiautomated and all automated segmentation methods used for cartilage qMRI studies segment both bones and their cartilage from the MR images. Thus, they already provide the basic data needed to process their 3D relationship. The segmented bones need to be surface rendered, after which they can be treated as 3D objects and their 3D spatial relationship can be calculated (Figure 12.30). The direction and the degree of the translation and possible rotation in OA are still under investigation. Furthermore, how these changes are associated or correlated to the progression of other OA changes or can serve as predictors for the disease progression still needs to be evaluated.

12.3 Quantitative MRI-Based Evaluation of OA: Procedures and Algorithms

José G. Tamez-Peña

12.3.1 Analytical Procedures for Use of Quantitative MRI in OA

During the last couple of decades, the progression of the OA has been studied using MRI and it has been found that the annual degree of cartilage loss is very small (Figure 12.31) [104–109]. These results and others indicate that precise measurements are required from the MRI analysis in order to achieve a good understanding of the OA progression. Not just good precision is required, but accurate measurements are required to compare results from different studies and to make valid conclusions from multicenter data. The desired accuracy and precision necessary in OA analysis require a good understanding of the process involved in getting the quantitative results. Therefore, it is important to map all the process in getting a quantitative measurement from patient preparation to the final calibrated measurement. This section will map the process involved in an OA measurement using a standardized approach that can be used to create robust operating procedures aimed to get valid OA measurements from current MRI equipment. Furthermore, this section will describe the different approaches

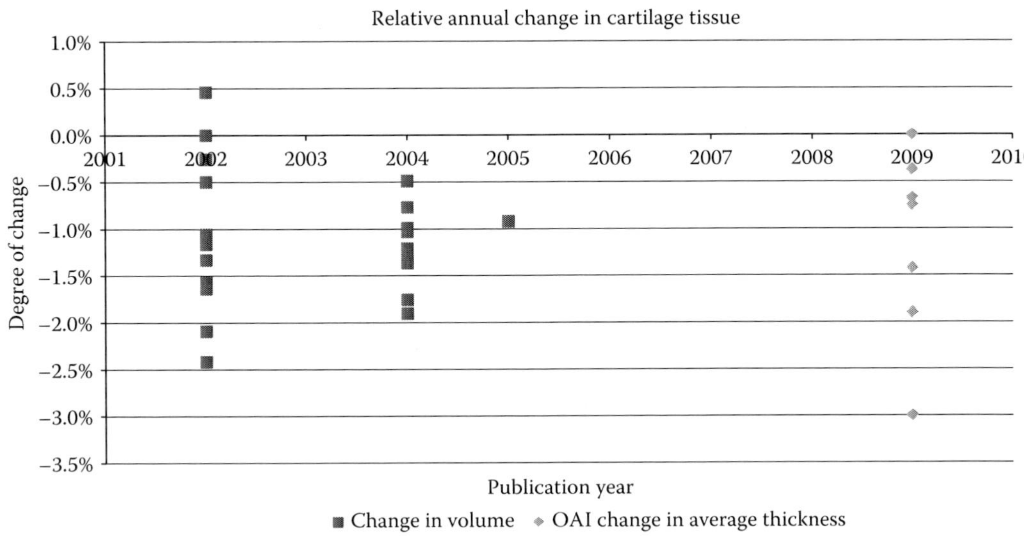

FIGURE 12.31 The slow progression of cartilage loss. The reported degree of cartilage loss by different studies (Eckstein et al. 2006) has been on average similar across time (1%). Therefore, the qMRI system must be able to measure small amounts of changes for year-long studies.

that have been used in getting quantitative information from MR images. Figure 12.32 shows a block diagram of the different processes involved in getting a measurement from the MRI system. These processes are

- Subject preparation
- Image acquisition
- Image analysis
- Quantification
- Phantom scanning
- MRI calibration

To get a valid measurement, each one of these processes must be standardized and the proper documentation must be created to aid in reproducibility of the results. Although all major OA studies have been following this recommendation, sometimes they miss taking into account some aspects that may affect the interpretation of the results. To mitigate the introduction of unwanted errors in the quantitative measurements, it is important to systematically map all the processes that may affect the measurements. In the following sections, we will summarize the different aspects involved in the quantification of the different OA features that can be analyzed in an MRI system. We will place special attention to the validity and reproducibility of the results in light of the nature of the OA progression: a slow degenerative disease.

By considering the quantitative MRI system as an analytical procedure for OA evaluation, the sources of variability and the uncertainties present in the measurement process can be mapped using current guidances. The mapping of uncertainties used in this chapter was done using the EURACHEM/CITAC Guide [110] or the National Institute of Standards and Technology (NIST) Technical note 1297 [111]. Once the variables that can affect the interpretation of the results have been identified, the quantitative procedures must be validated and it is recommended to follow the guidance on valid analytical procedures by the International Conference on Harmonization of Technical Requirements for Registration of Pharmaceuticals for Human Use (ICH) [112,113] and to use the philosophy of the Valid Analytical Methodology Programme of the UK (VAM: http://www.nmschembio.org.uk/).

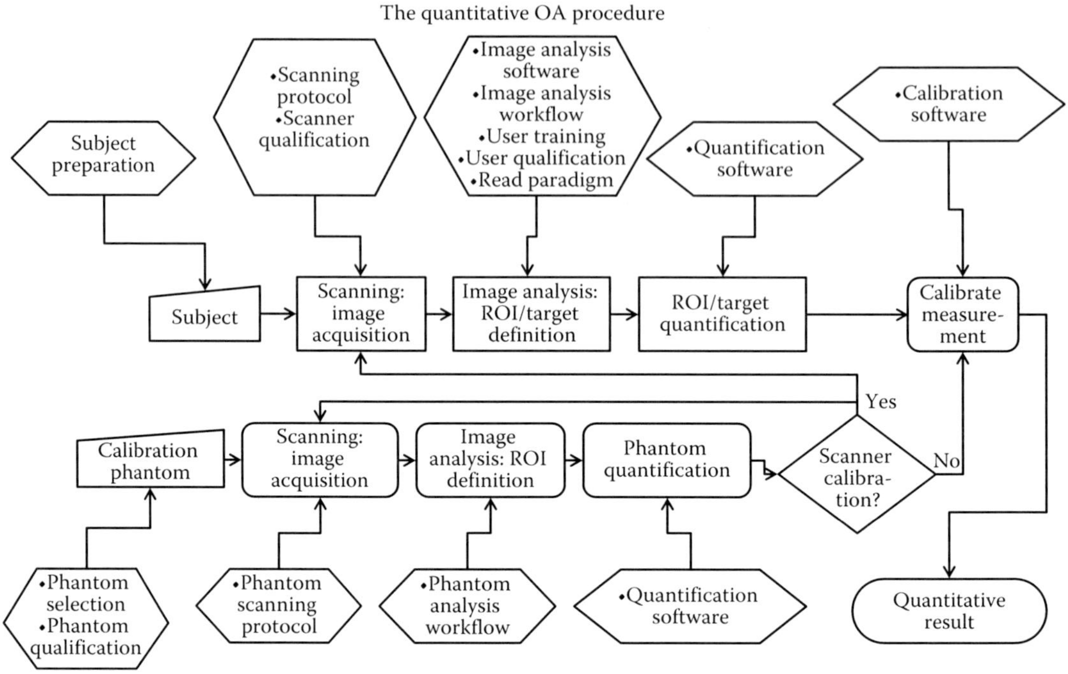

FIGURE 12.32 Procedures involved in obtaining a valid OA quantitative measure from an MRI equipment.

12.3.1.1 Uncertainties of OA Measurement

The mapping of uncertainties in OA is based on the EURACHEM/CITAC Guide CG 4 on the quantification of uncertainty in analytical measurements and the NIST Technical Note 1297 on the Guidelines for Evaluating and Expressing the Uncertainty of NIST Measurement Results [114]. The EURACHEM guide recommends drawing a cause–effect diagram to identify the major components of the system. Figure 12.33 shows a simple cause–effect diagram required to start mapping the possible sources of uncertainties that may play a role in the performance and accuracy of an OA quantitative measurement. The cause–effect diagram places the final measurement in the main horizontal axis and the diagonal arrows indicate different aspects of the quantitative process that may affect the measure via systematic errors (bias) or random errors. Figure 12.34 shows some of the aspects that may affect the quantitative measurements of OA systematically or randomly. As the diagram

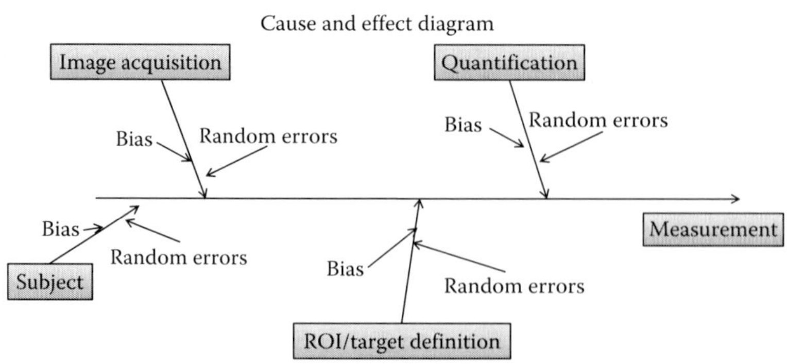

FIGURE 12.33 The identification of the sources of variation in an MRI-based quantitative system starts by drawing the cause–effect diagram of the different processes involved.

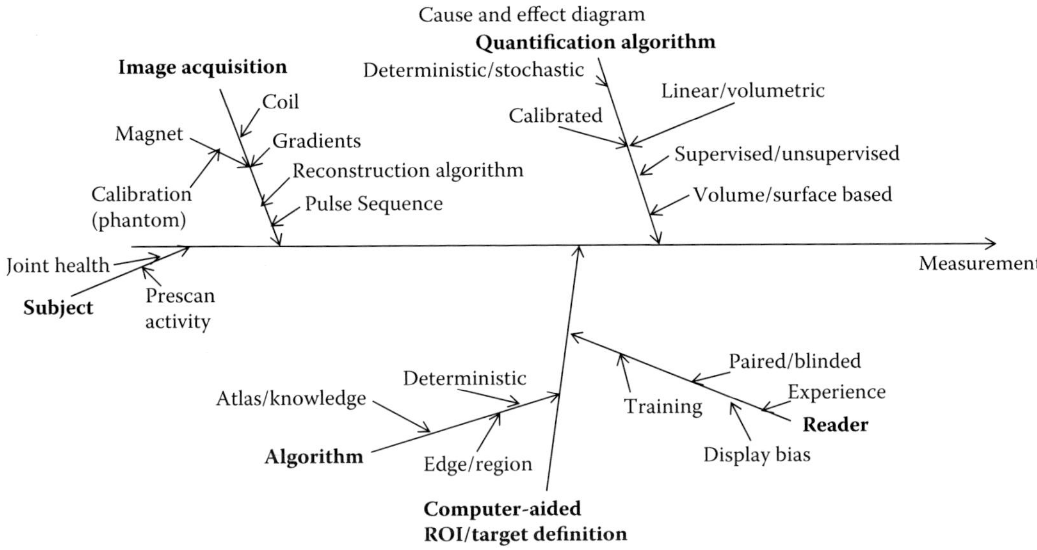

FIGURE 12.34 Different sources of systematic and random noise are present in the quantification of OA features. This fish-bone diagram identifies the different sources of errors that can affect the OA measurements.

shows, we have to consider the following aspects when evaluating the validity of an OA structural measurement:

- Subject:
  - Joint degradation stage: attrition, fibrillation, cartilage lesions, bone lesions
  - Prescan activity: anxiety, standing, walking, running, climbing stairs, and so on
  - magnetic susceptibility: tattoos, metal implants, blood clots, and so on
- Image acquisition:
  - Global distortions: B_0 drift, shim calibration, calibration phantoms
  - Local distortions and spatial sampling: gradient coils, pulse sequence
  - Signal: receiving coil, B_1 inhomogeneity, pulse sequence, flow, patient motion, truncation, partial volumes, and so on
  - Image reconstruction: reconstruction algorithm, geometric reconstruction algorithm, calibrations
- Image analysis—ROI definition:
  - Algorithm type:
    - Supervised
    - Edge/contour-based
    - Region-based
    - Unsupervised
    - Atlas-based: training set
    - Statistical knowledge-based
    - Stochastic/deterministic
  - Reader:
    - Expertise
    - Level of education
    - Years of experience

- Reading paradigm:
  - Blinded
  - Unblinded
- Graphical user interface and hardware:
  - Input device
  - Display
- Quantitation:
  - Preprocessing steps
  - Volume/voxel-based
  - 3D surface reconstructed
  - Stochastic/deterministic
  - Parametric model-based:
    - Linear/nonlinear
    - Fitting algorithm
  - Parameter calibration procedure

The former are some of the steps/issues involved in getting a measurement from an MRI image. Evaluating their individual impact on the final quantitative measurement will be a daunting task, but determining if the effect of each step is random or systemic is not. The systematic errors can be calibrated or randomly controlled and the stochastic errors can be controlled by increasing the number of samples. In the following sections, we will describe the most common algorithms used for the extraction of OA features, but first we will introduce the characteristics of a valid OA measure.

12.3.1.2 Quantitative Validation Characteristics

The quantitative OA measurement may be affected by any of the aspects described in the previous section; therefore, an OA/qMRI manual (imaging charter) will be required to treat the complex imaging procedure as a simple measurement unit that can be described with a unique validation characteristic. The OA/MRI manual procedure refers to the way of performing the quantitative analysis. It should describe the steps necessary to perform each test in detail. This may include but is not limited to the subject cohort, the phantoms, subject preparation, site preparations, use of the imaging equipment, generation of the calibration procedures, image analysis work-stream, and use of the algorithms for the calculation, among others.

We list the validation characteristics of a measurement in Table 12.1. Based on that, we recommend the quantification and documentation of the characteristics in order to have confidence in the final OA quantitative result. The characteristics are

- *Specificity:* The ability of the system to unequivocally assess the imaging endpoint in the presence of components which may be expected to be present. Typically, this may include imaging artifacts, lesions, and so on. The lack of specificity of an individual imaging endpoint can be compensated by other supporting imaging procedures. This definition has strong implications in the pulse sequence selection in OA studies: Not all OA characteristics can be evaluated in a single pulse sequence. Therefore, for structural cartilage evaluation, 3D high-resolution pulse sequences with fat suppression are preferred over 2D sequences. On the other hand, T1 and T2 cartilage evaluation requires sequences that have good SNR; therefore, 2D sequences are preferred over 3D sequences. At the end, the pulse sequence specificity plays an important role in the overall system specificity.
- *Accuracy:* Expresses the closeness of the agreement between the value which is accepted as either the conventional truth or an accepted reference value and the value found.

TABLE 12.1 ICH Q2A Requirements for the Identification of the Validation Characteristics of Any Analytical Procedure

Type of Image Analysis	Semiquantitative	Quantitative Continuous
Accuracy	NA[a]	+
Precision		
• Repeatability	+	+
• Intermediate precision	+	+
Specificity	+	+
Detection limit	–	–
Quantitation limit	–	–
Linearity	–	+
Range	–	+

Note: The quantitative analysis of OA can be seen as an analytical procedure.
[a] NA, not applicable.

The accuracy of a given measurement is composed by the bias and the precision.

- *Bias/linearity:* Expresses the average deviation between the value which is accepted as either the conventional true value or an accepted reference value and the value found using a series of measurements obtained from multiple sampling of the same subject(s) under the prescribed conditions.
- *Precision:* Expresses the closeness of agreement between a series of measurements obtained from multiple sampling of the same subject(s) under the prescribed conditions.
  - *Repeatability:* Expresses the precision under the same operating conditions over a short period of time. When a human observer is involved in the analysis process, this is also known as the short-term intraoperator precision.
  - *Intermediate precision:* Expresses the variability of the whole system under different analysts, different equipment, and so on. When only human observers are involved, this is known as the interobserver variability.
  - *Reproducibility:* Expresses the precision among laboratories (collaborative studies usually applied to standardization of methodology).

The specificity, accuracy, and precision in OA have been determined using pilot experiments usually involving TKA or cadaver studies [115–117]. Volume accuracy has been determined by water displacement, while area evaluation has been done using thin foil analysis [118,119]. The proper method to evaluate and report accuracy and precision is the one recommended by Bland and Altman. Bland and Altman define the statistical instruments to measure the agreement between the new method and the reference standard. Most of the time the pulse sequence defines the sensitivity of the OA measurements, the magnet/hardware calibration dominates the accuracy and the image analysis method is a key player in the precision [106,120,121]. The quantification algorithm also may play a role in the accuracy/precision of the measurement and this is especially true in indirect measurements, like area, thickness, T1, and T2 [122–125]. Therefore, the reported accuracy and precision reported in the literature must be then interpreted carefully to readily identify the source of the measured bias/accuracy/precision in the OA measurement. The specification of the specificity, accuracy, and precision should be accompanied by the following characteristics:

- *Linearity:* The ability (within a given range) to obtain test results which are directly proportional to the accurate value of the measurement.
- *Range:* The interval between the upper and lower value (amount) of the measurement for which it has been demonstrated that system has a suitable level of precision, accuracy, and linearity.

In OA, the linearity/range is important in all T2 and T1 measurements. The pulse sequence selection and the fitting algorithm will determine either the range at which the T1 measurements have a linear relationship between the T1 value and the GAG concentration or if the T2 measurement has this linear relationship to the true T2 relaxation.

For structural measurements, the range is also a very important parameter. Due to the volumetric sampling characteristics of the MRI, the volume precision is not homoscedastic as seen in Figure 12.35a. Therefore, it is important to specify the tested range of the volume, or apply the proper mathematical transform that may remove the dependency of the variance to the measurement level. For volume and area measurements, the cubic root and the square root transforms make a decent job in removing the dependency of the sampling variance from the measurement level as seen in Figure 12.35b and c.

Besides linearity and range, the following characteristics are useful in defining rules for removing questionable OA measurements.

- *Detection limit:* The detection limit is not usually reported or measured, but when reported, it represents the smallest size, signal, or amount in an individual subject that can be imaged by the acquisition device, but not necessarily quantitated as an exact value. In qMRI, the minimum detection limit is governed by the Nyquist–Shannon sampling theorem. Therefore, we can use this theorem to estimate the detection limit in all structural measurements. It is important to remember that the sampling rate will be given by the sampled voxel resolution, not the reconstructed voxel. It will be safe to set the detection limit to twice the acquisition slice thickness. Structures that are smaller than two slice thicknesses will not be properly measured and quantitative measurements from those structures cannot be trusted. For signal-based measurements, the detection limit can be placed at 3.3 standard deviations of the precision error [113].
- *Quantitation limit:* As well as in the detection limit, the quantitation limit is not usually reported or measured, but this characteristic represents the smallest size, signal, or amount that will be quantitated with the desired accuracy. For OA, the quantitation limit can be safely placed at 10

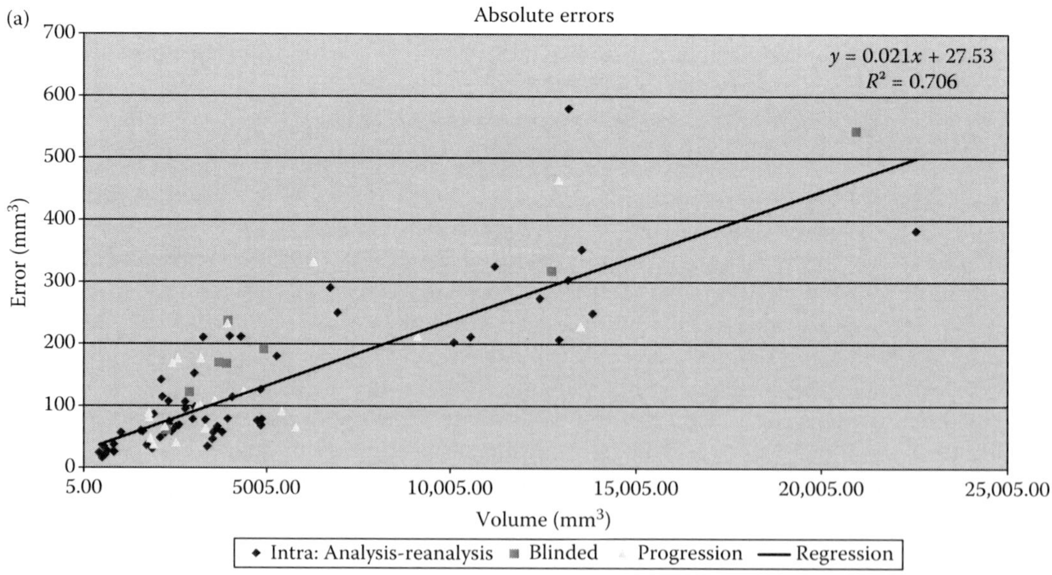

FIGURE 12.35 Reported random errors in the measurement of cartilage tissue. (a) The error in the quantification of cartilage volume from MRI is highly associated with the volume itself. This is not a desired characteristic of a measurement. (b) The reported CV values are also associated with the size of the cartilage tissue. There is a nonlinear relationship between CV values and cartilage size. (c) After transforming the cartilage volumes by a cubic root transform, the CV values of the measurements do not show a dependency to cartilage size. This transformation allows the cross-comparison of reported CV values across studies.

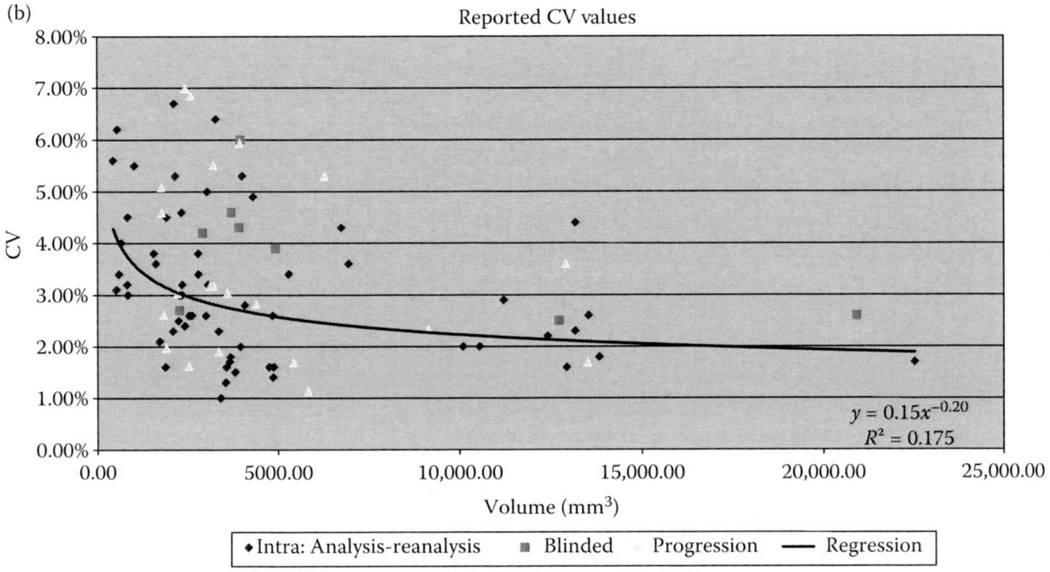

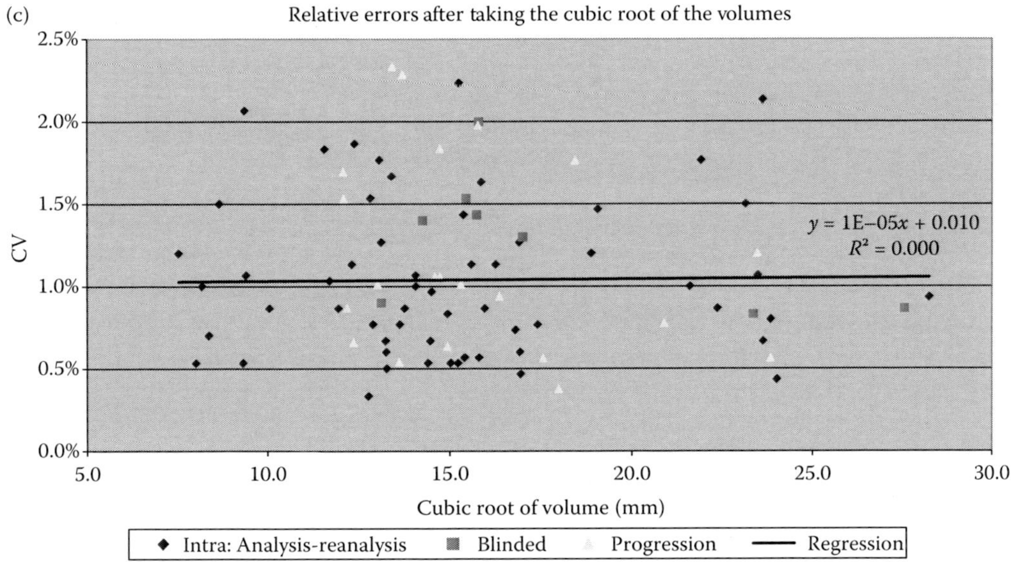

FIGURE 12.35 Continued.

standard deviations of the observed absolute measurement precision [113]. Knowing the signal quantitation limit is also very important in all T1 and T2 measurements especially when fitting the signal data to the nonlinear models of the signal formation.

Finally, it is recommended the whole system be tested for

- *Robustness or ruggedness:* It is a measure of the ability to remain unaffected by small but deliberate variations in method parameters and provides an indication of its reliability during normal usage. Ruggedness tests are used to determine the parameters that can affect the measurement.

Therefore, for any quantitative OA measurement, it is important to evaluate the roughness of the measure to Image Reader expertise, Paired-Read Order, Display Contrast, Image Analysis Algorithm Parameters, and so on.

12.3.1.3 Valid Quantitative Image Analysis Method in OA

The knowledge of the measurement characteristics of the MRI-based OA quantitative system does not guarantee that the system is producing valid measurements. The UK VAM Programme identified six principles of a valid analytical measurement (http://www.nmschembio.org.uk). We took the freedom to adjust them to fit the OA quantitative imaging requirements. Those principles are

1. Image measurements should be made to satisfy an agreed requirement.
2. Image measurements should be made using methods and equipment which have been tested to ensure that they are fit for purpose.
3. Staff making image measurements should be both qualified and competent to undertake the task.
4. There should be a regular independent assessment of the technical performance of the system.
5. Image measurements made in one location should be consistent with others made elsewhere.
6. Organizations making image measurements should have well-defined quality control and quality assurance procedures.

These principles are in line with the ICH2Qa guidance in validated analytical methods and in line with the FDA draft guidance in the use of statistical analysis plans, independent review charters (imaging charters or MRI manuals). The main purpose of the principles is to remind us that a good measurement device does not necessarily guarantee valid results. Quality control steps are required to guarantee that the numbers provided can be reproduced. Taking that recommendation into account, the simple OA procedure described in Figure 12.31 can be expanded to add the quality control and quality assurance procedures.

For example, in the OAI observational trail, there is an MRI manual that describes all the quality control steps [126,127]. The magnet is carefully monitored using special OA phantoms seen in Figure 12.36. These phantoms are quantitated using specific procedures and the result has been published. Figure 12.37a and b shows how one of the phantoms measurements slightly changes over time. These changes may have an impact on the measurement, especially when the expected degree of change in the measurements is in the order of magnitude of the natural drift of the magnet.

12.3.1.4 Revalidation

Revalidation of an OA measurement system may be necessary in the following circumstances:

- Changes in the composition of the system, that is,
  - Subject preparation guidelines
  - Imaging device
  - Changes in imaging phantoms and/or calibration criteria
  - Imaging protocol/pulse sequences
  - Change in readers
  - Reading paradigm: paired, blinded, sequential, and so on
- Image quantification algorithms
- Changes in the software versions

The degree of revalidation required depends on the nature of the changes. The most dramatic change will be modifications in the system itself. Changes in software versions must be carefully monitored because they can introduce new analysis algorithms, or may change the way a system performs. Careful software design controls will be required to verify that the changes in software meet the requirements. Do no change validated and calibrated image analysis algorithms in the middle of the study.

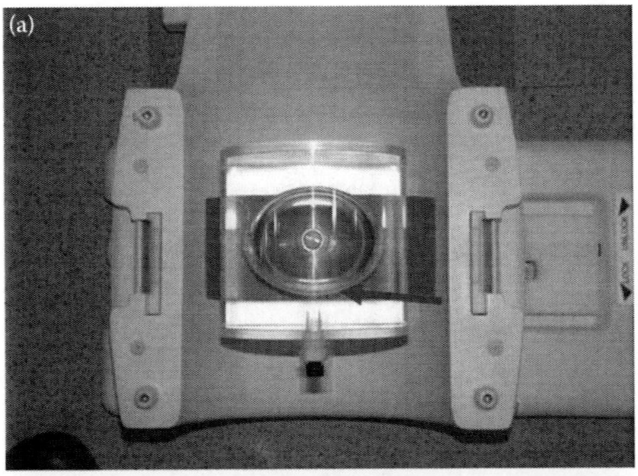

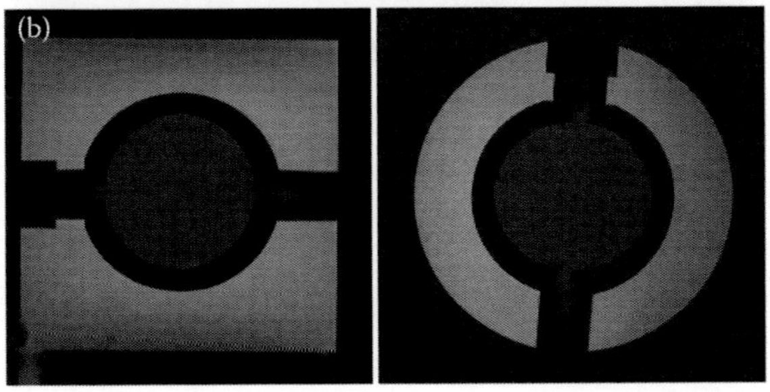

FIGURE 12.36 (a) The OIA using a cartilage phantom to monitor any deviation from the quantitation requirements of all the OAI equipment. (b) MRI imaging of the OAI phantom. (Courtesy of the OAI.)

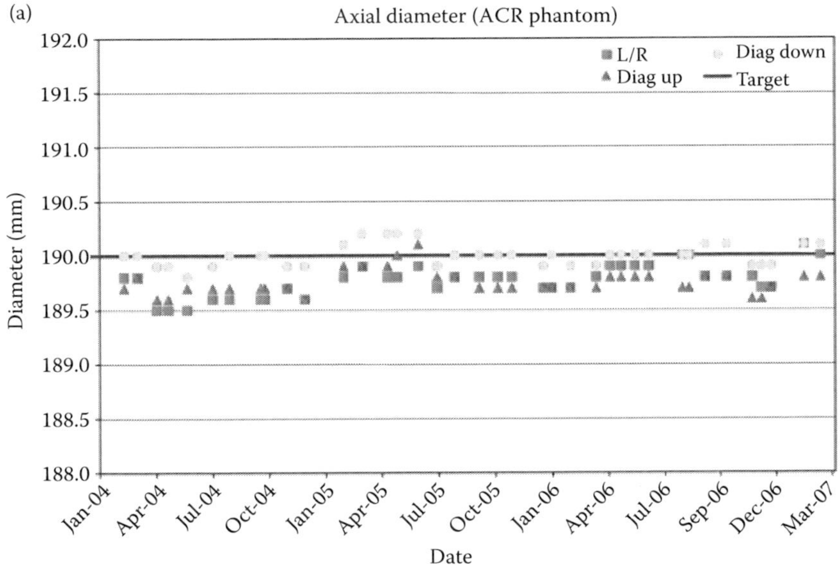

FIGURE 12.37 Quantitative analysis of the ACR and OAI phantom. (a,b) The ACR and OAI show small oscillations from the expected values. These variations are not large enough to drift measurements, but they show how the random variations of the MRI magnetic field are a source of linear noise on the order of 0.25%. (c) Example of longitudinal variation in OAI phantom spherical volume (b). The volume drift is from 210 to 211 mm^3. This 0.5% drift can yield to progression misinterpretations if ignored in the volumetric analysis of cartilage tissue.

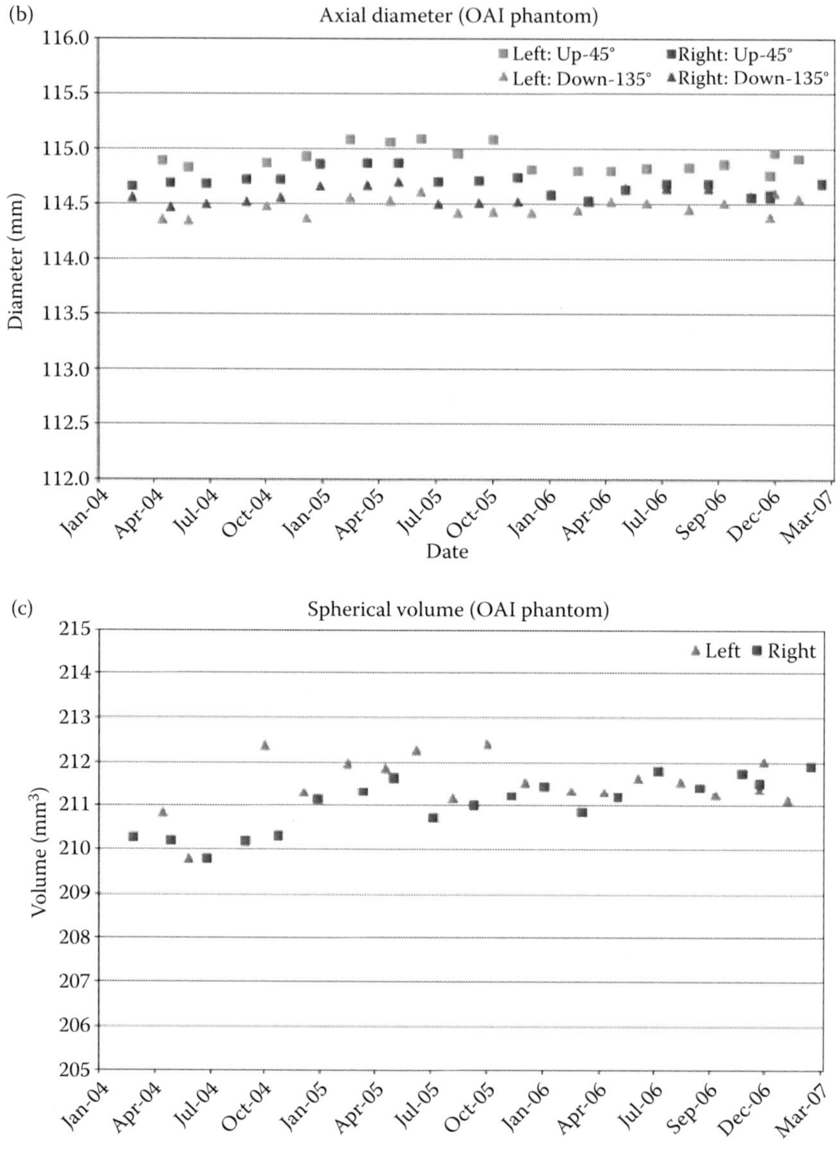

FIGURE 12.37 Continued.

12.3.2 MRI Image Analysis Algorithms in OA

As we have described in the former sections, an image analysis procedure is required to get quantitative data from the MRI. The most common algorithms used in OA are segmentation algorithms, registration algorithms, and quantification algorithms.

12.3.2.1 OA Segmentation Algorithms

As seen in Table 12.2, the OA segmentation algorithms can be classified into three major categories: supervised, atlas-based, and hybrid algorithms. Most of the OA quantitative work done up to date is based on supervised segmentation methods [120,125,128–131]. The reason behind this is the difficulty in accurately separating cartilage tissue from bone, synovium, or fat in MRI images with OA. The task

TABLE 12.2 Segmentation Methods with Different Sources of Random and Systematic Noise

Tissue Region of Interest	Reference Standard	Segmentation Method	Sources of Bias	Bias Control	Sources of Random Error
Cartilage	Expert manual trace	Manual/ semiautomated	Reader expertise Order of analysis	Training Blinding	Reader + acquisition noise
		Atlas	Atlas set	Multiple atlas	Acquisition noise/ registration
		Voxel classification	Training set/pulse sequence shift	Pulse sequence/ coil control	Acquisition noise
Bone marrow lesion	Expert manual trace	Manual/ semiautomated	Reader expertise Order of analysis	Training Blinding	Reader + acquisition noise
		Atlas/outlier detection	Outlier rules	Parameter calibration	Acquisition noise/ registration
		Voxel classification	Classification rules	Parameter calibration	Acquisition noise

Note: The degree of bias can be controlled by different procedures, but still, the human observer is considered the gold standard in all cartilage segmentation procedures.

is so challenging that even semiautomated approaches require trained users to properly identify the tissues. The nature of OA also makes the unsupervised segmentation methods fairly complex [132–134]. They are based in the use of anatomical atlases that are used to seed either a more complex classification algorithm or knowledge-based [135–141]. Figure 12.38 shows one of the early attempts in getting a fully automated segmentation from MRI images, and Figure 12.39 shows a more recent development of the use of multiple atlases to get accurate segmentations from MRI images. The main limitation of the fully automated approaches is the large variety of OA abnormalities that can be present in the joints: bone marrow lesions, synovities, inflammation, cartilage defect, full thickness defects, flow artifacts, and so on. Therefore, most of the fully automated approaches are limited to early stages of OA.

12.3.2.2 OA Registration Algorithms

The registration algorithms in OA are commonly used in atlas-based approaches to align the atlas to the incoming data set. These registration algorithms must accommodate to the articulating nature of

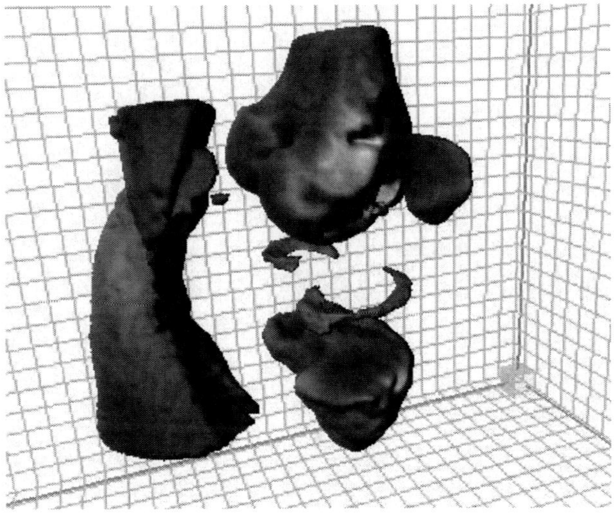

FIGURE 12.38 **(See color insert.)** The normal knee anatomy can be automatically segmented from MRI data sets using advanced statistical analysis tools.

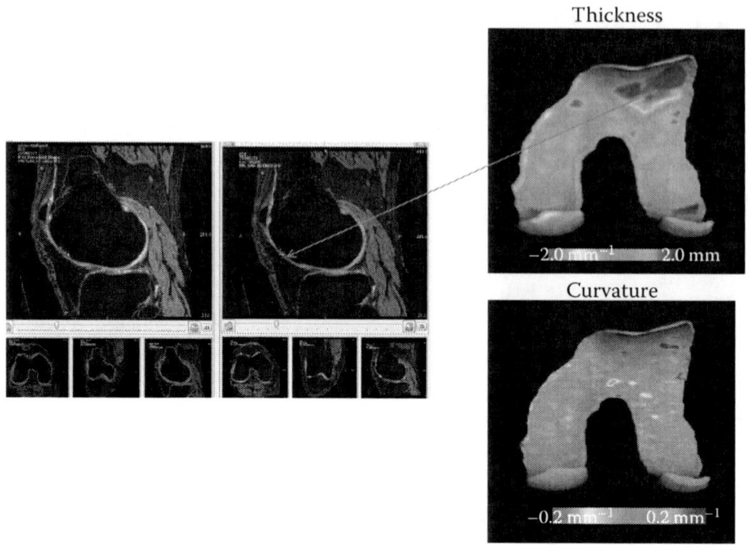

FIGURE 12.39 **(See color insert.)** In the presence of OA, atlas-based approaches are a practical alternative to the segmentation of thousands of MRI images. Left, slice through the DESS images. (Courtesy of OAI.) Top right, 3D reconstruction of the cartilage thickness in the atlas space. Red values represent areas of cartilage loss. Bottom right, the curvature of the cartilage surface can be computed and measured. Drastic changes in cartilage topography can be measured by the curvature values and are seen as red and blue spots in the curvature map.

the joints [142,143]. Therefore, masks are required to isolate each articulating bone from each other. Besides masks, the registration algorithms must deal with large deformations due to the development of bone deformities and osteophyte formation. The other challenge of an OA registration algorithm is the use of surface coils in joint imaging. These coils make the signal intensity fairly nonuniform across the imaging plane; therefore, robust algorithms are required to deal with signal fading or, preprocessing is required to mitigate the change in signal intensity.

A second use of image registration in OA is the intrasubject alignment of the different pulse sequences either for T1/T2 computations or for MRI pulse-sequence fusion required for multispectral-based segmentation. Although intraalignment is an easier task than interpatient alignment, it is still a challenge in OA analysis. Subjects tend to move between pulse sequence acquisitions. That movement may include a change in the relative position of bones, tendons, ligaments, and muscles. Therefore, articulated registration is required for the proper alignment of intrasubject MRI series. The final use of image registration algorithms in OA is the alignment of segmented data to get statistical maps of changes. This last registration algorithm is usually a surface-to-surface method, but atlas dense motion flow can be used to get the statistical maps. Once the surfaces are aligned, spatial statistics analysis can be done in the aligned surfaces [125,144]. Figure 12.40 shows an example of statistical maps generated after registering the quantitative thickness data.

12.3.2.3 Quantitation Algorithms

The final classes of algorithms used in OA analysis are quantitation algorithms as those listed in Table 12.3. Those algorithms can be as simple as voxel counting or as complex as the generation of statistical parametric maps that describe the voxel-by-voxel behavior of cartilage tissue. The three mayor classes of quantitation algorithms used in OA are the algorithms used for the estimation of cartilage thickness, the algorithms required to get T1 and T2 measurements, and the algorithms required to get statistical maps of the point-by-point measurements of thickness, curvature, T1, T2, and so on.

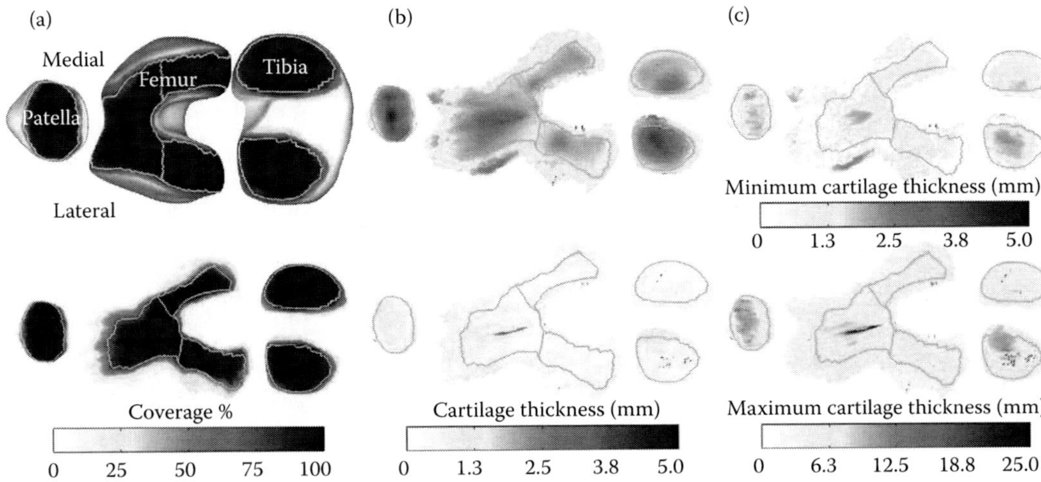

FIGURE 12.40 (**See color insert.**) Statistical maps of changes in cartilage thickness. The thickness maps are a very useful tool for the visualization of the localized changes and heterogeneity in the patterns of cartilage loss in subjects with OA. (a) Top, 3D rendering of knee bones with coverage analysis. (b) Mean (top) and standard deviation (bottom) of cartilage thickness. (c) Minimum and Maximum of the thickness. (This picture has copyrights by the IEEE TMI.)

TABLE 12.3 Quantification Procedures Calibrated to Avoid the Introduction of Biases into the Measurements

Quantitative Measurement	Reference Standard	Quantitation Method	Sources of Bias	Bias Control	Sources of Random Error
Volume	Phantom/ synthetic or real	Voxel count Surface volume	B_0 shift/gradients B_0 shift/gradients + surface reconstruction algorithm	Calibration	B_0 heterogeneties
Thickness	Phantom/ synthetic or real	Euclidian distance Laplacian distance Mid-surface	B_0 shift/ gradients + algorithm	Calibration	B_0 heterogeneties
Area	Phantom/ synthetic or real	Surface reconstruction Voxel edge count	B_0 shift/gradients + algortihm	Calibration	B_0 heterogeneties
T2	Phantom/ synthetic or real	Spin echo monoexponential fitting Constrained spin echo monoexponential fitting	B_1/pulse sequence + algorithm	Calibration	B_1 Heterogeneities + flow + partial volume
T1	Phantom/ synthetic or real	Nonlinear IR fitting Nonlinear multiple flip angle fitting	B_1/pulse sequence + registration + algorithm	Calibration	B_1 Heterogeneities + registration + flow + patial volume
T1 dGEMRIC	Phantom/ synthetic or real	Nonlinear IR fitting Nonlinear multiple flip angle fitting	B_1/pulse sequence + registration + algorithm	Calibration	B_1 heterogeneities + registration + flow + partial volume + Gd-DTPA(2−)/ subject

Note: The degree of bias depends heavily on the algorithm used to compute the measurement. Calibration and characterization in the range of the measurement are essential to avoid using those quantification algorithms outside the specification range.

The main issue in the quantitative algorithms is the validation and calibration of the quantitative data. To isolate image segmentation and image registration issues from the validation and calibration procedures, it is recommended to use synthetic phantoms or real phantoms for the calibration. Real phantoms are very important in the calibration of T1/T2 algorithms while synthetic phantoms are very useful in the validation of structural measurement algorithms [145]. The advantage of the synthetic phantoms is that the ground truth is known and that they can be designed to represent the whole range of conditions present in real life.

12.3.3 OA Image-Based Biomarkers

The FDA website (www.FDA.gov) states: "Biomarker, or biological marker, A characteristic that is objectively measured and evaluated as an indicator of normal biologic processes, pathogenic processes, or pharmacologic responses to a therapeutic intervention."

The final goal of the quantitative data generated from an MRI is to develop surrogated endpoints to stage the progression of the disease. The road to get image-based surrogated markers is very challenging and it requires the consensus of the OA community in a clear definition of a useful clinical endpoint for clinical diagnosis, for treatment decision-making, or for treatment monitoring. Presently, the most common criteria to stage the disease is the radiological evaluation of OA via the use of KL scores or radiological atlases like OARSIs. MRI-based evaluation of OA is still under development. In the meantime, WORMS and BLOCKS are used as tools to evaluate the findings and severity of OA-related features [146,147]. The other alternative is to use clinical evaluations via the use of either WOMAC scores or KOOS scores [148–150]. Both approaches are ill-defined and have not shown clear sensitivity or objectivity to the progression of OA.

Regardless of the state of the clinical endpoint in the OA community, the OA MRI-based biomarkers can be classified into structural biomarkers and imaging biomarkers. Structural biomarkers are single measurements of a specific tissue or region of interest, that is, total femoral cartilage volume, average medial tibia area, total bone marrow lesion volume, and so on. On the other hand, imaging biomarkers are maps or quantitative images of cartilage thickness, cartilage curvature, T1 dGEMRIC values, T2, and so on. Because the imaging biomarkers create maps or quantitative images, the imaging biomarkers can be described by statistical descriptors like mean, standard deviations, skewness, and so on. Therefore, there is an endless count of possible statistical descriptors that can be used as surrogated endpoints coming from the imaging biomarkers. Figure 12.41 shows an example of an interjoint distance map generated using MRI data. The map can be interpreted as an imaging biomarker of the distance between the tibia and the femur. On the other hand, functional biomarkers of the cartilage and surrounding tissues are possible with the use of loading devices as seen in Figure 12.42. The loading device places a load into the knee that allows the computation of stress and strain in the cartilage

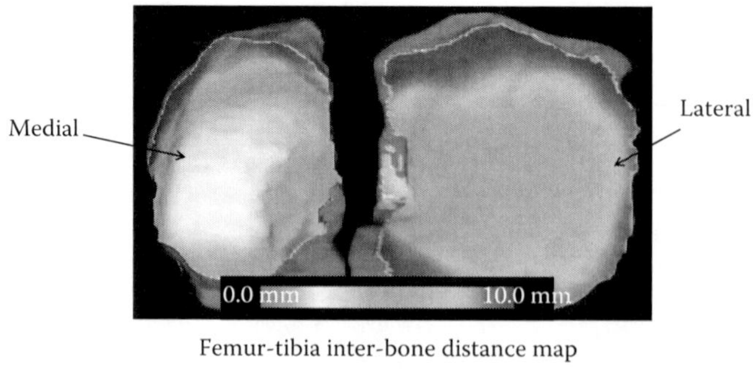

FIGURE 12.41 **(See color insert.)** The MRI images can be used to compute the interjoint distance. Distance maps can automatically be computed and they are an indirect measurement of changes in the soft tissues in joint anatomy.

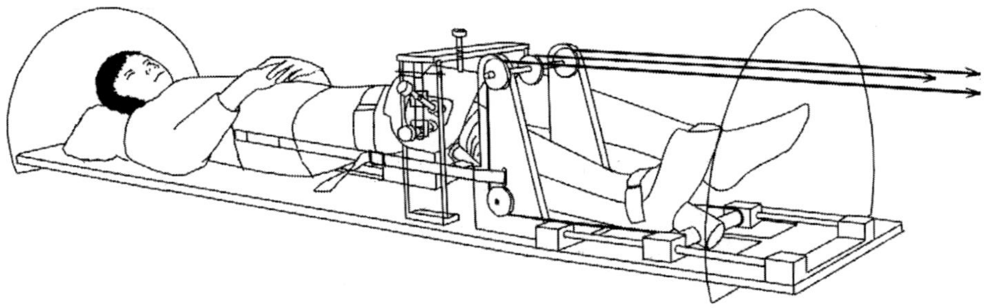

FIGURE 12.42 Functional joint analysis can be done by using lading devices. These devices allow the application of dynamic loads into the knee while the person is being imaged. The changes in cartilage deformation may be an indication of early cartilage degeneration.

and meniscus. Therefore, early stages of joint degradation can be inferred by changes in the mechanical properties estimated from those functional experiments.

Regardless of the type of image-based biomarker, the ideal image-based OA biomarker must predict the stage of the disease and must have good responsiveness to either disease progression or treatment intervention. The search of this biomarker has produced hundreds of candidates, some of them with good responsiveness but not clear association to the ill-defined OA stage. Others have a good association to some OA definition, but they do not show clear change over time, which makes them weak candidates for monitoring OA progression. At this stage of the OA biomarker development, only a few measurements have been able to be associated with the established radiological OA evaluation or clinical indexes [151–153].

A very important point in the validation of OA biomarker is the proper design and its association to disease stage and change responsiveness. Statistical models must be used to do these tasks and several time points (at least three) must be included to evaluate the responsiveness. The use of only two time points to evaluate biomarker progression rate has yielded many false estimations of progression rate. One of the main factors is that investigators forget to take into account that some biomarker or image analysis procedures are biased toward positive change even in the lack of real progression. The most common source of analysis bias is unblinded pair reading. Paired evaluation is biased by the natural expectation of the human observer toward symptoms worsening in OA. Besides analysis bias, there are biased biomarkers. Biased biomarkers are those which may have a nonzero value between repeated measurements. Order values, percentile analysis, and detection analysis are samples of biomarkers that will have that behavior in the presence of random noise. The estimation of the right rate of progression of those biased biomarkers is the inclusion of at least three time points and the use of a statistical model with bias term. Once the statistical model is fitted to the data, the bias and the time responsiveness can be estimated. The beta coefficient of the time-dependent variable can be used to estimate the rate of progression.

The number of biomarkers that is generated from imaging is very large, and the analysis of this data is very complex. Therefore, bioinformatics tools (see Figure 12.43) may be required to establish surrogated biomarker candidates that in a near future may be accepted by the OA community to be used as a diagnosis criteria, disease staging, or treatment monitoring.

12.3.4 Conclusion

The MRI has provided the OA community with an invaluable tool to observe and understand the progression of this chronic disease. Magnets, coils, and pulse sequences coupled with image analysis algorithms have made the quantitation of the OA phenomena possible. This has been very useful in understanding and redefining the disease progression. The final goal is to obtain a surrogate biomarker of OA that can be used to first diagnose the disease clinically, and then used to monitor treatment

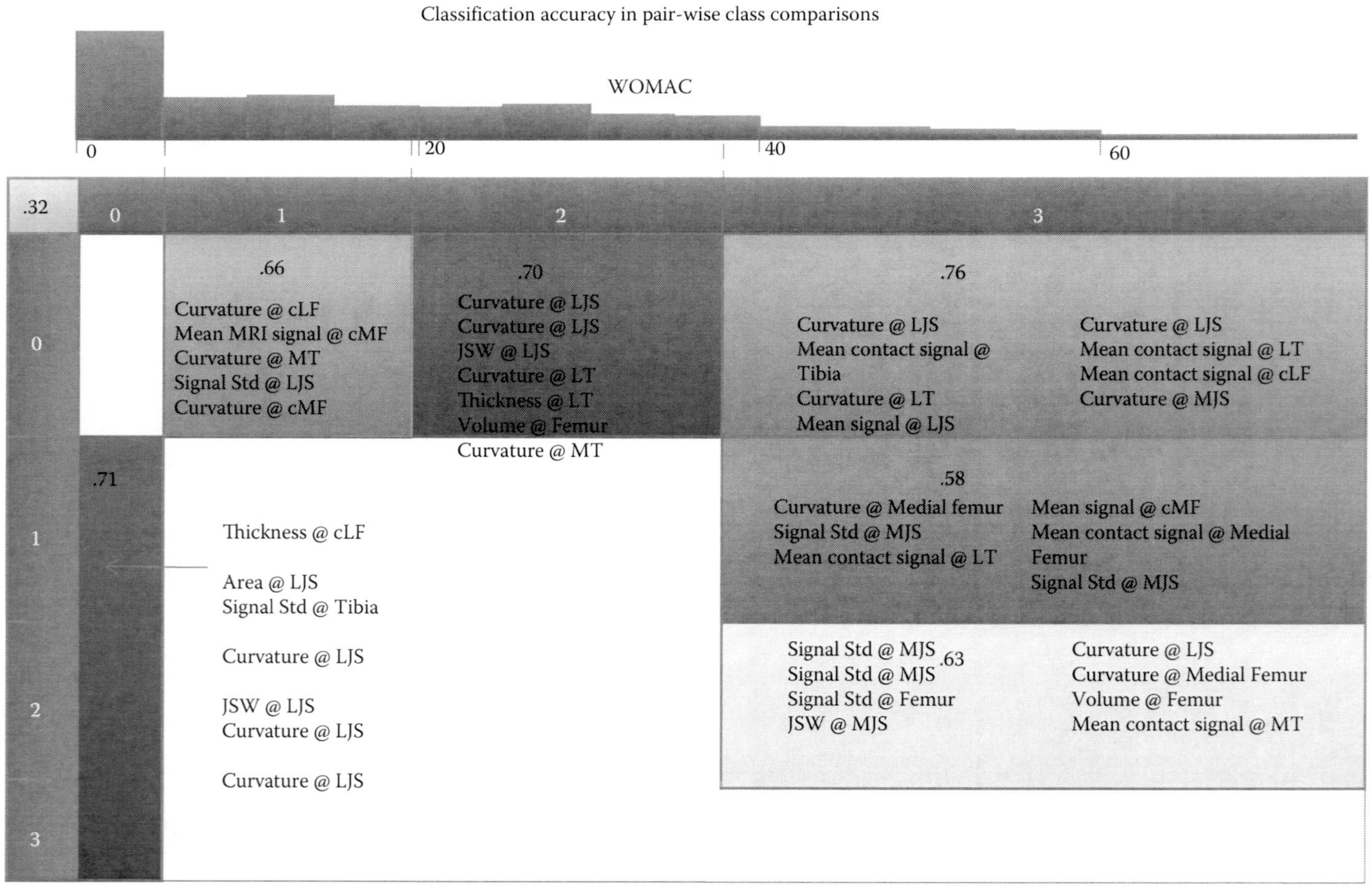

Classes 0,1,2, and 3 contain 0–25%, 25–50%, 50–75% and 75–100% respectively of the WOMAC population.

FIGURE 12.43 Advanced bioinformatic tools can be used to explore hundreds of variables and their association to clinical outcomes. In this example, GALGO was used to explore associations between MRI measurements and the total WOMAC score.

efficacy. We are not there yet, but this desire of the ultimate biomarker has pushed the scientific, pharmaceutical device, and regulatory community to work together to develop OA studies, algorithms, and standard methodologies aimed to get the biomarker that will enable the development of effective OA therapies.

Acknowledgment

The OAI is a public–private partnership comprised of five contracts (N01-AR-2-2258; N01-AR-2-2259; N01-AR-2-2260; N01-AR-2-2261; N01-AR-2-2262) funded by the National Institutes of Health, a branch of the Department of Health and Human Services, and conducted by the OAI Study Investigators. Private funding partners include Merck Research Laboratories, Novartis Pharmaceuticals Corporation, GlaxoSmithKline, and Pfizer, Inc. Private sector funding for the OAI is managed by the Foundation for the National Institutes of Health. This manuscript was prepared using an OAI public use data set and does not necessarily reflect the opinions or views of the OAI investigators, the NIH, or the private funding partners.

References

1. Brandt KD, Dieppe P, and Radin EL. Etiopathogenesis of osteoarthritis. *Rheum Dis Clin North Am.* 2008;34:531–559.
2. Brandt KD, Dieppe P, and Radin EL. Commentary: Is it useful to subset "primary" osteoarthritis? A critique based on evidence regarding the etiopathogenesis of osteoarthritis. *Semin Arthritis Rheum.* 2009;39:81–95.
3. Brandt K. Defining osteoarthritis: What it is and what it is not. *J Musculoskel Med.* 2010;27:338–350.
4. Felson DT, Lawrence RC, Dieppe PA, Hirsch R, Helmick CG, Jordan JM, Kington RS et al. Osteoarthritis: New insights. Part 1: The disease and its risk factors. [Review] [120 refs]. *Ann Internal Med.* 2000;133(8):635–646.
5. Frobell RB, Roos EM, Roos HP, Ranstam J, and Lohmander LS. A randomized trial of treatment for acute anterior cruciate ligament tears. *N Engl J Med.* 2010;363(4):331–342.
6. Hansen JT, and Lambert DR. *Netters Clinical Anatomy*, Saunders Elsevier, Philadelphia, PA, 2008: pp. 234–239.
7. Pritzker KPH, Gay S, Jimenez SA,Ostergaard K, Pelletier JP, Revell PA, Salter D, and van den Berg WB. Osteoarthritis cartilage histopathology: Grading and staging. *Osteoarthritis Cartilage.* 2006;14:13–29.
8. Jiang J, Leong NL, Mung JC, Hidaka C, and Lu HH. Interaction between zonal populations of articular chondrocytes suppresses chondrocyte mineralization and this process is mediated by PTHrP. *Osteoarthritis Cartilage.* 2008;16:70–82.
9. Modl JM, Sether LA, Haughton VM, and Kneeland JB. Articularcartilage: Correlation of histologic-zones with signal intensity at MR imaging. *Radiology.* 1991;181(3):853–855.
10. Uhl M, Ihling C, Allmann KH, Laubenberger J, Tauer U, Adler CP, and Langer M. Human articular cartilage: *In vitro* correlation of MRI and histologic findings. *Eur Radiol.* 1998;8(7):1123–1129.
11. Minns RJ, and Steven FS. The collagen fibril organization in human articular cartilage. *J Anat.* 1977;123:437–457.
12. Xia Y, Moody JB, Burton-Wurster N, and Lust G. Quantitative *in situ* correlation between microscopic MRI and polarized light microscopy studies of articular cartilage. *Osteoarthritis Cartilage.* 2001;9:393–406.
13. Nieminen MT, Rieppo J, Töyräs J, Hakumäki JM, Silvennoinen J, Hyttinen MM, Helminen HJ, and Jurvelin JS. T2 relaxation reveals spatial collagen architecture in articular cartilage: A comparative quantitative MRI and polarized light microscopic study. *Magn Reson Med.* 2001;46(3):487–493.
14. Bashir A, Gray ML, Boutin RD, and Burstein D. Glycosaminoglycan in articular cartilage: *In vivo* assessment with delayed Gd(DTPA)(2-)-enhanced MR imaging. *Radiology.* 1997;205(2):551–558.

15. Van Ginckel A, Baelde N, Almqvist KF, Roosen P, McNair P, and Witvrouw E. Functional adaptation of knee cartilage in asymptomatic female novice runners compared to sedentary controls. A longitudinal analysis using delayed Gadolinium Enhanced Magnetic Resonance Imaging of Cartilage (dGEMRIC). *Osteoarthritis Cartilage*. 2010;18(12):1564–1569.
16. Hauch KN, Villegas DF, and Haut Donahue TL. Geometry, time dependent and failure properties of human meniscal attachments. *J Biomech*. 201010;43(3):463.
17. Kaplan PA, Nelson NL, Garvin KL, and Brown DE. MR of the knee: The significance of high signal in the meniscus that does not clearly extend to the surface. *AJR Am J Roentgenol*. 1991;156(2):333–336.
18. Giuliani JR, Kilcoyne KG, and Rue JP. Anterior cruciate ligament anatomy: A review of the anteromedial and posterolateral bundles. *J Knee Surg*. 2009;22(2):148–154.
19. Chhabra A, Starman JS, Ferretti M, Vidal AF, Zantop T, and Fu FH. Anatomic, radiographic, biomechanical, and kinematic evaluation of the anterior cruciate ligament and its two functional bundles. *J Bone Joint Surg Am*. 2006;88 Suppl 4:2–10.
20. Roberts CC, Towers JD, Spangehl MJ, Carrino JA, and Morrison WB. Advanced MR imaging of the cruciate ligaments. *Magn Reson Imaging Clin N Am*. 2007;15(1):73–86.
21. Johal P, Williams A, Wragg P, Hunt D, and Gedroyc W. Tibio-femoral movement in the living knee. A study of weight bearing and non-weight bearing knee kinematics using "interventional" MRI. *J Biomech*. 2005;38(2):269–276.
22. Burr DB. Anatomy and physiology of the mineralized tissues: Role in the pathogenesis of osteoarthrosis. *Osteoarthritis Cartilage*. 2004;12:S20–S30.
23. Bullough PG. The role of joint architecture in the etiology of arthritis. *Osteoarthritis Cartilage*. 2004;12:S2–S9.
24. Walker PS, and Erkman MJ. The role of the menisci in force transmission across the knee. *Clin Orthop Relat Res*. 1975;109:184–192.
25. Fukuda Y, Takai S, Yoshino N, Murase K, Tsutsumi S, Ikeuchi K, and Hirasawa Y. Impact load transmission of the knee joint-influence of leg alignment and the role of meniscus and articular cartilage. *Clin Biomech (Bristol, Avon)*. 2000;15(7):516–521.
26. Hasler EM, Herzog W, Wu JZ, Müller W, and Wyss U. Articular cartilage biomechanics: Theoretical models, material properties, and biosynthetic response. *Crit Rev Biomed Eng*. 1999;27(6):415–488. Review.
27. Greis PE, Bardana DD, Holmstrom MC, and Burks RT. Meniscal injury: I. Basic science and evaluation. *Am Acad Orthop Surg*. 2002;10(3):168–176.
28. Aagaard H, and Verdonk R. Function of the normal meniscus and consequences of meniscal resection. *Scand J Med Sci Sports*. 1999;9(3):134–140.
29. Musahl V, Citak M, O'Loughlin PF, Choi D, Bedi A, and Pearle AD. The effect of medial versus lateral meniscectomy on the stability of the anterior cruciate ligament-deficient knee. *Am J Sports Med*. 2010;38(8):1591–1597.
30. Moglo KE, and Shirazi-Adl A. Biomechanics of passive knee joint in drawer: Load transmission in intact and ACL-deficient joints. *Knee*. 2003;10(3):265–276.
31. Burr DB, and Schaffler MB. The involvement of subchondral mineralized tissues in osteoarthrosis: Quantitative microscopic evidence. *Microsc Res Tech*. 1997;37(4):343–357.
32. Lane LB, Villacin A, and Bullough PG. The vascularity and remodelling of subchondrial bone and calcified cartilage in adult human femoral and humeral heads. An age- and stress-related phenomenon. *J Bone Joint Surg Br*. 1977;59:272–278.
33. Muir P, McCarthy J, Radtke CL, Markel MD, Santschi EM, Scollay MC, and Kalscheur VL. Role of endochondral ossification of articular cartilage and functional adaptation of the subchondral plate in the development of fatigue microcracking of joints. *Bone*. 2006;38:342–349.
34. Suri S, Gill SE, Massena de Camin S, Wilson D, McWilliams DF, and Walsh DA. Neurovascular invasion at the osteochondral junction and in osteophytes in osteoarthritis. *Ann Rheum Dis*. 2007;66:1423–1508.
35. Kellgren JH, and Lawrence JS. Radiological assessment of osteo-arthrosis. *Ann Rheum Dis*. 1957;16:494–502.

36. Wluka AE, Wang Y, Davis SR, and Cicuttini FM. Tibial plateau size is related to grade of joint space narrowing and osteophytes in healthy women and in women with osteoarthritis. *Ann Rheum Dis.* 2005;64:1033–1037.
37. Eckstein F, Hudelmaier M, Cahue S, Marshall M, and Sharma L. Medial-to-lateral ratio of tibiofemoral subchondral bone area is adapted to alignment and mechanical load. *Calcif Tissue Int.* 2009;84(3):186–194.
38. Bowes MA, Vincent GR, Williams TC, Hutchinson CE, Lacey T, and Brett A. Statistical shape models show differences in bone shape between progressors and non-progressors in OAI progression cohort. *Osteoarthritis Cartilage.* 2008;16(S4):S26–S27.
39. Roemer FW, Frobell R, Hunter DJ, Crema MD, Fischer W, Bohndorf K, and Guermazi A. MRI-detected subchondral bone marrow signal alterations of the knee joint: Terminology, imaging appearance, relevance and radiological differential diagnosis. *Osteoarthritis Cartilage.* 2009;17(9):1115–1131.
40. Davies-Tuck ML, Wluka AE, Forbes A, Wang Y, English DR, Giles GG, O'Sullivan R, and Cicuttini FM. Development of bone marrow lesions is associated with adverse effects on knee cartilage while resolution is associated with improvement—A potential target for prevention of knee osteoarthritis: A longitudinal study. *Arthritis Res Ther.* 2010;12(1):R10.
41. Wluka AE, Hanna F, Davies-Tuck M, Wang Y, Bell RJ, Davis SR, Adams J, and Cicuttini FM. Bone marrow lesions predict progression of cartilage defects and loss of cartilage volume in healthy middle-aged adults without knee pain over 2 yrs. *Rheumatology (Oxford).* 2008;47(9):1392–1396.
42. Messent EA, Ward RJ, Tonkin CJ, and Buckland-Wright C. Tibial cancellous bone changes in patients with knee osteoarthritis. A short-term longitudinal study using Fractal Signature Analysis. *Osteoarthritis Cartilage.* 2005;13(6):463–470.
43. Lindsey CT, Narasimhan A, Adolfo JM, Jin H, Steinbach LS, Link T, Ries M, and Majumdar S. Magnetic resonance evaluation of the interrelationship between articular cartilage and trabecular bone of the osteoarthritic knee. *Osteoarthritis Cartilage.* 2004;12(2):86–96.
44. Peterfy CG, Guermazi A, Zaim S, Tirman PF, Miaux Y, White D, Kothari M et al. Whole-Organ Magnetic Resonance Imaging Score (WORMS) of the knee in osteoarthritis. *Osteoarthritis Cartilage.* 2004;12(3):177–190.
45. Huebner JL, Williams JM, Deberg M, Henrotin Y, and Kraus VB. Collagen fibril disruption occurs early in primary guinea pig knee osteoarthritis. *Osteoarthritis Cartilage.* 2010;18(3):397–405. Epub 2009, October 1.
46. Calvo E, Palacios I, Delgado E, Sánchez-Pernaute O, Largo R, Egido J, and Herrero-Beaumont G. Histopathological correlation of cartilage swelling detected by magnetic resonance imaging in early experimental osteoarthritis. *Osteoarthritis Cartilage.* 2004;12:878–886.
47. Watson PJ, Carpenter TA, Hall LD, and Tyler JA. Cartilage swelling and loss in a spontaneous model of osteoarthritis visualized by magnetic resonance imaging. *Osteoarthritis Cartilage.* 1996;4:197–207.
48. Thambyah A, and Broom N. On new bone formation in the pre-osteoarthritic joint. *Osteoarthritis Cartilage.* 2009;17:456–463.
49. Bi X, Yang X, Bostrom MP, Bartusik D, Ramaswamy S, Fishbein KW, Spencer RG, and Camacho NP. Fourier transform infrared imaging and MR microscopy studies detect compositional and structural changes in cartilage in a rabbit model of osteoarthritis. *Anal Bioanal Chem.* 2007;387:1601–1612.
50. David-Vaudey E, Ghosh S, Ries M, and Majumdar S. T2 relaxation time measurements in osteoarthritis. *Magn Reson Imaging.* 2004;22:673–682.
51. Hannila I, Nieminen MT, Rauvala E, Tervonen, and Ojala, O. Patellar cartilage lesions: Comparison of magnetic resonance imaging and T2 relaxation-time mapping. *Acta Radiol.* 2007;48:444–448.
52. Ding C, Garnero P, Cicuttini F, Scott F, Cooley H, and Jones G. Knee cartilage defects: Association with early radiographic osteoarthritis, decreased cartilage volume, increased joint surface area and type II collagen breakdown. *Osteoarthritis Cartilage.* 2005;13:198–205.

53. McDermott ID, and Amis AA. The consequences of meniscectomy. *J Bone Joint Surg Br.* 2006;88(12):1549–1556.
54. Englund M, Guermazi A, and Lohmander SL. The role of the meniscus in knee osteoarthritis: A cause or consequence? *Radiol Clin North Am.* 2009;47(4):703–712.
55. Bhattacharyya T, Gale D, Dewire P, Totterman S, Gale ME, McLaughlin S, Einhorn TA, and Felson DT. The clinical importance of meniscal tears demonstrated by magnetic resonance imaging in osteoarthritis of the knee. *J Bone Joint Surg Am.* 2003;85-A:4–9.
56. Crema MD, Guermazi A, Li L, Nogueira-Barbosa MH, Marra MD, Roemer FW, Eckstein F, Le Graverand MP, Wyman BT, and Hunter DJ. The association of prevalent medial meniscalpathology with cartilage loss in the medial tibiofemoral compartment over a 2-year period. *Osteoarthritis Cartilage.* 2010;18(3):336–343.
57. Wang Y, Wluka AE, Pelletier JP, Martel-Pelletier J, Abram F, Ding C, and Cicuttini FM. Meniscal extrusion predicts increases in subchondral bone marrow lesions and bone cysts and expansion of subchondral bone in osteoarthritic knees. *Rheumatology (Oxford).* 2010;49(5):997–1004.
58. Roberts CC, Towers JD, Spangehl MJ, Carrino JA, and Morrison WB. Advanced MR imaging of the cruciate ligaments. *Magn Reson Imaging Clin N Am.* 2007;15(1):73–86.
59. Frobell RB, Le Graverand MP, Buck R, Roos EM, Roos HP, Tamez-Pena J, Totterman S, and Lohmander LS. The acutely ACL injured knee assessed by MRI: Changes in joint fluid, bone marrow lesions, and cartilage during the first year. *Osteoarthritis Cartilage.* 2009;17(2):161–167.
60. Li G, Moses JM, Papannagari R, Pathare NP, DeFrate E, and Gill TJ. Anterior cruciate ligament deficiency alters the *in vivo* motion of the tibiofemoral cartilage contact points in both the anteroposterior and mediolateral directions. *J Bone Joint Surg Am.* 2006;88:1826–1834.
61. Fernandez-Madrid F, Karvonen RL, Teitge RA, Miller PR, and Negendank WG. MR features of osteoarthritis of the knee. *Magn Reson Imaging.* 1994;12:703–709.
62. Crema MD, Roemer FW, Marra MD, Burstein D, Gold GE, Eckstein F, Baum T, Mosher TJ, Carrino JA, and Guermazi A. Articular cartilage in the knee: Current MR imaging techniques and applications in clinical practice and research. *Radiographics.* 2011;31(1):37–61.
63. Roemer FW, Eckstein F, and Guermazi A. Magnetic resonance imaging-based semiquantitative and quantitative assessment in osteoarthritis. *Rheum Dis Clin North Am.* 2009;35(3):521–555.
64. Tyler DJ, Robson MR, Henkelman M, Young IR, and Bydder GM. Magnetic resonance imaging with ultrashort TE (UTE) PULSE sequences: Technical considerations. *J Magn Reson Imaging.* 2007;25:279–289.
65. Chen CA, Kijowski R, Shapiro LM, Tuite MJ, Kirkland W, Davis KW, Klaers JL, Block WF, Reeder SB, and Gold GE. Cartilage morphology at 3.0T: Assessment of three-dimensional magnetic resonance imaging techniques. *J Magn Reson Imaging.* 2010;32:173–183.
66. Wirth W, Nevitt M, Hellio Le Graverand MP, Benichou O, Dreher D, Davies RY, Lee J et al. OAI investigators: Sensitivity to change of cartilage morphometry using coronal FLASH, sagittal DESS, and coronal MPR DESS protocols—Comparative data from the Osteoarthritis Initiative (OAI). *Osteoarthritis Cartilage.* 2010;18(4):547–554.
67. Eckstein F, Maschek S, Wirth W, Hudelmaier M, Hitzl W, Wyman B, Nevitt M, and Le Graverand MP. OAI Investigator Group: One year change of knee cartilage morphology in the first release of participants from the Osteoarthritis Initiative progression subcohort: Association with sex, body mass index, symptoms and radiographic osteoarthritis status. *Ann Rheum Dis.* 2009;68(5):674–679.
68. Cashman PM, Kitney RI, Gariba MA, and Carter ME. Automated techniques for visualization and mapping of articular cartilage in MR images of the osteoarthritic knee: A base technique for the assessment of microdamage and submicro damage. *IEEE Trans Nanobiosci.* 2002;1(1):42–51.
69. Hunter DJ, Li L, Zhang YQ, Totterman S, Tamez J, Kwoh CK, Eaton CB, Hellio Le Graverand M-P, and Beals CR. Region of interest analysis: By selecting regions with denuded areas can we detect greater amounts of change? *Osteoarthritis Cartilage.* 2010;18:175–183.

70. Frobell RB, Le Graverand MP, Buck R, Roos EM, Roos HP, Tamez-Pena J, Totterman S, and Lohmander LS. The acutely ACL injured knee assessed by MRI: Changes in joint fluid, bone marrow lesions, and cartilage during the first year. *Osteoarthritis Cartilage.* 2009;17(2):161–167.
71. Williams TG, Holmes AP, Waterton JC, Maciewicz RA, Hutchinson CE, Moots RJ, Nash AF, and Taylor CJ. Anatomically corresponded regional analysis of cartilage in asymptomatic and osteoarthritic knees by statistical shape modeling of the bone. *IEEE Trans Med Imaging.* 2010;29(8):1541–1559.
72. Williams TG, Holmes AP, Bowes M, Vincent G, Hutchinson CE, Waterton JC, Maciewicz RA, and Taylor CJ. Measurement and visualization of focal cartilage thickness change by MRI in a study of knee osteoarthritis using a novel image analysis tool. *Br J Radiol.* 2010;83(995):940–948.
73. Tamez-Pena JG, Jackson R, Yu J, Eaton CB, Schneider E, and Totterman S. Thickness delta maps: Methodology for the spatial detection and quantification longitudinal changes in cartilage thickness. *Osteoarthritis Cartilage.* 2007;15(Suppl C):C180–C181.
74. Tamez-Peña JG, Gonzalez PC, Schreyer EH, Farber JM, and Totterman SM. Atlas based standardized quantification of cartilage thickness maps: Data from osteoarthritis Initiative. *Osteoarthritis Cartilage.* 2010;18(Suppl 2):S64–S65.
75. Tamez-Peña J, González PC, Schreyer EH, Farber JM, and Totterman SM. Detection of early changes in subchondral bone plate curvature in OA: Data from Osteoarthritis initiative. *Osteoarthritis Cartilage.* 2010;18 S 2:60–61.
76. Wirth W, Buck R, Nevitt M, Le Graverand MP, Benichou O, Dreher D, Davies RY et al. MRI-based extended ordered values more efficiently differentiate cartilage loss in knees with and without joint space narrowing than region-specific approaches using MRI or radiography—Data from the OA initiative. *Osteoarthritis Cartilage.* 2011;19(6):689–699.
77. Muir P, McCarthy J, Radtke CL, Markel MD, Santschi EM, Scollay MC, and Kalscheur VL. Role of endochondral ossification of articular cartilage and functional adaptation of the subchondral plate in the development of fatigue microcracking of joints. *Bone.* 2006;38:342–349.
78. Thambyah A, and Broom N. On new bone formation in the pre-osteoarthritic joint. *Osteoarthritis Cartilage.* 2009;17:456–63.
79. Pritzker KPH, Gay S, Jimenez SA,Ostergaard K, Pelletier J-P, Revell PA, Salter D, and van den Berg WB. Osteoarthritis cartilage histopathology: Grading and staging. *Osteoarthritis Cartilage.* 2006;14:13–29.
80. Wirth W, Larroque S, Davies RY, Nevitt M, Gimona A, Baribaud F, Lee JH et al. OA Initiative Investigators Group: Comparison of 1-year vs 2-year change in regional cartilage thickness in osteoarthritis results from 346 participants from the Osteoarthritis Initiative. *Osteoarthritis Cartilage.* 2011;19(1):74–83.
81. David-Vaudey E, Ghosh S, Ries M, and Majumdar S. T2 relaxation time measurements in osteoarthritis. *Magn Reson Imaging.* 2004;22(5):673–682.
82. Riek JK, and Zhang. A comparison of T2 mapping sequences at 1.5 Tesla for use in clinical trial. *Osteoarthritis Cartilage.* 2009;17(Suppl 1):S231–S232.
83. Li X, Benjamin Ma C, Link TM, Castillo DD, Blumenkrantz G, Lozano J, Carballido-Gamio J, Ries M, and Majumdar S. *In vivo* T(1rho) and T(2) mapping of articular cartilage in osteoarthritis of the knee using 3 T MRI. *Osteoarthritis Cartilage.* 2007;15(7):789–797.
84. Williams A, Qian Y, and Chu CR. UTE-T2* mapping of human articular cartilage in vivo: A repeatability assessment. *Osteoarthritis Cartilage.* 2011;19(1):84–88.
85. Qazi AA, Dam EB, Nielsen M, Karsdal MA, Pettersen PC, and Christiansen C. Separation of healthy and early osteoarthritis by automatic quantification of cartilage homogeneity. *Osteoarthritis Cartilage.* 2007;15(10):1199–1206.
86. Kellgren JH, and Lawrence JS. Radiological assessment of osteo-arthrosis. *Ann Rheum Dis.* 1957;16:494–502.
87. Wluka AE, Wang Y, Davis SR, and Cicuttini FM. Tibial plateau size is related to grade of joint space narrowing and osteophytes in healthy women and in women with osteoarthritis. *Ann Rheum Dis.* 2005;64:1033–1037.

88. Roemer FW, Frobell R, Hunter DJ, Crema MD, Fischer W, Bohndorf K, and Guermazi A. MRI-detected subchondral bone marrow signal alterations of the knee joint: Terminology, imaging appearance, relevance and radiological differential diagnosis. *Osteoarthritis Cartilage.* 2009;17(9):1115–1131.
89. Tamez-Peña J, Farber J, González P, Schreyer E, and Totterman S. Unsupervised segmentation and quantification of knee features: Data from the osteoarthritis initiative, *IEEE Transactions on Biomedical Engineering.* 2012;59(4):1177–1186.
90. Frobell RB, Le Graverand MP, Buck R, Roos EM, Roos HP, Tamez-Pena J, Totterman S, and Lohmander LS. The acutely ACL injured knee assessed by MRI: Changes in joint fluid, bone marrow lesions, and cartilage during the first year. *Osteoarthritis Cartilage.* 2009;17(2):161–167.
91. Folkesson J, Goldenstein J, Carballido-Gamio J, Kazakia G, Burghardt AJ, Rodriguez A, Krug R, de Papp AE, Link TM, and Majumdar S. Longitudinal evaluation of the effects of alendronate on MRI bone microarchitecture in postmenopausal osteopenic women. *Bone.* 2011;48(3):611–621.
92. Peterfy CG, Guermazi A, Zaim S, Tirman PF, Miaux Y, White D, Kothari M et al. Whole-organ magnetic resonance imaging score (WORMS) of the knee in osteoarthritis. *Osteoarthritis Cartilage.* 2004;12(3):177–190.
93. Swanson MS, Prescott JW, Best TM, Powell K, Jackson RD, Haq F, and Gurcan MN. Semi-automated segmentation to assess the lateral meniscus in normal and osteoarthritic knees. *Osteoarthritis Cartilage.* 2010;18(3):344–353.
94. Wirth W, Frobell RB, Souza RB, Li X, Wyman BT, Le Graverand MP, Link TM, Majumdar S, and Eckstein F. A three-dimensional quantitative method to measure meniscus shape, position, and signal intensity using MR images: A pilot study and preliminary results in knee osteoarthritis. *Magn Reson Med.* 2010;63(5):1162–1171.
95. Zarins ZA, Bolbos RI, Pialat JB, Link TM, Li X, Souza RB, and Majumdar S. Cartilage and meniscus assessment using T1rho and T2 measurements in healthy subjects and patients with osteoarthritis. *Osteoarthritis Cartilage.* 2010;18(11):1408–1416.
96. Kraus VB, Vail TP, Worrell T, and McDaniel G. A comparative assessment of alignment angle of the knee by radiographic and physical examination methods. *Arthritis Rheum.* 2005;52(6):1730–1735.
97. Hsu RW, Himeno S, Coventry MB, and Chao EY. Normal axial alignment of the lower extremity and load-bearing distribution atthe knee. *Clin Orthop.* 1990;255:215–227.
98. Hunter DJ, Zhang Y, Niu J, Tu X, Amin S, Goggins J, Lavalley M, Guermazi A, Gale D, and Felson DT. Structural factors associated with malalignment in knee osteoarthritis: The Boston osteoarthritis knee study. *J Rheumatol.* 2005;32:2192–2199.
99. Allaire R, Muriuki M, Gilbertson L, Harner CD, Allaire R, Muriuki M, Gilbertson L, and Harner CD. Biomechanical consequences of a tear of the posterior root of the medial meniscus. Similar to total meniscectomy. *J Bone Joint Surg Am.* 2008;90(9):1922–1931.
100. Li G, Moses JM, Papannagari R, Pathare NP, DeFrate E, and Gill TJ. Anterior cruciate ligament deficiency alters the *in vivo* motion of the tibiofemoral cartilage contact points in both the anteroposterior and mediolateral directions. *J Bone Joint Surg Am.* 2006;88:1826–1834.
101. Papannagari R, Gill TJ, Defrate LE, Moses JM, Petruska AJ, and Li G. *In vivo* kinematics of the knee after anterior cruciate ligament reconstruction: A clinical and functional evaluation. *Am J Sports Med.* 2006;34(12):2006–2012.
102. Van de Velde SK, Gill TJ, and Li G. Evaluation of kinematics of anterior cruciate ligament-deficient knees with use of advanced imaging techniques, three-dimensional modeling techniques, and robotics. *J Bone Joint Surg Am.* 2009;91 Suppl 1:108–114.
103. Van de Velde SK, Bingham JT, Hosseini A, Kozanek M, DeFrate LE, Gill TJ, and Li G. Increased tibiofemoral cartilage contact deformation in patients with anterior cruciate ligament deficiency. *Arthritis Rheum.* 2009;60(12):3693–3702.
104. Eckstein F, Burstein D, and Link TM. Quantitative MRI of cartilage and bone: Degenerative changes in osteoarthritis. *Nmr in Biomed.* 2006; 19:822–854.

105. Burstein D. MRI for development osteoarthritis drugs of disease-modifying. *NMR Biomed.* 2006;19:669–680.
106. Wirth W, Nevitt M, Le Graverand MPH, Benichou O, Dreher D, Davies RY et al. Sensitivity to change of cartilage morphometry using coronal FLASH, sagittal DESS, and coronal MPR DESS protocols—Comparative data from the osteoarthritis initiative (OAI). *Osteoarthritis Cartilage.* 2010;18:547–554.
107. Eckstein F, Guermazi A, and Roemer FW. Quantitative MR Imaging of cartilage and trabecular bone in osteoarthritis. *Radiol Clin North Am.* 2009;47:655–+.
108. Hunter DJ, Niu J, Zhang Y, Totterman S, Tamez J, Dabrowski C et al. Change in cartilage morphometry: A sample of the progression cohort of the Osteoarthritis Initiative. *Ann Rheum Dis.* 2009;68:349–356.
109. Eckstein F, Wyman BT, Buck RJ, Wirth W, Maschek S, Hudelmaier M et al. Longitudinal quantitative MR imaging of cartilage morphology in the presence of gadopentetate dimeglumine (Gd-DTPA). *Magn Reson Med.* 2009;61:975–980.
110. Ellison SLR, Rosslein M, and Williams A. Quantifying uncertainty in analytical measurement. s.l; *Eurachem.* ii, 2000.
111. Taylor BN, and Kuyatt CE. Guidelines for Evaluating and Expressing the Uncertainty of NIST Measurement Results. In: NIST: NIST 1994.
112. FDA. ICH-Q2A: Text on Validation of Analytical Procedures. In: FDA CDER 1995.
113. FDA. ICH-Q2B: Validation of Analytical Procedures: Methodology. In: FDA CDER 1996.
114. Taylor BN, and Kuyatt CE. Technical Note 1297: Guidelines for Evaluating and Expressing the Uncertainty of NIST Measurement Results. In: NIST: NIST 1994.
115. Eckstein F, Sittek H, Milz S, Putz R, and Reiser M. The morphology of articular-cartilage assessed by magnetic-resonance-imaging (MRI)—Reproducibility and anatomical correlation. *Surgical Radiologic Anatomy.* 1994;16:429–438.
116. Schnier M, Eckstein F, Priebsch J, Haubner M, Sittek H, Becker C et al. Three-dimensional thickness and volume measurements of knee joint cartilage with MRI: Validation in anatomical specimens via CT arthrography. *Rofo-Fortschritte Auf Dem Gebiet Der Rontgenstrahlen Und Der Bildgebenden Verfahren.* 1997;167:521–526.
117. Eckstein F, Schnier M, Haubner M, Priebsch J, Glaser C, Englmeier KH et al. Accuracy of cartilage volume and thickness measurements with magnetic resonance imaging. *Clin Orthopaedics Related Res.* 1998:137–148.
118. Muensterer OJ, Eckstein F, Hahn D, and Putz R. Computer-aided three dimensional assessment of knee-joint cartilage with magnetic resonance imaging. *Clin Biomech.* 1996;11:260–266.
119. Hohe J, Ateshian G, Reiser M, Englmeier KH, and Eckstein F. Surface size, curvature analysis, and assessment of knee joint incongruity with MRI in vivo. *Magn Reson Med.* 2002;47:554–561.
120. Eckstein F, Westhoff J, Sittek H, Maag KP, Haubner M, Faber S, Englmeier KH, and Reiser M. In vivo reproducibility of three-dimensional cartilage volume and thickness measurements with MR imaging. *Am J Roentgenol.* 1998; 170:593–597.
121. Eckstein F, Buck RJ, Burstein D, Charles HC, Crim J, Hudelmaier M et al. Precision of 3.0 Tesla quantitative magnetic resonance imaging of cartilage morphology in a multicentre clinical trial. *Ann Rheum Dis.* 2008;67:1683–1688.
122. McKenzie CA, Williams A, Prasad PV, and Burstein D. Three-dimensional delayed gadolinium-enhanced MRI of cartilage (dGEMRIC) at 1.5T and 3.0T. *J Magn Reson Imaging.* 2006;24:928–933.
123. Williams A, Gillis A, McKenzie C, Po B, Sharma L, Micheli L et al. Glycosaminoglycan distribution in cartilage as determined by delayed gadolinium-enhanced MRI of cartilage (dGEMRIC): Potential clinical applications. *Am J Roentgenol.* 2004;182:167–172.
124. Liang ZR, Macfall JR, and Harrington DP. Parameter-estimation and tissue segmentation from multispectral MR-images. *IEEE Trans Med Imaging.* 1994;13:441–449.
125. Williams TG, Holmes AP, Waterton JC, Maciewicz RA, Hutchinson CE, Moots RJ et al. Anatomically corresponded regional analysis of cartilage in asymptomatic and osteoarthritic knees by statistical shape modelling of the bone. *IEEE Trans Med Imaging.* 2010;29:1541–1559.

126. Nevit MC, Felson DT, and Lester G. Protocol for the Cohort study. In: OAI 2002.
127. Synarc. MRI Procedure manual for examinations of the knee and thigh. In: OAI 2006.
128. Williams TG, Holmes AP, Waterton JC, Maciewicz RA, Nash AFP, Taylor CJ. Regional quantitative analysis of knee cartilage in a population study using MRI and model based correspondences. *In Proceedings of the 3rd IEEE International Symposium on Biomedical Imaging: From Nano to Macro.* Arlington, VA. 2006:311–314.
129. Raynauld JP, Kauffmann C, Beaudoin G, Berthiaume MJ, de Guise JA, Bloch DA et al. Reliability of a quantification imaging system using magnetic resonance images to measure cartilage thickness and volume in human normal and osteoarthritic knees. *Osteoarthritis Cartilage.* 2003;11:351–360.
130. Raynauld JP, Martel-Pelletier J, Berthiaume MJ, Beaudoin G, Choquette D, Haraoui B, Tannenbaum H et al. Long term evaluation of disease progression through the quantitative magnetic resonance imaging of symptomatic knee osteoarthritis patients: Correlation with clinical symptoms and radiographic changes. *Arthritis Res Therap.* 2006; 8:R21, Available at http://0-www.scopus.com.millenium.itesm.mx/inward/record.url?eid=2-s2.0-34247551198&partnerID=40&md5=412d7bb1fe07eaaae886387cac70501b
131. Wirth W, and Eckstein F. A technique for regional analysis of femorotibial cartilage thickness based on quantitative magnetic resonance Imaging. *IEEE Trans Med Imaging.* 2008;27:737–744.
132. Hudelmaier M, Wirth W, Wehr B, Kraus V, Wyman BT, Le Graverand MPH et al. Femorotibial cartilage morphology: Reproducibility of different metrics and femoral regions, and sensitivity to change in disease. *Cells Tissues Organs.* 2010;192:340–350.
133. Guermazi A, Burstein D, Conaghan P, Eckstein F, Le Graverand-Gastineau MPH, Keen H et al. Imaging in osteoarthritis. *Rheum Dis Clin North Am.* 2008; 34:645–+.
134. Eckstein F, Mosher T, and Hunter D. Imaging of knee osteoarthritis: Data beyond the beauty. *Curr Opin Rheumatol.* 2007;19:435–443.
135. Tamez-Pena JG, Gonzalez PC, Schreyer EH, Farber JM, and Totterman SM. Atlas-based standardized quantification of cartilage thickness maps: Data from the osteoarthritis initiative. *Osteoarthritis Cartilage.* 2010;18:S64–S65.
136. Pakin SK, Tamez-Pena JG, Totterman S, and Parker KJ. Segmentation, surface extraction and thickness computation of articular cartilage. *Medical Imaging 2002: Image Processing Vol 1–3* 2002; 4684:155–166.
137. Tamez-Pena JG, Barbu-McInnis M, and Totterman S. Knee cartilage extraction and bone-cartilage interface analysis from 3D MRI data sets. *Medical Imaging 2004: Image Processing,* Pts, 2004;5370:1774–1784.
138. Dodin P, Pelletier JP, Martel-Pelletier J, and Abram F. Automatic human knee cartilage segmentation from 3-D magnetic resonance images. *IEEE Trans Biomed Eng.* 2010;57:2699–2711.
139. Bowers ME, Trinh N, Tung GA, Crisco JJ, Kimia BB, and Fleming BC. Quantitative MR imaging using "LiveWire" to measure tibiofemoral articular cartilage thickness. *Osteoarthritis Cartilage.* 2008;16:1167–1173.
140. Folkesson J, Dam EB, Olsen OF, Pettersen PC, and Christiansen C. Segmenting articular cartilage automatically using a voxel classification approach. *IEEE Trans Med Imaging.* 2007;26:106–115.
141. Baldwin MA, Langenderfer JE, Rullkoetter PJ, and Laz PJ. Development of subject-specific and statistical shape models of the knee using an efficient segmentation and mesh-morphing approach. *Computer Methods Programs Biomed.* 2010;97:232–240.
142. Tamez-Pena JG, Parker KJ, and Totterman S. The integration of automatic segmentation and motion tracking for 4D reconstruction and visualization of musculoskeletal structures. In: Vemuri B Ed. *IEEE Workshop on Biomedical Image Analysis.* IEEE, Santa Barbara, CA, 1998: pp. 154–163.
143. Tamez-Pena JG, Totterman S, and Parker KJ. Kinematic analysis of musculoskeletal structures via volumetric MRI and unsupervised segmentation. *Medical Imaging 1999: Physiology and Function from Multidimensional Images.* 1999;3660:488–499.

144. Tamez-Pena JG, Barbu-McInnis M, and Totterman S. Structural quantification of cartilage changes using statistical parametric mapping—Art. no. 651248. In: Pluim JPW, Reinhardt JM Eds. *Medical Imaging 2007 Conference*. San Diego, CA, 2007: pp. 51248–51248.
145. Kauffmann C, Gravel P, Godbout B, Gravel A, Beaudoin G, Raynauld JP et al. Computer-aided, method for quantification of cartilage thickness and volume changes using MRL: Validation study using a synthetic model. *IEEE Trans Biomed Eng*. 2003;50:978–988.
146. Peterfy CG, Guermazi A, Zaim S, Tirman PFJ, Miaux Y, White D et al. Whole-organ magnetic resonance imaging score (WORMS) of the knee in osteoarthritis. *Osteoarthritis Cartilage*. 2004;12:177–190.
147. Felson DT, Lynch J, Guermazi A, Roemer FW, Niu J, McAlindon T et al. Comparison of BLOKS and WORMS scoring systems part II. Longitudinal assessment of knee MRIs for osteoarthritis and suggested approach based on their performance: Data from the Osteoarthritis Initiative. *Osteoarthritis Cartilage*. 2010;18:1402–1407.
148. Wolfe F, and Kong SX. Rasch analysis of the Western Ontario MacMaster Questionnaire (WOMAC) in 2205 patients with osteoarthritis, rheumatoid arthritis, and fibromyalgia. *Ann Rheum Dis*. 1999;58:563–568.
149. Thumboo J, Chew LH, and Soh CH. Validation of the Western Ontario and McMaster University Osteoarthritis Index in Asians with osteoarthritis in Singapore. *Osteoarthritis Cartilage*. 2001;9:440–446.
150. Roos EM, Roos HP, Lohmander LS, Ekdahl C, and Beynnon BD. Knee injury and osteoarthritis outcome score (KOOS)—Development of a self-administered outcome measure. *J Orthopaedic Sports Physical Therapy*. 1998;28:88–96.
151. Tamez-Pena JG, Gonzalez PC, Schreyer E, Farber J, Trevino V, and Totterman S. Advanced MRI-based measurements as surrogate markers of KOOS pain and KOOS other knee symptoms: Data from the osteoarthritis initiative. *Osteoarthritis Cartilage*. 2010;18:S187–S187.
152. Eckstein F, Wirth W, Hunter DJ, Guermazi A, Kwoh CK, Nelson DR et al. Magnitude and regional distribution of cartilage loss associated with grades of joint space narrowing in radiographic osteoarthritis—Data from the Osteoarthritis Initiative (OAI). *Osteoarthritis Cartilage*. 2010;18:760–768.
153. Eckstein F, Wirth W, Hudelmaier MI, Maschek S, Hitzl W, Wyman BT et al. Relationship of compartment-specific structural knee status at baseline with change in cartilage morphology: A prospective observational study using data from the osteoarthritis initiative. *Arthritis Res Therapy*. 2009; 11(3):R90, Available at http://0-www.scopus.com.millenium.itesm.mx/inward/record.url?eid=2-s2.0-67650095293&partnerID=40&md5=de3cf28d340df0839ad7c6919ff32767
154. Tamez-Pena JG, Lerner AL, Yao J, Salo AD, Totterman S. Evaluation of distance maps from fast GRE MRI as a tool to study the knee joint space. *Medical Imaging 2003: Physiology and Function: Methods, Systems, and Applications*. 2003;5031:551–562.

13

Utility of PET in Pharmaceutical Development

R. Boellaard
VU University Medical Center

M. Lubberink
Uppsala University

R.P. Maguire
Novartis Institutes for Biomedical Research

13.1 Introduction

Noninvasive imaging is a critical biotechnology tool for efficient pharmaceutical development. Precise measurement of pharmaceutical effects *in vivo*, in man, allows detailed study of target interaction and effects on pathology. Positron emission tomography (PET) enables measurement of uptake, distribution, and binding with high specificity. Optimal application of the technology requires critical selection of spatial and temporal resolution to optimize the sensitivity for the specific clinical question. From knowledge of the pharmacokinetics of the PET tracer and the statistical properties of the measurement system, data analysis methods for each radiotracer and application have been developed. In this way, the measurement sensitivity to radioligands and targets has been maximized and the signal response to potential confounding effects minimized. Both scientifically and from an operational and regulatory standpoint, PET imaging in pharmaceutical development has matured. Further research and work will be necessary to refine the technology, in order to reduce the number of study subjects required and increase the sensitivity to small changes.

PET may be useful in several different ways for clinical trials:

1. Longitudinal outcome biomarker. Changes in uptake are used as an index for treatment response.
2. Target engagement biomarker. Receptor expression or receptor occupancy studies using a specific radiotracer.
3. Staging and inclusion criteria for patient participation in clinical trials.
4. Stratification, that is, based on the PET findings patients may be assigned to a specific arm of the trial (e.g., high-risk vs. low-risk groups).
5. Biodistribution of a radiolabeled pharmaceutical.

Application of molecular imaging for receptor occupancy measurement and tumor metabolism is well understood in the pharmaceutical industry. The opportunity to derisk development programs by proving pathway engagement and efficacy early in clinical trials provides a critical advantage in pharmaceutical projects. Molecular imaging has also been used to study preclinical disease and therapy models. The main advantage of noninvasive measurement is clear, but it may not deliver the best sensitivity. Other methods such as *ex vivo* autoradiography may in fact be more appropriate to study pharmaceuticals in detail. The main advantages of *in vivo* translational imaging are the within-subject, longitudinal study designs allowing better precision, the ability to study processes *in vivo, in situ*, and the potential to reduce subject numbers, especially for precious animal models. Development of specific ligands for other organs and receptor systems will understandably increase the number of applications. As protein and antibody therapies are developed, PET imaging will be available to confirm biodistribution. Novel combined modality PET-MR devices will enable the measurement of function (pharmacodynamics) and target simultaneously. Development of noninvasive techniques, such as amyloid imaging in Alzheimer's disease, will permit earlier diagnosis and treatment. This chapter concentrates on current advances in technology that will permit advances in the next generation of vital treatments.

13.2 Phases of Pharmaceutical Development

As a pharmaceutical is developed from an initial concept and a pathway target, it is tested first in *in vitro* models and preclinical models to understand its interaction with the biological system in detail. The first clinical studies are designed to confirm the pharmacokinetic properties and safety profile of the pharmaceutical. As the new treatment moves into clinical trials in larger cohorts of patients, the clinical science, safety, and efficacy of the pharmaceutical will have more emphasis than the basic science of the pathway (Figure 13.1).

Molecular imaging can make contributions in all these phases. However, since it is related to the target, it is in the translational phase of early clinical development where the most obvious contribution can be made. Molecular imaging allows the noninvasive study of receptor and molecular level effects *in vivo* in man. This allows a more precise measurement and selection of dose, depending on the expected effect at the receptor/pathway level. In later stages of pharmaceutical development, the molecular imaging signal will be more directly related to disease outcome and the use of *non invasive* imaging can be used to monitor treatment effects on the disease. In this respect, the linkage between imaging signal change and the disease needs to be established and this will likely require that the imaging signal has been closely associated with disease outcome in previous treatment trials.

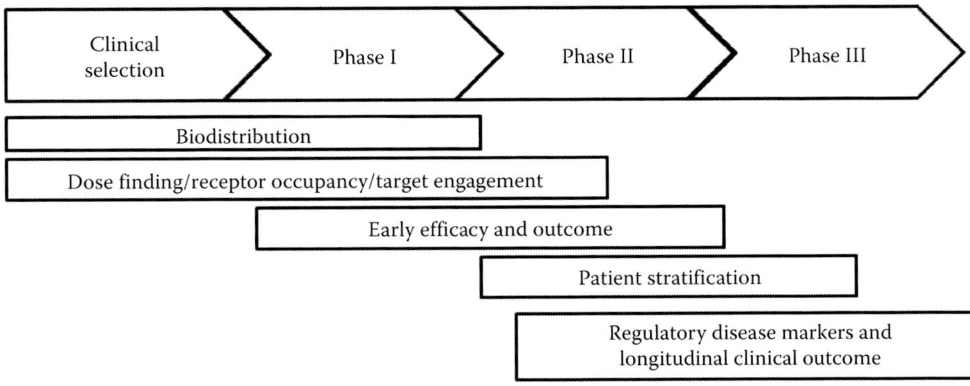

FIGURE 13.1 Applications of PET in different phases of pharmaceutical development.

13.3 State-of-the-Art Instrumentation for Clinical Trials

In this section, a brief summary of PET imaging principles, acquisition, and image reconstruction is given (Townsend 2004). Interested readers are referred to Townsend part 1 (Townsend 2004) for a more detailed explanation of PET imaging fundamentals.

13.3.1 Principles of PET

PET is a molecular imaging technique, which measures the distribution of a radioactive tracer *in vivo*. Upon administration of very small amounts (pico- or nanomoles) of a radiotracer to the patient, it distributes in the central compartment and in organs. The radioactive atom of the radiotracer (e.g., [F-18] FDG) emits positrons. The emitted positron combines with an electron after traveling a distance up to several mm in tissue. The positron and electron are then converted into two photons, each with an energy of 511 keV, which are emitted into near opposite directions. PET image acquisition is based on the simultaneous (coincidence) detection of these two photons. A PET scanner consists of many photon detectors surrounding the patient. A line connecting any two detectors that have detected two annihilation photons simultaneously is called the line of response (LOR) and ideally indicates the line along which a positron emission took place. In practice, a time window of a couple of ns (4–10 ns) is usually applied for the "simultaneous" detection of the annihilation photons. During a PET scan, millions of coincidence detections along LORs are being collected, providing information about the distribution of the radiotracer in tissue.

Unfortunately, not all coincidences contribute to the signal, that is, the "true" 3D distribution of the tracer. Background noise is added to the signal due to photons that are scattered before detection or by coincidence detection of two uncorrelated photons, that is, the so-called random coincidences. True coincidences arise from the simultaneous (coincident) detection of two annihilation photons generated by one positron emission. Ideally, only true counts are detected. A large fraction of the emitted photons (up to 50%) is scattered, mainly by Compton scattering, before leaving the patient. When one of the photons has been scattered, it will result in a dislocation of the "true" coincidence detection. Moreover, when two photons from two different positron emissions are accidentally (randomly) detected simultaneously, that is, within the duration of the time window (while the others are undetected), the PET camera will notice a random coincidence detection. It may be clear that these random coincidences result in image distortions (appears as an addition of a smooth background). Finally, multiple detections can occur when three or more photons are detected at the same time. These multiples are usually discarded.

In PET studies, it is necessary to account for the contribution of scattered and random coincidences. In most modern PET systems, scatter correction is usually performed using the single scatter simulation (SSS) method developed by Watson (2000). Randoms are usually measured by the so-called delayed coincidence time window technique. It is beyond the scope of this chapter to address these corrections in detail. Yet, it should be realized that especially accurate and precise scatter correction is one of the most challenging issues in quantitative PET imaging.

Moreover, due to attenuation (scatter and absorption) of photons in the patient, a large fraction of the emitted photons are not detected. Fortunately in PET, attenuation does not depend on the location of the positron emission along the LOR, but is determined by the total "radiological," or attenuating, thickness of the patient along the LOR. Consequently, by making specific (low dose) CT scans, the transmission through the subject along all LORs can be measured and thus correct exactly for the effects of attenuation. CT scans made for attenuation correction purposes are usually acquired using a low beam current (e.g., 30–70 mAs) to minimize the radiation dose for the patient. The (average) energy of the photons generated by the x-ray tube of the CT scanner is usually about 80 keV (actually it is a spectrum). In order to use these CT scans for attenuation correction of 511 keV annihilation photon, the CT image first needs to be converted into a 511 keV attenuation coefficient map. The latter step usually involves automated segmentation of the CT image in various tissue classes (lung, soft tissue, and bone) followed by (bi-)linear rescaling functions. After these conversions, an accurate attenuation correction can be derived from the CT images. In practice, however, attenuation correction is somewhat hampered by patient motion (e.g., breathing). An overview of potential pitfalls or artifacts when using CT-based attenuation correction is given by Sureshbabu and Mawlawi (2005).

Finally, with current state-of-the-art PET detectors, it is possible to measure the difference in arrival time of the two emitted annihilation photons with an accuracy of 300–600 ps. The technology that measures and uses this difference in arrival time is called "time of flight (TOF)." At present, with TOF (300–600 ps), it is possible to determine the location of the positron emission along the LOR with a spatial accuracy of about 5–10 cm. Clearly, this spatial accuracy is not high enough to exactly (within 1 mm) indicate the location of the positron emission. Yet, by taking TOF into account during image reconstruction, the image quality is improved and specifically the detection of small tumors nearby large high uptake organs (liver, heart, brain, and bladder) is enhanced.

13.3.2 Reconstruction Algorithms

PET or PET/CT systems measure are based on the coincidence detection of annihilation photons that are emitted from the patient; thus, a PET or PET/CT system measures coincident photons or projection data outside and around the patient. These projection or "raw" data are then used to calculate a 3D PET image representing the distribution of the radiotracer in the patient. The process that calculates the 3D PET image from the raw projection data while including all necessary corrections, such as, for example, scatter, randoms, and attenuation corrections, is called PET image reconstruction.

Up to the late 1990s, raw PET data were primarily reconstructed using the so-called filtered back-projection (FBP). The latter method is linear and quantitatively accurate but suffers from noise and streak artifacts. Nowadays, iterative reconstructions are almost exclusively used and PET/CT systems are equipped with dedicated reconstruction computer clusters to deal with the higher computational demand of iterative versus FBP reconstruction. Most commercially available iterative reconstruction methods are based on the maximum likelihood expectation maximization (MLEM) algorithm described by Shepp and Vardi (1982). Later, Hudson and Larkin (1994) invented a faster implementation of MLEM by taking the so-called ordered subsets of the raw data during the reconstruction process resulting in the ordered subset expectation maximization (OSEM) method. The main advantage of iterative reconstructions is that they provide enhanced image quality compared to FBP and thereby improving diagnostic accuracy. Another advantage is that various corrections can be applied or implemented more easily and/or in a more sophisticated way.

Recently, two new technologies have been introduced into PET image reconstruction technology. First of all, TOF information can be incorporated in the reconstruction process. By using TOF, the recovery of signal from small lesions, especially when located to larger high-uptake regions, is improved and this technique therefore enhances the detectability of small lesions. Secondly, PET/CT systems suffer from a relatively low resolution. Typically, the spatial resolution of a clinical PET system is about 4–6 mm FWHM. Yet, the PET system's point spread function can be measured and taken into account during the iterative image reconstruction process, the so-called recovery corrected image reconstruction (Brix et al. 1997). There are several terminologies being used for this reconstruction technique, such as resolution recovery, resolution modeling, "high definition," and PVC-OSEM reconstruction. Use of resolution recovery during reconstruction improves the spatial resolution of the reconstructed images, but the quantitative robustness of this technique has not yet been fully validated.

13.3.3 Multimodality Imaging Systems Useful for Clinical Trials

13.3.3.1 PET/CT

Nowadays, most PET systems are combined multimodality PET/CT systems. An excellent overview on PET/CT technology was recently published in the *European Journal of Nuclear Medicine and Molecular Imaging* by Mawlawi and Townsend (2009). To the best knowledge of these authors, there are at present five vendors offering PET/CT systems: Philips Healthcare, Siemens Medical Solutions, Hitachi Medical, Toshiba Medical Corporation, and GE Healthcare. All PET/CT systems have a sequential, but integrated, system design in which the CT scanner is placed in front of the PET part either within one large enclosure or in two separate enclosures, allowing the two systems to be moved apart. The latter may allow for a more easy patient access and/or improve patient comfort (e.g., claustrophobia). The main advantage of combined PET/CT image is the spatial correlation of functional information obtained with PET with the anatomical information collected with CT. The latter allows for more accurate localization of the PET tracer distribution, for example, the localization of PET tracer avid lesions. Yet, an important application for clinical trials may be the (near) simultaneous collection of structural and functional information. Specifically, the use of contrast-enhanced dynamic CT imaging may allow for assessment of (tumor or organ) perfusion, which can then be used in combination with both structural information (from the CT) and functional or metabolic information from the PET. Use of multimodality imaging may thus be helpful to understand to what extent changes in tracer uptake because of therapy may have been caused by changes in perfusion and/or structural changes. Finally, tumor response can now be assessed based on anatomical (tumor size) criteria as indicated by RECIST 1.1, and also on the basis of metabolic responses using PET at the same time (Wahl et al. 2009). In all causes, multimodality imaging may provide a more complete picture of (tumor) response to therapy.

13.3.3.2 PET/MR

At present, the first clinical PET/MRI systems are being built and/or have been installed. There are now PET/MRI systems with different designs: (a) A PET insert placed within the MRI thereby allowing for true simultaneous PET and MR acquisitions, but restricting its application to brain imaging due to the small FOV/gantry opening of the PET insert. (b) A fully integrated whole-body PET/MRI system in which a 3D PET system is fully integrated within the MRI system. (c) A design in which the MRI and PET are placed side by side, that is, in a similar arrangement as used for PET/CT systems. In the latter case, acquisitions will be near but not exact simultaneous, but these systems also allow for whole-body acquisitions. A possible advantage of this setup is that the PET system is distant (about 3–5 m) from the MRI system, thereby avoiding difficulties associated with the sensitivity of conventional PET detectors for the high magnetic fields. In other words, a "regular" state-of-the-art PET system can be used in such a sequential system design. Major challenges for integrated MRI/PET systems are the development of PET detectors that are insensitive to the magnetic field of the MR and, for any PET/MR system, use of MR data for attenuation (and scatter) correction of the PET data. High count rate capable detectors with good timing resolution (required for TOF acquisitions) that are insensitive to the magnetic field may allow for a fully integrated PET/MRI design for which the PET acquisition can be made while applying TOF technology to enhance the image quality. Much progress has been made and reported to address these issues (Pichler et al. 2008). Despite the technical challenges for optimizing PET/MR design and PET image quality and quantification, there may be several promising advantages for PET/MR. First, MR provides images with better soft tissue contrast, thereby enhancing the localization of areas with unusual tracer uptake. Moreover, MR can be used to provide various types of functional information, such as perfusion by use of contrast-enhanced or arterial spin labeling MR, MR spectroscopy to determine presence of specific molecules in tissue, and changes in blood flow by functional MRI.

13.4 Quantitation

PET is a clinical tool and radiological evaluation of PET images is important in clinical trial application. For subject inclusion and evaluation, this may be the most appropriate analysis approach. PET data also has critical spatial, temporal, and specific uptake and binding data and it is often appropriate to use quantitiative analysis.

13.4.1 Pharmacokinetics

The uptake and distribution of the PET ligand in the central blood compartment and side compartments, including the target organ, is called pharmacokinetics. In most PET applications, the tracer is given as an intravenous injection, although it can be given as an intravenous infusion. The ligand rapidly distributes itself in the blood and is taken up generally more slowly into other organs. In the target organ, the uptake is usually described in two phases—(1) extraction and nonspecific distribution into the organ and (2) binding (perhaps after enzymatic conversion). The binding phase can either be clearly reversible—for example, receptor binding—or irreversible—for example, trapping after incorporation.

Pharmacokinetic models for PET data analysis have been developed to address the irreversible and reversible-binding situations. A full kinetic analysis of the PET study might require extensive measurement of both the blood and the organ concentrations of the radiotracer. That requires PET measurement from the time of injection and during the uptake phase, which might take an hour or longer. At the same time, blood, venous, or arterial samples need to be measured. Nonlinear compartmental analysis to compute the kinetic rate constants of all exchange processes is one approach to estimate parameters (Gunn et al. 2001). In clinical trials, it is critical to minimize the patient burden whenever possible and it is often possible to fulfill the study objectives using simpler quantification approaches.

For [F-18]FDG and other irreversible tracers, the multiple time graphical analysis approach of Gjedde (Gjedde et al. 1985) and Patlak (Patlak et al. 1983) allows computation of the metabolic or incorporation rate K_m. This approach is a linearization which permits linear regression to compute the incorporation rate. The standardized uptake value is an even simpler methodology which only requires PET measurement of the organ activity at the end of the uptake phase.

Radiotracers which show reversible binding can best be analyzed by considering the partition between free and bound radiotracer at equilibrium. Although the equilibrium conditions are difficult to achieve in a PET clinical trial, it is possible to use the measured image and blood concentration data to compute the partition ratio. The partition ratio between free and bound is directly related to the affinity (K_d) and the available receptor density (B_{avail}). At least two methods, the simplified reference tissue method (SRTM) and the Logan plot, can be used to analyze PET data from reversible tracers. A standardized nomenclature associated with these concepts has been published (Innis et al. 2007).

13.5 Oncology Imaging

The application of PET in the clinic is most widespread in oncology. The glucose analog tracer [F-18]FDG is avidly metabolized by many tumor types because of their increased metabolism, and also because of changes in the metabolic pathways and transporters. The combination of functional PET and structural CT has allowed clinical imaging and staging of soft tissue tumors and has been a major driver for the expansion of PET technology in the clinic.

13.5.1 Oncology Quantitation

There are several degrees of quantification. To date, the most common method for quantifying tracer (mostly FDG) uptake is the use of standardized uptake values (SUV). This represents uptake in a tumor or organ at a fixed time after tracer administration, corrected for net injected dose of the tracer per patient weight. Limitations of SUV include its dependency on patient preparation, scanning procedure, image reconstruction, and image analysis procedures (Boellaard 2009; Weber 2005). These limitations, however, can to a large extent be overcome by proper standardizing procedures (Boellaard 2009; Boellaard et al. 2010; Shankar et al. 2006).

Apart from SUV, various other more complex and/or accurate quantitative measures can be used for the analysis of FDG-PET studies. Hoekstra et al. (2000, 2002) reviewed and compared various methods for quantifying FDG uptake in nonsmall-cell lung cancer (NSCLC). Nonlinear regression (NLR) of the data using a pharmacokinetic model could be used as gold standard. Yet, Patlak's analysis and various simplified methods, such as the simplified kinetic method (SKM) (Hunter et al. 1996), SUV normalized by body surface area and corrected for glucose (SUVBSAg), and SadatoBSA (Sadato et al. 1998), correlated well with NLR (R2 > 0.95). Lammertsma (Lammertsma et al. 2006) discussed the relationship between these simplified methods and full kinetic analysis for monitoring response to classic cytotoxic medicines. Both authors emphasized that care should be taken in using simplified methods (i.e., SUV-based methods) for evaluation of new medicines.

Potential discrepancies between SUV and full kinetic analysis results may be caused by changes in plasma glucose levels, changes in perfusion, or differences in FDG plasma clearance among scans (Cheebsumon et al. 2011). Therefore, full kinetic approaches are considered most quantitative as they use the entire (measured) arterial input and (tumor or organ) time–activity curves in combination with a tracer kinetic model to derive a more quantitative measure of tracer uptake or binding. However, one of the main limitations of full quantitative studies is the need for a dynamic scanning protocol as it requires information of tracer uptake as a function of time after tracer injection (typically up to 60–90 min or most tracers), thereby limiting the data acquisition to one bed position, that is, a restricted axial field of view or anatomical coverage. In addition, full quantitative kinetic analysis requires an input function, that is, the concentration of the radiotracer in plasma over time. The input function can be measured invasively by taking several or continuous arterial samples during the PET study. The latter procedure is invasive and causes patient discomfort. Therefore, when feasible, use of image-derived input function (IDIF) should be considered. IDIFs may be derived directly from the dynamic PET images in case large blood pool structures, such as the left ventricle or aorta ascendens, are in the field of view of the PET scanner. For some brain studies, an alternative approach is the use of reference tissue kinetic modeling, for which the time course of tracer uptake in a reference region in the brain can be used as input function. Typically, use of reference regions requires the presence of an anatomical region that does not show any specific uptake or binding of the radiotracer. Due to the various limitations of full kinetic analysis, simplified semiquantitative indices are preferred in a clinical setting and thus for larger clinical trials (phase 3), as they allow for whole-body scanning and require less complex scanning and data analysis procedures. On the other hand, they may be less accurate for reasons listed above.

13.6 Brain Imaging

13.6.1 Receptor Occupancy

PET using radiotracers that bind (reversibly) to a neuroreceptor can be used to assess the occupancy of a new drug as function of drug dosage (Figure 13.2). At the same time, it can be used to determine the optimal drug administration or intake schedule. The principle of measuring drug neuroreceptor occupancy is based on receptor availability. Usually, PET studies are performed by administering tracer (picomolar) amounts of a radiopharmaceutical. At these tracers levels, only a few percent of all available neuroreceptors in a brain region are occupied by the radiotracer and nearly 100% of neuroreceptors are freely available for radiotracer binding. Performing a PET study under tracer conditions thus results in highest uptake or (specific) binding of that tracer. Drug receptor occupancy can be measured when drug and radiotracer are binding to the same receptor. By giving a drug at various doses to a subject, different levels of neuroreceptors are occupied by that drug and consequently, the amount of freely available neuroreceptors for radiotracer binding is reduced. As a result, radiotracer uptake or binding measured by PET will be lower at increasing dosages of the drug. By quantifying tracer uptake or specific binding at various drug dosage levels, it is possible to determine the percentage neuroreceptor occupancy by that drug. For example, for an antagonist, when the PET signal is 50% of that measured under tracer conditions, 50% of the neuroreceptors are occupied by the drug. A drug-receptor occupancy of about 50–80% is usually considered to provide a (sufficient possible) therapeutic effect. Likewise, by making several PET studies after administration of a drug, the neuroreceptor occupancy by the drug can be measured as function of the postinjection time interval, which provides useful information for optimizing the drug dosage scheduling.

13.6.2 Amyloid Imaging

In the past decade, various tracers have been developed that bind to amyloid in the brain. Amyloid is a protein that accumulates in the brain of Alzheimer's disease (AD) patients. Already before clinical symptoms are present, increased levels of amyloid may be present in the brain. PET is a very sensitive technique that can detect the presence of amyloid in the brain at very low levels (Figure 13.3). Amyloid imaging by PET may thus be helpful for early identification of subjects with increasing levels of amyloid

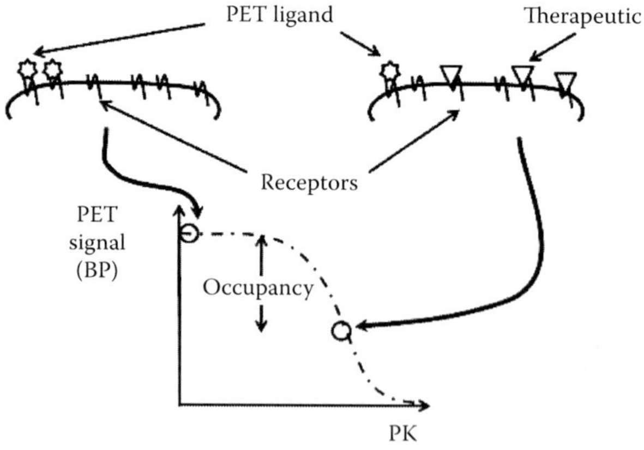

FIGURE 13.2 Receptor occupancy. Upper left shows a cell membrane with receptors crossing it. The PET ligand (stars) occupies the receptors in proportion to their availability. On the upper right, as the receptors are occupied by pharmaceutical (triangles), the PET ligand concentration reduces in proportion. The two conditions are plotted on a binding potential versus pharmaceutical concentration (PK) curve at the bottom. The receptor occupancy is the proportional change in binding potential.

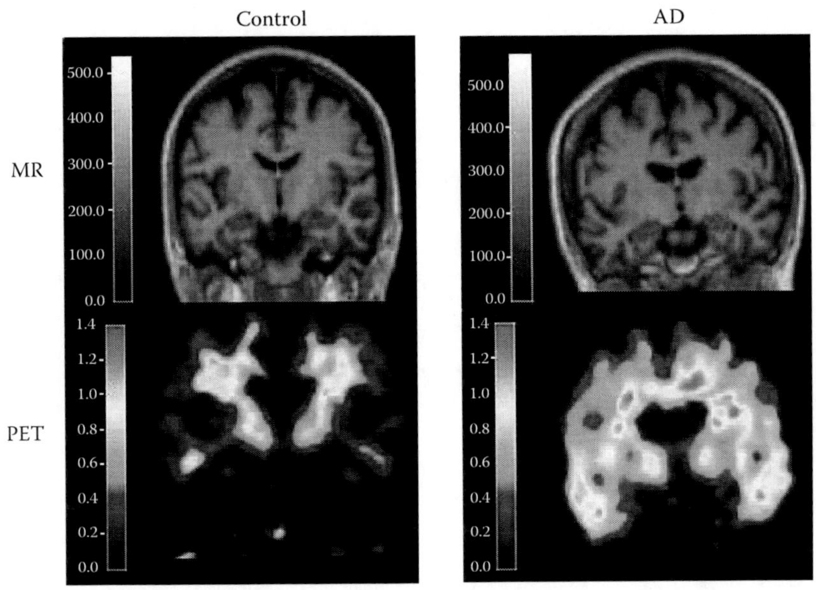

FIGURE 13.3 (**See color insert.**) MRI (top row) and PET tracer-binding images obtained with the amyloid tracers [11C]-PIB (bottom) row for a healthy control (first column) and AD patient (second column). A clear increase in tracer binding is seen in the AD subject compared to the healthy control.

in their brain, even before they show symptoms of dementia. Moreover, amyloid accumulation is characteristic for AD and PET can thus be supportive for improved differential diagnosis. Identification of Alzheimer's disease patients early in the disease spectrum is very important for clinical trials. Amyloid PET imaging could be used as an inclusion criterion in order to correctly identify AD subject eligible for trial participation. At present, the first response assessment studies using changes in PET tracer amyloid binding as index for drug efficacy are ongoing.

13.7 Cardiac PET

PET has been used in cardiology since the late 1970s, with its main application in the measurement of myocardial blood flow (MBF) and viability, but it has also been used extensively in basic research in the area of myocardial substrate metabolism. Since molecular cardiac imaging with PET is aimed at diagnosis at the earliest stage of disease, its use can be valuable in drug development. It provides well-defined end-points for clinical trials, for example, when measuring myocardial efficiency after therapies for heart failure or when imaging the effects of metabolic interventions (Bengel et al. 2009).

The advantage of using PET to measure MBF compared to other methods is that MBF can not only be visualized but also quantified, giving absolute MBF values in ml blood per gram tissue per minute. Several different tracers have been suggested for measuring MBF with PET, of which ^{13}N-ammonia, ^{82}Rb (rubidium), and ^{15}O-water have been used most widely. ^{15}O-Water is considered the gold standard for flow measurements because it is freely diffusible, metabolically inert, and has a 100% extraction. The extraction of ^{13}N-ammonia and ^{82}Rb is flow dependent and decreases for high flow levels. ^{13}N-Ammonia and ^{15}O-water require the presence of an onsite cyclotron because of the 9 and 2 min half-lives of both tracers, whereas ^{82}Rb is generator produced and hence more easily available where it has been approved for clinical use. ^{13}N-Ammonia and ^{82}Rb can give qualitative images showing the distribution of flow over the myocardium, whereas this is not possible with ^{15}O-water. Recent developments in scanner technology and image processing methods, however, have also enabled generation of quantitative MBF images with ^{15}O-water, showing MBF at the voxel level (Harms et al. 2011).

All of these three tracers have a very short radioactive half-life, but a number of promising F-18-labeled tracers are currently under development. One of these is ^{18}F-BMS747158, a novel pyridaven derivative which has a first-pass extraction of 94% and a myocardial uptake that is proportional to MBF up to high flow values (Nekolla et al. 2009). This compound is currently being tested in clinical studies and may allow for a more widespread use of MBF imaging with PET. The introduction of integrated PET-CT scanners allows for acquisition of functional data, CT angiography, and calcium score within a single session with quantitative MBF PET imaging (Knaapen et al. 2010).

FDG is considered the gold standard for imaging myocardial viability, with viable areas showing a mismatch with reduced MBF but normal FDG. Another marker of myocardial viability is the perfusable tissue index (PTI), measured using ^{15}O-water in the same procedure as MBF (Iida et al. 1995). Oxygen gas labeled with ^{15}O, as well as ^{11}C-acetate, can provide quantitative measures of myocardial oxygen consumption. ^{11}C-Palmitate can be used to measure fatty acid metabolism (Schelbert et al. 1986). Myocardial sympathetic innervation has been assessed with ^{11}C-hydroxyephidrine (HED) (Schäfers et al. 1998). A mismatch between MBF and HED retention is thought to be a predictor of ventricular arrhythmia in patients with previous heart failure. Recently, there has been an increased interest in the development of imaging markers for aortic aneurisms and especially vulnerable plaques before rupture.

13.8 Other Therapeutic Areas

13.8.1 Use of [F-18]FDG in Inflammatory Disease

The most common PET tracer, [F-18]FDG is used to measure the metabolism through the glucose pathway. Utilization of glucose for cellular energy metabolism means that this signal can be found in many cells and depends not only on the energy metabolism rate, but also on the density of cells. In brain, for example, a reduced [F-18]FDG uptake in a certain area of the brain can be indicative of either reduced glucose metabolism (e.g., due to a reduction in neuronal activity) or a loss of neurons, depending on the circumstances. In inflammation, the invasion of tissues by neutrophils increases the density (cells per unit volume) of high-energy-demand cells. This results in a strong PET signal which can be used as a marker of inflammation, for example, in COPD (Schuster 2007) and gastrointestinal disorders (Jacene et al. 2009).

13.8.2 Other Tracers for Inflammatory Disease

Joint inflammation is a complex process, which can be studied using functional markers. Leucocytes labeled with Tc-99m outside the body and reinjected to concentrate at the site of inflammation are a standard technique in nuclear medicine with SPECT (Basu et al. 2009). Because of the shorter half-life of the common PET isotopes, which is much shorter than the distribution time of labeled cells, this technique has not translated well to the PET technology. Tracers (Gotthardt et al. 2010; van der Bruggen et al. 2010) such as Ga-67-citrate have been used to study inflammation through perfusion changes. In the future, Ga-68-labeled peptides may provide a means for distribution of highly specific ligands to enable study of inflammatory pathways with PET (Rizzello et al. 2009). Gastrointestinal imaging with SPECT using Tc-99m and In-111 is well established (Wilding et al. 2001). These techniques have been used to study gastric emptying and gut motility.

13.9 Qualification of PET as a Biomarker

Processes and methods for qualification of imaging and other biomarkers are an area of current active development. Health authorities have an active interest in advancing new imaging tools that will allow testing of safe and effective medicines at all stages in development (Goodsaid and Mendrick 2010). Imaging biomarkers that are being used to understand pathway and receptor interactions can aid in advancing the basic science of programs and can be enabling tools to discover new targets to advance treatments for unmet medical needs.

13.9.1 Qualification of PET as Fit-for-Purpose

In qualifying a biomarker, the specific context and purpose of use have to be defined clearly. For example, use of PET imaging to monitor treatment-related change in a clinical study will have to be considered differently from a screening tool to select patients with rapidly progressing disease for inclusion in trials of neuroprotective agents. Both the FDA in the United States and the EMA in Europe are engaged in biomarker qualification initiatives and have invited interested parties to submit dossiers outlining potential biomarkers to be qualified for application. For PET imaging biomarkers, it will be critical to establish the technical validation of the instrument, outlining the minimum standards required to ensure instrument calibration. The performance of the specific PET tracer will also be important; What is the target? How is this related to the disease, pathology, or target? The promise of the validation process is that, having established a marker for a particular purpose with the health authority, a user can be confident that the health authority will accept the marker for this purpose. In this way, PET imaging biomarker investment can be derisked.

13.9.2 Using PET as a Surrogate Biomarker

In order to develop surrogate biomarker tools for late-stage development, we need to establish the linkage (Frank and Hargreaves 2003) between the signal and the disease outcome further, the change in signal under effective treatments needs to be established and correlated with outcome. This is a big hurdle. The number of PET techniques which can be considered to be surrogate markers is very limited. Some examples of previous use in this context are the use of F-18-FDOPA to measure dopaminergic neurodegeneration in Parkinson's disease (Oertel et al. 2006) and the use of F-18-FDG in Lymphoma oncology trials (Kelloff et al. 2007).

13.10 Antibody Imaging

There has been a large advance in the design of medicines targeting molecular targets on tumor cells, such as those involved in proliferation, differentiation, apoptosis, angiogenesis, and so on. Monoclonal antibodies (MAbs) form the largest category of targeted medicines. Targeted therapies are challenging to develop and will generally be beneficial to a subgroup of patients, depending, for example, on overexpression of the target and variability and heterogeneity of tumor uptake. The efficacy of a targeted medicine in an individual patient has to be evaluated by assessing its uptake in tumor and normal tissue during a scouting procedure prior to start of therapy or immediately upon onset of therapy to avoid unnecessary expensive treatments and delayed onset of effective treatment. Response, however, as addressed using RECIST criteria is usually not visible during earlier stages of therapy, and even the use of FDG-PET for response monitoring requires one or two cycles of therapy before efficacy can be evaluated.

Labeling of the targeted pharmaceutical with a positron-emitting nuclide, however, may provide a predictor of treatment efficacy even before the onset of therapy. Quantitative PET can image the distribution of the pharmaceutical both prior to and during therapy. Since the radioactive half-lives of the most common PET isotopes F-18 and C-11 are too short to measure the kinetics of MAbs, isotopes with longer half-lives such as I-124 (100 h) and Zr-89 (78 h) have been used for labeling MAbs. Methods for large-scale production of highly pure Zr-89 and I-24I, and for facile and stabling coupling of these isotopes with MAbs, have been developed (Perk et al. 2010; Vosjan et al. 2010, 2011), and the radioactive half-life of Zr-89 and I-124 allows for their worldwide supply.

A microdosing PET study with a labeled targeted medicine can be used to understand the biodistribution and target engagement before starting the treatment. This, of course, assumes that target organ distribution of the pharmaceutical is a good predictor for treatment efficacy, and that a tracer study is representative of the uptake of the medicine at efficacious concentrations. In pharmaceutical development, quantitative imaging of labeled-targeted pharmaceuticals with PET can provide information on optimal dosage, uptake in critical organs, and interpatient variations in kinetics and targeting. Pharmaceutical imaging provides this information in an efficient and safe way, with fewer patients treated at suboptimal doses. This approach is especially attractive when the pharmaceutical of interest is directed against a novel tumor target that has not been previously validated. Thus, use of labeled pharmaceuticals allows for early selection of promising pharmaceutical candidates and reduction of development costs.

13.11 Precompetative Consortia

Health authorities have recognized (Altar 2008; Hunter 2008; Kamel et al. 2008; Woodcock 2009) the need to advance imaging and other biomarkers as decision-making tools in pharmaceutical development and a number of interconnected public–private consortia partnerships have been established in order to tackle these issues. The Innovative Medicines Initiative (IMI) is a partnership between European Pharmaceutical Industry partners represented by EFPIA and the European Union. Two of the current projects involve molecular imaging: IMIDIA which is studying Beta-cell demise in diabetes and NEWMEDS which is developing novel biomarkers for psychiatry. The Foundation of the National Institutes of Health Biomarkers Consortium has a broad stakeholder base of for-profit and nonprofit organizations, including the health agency and the pharmaceutical industry. The FNIH biomarkers consortium has identified multiple projects with a molecular imaging component: I-SPY 2 studying personalized medicine in oncology; ADNI developing imaging biomarkers, including amyloid imaging for Alzheimer's disease; and OAI, which has studied imaging technologies in osteoarthritis.

This is an opportunity to bring academic and industry scientists closer, where there is potential impact of a new technology on pharmaceutical development. The framework for these consortia provides a practical way to work on precompetitive projects and share resources and expertise.

One area which is perhaps underutilized in the current consortia is the development of novel radiopharmaceuticals for all disease areas. Although there are focused and fruitful efforts within specific consortia, this is such a critical area in molecular medicine with specific challenges on intellectual property and licensing that it is in need of a focused consortium activity.

13.12 Future Outlook

PET will continue to be a valuable tool to understand the *in vivo* action of pharmaceuticals on target, mechanisms, and disease outcomes. The contribution of PET imaging in oncology and neurology will continue and the use in decision making in clinical trials will be even clearer and more refined. At the same time, there is scope to consider how to develop PET agents and techniques as readouts that can be used in personalized medicine paradigms to select the treatments most appropriate for specific patients and to monitor the treatment efficacy. The use of PET technology in other disease areas and organ systems is gaining ground. Focused broad scientific understanding of the specific contributions PET can bring in decision making is critical to ensuring that the method is used efficiently.

Rapid development and characterization of specific radiotracers for target engagement are one area which would benefit from new methods and clear selection process, for example, high specific activity ligand synthesis techniques and microfluidic systems may improve radiochemical process yields. However, early *in vitro* test systems and screening tools to quickly identify the *in vivo* clinical performance of PET ligands would be an important step.

Further interdisciplinary integration of research and development of PET for clinical trials across disciplines combined with standardization of instrumentation and methods will increase future efficient application. Precompetitive consortia will allow academic and industry collaboration to develop an appropriate set of tools and to establish the validation of those methods for clinical trial application.

The application of PET in clinical trials has reached a stage of maturity that will now allow more refined early-development trial designs. This will enable the next generation of treatments to be selected and advanced in clinical development.

References

Altar, C.A. 2008. The Biomarkers Consortium: On the critical path of drug discovery. *Clin. Pharmacol. Ther.*, 83, (2) 361–364. Available from: PM:18183037.

Basu, S., Zhuang, H., Torigian, D.A., Rosenbaum, J., Chen, W., and Alavi, A. 2009. Functional imaging of inflammatory diseases using nuclear medicine techniques. *Semin. Nucl. Med.*, 39, (2) 124–145. Available from: PM:19187805.

Bengel, F.M., Higuchi, T., Javadi, M.S., and Lautamaki, R. 2009. Cardiac positron emission tomography. *J. Am. Coll. Cardiol.*, 54, (1) 1–15. Available from: PM:19555834.

Boellaard, R. 2009. Standards for PET image acquisition and quantitative data analysis. *J. Nucl. Med.*, 50 Suppl 1, 11S–20S. Available from: PM:19380405.

Boellaard, R., O'Doherty, M.J., Weber, W.A., Mottaghy, F.M., Lonsdale, M.N., Stroobants, S.G., Oyen, W.J. et al. 2010. FDG PET and PET/CT: EANM procedure guidelines for tumour PET imaging: Version 1.0. *Eur. J. Nucl. Med. Mol. Imaging*, 37, (1) 181–200. Available from: PM:19915839.

Brix, G., Doll, J., Bellemann, M.E., Trojan, H., Haberkorn, U., Schmidlin, P., and Ostertag, H. 1997. Use of scanner characteristics in iterative image reconstruction for high-resolution positron emission tomography studies of small animals. *Eur. J. Nucl. Med.*, 24, (7) 779–786. Available from: PM:9211765.

Cheebsumon, P., Velasquez, L.M., Hoekstra, C.J., Hayes, W., Kloet, R.W., Hoetjes, N.J., Smit, E.F., Hoekstra, O.S., Lammertsma, A.A., and Boellaard, R. 2011. Measuring response to therapy using FDG PET: Semi-quantitative and full kinetic analysis. *Eur. J. Nucl. Med. Mol. Imaging.*, 38, (5) 832–842. Available from: PM:21210109.

Frank, R. and Hargreaves, R. 2003. Clinical biomarkers in drug discovery and development. *Nat. Rev. Drug Discov.*, 2, (7) 566–580. Available from: PM:12838269.

Gjedde, A., Wienhard, K., Heiss, W.D., Kloster, G., Diemer, N.H., Herholz, K., and Pawlik, G. 1985. Comparative regional analysis of 2-fluorodeoxyglucose and methylglucose uptake in brain of four stroke patients. With special reference to the regional estimation of the lumped constant. *J. Cereb. Blood Flow Metab.*, 5, (2) 163–178. Available from: PM:3872872.

Goodsaid, F.M. and Mendrick, D.L. 2010. Translational medicine and the value of biomarker qualification. *Sci. Transl. Med.*, 2, (47) 47–44. Available from: PM:20811041.

Gotthardt, M., Bleeker-Rovers, C.P., Boerman, O.C., and Oyen, W.J. 2010. Imaging of inflammation by PET, conventional scintigraphy, and other imaging techniques. *J. Nucl. Med.*, 51, (12) 1937–1949. Available from: PM:21078798.

Gunn, R.N., Gunn, S.R., and Cunningham, V.J. 2001. Positron emission tomography compartmental models. *J. Cereb. Blood Flow Metab.*, 21, (6) 635–652. Available from: PM:11488533.

Harms, H.J., Knaapen, P., de, H.S., Halbmeijer, R., Lammertsma, A.A., and Lubberink, M. 2011. Automatic generation of absolute myocardial blood flow images using [(15)O]H (2)O and a clinical PET/CT scanner. *Eur. J. Nucl. Med. Mol. Imaging.*, 38, (5) 930–939. Available from: PM:21271246.

Hoekstra, C.J., Hoekstra, O.S., Stroobants, S.G., Vansteenkiste, J., Nuyts, J., Smit, E.F., Boers, M., Twisk, J.W.R., and Lammertsma, A.A. 2002. Methods to monitor response to chemotherapy in non-small cell lung cancer with F-18-FDG PET. *J. Nucl. Med.*, 43, (10) 1304–1309. Available from: ISI:000178393900011.

Hoekstra, C.J., Paglianiti, I., Hoekstra, O.S., Smit, E.F., Postmus, P.E., Teule, G.J.J., and Lammertsma, A.A. 2000. Monitoring response to therapy in cancer using [F-18]-2-fluouo-2-deoxy-D-glucose and positron emission tomography: An overview of different analytical methods. *Eur. J. Nucl. Med.*, 27, (6) 731–743. Available from: ISI:000087883200017.

Hudson, H.M. and Larkin, R.S. 1994. Accelerated image reconstruction using ordered subsets of projection data. *IEEE Trans. Med. Imaging*, 13, (4) 601–609. Available from: PM:18218538.

Hunter, A.J. 2008. The Innovative Medicines Initiative: A pre-competitive initiative to enhance the biomedical science base of Europe to expedite the development of new medicines for patients. *Drug Discov. Today.*, 13, (9–10) 371–373. Available from: PM:18468553.

Hunter, G.J., Hamberg, L.M., Alpert, N.M., Choi, N.C., and Fischman, A.J. 1996. Simplified measurement of deoxyglucose utilization rate. *J. Nucl. Med.*, 37, (6) 950–955. Available from: PM:8683318.

Iida, H., Takahashi, A., Tamura, Y., Ono, Y., and Lammertsma, A.A. 1995. Myocardial blood flow: Comparison of oxygen-15-water bolus injection, slow infusion and oxygen-15-carbon dioxide slow inhalation. *J. Nucl. Med.*, 36, (1) 78–85. Available from: PM:7799088.

Innis, R.B., Cunningham, V.J., Delforge, J., Fujita, M., Gjedde, A., Gunn, R.N., Holden, J. et al. 2007. Consensus nomenclature for *in vivo* imaging of reversibly binding radioligands. *J. Cereb. Blood Flow Metab.*, 27, (9) 1533–1539. Available from: PM:17519979.

Jacene, H.A., Ginsburg, P., Kwon, J., Nguyen, G.C., Montgomery, E.A., Bayless, T.M., and Wahl, R.L. 2009. Prediction of the need for surgical intervention in obstructive Crohn's disease by 18F-FDG PET/CT. *J. Nucl. Med.*, 50, (11) 1751–1759. Available from: PM:19837758.

Kamel, N., Compton, C., Middelveld, R., Higenbottam, T., and Dahlen, S.E. 2008. The Innovative Medicines Initiative (IMI): A new opportunity for scientific collaboration between academia and industry at the European level. *Eur. Respir. J.*, 31, (5) 924–926. Available from: PM:18448501.

Kelloff, G.J., Sullivan, D.M., Wilson, W., Cheson, B., Juweid, M., Mills, G.Q., Zelenetz, A.D. et al. 2007. FDG-PET lymphoma demonstration project invitational workshop. *Acad. Radiol.*, 14, (3) 330–339. Available from: PM:17307666.

Knaapen, P., de Haan, S., Hoekstra, O.S., Halbmeijer, R., Appelman, Y.E., Groothuis, J.G., Comans, E.F. et al. 2010. Cardiac PET-CT: Advanced hybrid imaging for the detection of coronary artery disease. *Neth. Heart J.*, 8, (2) 90-8. PubMed PMID: 20200615; PubMed Central PMCID: PMC2828569.

Lammertsma, A.A., Hoekstra, C.J., Giaccone, G., and Hoekstra, O.S. 2006. How should we analyse FDG PET studies for monitoring tumour response?. *Eur. J. Nucl. Med. Mol. Imaging.*, 33 Suppl 1, 16–21. Available from: PM:16763817.

Mawlawi, O. and Townsend, D.W. 2009. Multimodality imaging: An update on PET/CT technology. *Eur. J. Nucl. Med. Mol. Imaging.*, 36 Suppl 1, S15–S29. Available from: PM:19104808.

Nekolla, S.G., Reder, S., Saraste, A., Higuchi, T., Dzewas, G., Preissel, A., Huisman, M. et al. 2009. Evaluation of the novel myocardial perfusion positron-emission tomography tracer 18F-BMS-747158-02: Comparison to 13N-ammonia and validation with microspheres in a pig model. *Circulation*, 119, (17) 2333–2342. Epub 2009 Apr 20. PubMed PMID: 19380625.

Oertel, W.H., Wolters, E., Sampaio, C., Gimenez-Roldan, S., Bergamasco, B., Dujardin, M., Grosset, D.G. et al. 2006. Pergolide versus levodopa monotherapy in early Parkinson's disease patients: The PELMOPET study. *Mov. Disord.*, 21, (3) 343–353. Available from: PM:16211594.

Patlak, C.S., Blasberg, R.G., and Fenstermacher, J.D. 1983. Graphical evaluation of blood-to-brain transfer constants from multiple-time uptake data. *J. Cereb. Blood Flow Metab.*, 3, (1) 1–7. Available from: PM:6822610.

Perk, L.R., Vosjan, M.J., Visser, G.W., Budde, M., Jurek, P., Kiefer, G.E., and van Dongen, G.A. 2010. p-Isothiocyanatobenzyl-desferrioxamine: A new bifunctional chelate for facile radiolabeling of monoclonal antibodies with zirconium-89 for immuno-PET imaging. *Eur. J. Nucl. Med. Mol. Imaging.*, 37, (2) 250–259. Available from: PM:19763566.

Pichler, B.J., Judenhofer, M.S., and Wehrl, H.F. 2008. PET/MRI hybrid imaging: Devices and initial results. *Eur. Radiol.*, 18, (6) 1077–1086. Available from: PM:18357456.

Rizzello, A., Di, P.D., Lodi, F., Trespidi, S., Cicoria, G., Pancaldi, D., Nanni, C. et al. 2009. Synthesis and quality control of 68Ga citrate for routine clinical PET. *Nucl. Med. Commun.*, 30, (7) 542–545. Available from: PM:19424101.

Sadato, N., Tsuchida, T., Nakaumra, S., Waki, A., Uematsu, H., Takahashi, N., Hayashi, N., Yonekura, Y., and Ishii, Y. 1998. Non-invasive estimation of the net influx constant using the standardized uptake value for quantification of FDG uptake of tumours. *Eur. J. Nucl. Med.*, 25, (6) 559–564. Available from: PM:9618569.

Schäfers, M., Lerch, H., Wichter, T., Rhodes, C.G., Lammertsma, A.A., Borggrefe, M., Hermansen, F., Schober, O., Breithardt, G., Camici, P.G. 1998. Cardiac sympathetic innervation in patients with idiopathic right ventricular outflow tract tachycardia. *J. Am. Coll. Cardiol.*, 32, (1) 181–186. PubMed PMID: 9669268.

Schelbert, H.R., Henze, E., Sochor, H., Grossman, R.G., Huang, S.C., Barrio, J.R., Schwaiger, M., Phelps, M.E. 1986. Effects of substrate availability on myocardial C-11 palmitate kinetics by positron emission tomography in normal subjects and patients with ventricular dysfunction. *Am. Heart J.*, 111, (6) 1055–1064. PubMed PMID: 3487240.

Schuster, D.P. 2007. The opportunities and challenges of developing imaging biomarkers to study lung function and disease. *Am. J. Respir. Crit Care Med.*, 176, (3) 224–230. Available from: PM:17478617.

Shankar, L.K., Hoffman, J.M., Bacharach, S., Graham, M.M., Karp, J., Lammertsma, A.A., Larson, S., Mankoff, D.A., Siegel, B.A., den Abbeele, A., Yap, J., and Sullivan, D. 2006. Consensus recommendations for the use of F-18-FDG PET as an indicator of therapeutic response in patients in national cancer institute trials. *J. Nucl. Med.*, 47, (6) 1059–1066. Available from: ISI:000238104800028.

Shepp, L.A. and Vardi, Y. 1982. Maximum likelihood reconstruction for emission tomography. *IEEE Trans. Med. Imaging.*, 1, (2) 113–122. Available from: PM:18238264.

Sureshbabu, W., and Mawlawi, O. 2005. PET/CT imaging artifacts. *J. Nucl. Med. Technol.*, 33, (3) 156–161. Available from: PM:16145223.

Townsend, D.W. 2004. Physical principles and technology of clinical PET imaging. *Ann. Acad. Med. Singapore.*, 33, (2) 133–145. Available from: PM:15098626.

van der Bruggen, W., Bleeker-Rovers, C.P., Boerman, O.C., Gotthardt, M., and Oyen, W.J. 2010. PET and SPECT in osteomyelitis and prosthetic bone and joint infections: A systematic review. *Semin. Nucl. Med.*, 40, (1) 3–15. Available from: PM:19958846.

Vosjan, M.J., Perk, L.R., Roovers, R.C., Visser, G.W., Stigter-van, W.M., van Bergen En Henegouwen, P.M., and van Dongen, G.A. 2011. Facile labelling of an anti-epidermal growth factor receptor Nanobody with 68Ga via a novel bifunctional desferal chelate for immuno-PET. *Eur. J. Nucl. Med. Mol. Imaging.*, 38, (4) 753–763. Available from: PM:21210114.

Vosjan, M.J., Perk, L.R., Visser, G.W., Budde, M., Jurek, P., Kiefer, G.E., and van Dongen, G.A. 2010. Conjugation and radiolabeling of monoclonal antibodies with zirconium-89 for PET imaging using the bifunctional chelate p-isothiocyanatobenzyl-desferrioxamine. *Nat. Protoc.*, 5, (4) 739–743. Available from: PM:20360768.

Wahl, R.L., Jacene, H., Kasamon, Y., and Lodge, M.A. 2009. From RECIST to PERCIST: Evolving Considerations for PET response criteria in solid tumors. *J. Nucl. Med.*, 50 Suppl 1, 122S-150S. Available from: PM:19403881.

Watson, C.C. 2000. New, faster, image-based scatter correction for 3D PET. *IEEE Trans. Nucl. Sci.*, 47, (4) 1587–1594. Available from: ISI:000089576400009.

Weber, W.A. 2005. PET for response assessment in oncology: Radiotherapy and chemotherapy. *Br. J. Radiol.*, 78, 42–49. Available from: ISI:000234894100006.

Wilding, I.R., Coupe, A.J., and Davis, S.S. 2001. The role of gamma-scintigraphy in oral drug delivery. *Adv. Drug Deliv. Rev.*, 46, (1–3) 103–124. Available from: PM:11259836.

Woodcock, J. 2009. Chutes and ladders on the critical path: Comparative effectiveness, product value, and the use of biomarkers in drug development. *Clin. Pharmacol. Ther.*, 86, (1) 12–14. Available from: PM:19536116.

14

Medical Image Search

Thomas Deserno
Aachen University

14.1 Introduction

The search for images is a well-known problem to all of us using digital cameras: while image data is usually ordered on your hard disk by trips, events, or dates (alphanumerical metainformation), you rather remember a certain landscape, scene, or person shown in an image (pattern and content). In this chapter, we will explore the state of the art in medical image search by image content. While image mining denotes the analysis of (often large) observational data sets in order to find unsuspected relationships and to summarize the data in ways that are better understandable to human observers (Hand et al. 2001), content-based image retrieval (CBIR) or content-based visual image retrieval (CBVIR) aims at searching image databases for specific images that are similar to a given query image (Smeulders et al. 2000, Müller and Deserno 2011).

14.1.1 Relevance of Images

Images are the main carriers of information as processed in the human brain. It has been estimated that more than 90% of experience and knowledge of humans are learned via visual perception. This may also explain that best-working mnemonics are those where the numbers or words to remember are mapped to images, and their order is learned by composing a story as a virtual moving picture. For many of us,

FIGURE 14.1 Image pattern. (From Gordon IE. *Theories of Visual Perception*. 1989. Copyright Wiley-VCH Verlag GmbH & Co. KGaA.)

a face is recognized much easier than a name, although the visual pattern is by far more complex than a couple of letters.

In the beginning of the twentieth century, scientists discovered the steps of image pattern reception and perception in the human brain. According to the Gestalt theory (Wertheimer 1925), edges and color/texture information are extracted locally, and image components are formed, grouped, and recognized with access to the mind and its memory. Two of the gestalt paradigms are worth mentioning:

- *Emergence* is the process of complex pattern formation from simpler rules. However, objects are recognized as a whole, and not by parts. For example, Figure 14.1 depicts a Dalmatian dog sniffing the ground in the shade of overhanging trees. The dog is perceived as a whole, all at once, and is not recognized by first identifying its parts (feet, ears, nose, tail, etc.), and then inferring the dog from those component parts.
- *Reification* is the constructive or generative aspect of perception, by which the experienced percept contains more explicit spatial information than the sensory stimulus on which it is based. Again referring to the dog in Figure 14.1, the dog's shape is clearly seen as closed contour, although several parts of this imaginary line are not present in the image at all. Obviously, illusory contours are composed and treated by the human visual system as real contours.

Since then, computer vision aims at finding algorithms for segmentation and pattern recognition that are capable of performing like humans. When referring to Figure 14.1, a human may immediately recognize the "dog" or "Dalmatian" on a conceptual level, while today's computer programs are far away from responding to such words when fed by the binary pattern of Figure 14.1.

14.1.2 Relevance of Medical Images

Starting with the discovery of x-rays by Wilhelm Conrad Röntgen in 1896, images have been established in the medical domain as the main resource of diagnostics, therapy planning, and intervention. In the first part of this book, the most relevant imaging modalities have been described, such as radiography (plain x-ray imaging), computed tomography (CT), magnetic resonance imaging (MRI), positron emission tomography (PET), single photon emission computed tomography (SPECT), and ultrasound

(US). Furthermore, optical imaging as used in endoscopy and microscopy is frequently applied in medicine.

Nowadays, almost all of these imaging modalities are operated direct digitally. Picture archiving and communication systems (PACS) are installed in a hospital to manage (store, retrieve, and communicate) the image information. Easily, several terabytes of pixel data are acquired in a hospital each year (Müller and Deserno 2011). And again, however, this huge amount of data is managed solely by alphanumerical metainformation such as the patient's name, the date of exposure, or additional study information. The same holds for the radiologists reading images. He reports a diagnosis in natural language, which is coded additionally, for instance, using the international classification of diseases (ICD) nomenclature as nonvisual, that is, textual metainformation.

14.1.3 Idea of Content-Based Image Retrieval

Contrarily, CBIR aims at managing images by pattern and retrieving images from an (PACS) archive by means of visual similarity rather than comparing the alphanumerical metainformation. The advantages of this paradigm arise directly from the limitations of describing pictures by words. The saying "A picture is worth a thousand words" depicts the inherent incompleteness of any natural language description of an image—a problem well known in the medical field too.

Hence, any system for content-based image data mining usually is composed of the following four steps:

- *Feature extraction*: based on the image pattern, numerical features, which are also referred to as signature, are extracted. The features may describe color, texture, and shape of the entire image or image parts. They are stored separately to the image in a feature database.
- *Similarity computation*: According to the signature, appropriate similarity measures are defined to quantify the visual distance of two images or image patterns.
- *Signature indexing*: Assuming a large number of images hosted in the archive, processing a request requires many feature comparisons. Hence, the feature vectors are indexed to speed up the response time of the CBIR system.
- *Image retrieval*: On time of retrieval, a pattern is presented to the system, and its signature is extracted and compared to those indexed in the CBIR archive. According to the most similar signatures, the corresponding set of response images is presented to the user.

In this chapter, we focus on medical image search by means of CBIR. We will briefly discover the history of medical and nonmedical CBIR techniques and discuss its potential in medical use in general regarding diagnostics, research, and medical teaching. In Section 14.4, we explore the engineering fields in medical CBIR, and a state of the art of all important techniques is given in Section 14.5. As an example, three outstanding systems and ideas of medical CBIR applications are presented in Section 14.6, before we conclude this chapter with a prospective view on future developments.

14.2 History of CBIR

The emerging development of the World Wide Web holding millions of images has majorly influenced the development of visual image retrieval technologies. The first CBIR systems had already been discussed in the early 1980s (Chang and Kunii 1981). Hence, this section briefly reports on the nonmedical CBIR systems and discusses the fundamental differences when retrieving medical imaging data by content. A more comprehensive review on nonmedical CBIR approaches is given by Smeulders et al. (2000).

14.2.1 Nonmedical CBIR

In the early 1990s, the query by image content (QBIC) system of IBM was one of the first approaches to CBIR (Niblack et al. 1993). The QBIC signature is composed of color, texture, and edge orientation. In particular, the three-dimensional (3D) average color vector of an object or the whole image is computed. The texture features used in QBIC are modified versions of the coarseness, contrast, and directionality, as they have been proposed by Tamura et al. (1978). The shape features consist of shape area, circularity, eccentricity, major axis orientation, and a set of algebraic moment invariants. They are based on a reduced binary map of edge points, which is obtained applying a Canny edge detector to the gray-scale image (Canny 1986).

The query by image example (QBE) paradigm has also been proposed by the authors. QBIC allows queries based on (i) example images, (ii) user-constructed sketches, and (iii) color and texture patterns selected from a sampler. In the latter cases, the percentage of a desired color in the QBE is adjusted manually by moving sliders.

Since color has been identified as most relevant structure for CBIR, the *semantic gap* has been recognized within the first approaches to CBIR systems. It describes the differences between image similarity on the high level of human perception, which is based on emergence and reification (cf. Section 14.1.1), and on the low level of a few numerical numbers describing a mean color. Accordingly, first approaches to bridge the semantic gap are based on local image signatures.

The Blobworld system was one of the first CBIR systems that applied a local concept of image representation (Belongie et al. 1998, Carson et al. 2002). Again, color, texture, and edge features were extracted from the images. Then, the expectation-maximization (EM) algorithm was applied for clustering the feature space. The image pixel coordinates were added to the feature space representing the local connectedness. Prominent regions were described by best-fitting Gaussians and visualized with ellipses, which have been called "blobs" (Figure 14.2).

Proving its value for image search, nonmedical CBIR systems have already been commercialized. For instance, MICHELscope identifies stamps using CBIR technology (Siggelkow 2002). The first edition of the software product was released in 2002. At that time, it was based on a reference dataset of 12,000 scanned stamps from Germany. Nowadays, commercial CBIR products are also available for paintings, industrial components, textile patterns, and catalogs in general.

14.2.2 Differences to Medical Images

Images from the Internet or large sets of photographs or scans (e.g., stamps) are rather unstructured image collections, usually containing color as a most discriminate feature and usually differing significantly in

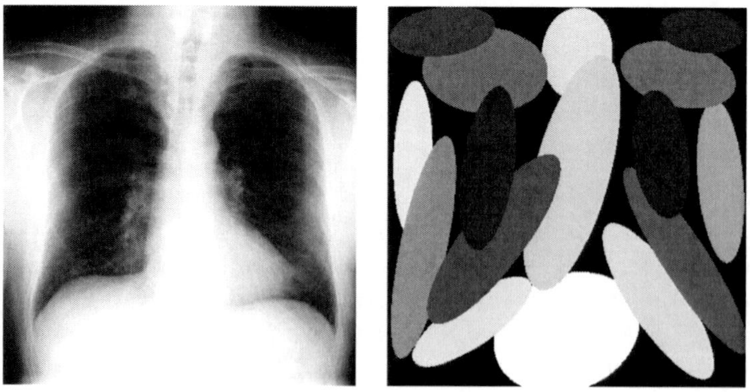

FIGURE 14.2 Blob representation of a chest radiograph. (From Greenspan H and Pinhas AT. *IEEE Trans Inf Technol Biomed.* 11(2): 190–202 © 2007 IEEE.)

their visual content. Hence, simple signatures may be suitable to describe them. However, the characteristic of medical images differs substantially:

- *Noncolor*: In medical imaging modalities, color is less relevant. For instance, x-ray imaging (radiography, fluoroscopy), CT, MRI, ultrasound, and nuclear medicine all form gray-scale images. Hence, color cannot be used to describe the images appropriately.
- *Structure*: The pathology that is visualized in medical images may result from only subtle or small alterations in the image pattern—if a history of images is available for comparison at all. Usually, intraindividual differences in size, shape, and appearance of biological tissue are more prominent than the diagnostically relevant alterations.
- *Multidimensionality*: Medical images are often 3D, for example, representing a volume composed of two-dimensional (2D) slices or a temporal sequence of 2D images (moving picture), or even four-dimensional (4D), such as MRI sequences of the beating heart. Hence, more complex features must be used in the medical domain.
- *Mutlimodality*: Furthermore, medical diagnostics frequently is based on more than only one set of image data, and imaging data is usually obtained from different modalities. Comparing ultrasound with radiography or MRI exemplifies that the different nature of medical modalities does not support representing all such images using the same numerical features.
- *Metainformation*: Medical images are only a part of the information that is relevant in a medical case. Gender, age, and other data are required to compare diagnostics in medical cases. Consequently, combining CBIR with natural language processing (NLP) is superior to medical document retrieval.

Anyway, in this chapter, we will focus on 2D gray-scale images. Starting in 2000, medical CBIR systems have been described in the literature. A brief history is given in the next section.

14.2.3 Medical CBIR Systems

A comprehensive review of CBIR systems in medical applications is given by Müller et al. (2004a). From the previous section, it is obvious that medical CBIR systems have been focused on certain imaging modalities, body regions, and diagnostic tasks (Table 14.1). The ASSERT system, for instance, is a physician-in-the-loop content-based retrieval system for high-resolution CT (HRCT) data. The gap of lacking appropriate signatures and similarity measures for such complex semantics is bridged with user interaction: the physician manually marks pathology-bearing regions of interest (ROIs) in selected 2D

TABLE 14.1 Various Images Types and the Medical CBIR Systems That are Using These Images

Images Used	Names of the Systems
HRCTs of the Lung	ASSERT
Functional PET	FICBDS
Spine x-rays	CBIR2, MIRS
Pathologic images	IDEM, I-Browse, PathFinder, PathMaster
CTs of the head	MIMS
Mammographies	APKS
Images from biology	BioImage, BIRN
Dermatology	MELDOQ, MEDS
Breast cancer biopsies	BASS
Varied images	I²C, IRMA, KMed, COBRA, MedGIFT, ImageEngine

Source: Adapted from Müller H et al. *Int J Med Inform.* 2004a; 73(1): 1–23.

slices (Shyu et al. 1999). Some approaches, however, try to address the medical imagery in general. Two of these approaches are depicted in the following sections.

14.2.3.1 MedGIFT

The MedGIFT project aims at providing an open-source framework of reusable components for a variety of medical applications. It is based strongly on the GNU image finding tool (GIFT)* as its central piece. Main developments are on the integration of various new components to create a domain-specific search and navigation tool (Müller et al. 2003, 2005).

GIFT supports retrieval of images by their visual content. Scripts permit indexing entire directory trees of images and build inverted file indices. Supported visual features are aimed at color photography, such as color histograms applied on the entire image and on block partitions at various scales. Gabor texture features are also computed and stored. Query and retrieval are done using a specialized extensible markup language (XML)-type language called multimedia retrieval markup language (MRML), which facilitates communication between the query interface and the retrieval engine.

MedGIFT supports some basic keyword-based retrieval by including search of patient records along with images. It should be noted that MedGIFT is not a fixed system but rather a framework upon which to base various image retrieval systems/experiments. For feature selection, MedGIFT applies general concepts from information retrieval and text mining:

- *Term frequency (TF)*: a feature frequent in an image describes this image well.
- *Inverse document frequency (IDF)*: a feature frequent in the collection is a weak indicator to distinguish images.

A weighted TF-IDF function is used to determine the feature weight, which is then applied to each feature computing an overall similarity score. The images are segmented to the extent of background removal and some object localization through thresholding or other high-level methods. Simple relevance feedback is also implemented.

14.2.3.2 IRMA

The Image Retrieval in Medical Applications (IRMA) project aims at developing and implementing high-level methods for CBIR, including prototypical application to medico-diagnostic tasks on a radiological image archive (Lehmann et al. 2000, 2004). The IRMA team wants to perform semantic and formalized queries on the medical image database, which includes intra- and interindividual variance and diseases. Three levels of image content similarity are modeled:

- *Global* features are linked to the entire images and used to automatically classify an image according to the (a) anatomic region and (b) biosystem that is captured, the imaging modality (c, creation) that is used, and the relative (d) direction of imaging device and patient, including the pose (ABCD code) (Lehmann et al. 2003).
- *Local* features are linked to prominent image regions and used for object recognition.
- *Structural* features are linked to spatial and/or temporal relations between the objects. They are used for high-level image interpretation.

A pipeline of image processing is suggested. Iterative region merging builds up a hierarchical attributed region adjacency graph (HARAG), the data structure that is used to represent images. Hence, image retrieval is converted to graph matching. Object comparison operates on the HARAG nodes, while scenes are modeled by rather structural graph to subgraph comparison.

The system is set up on a Unix platform, with a central server, relational database, and shared computing in the cluster of connected workstations for distributed computing. All interfaces are based

* Available at http://www.gnu.org/software/gift.

on HTML and recorded in the database too. The coupling of CBIR and PACS is based on the Digital Imaging and Communications in Medicine (DICOM) protocol.

14.3 Medical Use of CBIR

Using such generic frameworks, medical CBIR applications have been explored and developed recently. They cover a wide range of applications, from medical education to clinical routine and research. In this section, we will describe in general the possible use of CBIR in medicine.

Figure 14.3 emphasizes the radiological process (Greenes 1989). The outer loop refers to diagnostics, while quality assessment and performance monitoring are rather research-oriented. The third complex in using medical imagery addresses medical education and training.

14.3.1 Diagnostics

As mentioned before, medical images are acquired primarily for diagnostics, and hence CBIR systems are designed to assist diagnosis. In any case, the major goal is to retrieve similar images from the archive, since these images are annotated with metainformation that is stored in the medical record and may be helpful for the current case.

- *Case-based reasoning (CBR)*, broadly construed, is the process of solving new problems based on the solutions of similar past problems (Macura and Macura 1997). According to the radiological process, The CBR cycle (retrieve, reuse, revise, retain, Figure 14.4) requires access to previous cases (Aamodt and Plaza 1994). Usually, such cases are retrieved from textbooks and medical atlases (general knowledge). In particular, the radiologist may easily recognize an irregular structure in

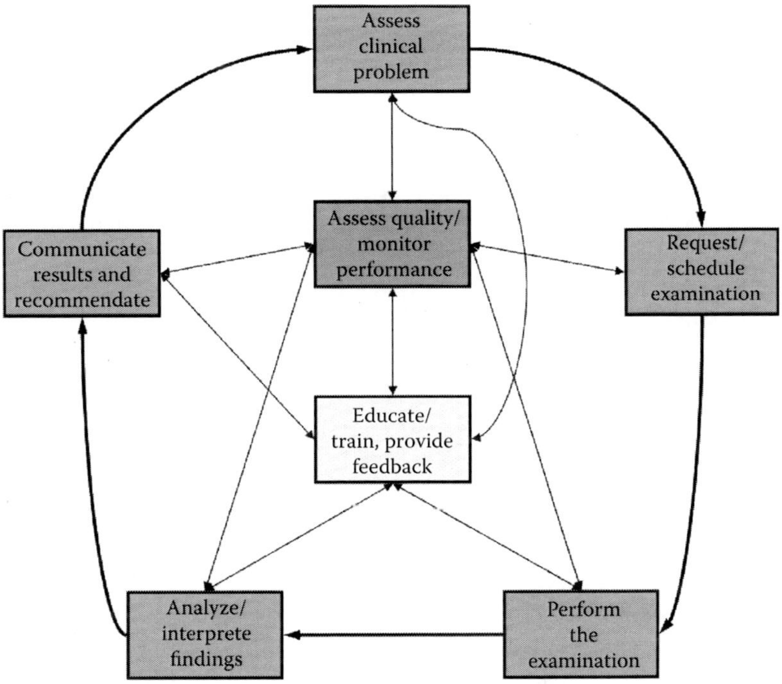

FIGURE 14.3 The radiological process of diagnostics, research, and education. (From Lehmann TM. Digitale Bildverarbeitung für Routineanwendungen. Evaluierung und Integration am Beispiel der Medizin. Deutscher Universitäts-Verlag, Wiesbaden 2005. ISBN 3-8244-2191-7 (in German).)

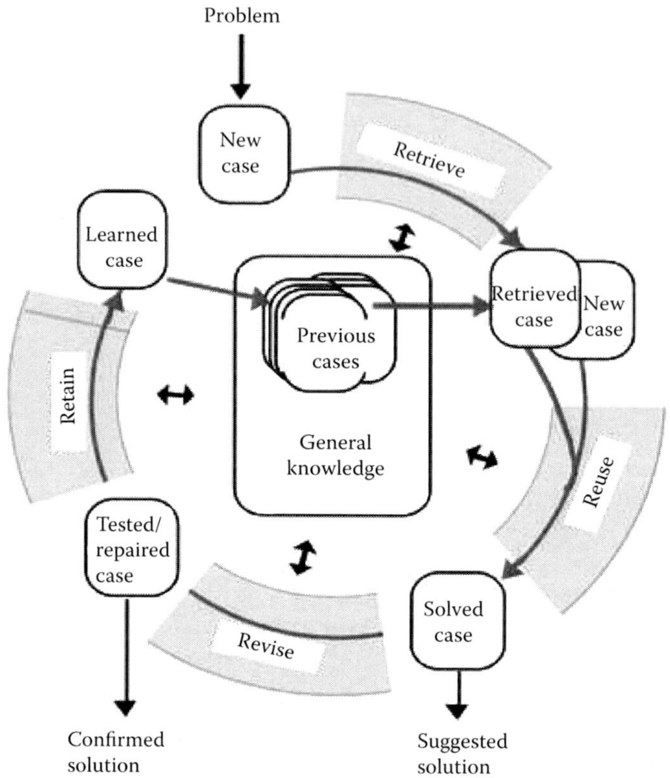

FIGURE 14.4 The CBR cycle transforms a problem into a confirmed solution. (From Aamodt A and Plaza E. *AICom—Artif Intell Commun.* 1994; 7(1): 39–59.)

an image, but deciding on its reason is more challenging. Using his knowledge, he may form some hypotheses, which are then evaluated step by step. For each hypothesis, corresponding images are sought in the printed resources by textual labels rather than image pattern. If the actual pathology is not among the initial hypotheses, relevant images will not be compared to the current case and the CBR process may fail. Within a CBIR-based CBR scenario, the physician may mark the ROI in the image, and the selected pattern is used to retrieve similar images by contained pattern. In a second step, the radiologist is put into the loop again, when selecting relevant images from the CBIR response. Accessing the medical records that are linked to these cases may expose the correct diagnosis even if this particular hypothesis has not been within the initial guess of the radiologist.

- *Computer-aided diagnosis*: The CBR process may be further automated by taking the physician out of the loop. This, however, requires a more reliable ground truth associated with the reference images in the archive. In a certain context, for example, for tumor staging, each class of diagnostics (e.g., benign vs. malignant) is represented by several images. Again, the current examination forms the QBE, and a set of similar images is retrieved automatically from the archive. Using appropriate features and similarity measures, which of course must be carefully selected and parameterized, which is still a field of ongoing research, the set of returned images is majorly composed of a particular diagnostic class, which is then transferred automatically to the current case. More sophisticated methods may construct a measurement automatically using similarity scores and ordinal measurements that are linked to the retrieved images in their accompanying metadata (ground truth).

14.3.2 Research

- *Evidence-based medicine (EBM)* aims at applying the best available evidence gained from the scientific method to clinical decision making (Guyatt et al. 1992, Sackett et al. 1996). It seeks to assess the strength of evidence of the risks and benefits of treatments (including lack of treatment) and diagnostic tests. The quality of evidence can range from meta-analyses and systematic reviews of double-blind, placebo-controlled clinical trials at the top end, down to conventional wisdom at the bottom. In contrast to CBR, EBM usually refers to more than only one similar case, and content-based methods may be applied to retrieve such cases. However, EBM rather implies image mining (identifying classes of images) than image retrieval (identifying images from a given class or type).
- *Drug development*: Another field of research-driven CBIR applications is the development process of drugs. Here, clinical trials are performed in order to assess safety and efficacy of a new drug:
    - *Phase I* trials are the first stage of testing in human subjects. A small group of up to 100 healthy volunteers is selected to assess the tolerability and safety (pharmacovigilance), pharmacokinetics, and pharmacodynamics of a drug.
    - *Phase II* trials are performed on larger groups of up to 300 subjects and designed to assess how well the drug is working, for example, determining dosing requirements (*Phase IIa*) and efficacy (*Phase IIb*).
    - *Phase III* trials are randomized controlled multicenter trials on large patient groups (3000 or more) aiming at assessing effectiveness of the drug in comparison to the current gold standard treatment.
    - *Phase IV* trials (postmarketing surveillance trial) involve safety surveillance (pharmacovigilance) and technical support after the drug has received permission to be sold. Involving up to a million of individuals, they may also aim at finding a new market or indication for the drug.

In each phase, a so-called endpoint is defined reflecting how a subject feels, functions, or survives. Based on a hypothetical new drug for treating congestive heart failure (CHF), Figure 14.5 emphasizes the four different phases of clinical trials and it is corresponding clinical endpoints.

Sometimes, however, the endpoint in a clinical trial cannot be assessed that easily by a simple parameter as, for instance, the heart's ejection rate. Then, a surrogate may be used. A surrogate endpoint is defined as a biomarker (objectively measured and evaluated indicator of normal biological processes,

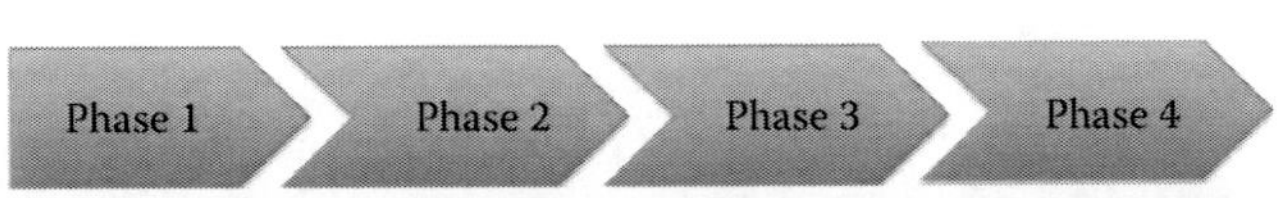

Stage of Development	Phase 1	Phase 2	Phase 3	Phase 4
End point	Safety	Efficacy	Efficacy	Efficacy
Specific end point	Safety profile	Cardiac output	Reduction in mortality rate	Reduction in mortality rate
Types of studies	Different indications; Single or multiple dose	Placebo controlled; Dose Escalation	Placebo controlled; Long-term follow up	Comparative; New indications

FIGURE 14.5 Specific endpoints in clinical trials for drug development. (From Silverman M. Clinical trial end points at different phases of clinical development. Technical Report, BioStrategics Consulting Ltd, 2011.)

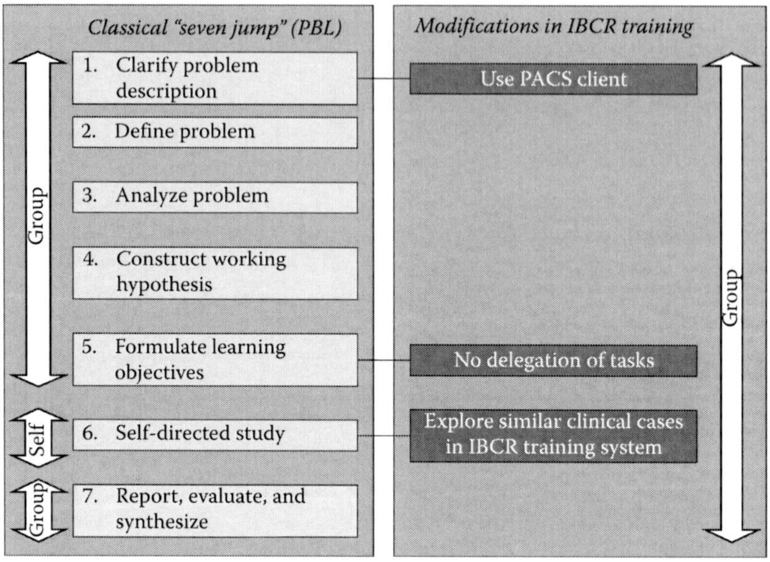

FIGURE 14.6 The Seven Jumps for PBL (left) and IBCR support (right). (From Welter P et al. *BMC*. 2011a; 11(68).)

pathogenic processes, or pharmacologic responses to a therapeutic intervention) that is intended to substitute for a clinical endpoint. Image-based surrogates, for instance, are computed from almost all of the imaging modalities that have been addressed in this book, including ultrasound (de Groot et al. 2004), CT (Jaffe 2008), MRI (Warach 2001), or molecular imaging methods such as PET (de Geus-Oei et al. 2007). It is obvious that image-based biomarkers require comprehensive evaluation (Analoui 2011), which shall rely on a large set of images. Hence, CBIR may become the key technology in further developments of surrogate endpoints.

14.3.3 Education

Medical education relies on practical experience within health care institutions such as hospitals. The majority of problems physicians have to deal with are directly concerning the patient. Therefore, the presentation of clinical cases in medical education is essential. Problem-based learning (PBL) is known to provide positive effects on medical education, especially pertaining to social and cognitive aspects (Koh et al. 2008). A clinical case is a stimulus for learning, while fostering the acquisition of knowledge related to the problem together with an improvement of problem-solving skills.

More formalized, a seven-step procedure usually is applied in PBL, which is referred to as Seven Jump (Gijselaers 1995, Wood and Diana 2003). Access to the case collection supports all of the steps but is essentially required at least in Steps 1 and 6. It is eased by integrating CBIR technology with the PACS (Welter et al. 2011a). If image-based case retrieval (IBCR) is applied to PBL, the PACS client is used to select the learning material, to clarify the problem description, and to explore similar cases in a self-directed study (Figure 14.6).

14.4 Engineering Fields of Medical CBIR Systems

Within the broad scope of CBIR in medical diagnosis, education, and research, there are plenty of problems that still have to be addressed by biomedical engineers. For instance, in Section 14.2.1, we have already discussed the semantic gap between simple numerical features that are applied to represent the meaning of an image for the human observer, and engineers are trying to bridge this gap defining advanced features, for instance, a HARAG (Fischer et al. 2004).

However, there are more essential gaps still hindering the clinical application of content-based image search (Deserno et al. 2009a). Figure 14.7 visualizes a classification scheme for such gaps and additional characteristics of medical CBIR systems. According to this scheme, the most relevant fields of medical CBIR engineering are addressed in the following sections.

14.4.1 Content and Features

The content gaps address the level of image understanding (semantic gap) as well as the imaging and/or clinical context in which a CBIR system may be used (use context gap). Obviously, designing a medical CBIR system for a broad use is more challenging, since the level of image details relevant for the retrieval, the type of image data (modality) to handle, and other system preferences are highly variable.

The feature gaps address the automation of feature extraction (extraction gap), the granularity of structure of image objects recognized by the system (structure gap), the granularity of visual details in the image processed by the system (scale gap), the dimensionality of spatial and time inputs actually used to compute the signature (space and time dimension gap), and the dimensionality of channel inputs actually used to compute the signature (channel dimension gap).

Biomedical engineers trying to establish clinical CBIR systems must face these problems. Signatures must be developed that are calculated automatically rather than manually from the entire dimensions and range of image data: 3D and 4D image data cannot be represented appropriately by features that have been extracted from 2D slices, and multichannel data (e.g., true color) cannot be completely represented using only singular channel (e.g., gray scale). Also, local and structural signatures must become available for clinical applications of CBIR.

14.4.2 Performance

Another field requiring customized engineering approaches is the performance of CBIR systems. In the medical field, images, and especially digital images, are produced in an ever-increasing quantity. For instance, the Radiology Department of the University Hospitals of Geneva (HUG), Geneva, Switzerland has produced more than 114,000 images a day in 2009, risen form 12,000 in 2002. Large hospital groups such as Kaiser Permanente that manage several hospitals have reported about 700 TB of data stored in the institutional archives by early 2009, and very large hospitals such as the University Hospital of Vienna, Austria currently produce more than 100 GB of image data per day (Müller and Deserno 2011).

For CBIR, numerous comparisons are required, and hence, the response time of a retrieval system becomes a serious parameter. The level of support for fast database searching must be increased to bridge the indexing gap. Hardware and software solutions are engineered, including parallel computing and extensive use of field programmable gate arrays (FPGAs) and graphics processing units (GPUs) (Scholl et al. 2011).

Currently, performance engineering also addresses the efficiency of medical CBIR systems. When applied in clinical routine, the performance of CBIR turns out to be rather low. Frequently, the images returned may be inappropriate with respect to the query semantics or even wrong. This fact results from insufficient evaluation of the algorithms, and the evaluation gap addresses the level to which the system validity of retrieval has been evaluated. Unavailability of large image repositories reliably annotated with ground truth information is one of the major problems to face.

14.4.3 Usability

However, deploying a valid and well-performing CBIR system does not guarantee its successful establishment in clinical practice. Graphical user interfaces (GUIs) and input/output (I/O) devices must

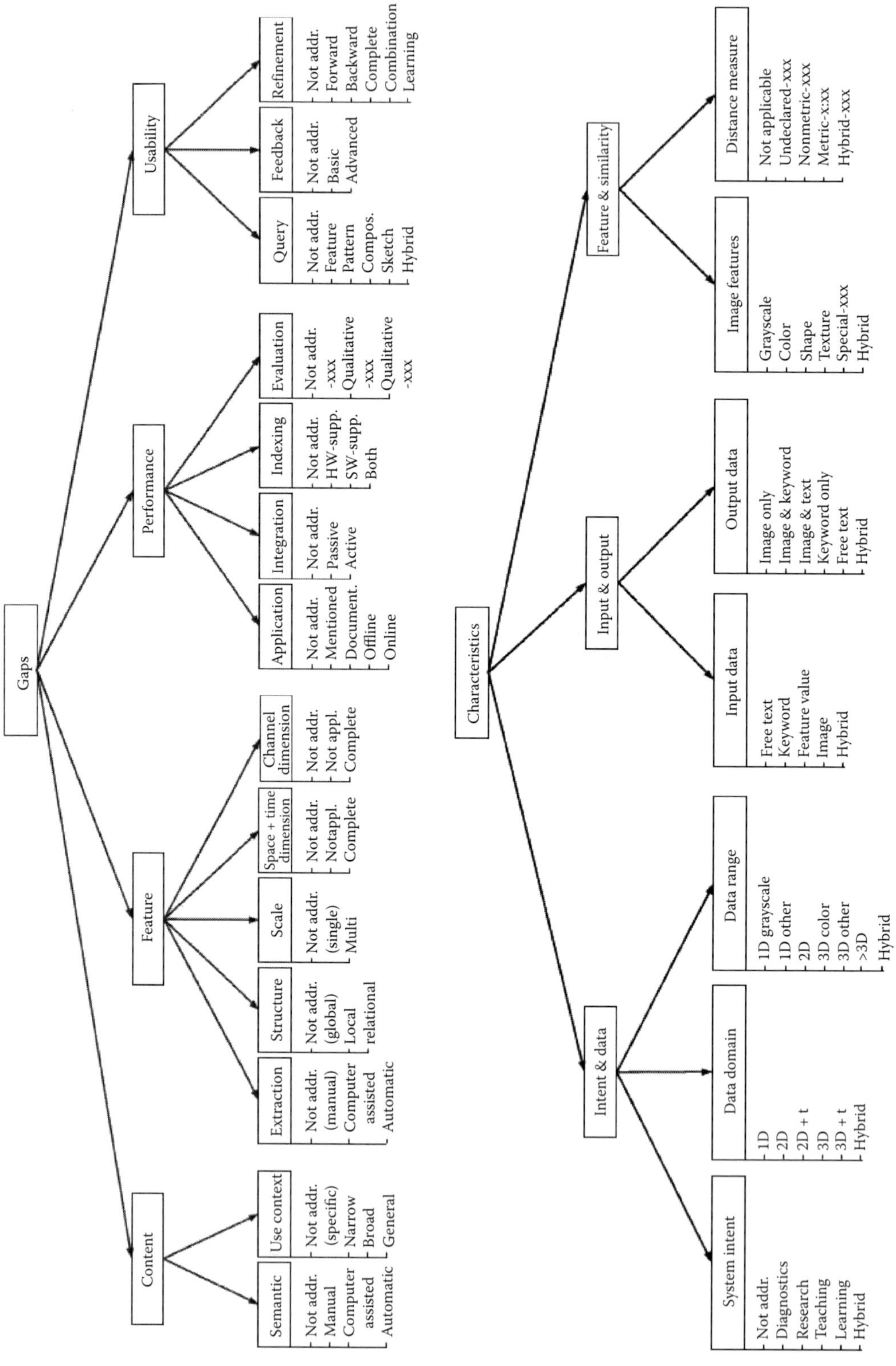

FIGURE 14.7 Gaps and characteristics of medical CBIR systems. (From Deserno TM, Antani S, and Long R. *J Digit Imaging*. 2009; 22(2): 202–215.)

be carefully designed, and the workflow within the application must be intuitive and easy to follow. Relevant fields of engineering address the levels to which a user may use and combine visual as well as textual (alphanumerical) queries (query gap), to which the system helps the user to understand the query results (feedback gap), and to which the system helps the user to refine and improve the query results (refinement gap).

Engineering approaches to bridge the visual query gap include 3D interaction such as marking a partial volume of interest (VOI). While in 2D, a surface may be sketched quickly upon an image, annotating a 3D surface into a voxel-based volume is challenging. Obviously, the 2D mouse is inappropriate. 3D input devices such as the gloves or optical scanners may be useful. Furthermore, the image data must be positioned appropriately in 3D space, which may require two rather than only one sophisticated input device. Even more difficult is user interaction for retrieval of volume sequences over time (4D data).

14.4.4 Integration

In Figure 14.7, the integration gap is defined within the group of performance gaps (see Section 14.4.2). However, integration of medical image processing technology into the clinical workflow is of fundamental prominence to justify paying a special attention.

More than 10 years ago, Eakins and Graham have already stated that the experience of all commercial vendors of CBIR software is that system acceptability is heavily influenced by the extent to which image retrieval capabilities can be embedded within user's overall work tasks (Eakins and Graham 1999). Such work tasks in medicine have already been visualized in Figure 14.3. They require seamless integration on different levels (Leisch et al. 1997, Winter et al. 2010):

- *Data integration* is achieved if any data stored in the modules must not be entered more than once.
- *Functional integration* is obtained if any services provided by a certain module can be used from any other module requiring this particular service within the entire system.
- *Presentation integration* is obtained if different modules present their data in a unified way and style.
- *Context integration* is present if specific setting such as the selection of a certain patient or image, which is done in one module, is passed automatically to another module when this is called.

This short list emphasizes the challenges of integration of CBIR technology into medical diagnosis, research, and education. Accordingly, research and development in medical image search must include communication interfaces such as DICOM and Health Level Seven (HL7).

14.5 State of the Art in Medical CBIR Technology

In this section, we will briefly describe the state of the art in medical content-based image search regarding features and similarity measures, user interfaces, workflow integration, combined visual and textual retrieval, and evaluation.

14.5.1 Features Extraction and Selection

As mentioned before, features or signatures may be computed on different levels. The following description of global and local features is according to Müller and Deserno (2011).

14.5.1.1 Global Features

Visual features have been classified into (i) primitive features such as color or shape, (ii) logical features such as identity of objects shown, and (iii) abstract features such as significance of depicted scenes

(Eakins and Graham 1999). However, basically, all currently available systems exclusively rely on primitive features, such as

- *Color*: In stock photography (large, varied databases used by artists, advertisers, and journalists), color has been the most effective feature. The red, green, blue (RGB) color space is only rarely used, as it does not correspond sufficiently with human color perception. Other color models such as hue, saturation, value (HSV), or the International Commission on Illumination (Commission internationale de l'éclairage, CIE) Lab and Luv spaces perform superiorly because a metric distance in such a color model is more similar to the difference between the colors that a human observer perceives. Much effort has also been spent on creating color spaces that are optimal with respect to lighting conditions or that are invariant to shades and other influences such as viewing position (Geusebroek et al. 2001).
- *Texture*: Texture measures try to capture the characteristics of the image with respect to changes in certain directions and the scale of the changes. This is most useful for images with homogeneous texture. Some of the most common measures are wavelets and Gabor filters. Invariances with respect to rotation, shift, or scale can be included into the feature space but information on the texture may get lost in this process (Milanese and Cherbuliez 1999). Other popular texture descriptors contain features derived from co-occurrence matrices (Haralick et al. 1973, Kuo et al. 2002), the Fourier transform (Milanese and Cherbuliez 1999), and the so-called wold features, which represent image texture combining wavelets that have been certainly tuned to various scales and rotations (deterministic component) with an autoregressive model (nondeterministic component) (Lu and Chung 1998).

14.5.1.2 Local Features

Both color and texture features can also be used on a local or regional level, that is, on parts of the image, disregarding whether such parts have been obtained from image-driven segmentation or geometric partitioning. The easiest way to employing regional features is to produce blocks of fixed size. However, such blocks do not reflect any semantics in the image (Squire et al. 2000). When allowing the user to choose ROIs (Comaniciu et al. 1998), or when segmenting the image into areas with similar properties (Winter and Nastar 1999), local features may capture more information about relevant image structures. The following local features are commonly used:

- *Shape*: Fully automated segmentation of images into objects itself is an unsolved problem. Even in fairly specialized domains, automated segmentation causes many problems. In image retrieval, several systems attempt to perform an automatic segmentation for feature extraction (Lucchese and Mitra 1999). The segmentation process should be based on color and texture properties of the image regions. Then, the segments can be described by shape features, usually being invariant to shift, rotation, and scaling (Loncaric 1998). Medical image segmentation for browsing large image repositories is frequently addressed in the literature (Lapeer et al. 2002).
- *Salient points*: In recent years, salient point-based features have best performed in most of the image retrieval and object classification tasks (Mikolajczyk and Schmid 2005). The idea is to find representative points (or points that attract visual attention) in the images and then analyze the relationships of the points. This permits to extract features that possess several invariants such as invariance to shift, rotation, scale, and even the point of view. A large number of such techniques exist for detecting as well as extracting features from the salient points. A prominent example is the scale-invariant feature transform (SIFT) (Lowe 2004).
- *Patches and visual words*: Patches and visual words are closely linked to salient point-based features, because they are usually extracted from regions in the images that were identified as salient point. Then, features are extracted regionally from these regions. Alternatively, a regular grid is placed in the image and the patches are extracted around the center points of the grid. The term "visual word" stems from text retrieval. Visual features are clustered into a limited number

of homogeneous characteristics that can have distributions similar to the distribution of words in text. This allows us to apply techniques that are well known from text retrieval (Sivic and Zisserman 2006). The image annotation task in the cross-language evaluation forum (CLEF) (Deselaers et al. 2007, 2008) has shown that x-ray categorization and retrieval on the organ and pathology level is performed best using patch-based visual words (Avni et al. 2011).

On the one hand, all of these features provide benefits and have confirmed application domains, but on the other hand, all of them are low-level visual features not corresponding to semantic categories. For this reason, text, whenever available, should be used additionally for medical image search, as semantic information is conveyed inherently with natural language. All benchmarks show that text has a superior performance compared to visual characteristics, but is complemented effectively by visual retrieval.

14.5.1.3 Structural Features

In medical image processing, scene-based approaches are used successfully for segmentation (Deserno 2011). For instance, the commercial software BoneXpert® (http://www.bonexpert.com) for automated determination of skeletal maturity is segmenting the relevant bones from left hand's radiographs using an active shape model (Thodberg et al. 2009).

The use of structural features for image retrieval is even more advanced and has not yet been implemented in any system that is used in routine. The challenging idea is to capture not only the relevant objects in an image but also their constellation in a spatial and/or temporal scene. Fischer et al. (2008) have published an approach where the so-called structural prototypes are trained from reference images. The prototypes are attributed relational graphs (ARG), where nodes and edges capture the object and relation properties, respectively (Figure 14.8). To cope with the variety in shape, appearance, and constellation of living biological tissue, node and edge annotations are represented by unimodal Gaussian distributions of the corresponding feature. On time of retrieval, graph matching is computed to assess

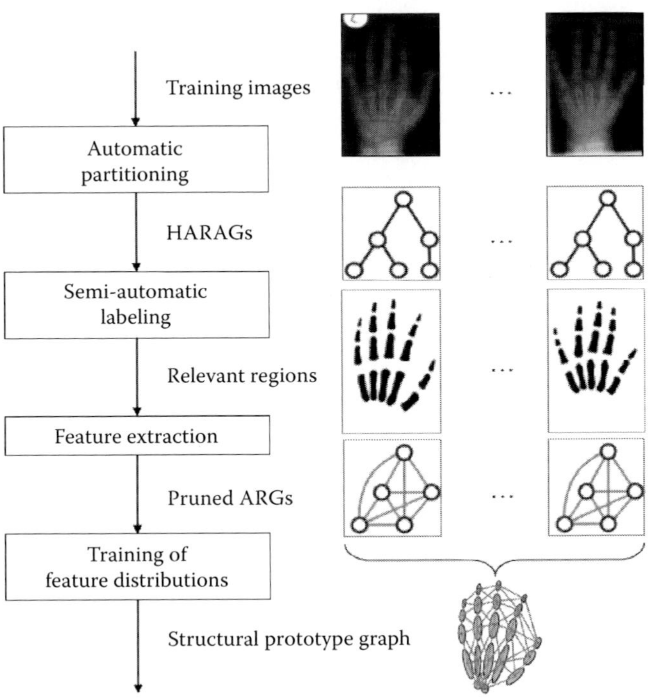

FIGURE 14.8 Structural prototypes for scene-based image retrieval. (From Fischer B et al. *Proc SPIE*. 2008; 6914: 1X1–1X9.)

the similarity of two scenes that can cope with missing objects. The similarity computed based on the signature is discussed in detail in the next section.

14.5.2 Distance and Similarity Measures

To perform a similarity query, we can define in general two main operators for comparison (Traina et al. 2011):

- *Range query*: Given a dataset, a query center, and a radius, each element is selected for which the distance between the query center and the element is below the radius; for example, retrieve all images that differ five units of similarity to the QBE.
- *Nearest-neighbor query*: Given a dataset, a query center, and an integer value $k \geq 1$, the k-nearest neighbor (k-NN) query selects the k elements of the dataset that are at the shortest distances from the query center; for example, retrieve the five most similar images to the QBE.

For both types of queries, one needs to define a distance or similarity measure as well as one or more reference elements (QBE), which also have been referred to as the query center. Basically, all systems for image retrieval assume equivalence of an image and its representation in the feature space, and the Euclidean vector space model is used for measuring the distances between a query image (represented by its features) and possible results representing all images as feature vectors in an n-dimensional (nD) vector space. Although such metrics have been shown to disagree with the visual perception of human (Tversky 1977), they are broadly used and formally defined in the following section.

14.5.2.1 Metrics

A metric space is defined as a pair $\mathbf{M} = \langle \mathbf{S}, d \rangle$, where $\mathbf{S}$ denotes the universe of valid elements and d is the function d: $\mathbf{S} \times \mathbf{S} \rightarrow \mathbf{R}^+$ called a metric that expresses the distance between elements in $\mathbf{S}$. To be a metric, d must satisfy the following properties for every $s_1, s_2, s_3 \in \mathbf{S}$:

1. *Symmetry*: $d(s_1, s_2) = d(s_2, s_1)$
2. *Nonnegativity*: $0 < d(s_1, s_2) < \infty$ if $s_1 \neq s_2$ and $d(s_1, s_1) = 0$
3. *Triangular inequality*: $d(s_1, s_3) \leq d(s_1, s_2) + d(s_2, s_3)$

Using a metric to compare elements, a similarity query returns the stored elements that satisfy a given similarity criterion, usually expressed in terms of query centers (QBE parameter in feature space).

In addition to the Euclidian distance, several other metrics are known for the vector space model such as the city-block distance, the Mahalanobis distance, histogram intersection, or other histogram-based metrics that can be used for histogram-based features such as SIFT (Pele and Werman 2008). However, the use of high-dimensional feature spaces bears the *curse of dimensionality* (increasing the number of features is decreasing the performance of the classifier), and appropriate features and distance measures must be selected carefully (Aggarwal et al. 2001).

14.5.2.2 Other Measures

Another approach is a probabilistic framework to measure the likelihood that an image is relevant (Jain et al. 2000). Some probabilistic approaches use the support vector machine (SVM) for grouping of images into classes of relevance and nonrelevance (de Oliveira et al. 2010). In most visual classification tasks, SVM reaches the overall best performance.

Various systems rely on methods that are well known from the field of text retrieval. Adapting them to visual features that are corresponding roughly to words in the text is based on the TF-IDF principle (cf. Section 14.2.3.1). Several weighting schemes for text retrieval have also been used in image retrieval. Accordingly, the TF-IDF method perfectly fits to features extracted from patches and visual words (cf. Section 14.5.1.2).

14.5.3 User Interaction

Designing ergonomic user interfaces is a challenging task of engineering in general, and becomes even more sophisticated if the task is complex and the desired user group is specialists from a nontechnical domain, such as in medicine and biology. Since the early 1990s, the International Organization of Standardization (ISO) is publishing a series of standards (ISO 9241) initially titled "Ergonomic requirements for office work with visual display terminals (VDTs)." Part 10 (nowadays part 110) of this series defines seven basic principles: (i) suitability for the task, (ii) self-descriptiveness, (iii) controllability, (iv) conformity with user expectation, (v) error tolerance, (vi) suitability for individualization, and (vii) suitability for learning, which in summary support (ISO 9241-11):

- *Effectiveness*: With respect to the ISO standard, effectiveness means that the software performs all the task it has been designed for, and is not interrupted by errors, malfunctions, or system failures of any kind. In addition, the results obtained using the software must have a minimum (well-defined) quality.
- *Efficiency*: All tasks can be performed with minimal efforts. In particular, required user interaction such as keyboard hits, mouse clicks, and so on must be minimized and redundancy shall be avoided.
- *User satisfaction*: In general, user satisfaction is defined as the opinion of the user about a specific computer application, which they use (Doll and Torkzadeh 1988, 1991). The user subjectively must "feel" effectiveness and efficiency, disregarding their objective measurements. We shall realize that user satisfaction is regarded as *the* general key measure of computer system success at all (de Lone and Mc Lean 1992, 2003).

With this in mind, important functionality for medical image search has been identified. In particular, mechanisms for query refinement and relevance feedback are required (Smeulders et al. 2000).

14.5.3.1 Query Refinement

Query refinement is a common task in document retrieval. For example, retrieving literature from bibliographic databases such as PubMed, the resulting set of papers matching a query is reduced in the numbers of entities by additional constraints, such as determining the language of the paper, its publication date, or simply requiring the presented key words to occur in the title of the paper rather than within its abstract. This process, called query refinement, is performed by the user until the set of retrieved documents is suitable for manual inspection.

In image retrieval, in particular, performing a range query, the user also needs to lower the query range. Here, however, the task is more sophisticated, since the range is defined by similarity rather than exact match, and the features are numerical vectors computed from color, texture, and shape as compared to simple words that are applied in text-based document retrieval. Hence, the process of query specification and display is iterated, where, in each step, the user revises the query. Considering such an interactive session in its entirety, the system updates the query space, attempting to learn the goals from the user's feedback. Accordingly, an interactive query session is defined as a sequence of query spaces Q visualized in display space $V(Q)$ (Smeulders et al. 2000):

$$\{Q^0, Q^1, \ldots, Q^{n-1}, Q^n\} \text{ with } A^n(q) = V(Q^n)$$

In a truly successful session, the system's answer $A^n(q)$ equals the user's search goal. The interaction process is indicated schematically in Figure 14.9. The framework is suitable to model different types of searches, such as the

- *Associative search* aiming at finding relevant images by a range or nearest-neighbor query
- *Category search* aiming at identifying a certain category the query image is belonging to, for example, in CAD applications, where each category represents a certain pathology or tumor stage
- *Target search* aiming at retrieving only a certain image

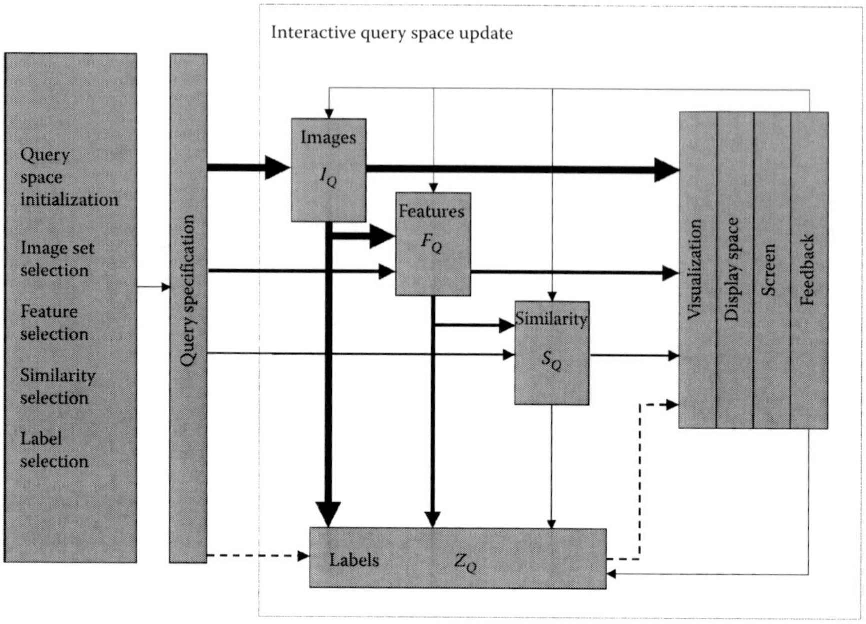

FIGURE 14.9 Framework for interactive CBIR sessions. (From Smeulders AWM et al. *IEEE Trans Pattern Anal Mach Intell.* 22(12): 1349–1380 © 2000 IEEE.)

14.5.3.2 Relevance Feedback

The interaction of the user yields a relevance feedback RF_i in every iteration i of the session. The transition from Q^i to Q^{i+1} materializes this feedback of the user. For all types of search, various ways of user feedback have been considered in the literature (Smeulders et al. 2000). All are balancing between obtaining as much information from the user as possible (cf. effectiveness) and keeping the burden on the user minimal (cf. efficiency). The simplest form of feedback is to indicate which of the retrieved images are relevant. More sophisticated approaches allow continuous-valued feedback between relevance and nonrelevance.

With respect to the framework defined in Figure 14.9, user feedback in general leads to an update of the query space

$$\{I_Q^i, F_Q^i, S_Q^i, Z_Q^i\} \rightarrow RF_i \rightarrow \{I_Q^{i+1}, F_Q^{i+1}, S_Q^{i+1}, Z_Q^{i+1}\}$$

Marking relevant images refines the query space on elements I_Q. Here, the user zooms on a target or category. More sophisticated, feedback from the user may also yield an optimization of the features used for retrieval. For instance, the user is allowed to select between color, texture, and shape-based features. Then, relevance feedback addresses the elements F_Q. For associative search, users typically interact to teach the system the right associations. Hence, the system updates the similarity function S_Q. The last set of systems operates on grouping or partitioning that is based on the similarity for each individual feature. The feedback from the user is employed to create compound groupings corresponding to a user given $z \in Z_Q$. Usually, the compound groupings are such that they include the entire positive and none of the negative examples.

14.5.3.3 User Interfaces

Aiming at full user satisfaction, the challenge in GUI engineering lies in integrating all required types of search and interaction into a simple, preferably web-based design, where medical CBIR can be used via a web browser that is already installed on the user's system. Furthermore, it cannot be assumed that query refinement is always successful, and previous query spaces Q^{n-i} shall stay available within a session. Based on only four types of interface modules, an award-winning approach has been proposed in Deserno et al. (2008):

1. *Output modules* contain all functionality regarding the visualization of information such as images, descriptions, or numerical parameters. In particular, "text," "image," "line," "shadow," "frame," and "table" are basic output modules. These modules are used to initialize the query, to visualize the query result, and to display the relevance facts.
2. *Parameter modules* allow the user to interact with the system. "Input field," "radio button," "slider," or "selection box" are some prominent examples. Furthermore, tools for selecting an ROI within an image or drawing a sketch belong to this class of modules. The parameter modules are used whenever a query is initialized or (re)submitted. In particular, they support relevance feedback.
3. *Transaction modules* include all functions that allow the user to step back and forward (UNDO or REDO functions) and to restore any "steady state" that the system has had within the current session, and also within already closed sessions (HISTORY functions). Using web-based GUIs, all output and parameter modules can be embedded in a hypertext form sheet, and the data transfer between server and client (the browser) is captured and stored. Then, previous stages are easily restored by "updating" the hypertext form with information from older stages.
4. *Process modules* are defined as union or intersection of intermediate sets of query results, which have already been refined by relevance feedback loops. The image lists and all corresponding relevant facts are stored, when process modules are activated during a session. Boolean AND/OR operations are available to operate on these lists for advanced query refinement.

Based on these module classes, Figure 14.10 diagrams the flow of a retrieval session. Parameter or function modules are used to initialize the query. After performing the search, which compares the query with all images in the system by means of the selected features, resulting images are displayed and relevance facts are given by means of the output or function modules.

In Figure 14.10, four nested loops are indicated. Without any query refinement, the straight line from start to end is followed. If the system offers relevance feedback, a central loop from output to parameter modules is added (Loop 1). The sequence of performing relevance feedback, computing a revised search, and presenting relevant facts of updated results is repeated until the user accepts the current response to his query. The transaction modules add a second inner loop to the flowchart intra-connecting the output modules (Loop 2). This loop is supported by the system's query logging, which also reads from the parameter and the output modules. Finally, a third loop is added to provide extended query refinement by combining successive but independent queries over the same database (Loop 3). If the system is currently not within a Boolean loop, such a loop can be opened to extend (OR relationship) or refine (AND relationship) the resulting image list, which is stored as pending in the database via the query logging mechanisms. The Boolean loop is closed automatically by merging the pending with the current response list of images if the user accepts the current result (Loop 4). Again, this loop can be iterated until the user agrees to the final result.

14.5.4 CBIR Workflow Integration

In Section 14.4, we have already emphasized the relevance of seamless workflow integration. In medical applications, CBIR-based CAD may assist the radiologists, and all levels of integration from data to context are equally important. In this section, we focus on data and functional integration. Data integration, however, needs to be considered in both ways, PACS to CBIR and CBIR to PACS.

14.5.4.1 CBIR-Based Access to PACS Images

Within a DICOM-based PACS, any DICOM-conformant system may be integrated and retrieves images from the PACS archive using the DICOM Query/Retrieve (Q/R) services (Onken et al. 2011).

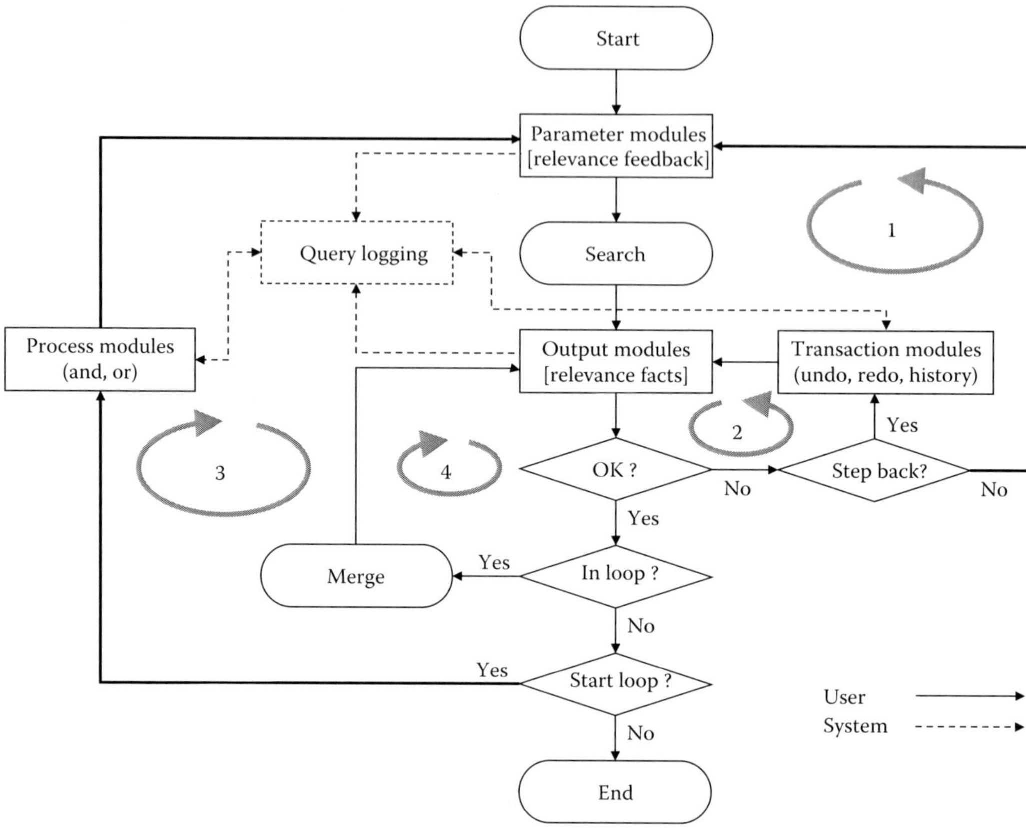

FIGURE 14.10 Flowchart of extended query refinement. (From Deserno TM et al. *J Digit Imaging.* 2008; 21(3): 280–289.)

Q/R offers different service object pair (SOP) classes for searching and retrieving, which are varying in what information can be searched and how transmission of objects is performed. In the query phase, an archive can be searched for images based on filter mechanisms, for example, it is possible to ask for a list of studies based on a defined patient name. If the service class user (SCU) decides to download any DICOM objects (retrieve phase), the Q/R protocol is used to initiate the transfer. However, the transmission itself is done via the DICOM Storage SOP classes. It is also possible for an SCU to tell the service class provider (SCP) not to transfer the DICOM objects to the SCU itself but to a third system. In this way, the reading application may also invoke a CBIR process by sending the QBE (functional integration).

The first approaches of CBIR-PACS integration based on DICOM have been proposed in 1999 (Qi and Snyder 1999). Le Bozec et al. have described new representational and retrieval models allowing hospital-wide access to the images based on demographic and procedure-type information and fostering CBIR to enhance the medical impact of image retrieval in daily practice. The approach was applied in the Gerorges Pompiduo Hospital in Paris, France (Le Bozec et al. 2000). More recently, Traina et al. have presented a novel PACS approach, called cbPACS, which provides content-based capabilities for image retrieval. The cbPACS answers range and k-NN similarity queries. By now, the system implemented works on features based on color distribution of the images through normalized histograms as well as metric histograms (Traina et al. 2005). Nonetheless, a recent review demonstrates the remaining gap between intensive CBIR research and development, and its effective use in clinical practice (Depeursinge et al. 2011).

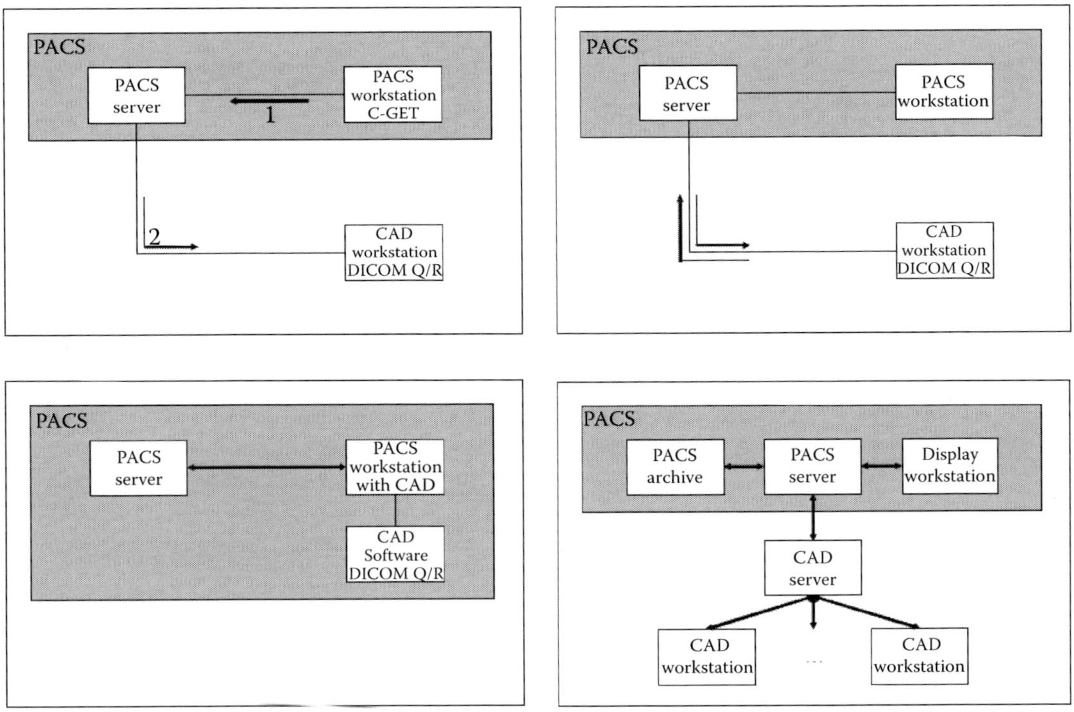

FIGURE 14.11 Conceptual methods of integration CAD with PACS. (From Huang HK et al. PACS-based computer-aided detection and diagnosis. Chapter 18 in Deserno, 2011; pp. 455–470.)

More methodically, data integration of CAD (including CBIR-based CAD) with DICOM PACS can have four approaches, which differ in the systems performing the query and retrieve commands (Huang et al. 2011). In panels (a), (b), and (c) of Figure 14.11, the CAD is connected directly to the PACS, while the fourth approach is to use a CAD server to connect with the PACS (Figure 14.11d). This is seen advantageous since DICOM must not be installed on the CBIR engine. The CAD server can automatically manage the clinical workflow of image studies to be processed and can archive the CAD results back to PACS, for the clinicians to review directly on the PACS workstation. This also eliminates the need for both the PACS and the CBIR manufacturer to open up their respective software platforms for installing an application programming interface (API).

Accordingly, a workflow management of CBIR for CAD support in PACS environments has been proposed recently by Welter et al. (2010). It is based on concepts developed by the Integrating the Healthcare Analysis (IHE) initiative, a consortium of health care professionals from industry and research that has been founded to establish unified and generally accepted process flows between information and communication technology (ICT) systems for medical applications improving interoperability. Information exchange is based on established standards like DICOM and Health Level 7 (HL7). IHE provides a framework that accumulates requirements in use cases and defines guidelines, called Integration Profiles, which represent scenarios and prescribe how well-known standards like DICOM shall be applied. Each scenario involves a set of actors, which represent components or modules that occur in medical information systems (MIS) such as the hospital information system (HIS).

14.5.4.2 Persistent Storing CBIR Results in PACS

However, CBIR-based CAD usually does not end up only with an image and some additional descriptive measures that have been created by the CAD software. Instead, it delivers a list of DICOM study unique

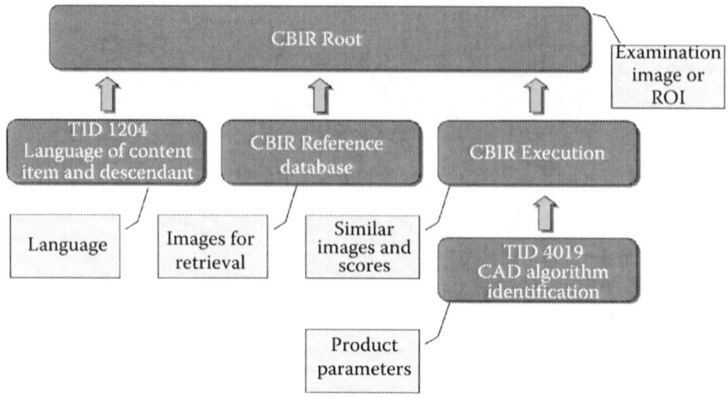

FIGURE 14.12 Composition of the CBIR SR template. (From Welter P et al. *J Am Med Informatics Assoc (JAMIA)*. 2011b; 18(4): 506–510.)

identifiers (UID) containing relevant images and pointing to relevant cases (electronic health record, EHR), which are already included within the HIS. This type of information cannot be stored back to the PACS by means of novel DICOM objects.

Here, DICOM Structured Reporting (SR) is an appropriate tool for designing persistent storage of CBIR results (Welter et al. 2011b). A CBIR-CAD DICOM-SR document is built from SR templates and includes information on the CBIR system, the QBE image or ROI of query, the reference database on which CBIR was performed, and a list of resulting image UIDs associated with similarity scores. In Figure 14.12, the template identifier (TID) refers to a predefined template from the DICOM standard.

14.5.4.3 Functional CAD-PACS Integration

In 2009, DICOM supplement 118 titled "Application Hosting" has become part of the standard. It may be regarded as the future of CAD-PACS coupling, including CBIR integration to the hospitals and clinical practice. It defines a generic API between the hosting systems and the hosted application that supports full integration on all the data, function, context, and presentation levels, since the application is now simply plugged into the clinical workstation, and all user interfaces remain from the hosting program. The API supports the entire software life cycle management. The hosting system is able to launch and terminate the hosted application. While running, hosting system and hosted application exchange input, processing, and output data, and communicate status information. This is done by the hosted application switching between several states (Onken et al. 2011), which are visualized in the state chart (Figure 14.13):

- *Idle*: waiting for a new task assignment from the hosting system. This is the initial state when the hosted application is started.
- *Inprogress*: performing the assigned task.
- *Suspended*: stopping processing and releasing as many resources as possible, while still preserving enough state to be able to continue processing.
- *Completed*: completing processing and waiting for the hosting system to access and release any output data.
- *Canceled*: stopping processing and releasing all resources giving up any chance to resume.
- *Exit*: terminating the service. This is the terminal state of the hosted application.

14.5.5 Combined Visual and Textual Retrieval

Disregarding all efforts on CBIR technology, it is important to realize that medical images are never created isolated. They are always embedded with rich annotations and metainformation, which may be

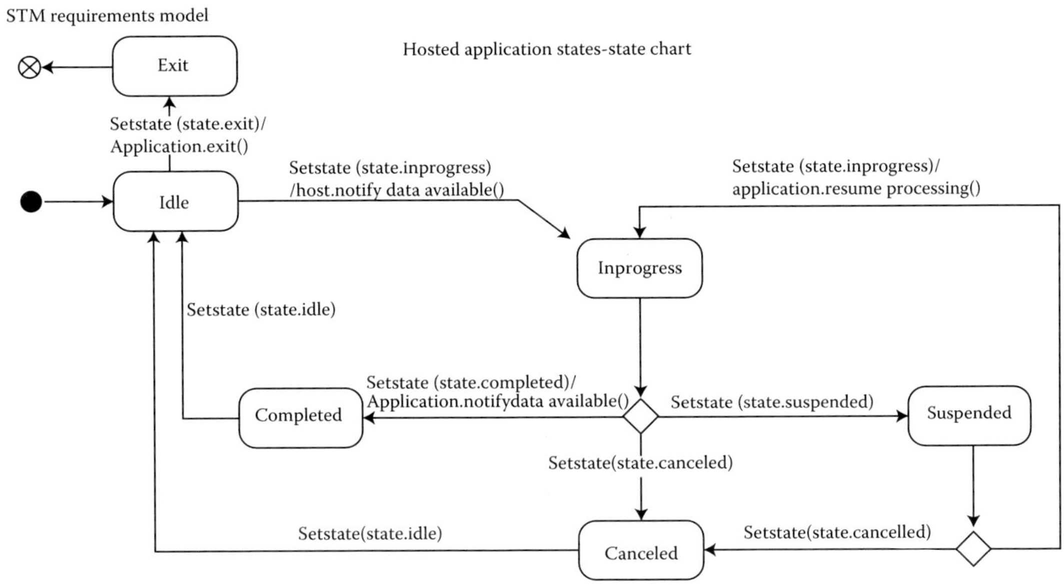

FIGURE 14.13 State chart of DICOM Application Hosting. (From DICOM Working Group 23. (ed). Digital Imaging and Communications in Medicine (DICOM) supplement 118 application hosting. Technical Report, DICOM Work Item 2004-09-E, 2010/09/30. Rosslyn, VA, USA.)

composed of either structured or unstructured text. Hence, CBIR shall always be combined with text-based retrieval methods based on either alphanumerical matches or natural language processing (NLP), respectively. Sinha et al. have reviewed methods to present imaging and associated data without causing an overload, including image study summarization using CBIR and NLP (Sinha et al. 2002).

Some examples are the works of Ruiz, combining image features, case descriptions, and unified medical language system (UMLS) concepts to improve retrieval of medical image (Ruiz 2006), and integrating this automatic classification method into the medical image retrieval process (Uwimana and Ruiz 2008). Three publicly available tools were combined: GIFT (cf. Section 14.2.3.1), a text retrieval system, and a tool for mapping free text to UMLS concepts. Data from the CLEF campaign was used for evaluation. Névéol et al. applied a set of 180 medical documents from a standard reference set in French language, combining the IRMA (cf. Section 14.2.3.2) engine with NLP and Medical Subject Heading (MeSH) mapping (Névéol et al. 2009). Another example of visual and textual combined retrieval is the cbPACS system (cf. Section 14.5.4.1). Here, the standard query language (SQL) was extended to seamlessly integrating similarity queries based on visual features resulting in a command language interpreter called Similarity Retrieval Engine (SiREn) (Barioni et al. 2009).

14.5.6 Evaluation

One of the most challenging tasks in medical image search is its comprehensive evaluation. Obviously, large databases with well-known references are required. As a response to the need for standardized test collections and evaluation forums, ImageCLEF was initiated in 2003. It has grown to become a preeminent venue today for image retrieval evaluation (Kalpathy-Cramer and Müller 2011). ImageCLEF itself also includes several subtracks dealing with various aspects of image retrieval (Müller et al. 2010).

One of these tracks is the IRMA image annotation task, where CBIR systems perform a categorization of images. In 2005 and 2006, the goal was a flat classification into 57 and 116 unique classes, respectively, and error rates based on the number of misclassified images have been used as evaluation metric (Deselaers et al. 2007). In 2007 and 2008, the hierarchical IRMA code (Lehmann et al. 2003) was used, and errors were penalized depending on the level of the hierarchy at which they occurred (Deselaers et al. 2008). Typically, participants were provided about 10,000 training images and were to submit classification for 1000 test images. All radiographs have been transferred into the IRMA system from the PACS of the Aachen University Hospital, Aachen, Germany, anonymized, and semiautomatically categorized by experienced physicians and radiologists providing the ground truth. Both secondarily scanned x-ray films and direct digitally acquired radiographs were included in the IRMA repository. In 2009, the goal was to classify 2000 test images using the different classification schemes used in 2005–2008, given a set of about 12,000 training images.

In ImageCLEF, the Medical Image Retrieval task's test collection has began with a rather small database of 8700 images as an amalgamation of several teaching case files in English, French, and German (Müller et al. 2004b). By 2007, it had grown to a collection of over 66,000 images from several teaching collections, as well as a set of topics that were known to be well suited for textual, visual, or mixed retrieval methods. The 2009 database already contained a total of 74,902 images (Kalpathy-Cramer and Müller 2011). However, such large collections cannot be annotated manually to provide a reliable ground truth. Hence, the results obtained by different CBIR methods in the campaign are merged and evaluated retrospectively by the experts. This method, however, leaves large parts of the repositories unexplored.

14.6 Examples

It is worth exemplarily describing outstanding CBIR approaches in more detail. Within the numerous fields of medical CBIR application that we have discussed already, case retrieval and CAD in diagnostics and research as well as medical document retrieval in research and education are focused in the following sections.

14.6.1 CBIR-Based Access to Medical Literature

It is self-suggesting to apply CBIR techniques to medical image repositories, and the medical literature itself is a prominent example. For instance, the 2005 volumes of (i) *New England Journal of Medicine*, (ii) *Radiology*, (iii) *Journal of Dental Research*, and (iv) *Journal of the American Medical Informatics Association* (JAMIA) in total contain about 11,753 journal pages and 11,238 figure panels (Deserno et al. 2009b).

Figure 14.14 demonstrates the variety of illustrations that can be found in the medical literature. On the top level, 12 categories have been defined:

1. *Diagnostic image*, that is, an original image as obtained from any medical imaging modality (e.g., radiography, microscopy, endoscopy, sonography) that may be color or grayscale and annotated.
2. *Diagnostic visualization*, that is, a color or grayscale computed visualization of medical image data, such as a 3D direct volume rendering of CT or MRI data.
3. *Photograph*, that is, any type of an optical static image, which, again, may be in color or grayscale and show devices, medical objects or situations, persons, or portraits.
4. *Screen shot*, that is, any illustration showing a computer screen, window, or a part thereof.
5. *Graph*, that is, any visualization of numerical data such as plots, curves, as well as block or pie charts.
6. *Diagram*, that is, any kind of functional or block diagram, scheme, or mind map.
7. *Drawing*, that is, any type of manual drawings.
8. *Multipanel figure*, that is, a composition of different parts, which may be composed of *strictly medical*.
9. *Nonmedical*, or.
10. *Mixed panels*, and may be presented in color or grayscale.
11. *Protein spot*, that is, a special type of multipanel figures, where the high number of spots (panels) frequently is ambiguous, and therefore, not countable.
12. *Other*, for instance, equations that are embedded in the PDF file as figure.

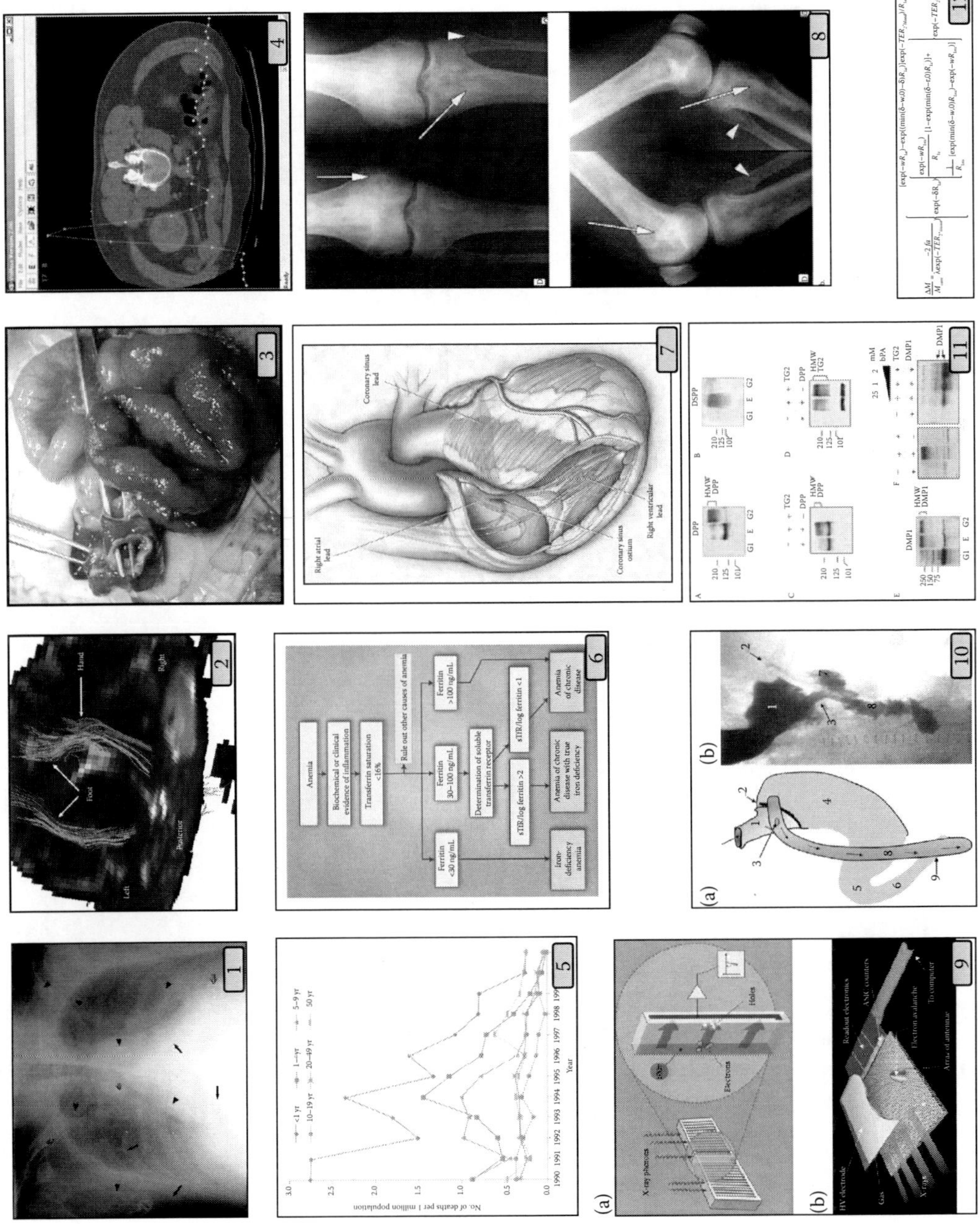

FIGURE 14.14 The 12 categories of images in medical scientific literature. (From Deserno TM, Antani S, and Long RL. *Methods Inf Med.* 2009b; 48(4): 371–380.)

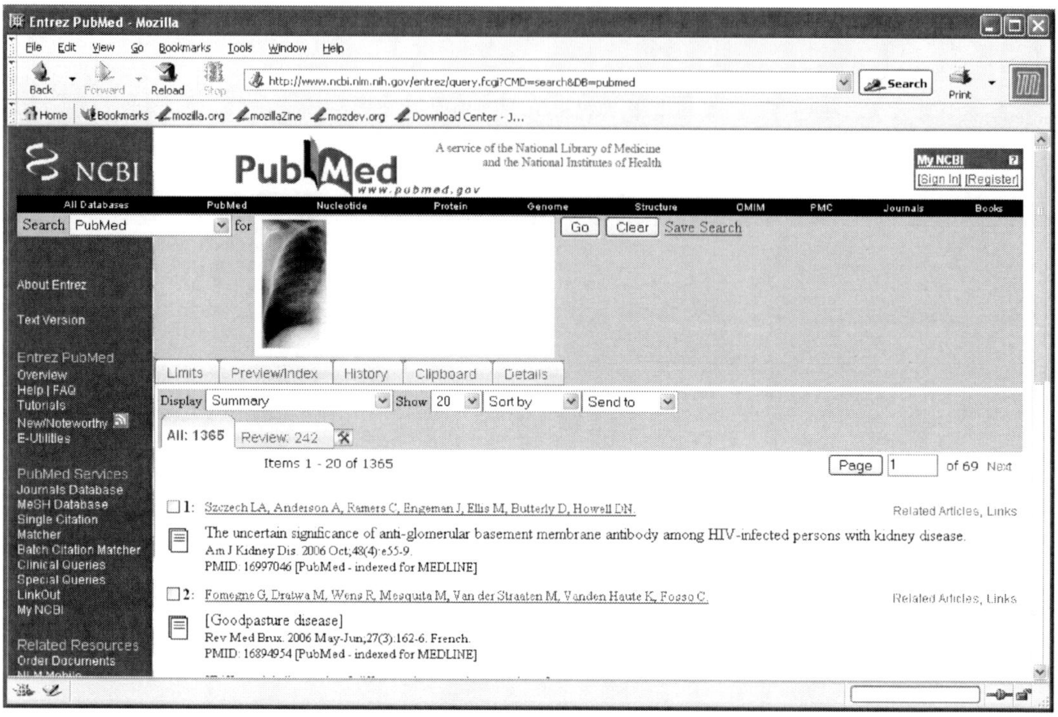

FIGURE 14.15 A vision on medical literature retrieval using CBIR technology.

Figure 14.15 visualizes the well-known PubMed Entrez screen, which has been expanded virtually to support CBIR-based queries. Deserno et al. (2009b) have shown exemplarily that combining textual and visual retrieval methods may improve the query completion when seeking scientific medical literature.

Accordingly, Müller has emphasized the enormous potential from content-based access to images in the scientific literature as physicians can directly compare a current case with similar existing cases in the peer-reviewed literature (Müller 2008). However, such applications still need to be implemented and integrated with the clinical patient record and large knowledge archives of literature.

14.6.2 CBIR-Based CBR for Surgery Planning

A case-based CBIR system for surgical planning of bone fractures has been developed at HUG (Zhou et al. 2011) and demonstrated live at the CADdemo@CARS 2010 (Depeursinge et al. 2011). It is using an image database built at the surgery department of HUG containing 2693 fracture cases associated with 43 different fracture types. Beside images, a few clinical attributes such as age, gender, implant type, and exact diagnosis are available in XML files. The fracture retrieval engine is based purely on visual information extracted from image content and the diagnosis information is used for evaluation only. Images are indexed using a bag of visual words strategy, where local descriptors based on SIFT are obtained at fixed positions (cf. Section 14.5.1.2). Finally, each image is represented by a histogram of 1000 numerical features.

Figure 14.16 shows the case-based query interface. Similarity measurements based on histogram intersection are used to rank the returned cases. Both query and results are case-based and contain multiple images. A fusion strategy based on a mix of sum and max operators is used (Zhou et al. 2011).

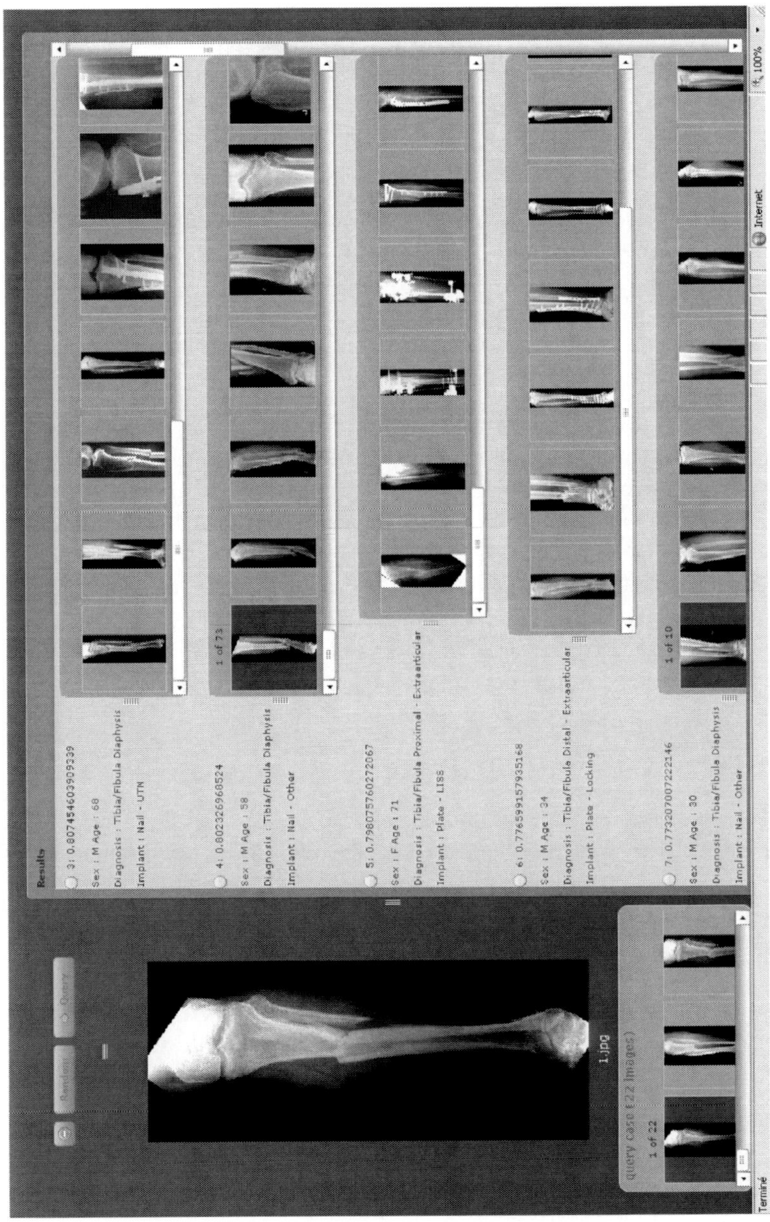

FIGURE 14.16 GUI for visual retrieval of fracture cases. (From Depeursinge A et al. *Open Med Imaging J*. 2011; 5 Suppl 1-M7: 58–72.)

14.6.3 CBIR-Based CAD for Bone Age Assessment

Recently, Depeursinge et al. have identified four CBIR-based CAD systems with a high degree of "user readiness" (Depeursinge et al. 2011). For example, with its underlying flexible structure of image processing and image retrieval algorithms, the IRMA framework has been adjusted to enrich CAD in the context of bone age assessment (BAA) (Fischer et al. 2011). Similar radiographs with validated ages are retrieved from a reference database and presented to the radiologist along with a suggested bone age. The reference database contains about 1000 hand radiographs, corresponding epiphysial ROIs, and metainformation such as gender, ethnic origin, chronological age, and the validated bone age from expert's readings.

For a new hand radiograph, the processing pipeline consists of four steps:

1. *Center localization*: At first, the centers of the epiphyses are localized. Optionally, manual and automatic methods can be applied.
2. *Region extraction*: A bounding box is adjusted automatically around these centers, scaled and extracted, yielding epiphysial ROIs in standard vertical orientation for the CBIR part of IRMA-BAA.
3. *CBIR query*: With each extracted ROI, a QBE *k*-NN query is performed (cf. Section 14.5.2). Similar images are returned with similarity scoring and validated bone age.
4. *Age assessment*: The overall bone age is predicted by a similarity score-weighted mean of ages associated to the response images.

The IRMA framework provides several modules for integration of CBIR into web-based interfaces (cf. Section 14.5.3.3). Using this technology, the result of CBIR-based age estimation is presented to the user (Figure 14.17). The query image and its extracted ROIs are shown at the top-most area of the interface. Their most similar counterparts retrieved from the database are shown below (scrollable) in decreasing similarity and with the validated bone age. The estimated bone age is shown below the query image. If the query image is contained in the demo database, the validated bone age is also provided. A click on one of the thumbnails opens the full-resolution image. The display mode can also be switched to show the hands belonging to the retrieved ROIs. In leaving-one-out experiments, a mean absolute error of 0.97 years and a variance of 0.63 were observed over all ages and regardless of gender (Fischer et al. 2011).

14.7 Summary and Conclusions

In this chapter, we have explored the relevance and state of the art of medical image search by means of CBIR techniques. Initially focused on nonmedical image repositories, and using color as most relevant feature, CBIR has been established in the commercial markets in the early 2000s. Deducted from the relevance of images in general and the relevance of medical images in particular, we have explored general fields of biomedical engineering, where CBIR methodology can be applied meaningfully in medical practice, including diagnostic, research, and medical education. The state of the art has been described regarding features and similarity measures, user interaction, workflow integration, combined visual and textual retrieval, and evaluation of medical CBIR systems. Exemplarily, three outstanding approaches have been described in detail, regarding CBIR-based access to medical literature, CBIR-based CBR for surgery planning, and CBIR-based CAD for maturity measurements.

Despite such user-ready implementations, CBIR technology in general has not yet been established in clinical practice. With respect to the concept of gaps that we have described in Section 14.4, current systems adequately narrow the semantic gap focusing on a particular imaging and application domain. However, this specific use context is hindering the transfer into other application domains in clinical practice. Future CBIR applications in the medical domain shall aim at more generic architectures, principles, and interfaces to support rapid prototyping and implementation in other fields of medical applications.

Nonetheless, today's prototype systems have bridged the various feature gaps, and efficiency has been proven on relatively large databases of reference and ground truth images. This can be seen as a prerequisite for any system in the field of CBIR-based CAD. However, user interaction and system integration still

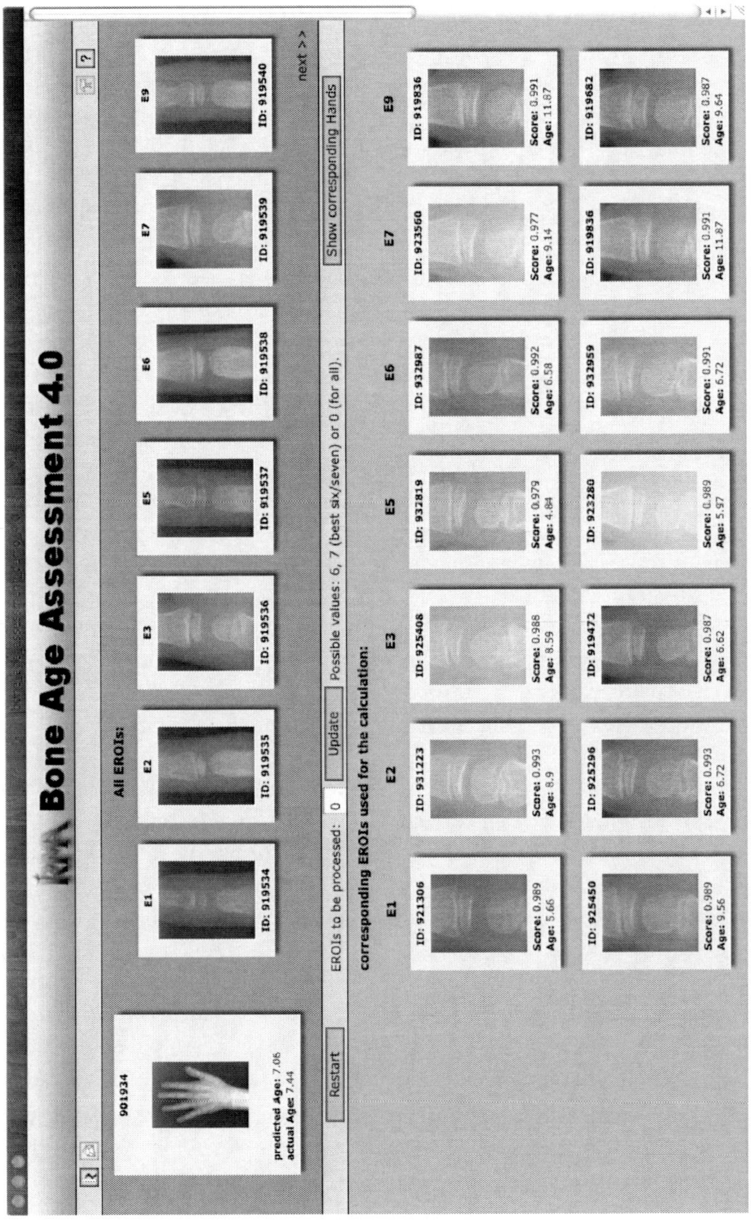

FIGURE 14.17 IRMA bone age assessment. (From Fischer B et al. *Int J Computer Assisted Radiol Surgery*. 2012; 7(3): 389–399.)

need improvement. So far, stand-alone systems are demonstrated, which are inappropriately integrated in the hospital information system environments. DICOM Hosted Application and DICOM Structured Reporting might be suitable concepts in bridging the integration gap (Welter et al. 2011b).

Müller has suggested establishing small modules or blocks, developed in open-source projects and made available to a large community free of charge, to foster the development of CBIR in medicine (Müller 2008). Plug-in technology, as it has been established in the Internet, is a good example of the impact obtained from open interfaces. Google Earth, for instance, is easily interconnected with any website, and the web browser itself offers various options to plug-in and combine other pieces of software.

However, the large reference databases that are needed for development and evaluation of medical CBIR systems are seen more crucial. Until now, sufficient data is unavailable, and the ground truth provided along with the images cannot always be trusted. Currently, legal affairs, data privacy policies, and other restrictions must be coped with laboriously to build a medical reference image database. In particular, written consent is needed for academia research if patient images are used else than personal diagnostic. On the other hand, social networks such as Facebook, Twitter, and Flickr have shown the intrinsic vigilance of the entire human population to share personal details. Soon, technology will be available to host medical images in the Internet, and educated individuals will take over the responsibility of their EHR. This will foster medical image sharing, and preparation of large medical image repositories including images and metainformation. Then, the need to navigate in databases of billions of medical images in a meaningful way, allowing efficient and effective retrieval satisfying the user, will become even more prominent. Here, it also seems necessary to better integrate visual, textual, and structured data retrieval into unique systems (Müller and Deserno 2011).

Anyway, reference annotations and reliable ground truth still need trusted establishment. For example, Horsch et al. (2011) have assessed the needs for the next-generation computer-aided mammography reference image databases and evaluation studies. The authors developed 13 recommendations concerning the construction and usage of a test database, as well as the application of statistical evaluation methods. In particular, a test database should ultimately reflect realistic clinical conditions, in order to allow for significant conclusions on the performance of the method or system under evaluation. Furthermore, a clear picture of the variability of a measure shall be given by reporting either the corresponding standard deviation or standard error, or the confidence interval, that is, the range of values which are likely.

As we have emphasized already, the 3D and 4D nature of the medical imagery is not yet sufficiently coped in both medical CBIR user interfaces and system's data processing. Interactive navigation in 3D and 4D data, its manual annotation, and the process of fast and efficient browsing according databases requires novel human machine interfaces and fast processing methods. Again, consumer industry might have a large impact on medical applications. Game industry already has developed powerful but—since huge numbers of pieces are being sold—inexpensive devices for 3D computer interaction. Parallel computing on GPU is already used successfully in medical image processing, and tablet computers (iPad) have been established in the operation theatre supporting augmented reality in surgery (Volonté et al. 2011). Furthermore, Mezzana et al. have introduced augmented reality in oculoplastic and orbit surgery through a simple iPhone application evaluating the position of the lateral canthal ligament (Mezzana et al. 2011). Such examples show how the medical field takes advantage of the lower costs of technology adaptation than technology development. The consumer has paid already for device development that is now used in medicine.

Another cost-intensive field is drug development, where a paradigm change has been noticed. Individualized medicine and personalized drug development are fostered, and medical imaging is expected to provide surrogate endpoints in clinical trials (Rosenkranz 2003). Pharmaceutical industry will face the need for large reference databases and efficient techniques for image-based data search, providing much more financial power than medical technology industry, which is reimbursed rather slowly by the public health insurance systems. 3D and 4D ultrasound visualizations in prenatal imaging or full-body CT scans are other examples emphasizing the impact of economic interests in medical imaging and image management technology.

Another point that is particular important for medical image search using CBIR technology is that medical image processing has not yet reached the level of scene interpretation. As we have emphasized in the introduction, computer-based image analysis still is by far less powerful than the human visual system. We have presented approaches for graph representations, where the nodes correspond to objects in the image and the edges in the graph describe their constellation in temporal and/or spatial domains. Such annotated graphs will form the basis of Web 2.0 technology, which is about to be delivered. Annotating hyperlinks with certain characteristics will enable search engines such as Google to retrieve information more selectively and provide a much better query completion. Such novel technologies will boost medical image search on both levels: scene analysis within an image and CBIR-based CBR within the entire medical databases.

All in all, medical image search is active field of biomedical engineering, which is expected to develop quickly by adopting technology and processing knowledge from other fields such as consumer industry, the Internet, and Web 2.0 technology.

References

Aamodt A and Plaza E. Case-based reasoning: Foundational issues, methodological variations, and system approaches. *AICom—Artif Intell Commun.* 1994; 7(1): 39–59.

Aggarwal CC, Hinneburg A, and Keim DA. On the surprising behavior of distance metrics in high dimensional space. *Lect Notes Computer Sci.* 2001; 1973: 420–434.

Analoui M. Quantitative medical image analysis for clinical development of therapeutics. Chapter 14 in Deserno, 2011; pp. 359–374.

Avni U, Greenspan H, Konen E, Sharon M, and Goldberger J. X-ray categorization and retrieval on the organ and pathology level using patch-based visual words. *IEEE Trans Med Imaging.* 2011; 30(3): 733–746.

Barioni MCN, Razente HL, Traina AJM, and Traina C, Jr. Seamlessly integrating similarity queries in SQL. *Software: Practice Experience.* 2009; 39(4): 355–384.

Belongie S, Carson C, Greenspan H, and Malik J. Color- and texture-based image segmentation using EM and its application to content-based image retrieval. *Sixth International Conference on Computer Vision (IEEE Cat. No.98CH36271).* Narosa Publishing House, New Delhi, India, 1998; pp. 675–682.

Canny JF. A computational approach to edge detection. *IEEE Trans Pattern Anal Mach Intell.* 1986; 8(6): 679–698.

Carson C, Belongie S, Greenspan H, and Malik J. Blobworld: Image segmentation using expectation-maximization and its application to image querying. *IEEE Trans Pattern Anal Mach Intell.* 2002; 24(8): 1026–1038.

Chang SK and Kunii T. Pictorial database systems. *IEEE Comput.* 1981; 14(11): 13–21.

Comaniciu D, Meer P, Foran D et al. Bimodal system for interactive indexing and retrieval of pathology images. *Proc IEEE Workshop Appl Comput.* 1998; 76–81.

de Geus-Oei LF, van der Heijden HFM, Visser EP, Hermsen R, van Hoorn BA, Timmer-Bonte JNH, Willemsen AT, Pruim J, Corstens FHM, Krabbe PFM, and Oyen WJG. Chemotherapy response evaluation with 18F-FDG PET in patients with non-small cell lung cancer. *J Nucl Med.* 2007; 48: 1592–1598.

de Groot E, Hovingh GK, Wiegman A, Duriez P, Smit AJ, Fruchart J-C, and Kastelein JJP. Measurement of arterial wall thickness as a surrogate marker for atherosclerosis. *Circulation.* 2004; 109: 33–38.

De Lone WH and Mc Lean ER. Information systems success: The quest for the dependent variable. *Inf Syst Res.* 1992; 3(1): 60–95.

De Lone WH and Mc Lean ER. The DeLone and McLean Model of information systems success: A ten-year update. *J Manage Inf Syst.* 2003; 19(4): 9–30.

De Oliveira JEE, Machado AMC, Chavez GC et al. MammoSys: A content-based image retrieval system using breast density patterns. *Comput Methods Programs Biomed.* 2010; 99(3): 289–297.

Depeursinge A, Fischer B, Müller H, and Deserno TM. Prototypes for content-based image retrieval in clinical practice. *Open Med Imaging J.* 2011; 5 Suppl 1-M7: 58–72.

Deselaers T, Müller H, Clough P, Ney H, and Lehmann TM. The CLEF 2005 automatic medical image annotation task. *Int J Computer Vis.* 2007; 74(1): 51–58.

Deselaers T, Deserno TM, and Müller H. Automatic medical image annotation in ImageCLEF 2007. Overview, results, and discussion. *Pattern Recognit Lett.* 2008; 29(15): 1988–1995.

Deserno TM, Güld MO, Plodowski B, Spitzer K, Wein BB, Schubert H, Ney H, and Seidl T. Extended query refinement for medical image retrieval. *J Digit Imaging.* 2008; 21(3): 280–289.

Deserno TM, Antani S, and Long R. Ontology of gaps in content-based image retrieval. *J Digit Imaging.* 2009a; 22(2): 202–215.

Deserno TM, Antani S, and Long RL. Content-based image retrieval for scientific literature access. *Methods Inf Med.* 2009b; 48(4): 371–380.

Deserno TM (ed). *Biomedical Image Processing. Series 3740: Biological and Medical Physics, Biomedical Engineering.* Springer, Berlin, 2011.

Deserno TM. Fundamentals of biomedical image processing. Chapter 1 in Deserno, 2011; pp. 1–51.

DICOM Working Group 23. (ed). Digital Imaging and Communications in Medicine (DICOM) supplement 118 application hosting. Technical Report, DICOM Work Item 2004-09-E, 2010/09/30. Rosslyn, VA, USA.

Doll WJ and Torkzadeh G. The measurement of end-user computing satisfaction. *MIS Quarterly.* 1988; 12(2): 259–274.

Doll WJ and Torkzadeh G. The measurement of end-user computing satisfaction: Theoretical considerations. *MIS Quarterly.* 1991; 15(1): 5–10.

Eakins J and Graham M. Content-based image retrieval. Technical Report No 39, JISC Technology Applications Programme, University of Northumbria at Newcastle, October 1999.

Fischer B, Thies C, Güld MO, and Lehmann TM. Content-based retrieval of medical images by matching hierarchical attributed region adjaciency graphs. *Procs SPIE.* 2004; 5370: 598–606.

Fischer B, Sauren M, Güld MO, and Deserno TD. Scene analysis with structural prototypes for content-based image retrieval in medicine. *Procs SPIE.* 2008; 6914: 1X1–1X9.

Fischer B, Welter P, Günther RW, and Deserno TM. Web-based bone age assessment by content-based image retrieval for case-based reasoning. *Int J Computer Assisted Radiol Surgery.* 2011; 7(3): 389–399.

Geusebroek JM, van den Boogaard R, Smeulders AWM et al. Color invariance. *IEEE Trans Pattern Anal Mach Intell.* 2001; 23(12): 1338–1350.

Gijselaers W. Perspectives on problem-based learning; Chapter 5 in Gijselaers W, Tempelaar D, Keizer P, Blommaert J, Bernard E, and Kapser H (eds). *Educational Innovation in Economics and Business Administration: The Case of Problem-Based Learning.* Kluwer, Dordrecht, 1995; pp. 39–52.

Gordon IE. *Theories of Visual Perception.* Wiley, Chichester, 1989.

Greenes RA. The radiologist as clinical activist—A time to focus outward. *Proceedings of the First International Conference on Image Management and Communication in Patient Care—Implementation and Impact.* IEEE Computer Society Press, Washington, 1989; pp. 136–140.

Greenspan H and Pinhas AT. Medical image categorization and retrieval for PACS using the GMM-KL framework. *IEEE Trans Inf Technol Biomed.* 2007; 11(2): 190–202.

Guyatt G, Cairns J, Churchill D et al. Evidence-based medicine. A new approach to teaching the practice of medicine. *J Am Med Assoc.* 1992; 268(17): 2420–2425.

Hand D, Manila H, and Smyth P. *Principles of Data Mining.* MIT Press, Cambridge, MA, 2001.

Haralick RM, Shanmugam K, and Dinstein I. Textural features for image classification. *IEEE Trans Syst Man Cybern.* 1973; 3(6): 610–621.

Horsch A, Hapfelmeier A, and Elter M. Needs assessment for next generation computer-aided mammography reference image databases and evaluation studies. *Int J CARS.* 2011; 6(6):749–767.

Huang HK, Liu BJ, Le AH, and Documet J. PACS-based computer-aided detection and diagnosis. Chapter 18 in Deserno, 2011; pp. 455–470.

Jaffe CC. Response assessment in clinical trials: Implications for sarcoma clinical trial design. *The Oncologist*. 2008; 13(Suppl 2): 14–18.

Jain AK, Duin RPW, and Mao J. Statistical pattern recognition: A review. *IEEE Trans Pattern Anal Mach Intell.* 2000; 22(1): 4–37.

Kalpathy-Cramer J and Müller H. Systematic evaluations and ground truth. Chapter 20 in Deserno, 2011; pp. 497–520.

Koh GC, Khoo HE, Wong ML, and Koh D. The effects of problem-based learning during medical school on physician competency: A systematic review. *CMAJ.* 2008; 178(1): 34–41.

Kuo WJ, Chang RF, Lee CC et al. Retrieval technique for the diagnosis of solid breast tumors on sonogram. *Ultrasound Med Biol.* 2002; 28(7): 903–909.

Lapeer RJ, Tan AC, and Aldridge R. A combined approach to 3D medical image segmentation using marker-based watersheds and active contours: The active watershed method. *Lect Notes Computer Sci.* 2002; 2488: 596–603.

Le Bozec C, Zapletal E, Jaulent MC, Heudes D, and Degoulet P. Towards content-based image retrieval in a HIS-integrated PACS. *Proc AMIA Symp*. 2000; 477–481. 1

Lehmann TM, Wein B, Dahmen J, Bredno J, Vogelsang F, and Kohnen M. Content-based image retrieval in medical applications. A novel multi-step approach. *Procs SPIE.* 2000; 3972: 312–320.

Lehmann TM, Schubert H, Keysers D, Kohnen M, and Wein BB. The IRMA code for unique classification of medical images. *Procs SPIE.* 2003; 5033: 440–451.

Lehmann TM, Güld MO, Thies C, Fischer B, Spitzer K, Keysers D, Ney H, Kohnen M, Schubert H, and Wein BB. Content-based image retrieval in medical applications. *Methods Inf Med.* 2004; 43(4): 354–361.

Lehmann TM. Digitale Bildverarbeitung für Routineanwendungen. Evaluierung und Integration am Beispiel der Medizin. Deutscher Universitäts-Verlag, Wiesbaden 2005. ISBN 3-8244-2191-7 (in German).

Leisch E, Sartzetakis S, Tsiknakis M, and Orphanoudakis SC. A framework for the integration of distributed autonomous healthcare information systems. *Medical Informatics, Special Issue.* 1997; 22(4): 325–335.

Loncaric S. A survey of shape analysis techniques. *Pattern Recognit.* 1998; 31(8): 983–1001.

Lowe DG. Distinctive image features from scale-invariant keypoints. *Int J Computer Vision.* 2004; 60(2): 91–110.

Lu CS and Chung PC. Wold features for unsupervised texture segmentation. *Proc Int Conf Pattern Recognit Lett.* 1998; 2: 1689–1693.

Lucchese L and Mitra SK. Unsupervised segmentation of color images based on *k*-means clustering in the chromaticity plane. *Proc Content Based Access of Image and Video Libraries.* 1999; 74–78.

Macura RT and Macura K. Case-based reasoning: Opportunities and applications in health care. *Artif Intell Med.* 1997; 9(1): 1–4.

Mezzana P, Scarinci F, and Marabottini N. Augmented reality in oculoplastic surgery: First iPhone application. *Plast Reconstr Surg.* 2011; 127(3): 57e–58e.

Mikolajczyk K and Schmid C. A performance evaluation of local descriptors. *IEEE Trans Pattern Anal Mach Intell.* 2005; 27(10): 1615–1630.

Milanese R and Cherbuliez M. A rotation, translation and scale-invariant approach to content-based image retrieval. *J Vis Commun Image Represent.* 1999; 10: 186–196.

Müller H, Fabry P, Lovis C, and Geissbuhler A. medGIFT—Retrieving medical image by their visual content, in *World Summit of the Information Society, Forum Science and Society*, 2003.

Müller H, Michoux N, Bandon D, and Geissbuhler A. A review of content-based image retrieval systems in medical applications. Clinical benefits and future directions. *Int J Med Inform.* 2004a; 73(1): 1–23.

Müller H, Rosset A, Vallée JP, Terrier F, and Geissbuhler A. A reference data set for the evaluation of medical image retrieval systems. *Comput Med Imaging Graph.* 2004b; 28(6): 295–305.

Müller H, Lovis C, and Geissbuhler A. Medical image retrieval and the medGIFT project, in *Medical Imaging and Telemedicine (MIT 2005)*, pp. 2–7, 2005.

Müller H. Medical multimedia retrieval 2.0. *Yearb Med Inform.* 2008; 55–63.

Müller H, Clough P, Deselaers T, and Caputo B (eds). *ImageCLEF: Experimental Evaluation in Visual Information Retrieval.* Springer, Berlin, 2010.

Müller H and Deserno TM. Content-based medical image retrieval. Chapter 19 in Deserno, 2011; pp. 471–494.

Névéol A, Deserno TM, Darmoni SJ, Güld MO, and Aronson AR. Natural language processing versus content-based image analysis for medical document retrieval. *J Am Soc Inf Sci Technol.* 2009; 60(1): 123–34.

Niblack W, Barber R, Equitz W, Flickner MD, Glasman EH, Petkovic D, Yanke, P, Faloutsos C, and Taubin G. The QBIC Project: Querying images by content, using color, texture, and shape. *Proc. Storage Retrieval Image Video Databases (SPIE).* 1993; 1908: 173–187.

Onken M, Eichelberg M, Riesmeier J, and Jentsch P. Digital imaging and communications in medicine. Chapter 17 in Deserno, 2011; pp. 427–454.

Pele O and Werman M. A linear time histogram metric for improved SIFT matching. *Lect Notes Computer Sci.* 2008; 5304: 495–508.

Qi H and Snyder WE. Content-based image retrieval in picture archiving and communications systems. *J Digit Imaging.* 1999; 12(2 Suppl 1): 81–83.

Rosenkranz B. Biomarkers and surrogate endpoints in clinical drug development. *Appl Clin Trials.* 2003; 7: 30–40.

Ruiz ME. Combining image features, case descriptions and UMLS concepts to improve retrieval of medical images. *AMIA Annu Symp Proc.* 2006; 674–678.

Sackett DL, Rosenberg WM, Gray JA, Haynes RB, and Richardson WS. Evidence based medicine: What it is and what it isn't. *BMJ.* 1996 13; 312(7023): 71–72.

Scholl I, Aach T, Deserno TM, and Kuhlen T. Challenges of medical image processing: From kilo- to terabyte. *Computer Sci Res Dev.* 2011; 26(1): 5–13.

Shyu CR, Brodley CE, Kak AC et al. ASSERT: A physician-in-the-loop content-based retrieval system for HRCT image databases. *Comput Vis Image Underst.* 1999; 75(1–2): 111–132.

Siggelkow S. *Feature Histograms for Content-Based Image Retrieval.* PhD Thesis. University of Freiburg, Freiburg im Breisgau, Germany, 2002.

Silverman M. Clinical trial end points at different phases of clinical development. Technical Report, BioStrategics Consulting Ltd, 2011.

Sinha U, Bui A, Taira R, Dionisio J, Morioka C, Johnson D, and Kangarloo H. A review of medical imaging informatics. *Ann N Y Acad Sci.* 2002; 980: 168–197.

Sivic J and Zisserman A. Video Google: Efficient visual search of videos. *Lecture Notes in Computer Science* 2006; 4170: 127–14.

Smeulders AWM, Worring M, Santini S, Gupta A, and Jain R. Content-based image retrieval at the end of the early years. *IEEE Trans Pattern Anal Mach Intell.* 2000; 22(12): 1349–1380.

Squire DM, Müller W, Müller H et al. Content-based query of image databases: Inspirations from text retrieval. *Pattern Recognit Lett.* 2000; 21(13–14): 1193–1198.

Tamura H, Mori S, and Yamawaki T. Texture features corresponding to visual perception. *IEEE Trans Systems, Man Cybernetics.* 1978; 8(6): 460–473.

Thodberg HH, Kreiborg S, Juul A, and Pedersen KD. The BoneXpert method for automated determination of skeletal maturity. *IEEE Trans Med Imaging.* 2009; 28(1): 52–66.

Traina AJM, Traina C Jr, Balan AGR, Ribeiro MX, Bugatti PH, Watanabe CYVW, and Paulo M. Feature extraction and selection for decision making. Chapter 8 in Deserno, 2011; pp. 197–223.

Traina C Jr, Traina AJ, Araújo MR, Bueno JM, Chino FJ, Razente H, and Azevedo-Marques PM. Using an image-extended relational database to support content-based image retrieval in a PACS. *Comput Methods Programs Biomed.* 2005; 80(Suppl 1): S71–S83.

Tversky A. Features of similarity. *Psychol Rev.* 1977; 84(4): 327–352.

Uwimana E and Ruiz ME. Integrating an automatic classification method into the medical image retrieval process. *AMIA Annu Symp Proc.* 2008; 747–751.

Volonté F, Robert JH, Ratib O, and Triponez FA. Lung segmentectomy performed with 3D reconstruction images available on the operating table with an iPad. *Interact Cardiovasc Thorac Surg.* 2011; March 8; 12(6): 1066–1068 (Epub ahead of print).

Warach S. Use of diffusion and perfusion magnetic resonance imaging as a tool in acute stroke clinical trials. *Curr Control Trials Cardiovasc Med.* 2001; 2: 38–44.

Welter P, Hocken C, Deserno TM, Grouls C, and Günther RW. Workflow management of content-based image retrieval for CAD support in PACS environments based on IHE. *Int J Comput Assist Radiol Surg.* 2010; 5(4): 393–400.

Welter P, Spreckelsen C, Fischer B, Günther RW, and Deserno TM. Case-based medical learning in radiological decision making using content-based image retrieval. *BMC.* 2011a; 11(68).

Welter P, Riesmeier J, Fischer B, Grouls C, Kuhl C, and Deserno TM. Bridging the integration gap from imaging to information systems: A uniform data concept for content-based image retrieval in computer-aided diagnosis. *J Am Med Informatics Assoc (JAMIA).* 2011b 18(4): 506–510.

Wertheimer M. Über Gestalttheorie. Vortrag vor der Kant-Gesellschaft, Berlin am 17. Dezember 1924. Verlag der Philosophischen Akademie, Erlangen 1925, in German.

Winter A and Nastar C. Differential feature distribution maps for image segmentation and region queries in image databases. *Proc Content-Based Access of Image and Video Libraries.* 1999; 9–17.

Winter A, Haux R, Ammenwerth E, Brigl B, Hellrung N, and Jahn F. *Health Information Systems: Architectures and Strategies.* Springer, Berlin 2010.

Wood DF. ABC of learning and teaching in medicine. Problem based learning. *BMJ.* 2003; 326(7384): 328–330.

Zhou X, Stern R, and Müller H. Multi-scale salient point-based retrieval of fracture cases. *Proc SPIE.* 2011; 7967: 06.

Index

B

C

E

G

H

I

J

K

L

M

N

O

Q

V

W

X